Birkhäuser

Trends in Mathematics

Trends in Mathematics is a series devoted to the publication of volumes arising from conferences and lecture series focusing on a particular topic from any area of mathematics. Its aim is to make current developments available to the community as rapidly as possible without compromise to quality and to archive these for reference.

Proposals for volumes can be submitted using the Online Book Project Submission Form at our website www.birkhauser-science.com.

Material submitted for publication must be screened and prepared as follows:

All contributions should undergo a reviewing process similar to that carried out by journals and be checked for correct use of language which, as a rule, is English. Articles without proofs, or which do not contain any significantly new results, should be rejected. High quality survey papers, however, are welcome.

We expect the organizers to deliver manuscripts in a form that is essentially ready for direct reproduction. Any version of TEX is acceptable, but the entire collection of files must be in one particular dialect of TEX and unified according to simple instructions available from Birkhäuser.

Furthermore, in order to guarantee the timely appearance of the proceedings it is essential that the final version of the entire material be submitted no later than one year after the conference.

Anna Maria Candela •
Mirella Cappelletti Montano • Elisabetta Mangino
Editors

Recent Advances in Mathematical Analysis

Celebrating the 70th Anniversary of Francesco Altomare

Birkhäuser

Editors
Anna Maria Candela
Dipartimento di Matematica
Università degli Studi di Bari Aldo Moro
Bari, Italy

Mirella Cappelletti Montano
Dipartimento di Matematica
Università degli Studi di Bari Aldo Moro
Bari, Italy

Elisabetta Mangino
Dipartimento di Matematica e Fisica
"Ennio De Giorgi"
Università del Salento
Lecce, Italy

ISSN 2297-0215 ISSN 2297-024X (electronic)
Trends in Mathematics
ISBN 978-3-031-20020-5 ISBN 978-3-031-20021-2 (eBook)
https://doi.org/10.1007/978-3-031-20021-2

Mathematics Subject Classification: 41A36, 41A60, 31C05, 26B25, 35J25, 35J15, 47D06, 35J92, 35P30, 47J30, 58E05, 41A65, 41A36, 41A25, 41A63, 41A30, 41A25, 41A36, 41A25, 35K65, 35K92, 35B65, 35B38, 35J20, 35Q55, 58E05, 35B09, 35C99, 35G10, 47D06, 47F05, 34N05, 39A10, 35J62, 35J92, 35Q55, 47J30, 58E05, 58E30, 35K67, 35B45, 47D07, 35J70, 35J75, 35J20, 35J60, 26A33, 34G20; Primary: 47A10, 47A11, 46E10, 46F05, 46H35, 35J30, 30H30, 47B33, 34B27, 34B08, 47D06, 35K65, 47D06; Seconday: 47A53, 47A55, 47B38, 47LXX, 35B65, 35R05, 46E30, 31A05, 46A04, 45A13, 46E15, 47B07, 47B38, 34B05, 34B15, 34B18, 47B25, 47N20, 35K65, 35Q91

This book is published under the imprint Birkhäuser, www.birkhauser-science.com by the registered company Springer Nature Switzerland AG
The registered company address is: Gewerbestrasse 11, 6330 Cham, Switzerland

Rector's Foreword

Francesco is a name that fits him perfectly. Professor Francesco Altomare has a kind and an accurate personality. This is the hallmark of his character.

I met him during those grueling meetings of the Academic Senate when with his speech he would illuminate the details, making them appear in front of everyone's eyes. He used to listen carefully and after that he used to argue and debate. Calm and prudent, scrupulous and acute, his language was marked by his ability to eliminate gleanings and controversies in favor of an overall vision.

One day we were in his studio, and we were engaged in a conversation about the intertwined trajectories of our only seemingly distant disciplines. With calm and tranquility, he considered them to the point of overlapping methods and approaches of our areas so that, to my eyes, they were less and less distant. It was afternoon and, without my perceiving, the light in that study dimmed. The progress of his reasoning had captured me so deeply that I did not realize that a long time had passed.

Francesco Altomare was not one of my teachers; I used not to attend his classes. I was with him in other places of the University system such as the Academic Senate or the long committee meetings. Yet, I consider him as a point of reference that, with an ancient and precise style, was able to make himself listened by giving shape to the reasoning and thus, even if I have been one of his colleagues, I was captured by him as one of his students. During those meetings, I discovered his elegance supported by a noble idea of the role of the professor. I must confess his friendship made me proud.

The University of Bari Aldo Moro has been very lucky to count Francesco Altomare among its professors. I say this well knowing how the cast of that civil humanity has left its mark over generations and generations not only of mathematics graduates.

Thanks Professor Emeritus Altomare, thanks Francesco.

Bari, Italy
August 3, 2022

Stefano Bronzini

Preface

The idea of the Special Volume *Recent Advances in Mathematical Analysis* was conceived during the International Conference on

Recent Developments in Mathematical Analysis
ReDiMA 2021,

organized by the Department of Mathematics of *Università degli Studi di Bari Aldo Moro* and the Department of Mathematics and Physics "Ennio De Giorgi" of *Università del Salento*, which was held in Bari on September 23–24, 2021, on the occasion of Francesco Altomare's 70th birthday.

Taking Francesco Altomare's large breadth of mathematical interests into account, the conference brought together mathematicians coming from different areas of Mathematical Analysis as well as experts in the deep relationship between Mathematics and other Sciences, including the Humanistic ones. In fact, one of the greatest motivations in Francesco's work, as well as an inspiration to his collaborators, has always been to build unexpected bridges among heterogeneous aspects of Mathematics.

In such a spirit, this Volume is meant to be a space where a dialogue among mathematicians is possible and, at the same time, it attests our deep esteem and affection towards Francesco Altomare.

This Volume is divided into two parts.

The first part is devoted to celebrate Francesco and his achievements.

During his very long and prolific career, Francesco Altomare has not only been a brilliant and pioneering mathematician, but he has also been appointed on several official academic responsibilities, that he has honored with commendable sense of duty. This is why the foreword of this Volume is written by the Rector of *Università degli Studi di Bari Aldo Moro*, Professor Stefano Bronzini, who has been acquainted with Francesco for many years.

We have also asked Francesco to share the speech he gave at the Opening Ceremony of ReDiMA 2021, since in our opinion his own words are the best way to express his scientific legacy as well as the spirit of this Volume.

The second part consists of 22 peer-reviewed articles, of very high scientific level, written by close friends and colleagues of Francesco. Their topics range from Approximation Theory, Semigroup Theory, and Banach Spaces to Elliptic Differential Equations and Nonlinear Analysis, with a wide array of applications.

Finally, we take this opportunity to thank all the contributors and the reviewers for their commitment to this project and to express our gratitude to Birkäuser for agreeing to our proposal and for supporting us during the preparation of this Volume.

Bari, Italy Anna Maria Candela
Lecce, Italy Mirella Cappelletti Montano
August 10, 2022 Elisabetta Mangino

From ReDiMA 2021 Opening Ceremony

Dear friends and colleagues,

I sincerely thank you for your presence and for the affectionate words towards me. I would like to express special thanks to the Vice-Chancellor of the University of Bari, Professor Anna Maria Candela, who, despite her numerous academic commitments, still found the time to deal with several aspects of the Conference together with all the other friends of the Organizing Committee, Professors Mirella Cappelletti Montano, Silvia Cingolani, Lorenzo D'Ambrosio, Elisabetta Mangino and Dora Salvatore.

I am very grateful indeed to the Organizers. They have committed themselves to organize the Conference, despite the objective difficulties caused by the Covid emergency, which unfortunately is still in progress.

Special thank also go to the coordinators of the Workgroup "Research Italian Network on Approximation" as well as of the group "Approximation Theory and Applications" of the Italian Mathematical Union for their desire to express their closeness to me and to the present Conference.

I also wish to thank the main speakers of the Conference, who with their participation and their lectures will highlight the high scientific and cultural quality of the Conference. Similarly, it is my pleasure to thank the President of the School of Science and Technology, Professor Domenico Di Bari, whom I also congratulate for his commitment and contagious enthusiasm for the activities of the School of Science and Technology.

I would like to express my special thanks to the Department of Mathematics of the University of Bari and the Department of Mathematics and Physics of the University of Salento, represented here by Professor Dora Salvatore and Professor Elisabetta Mangino, respectively, for partially supporting the Conference and for their warm participation into it.

Not without emotion, I greet Professor Luciano Lopez, coordinator of the Graduate School in Mathematics, for which I spent and developed most of my teaching activity.

I would also like to express my heartfelt thanks to Dr. Thomas Hempfling, Executive Director for Mathematics, Birkhauser, and Editorial Director for Mathematics of Springer, and to Mr. Clemens Heine, Executive Editor Mathematics for Birkhauser, for confirming, also on this occasion, a witness of the fruitful collaboration with the Department of Mathematics in the management of the *Mediterranean Journal of Mathematics* which I am honored to manage for more than 18 years.

I'm especially grateful to both as well as to the past Editor Mathematics of Birkhauser, Dr. Dorothy Mazlum, for having given credit to the scientific and cultural project which characterizes the journal, and which seeks to improve and encourage scientific collaborations among mathematicians of the Mediterranean area. This aim is indeed reflected in the choice of the title "Mediterranean Journal of Mathematics" and in the composition of the Editorial Board, which includes mathematicians from Mediterranean countries.

My retirement and the age I have reached inevitably lead to a general consideration on one's professional and family life as well as one's human existence. As you probably already know, my professional activity has been developed over 46 years, most of which at the University of Bari apart from a brief 3-year interlude at the University of Basilicata, of which I still have a nostalgic memory.

In carrying out my wide-ranging teaching activities, I have endeavored to give a high noble dignity to the teaching. The Italian translation of the word "teaching" is "insegnamento" which finds its roots in the Latin word "in-signum", namely, to give a sign, a guide, an orientation. The relationship between teaching, training and education has never been easy and yet I think that this trinomial must always be kept in mind; not just teaching without also training and educating but, rather, transmitting knowledge by offering cues and thoughts which bring out scenarios of broad cultural dimensions.

It is important to teach to ask questions to themselves too, stimulating students to solve problems which, of course, should be chosen at a suitable level to their potential knowledge. I think it could be very fruitful to tell the story of some mathematicians' lives, even non-eminent ones, when illustrating some of their important theorems, setting their lives in their own historical and cultural context, making use of films and audiovisual supports, when available. Furthermore, in developing the main aspects of a theory, it would be appropriate to dwell on the historical aspects and on the efforts and contributions developed before arriving to the most modern formulation of it.

It seems that the time is ripe for discussing about a challenge, which, if it was taken up by the younger generations of teachers, would bring an important, modern, turning point in the teaching of mathematics.

The challenge could consist in training and educating students to consider mathematics not only as the science of calculus, theorems, and algorithms, but also, as the same Galileo Galilei states, as a science that can make a contribution to the noblest cultural debates, to the understanding of the Nature, the Universe, the infinitely small, and the infinity.

In the common effort to research the ultimate truths, Mathematics interacts with other sciences, including the humanistic sciences, with which Mathematics, in particular, shares the goal of grasping harmony and beauty in what is studied, and in the results and applications which are discovered. Mathematics can help to deepen the sense of the interiority of the soul and to raise the own spirituality. Professor Paolo Maroscia will highlight these aspects with more details in his afternoon talk.

Scientific research and post-doc training for research initiation of young researchers have also occupied a prominent place in my professional activity. I have always been fascinated by the interrelationships between different fields and I have tried to study and understand them within the limits of my abilities. So, often I have dealt with issues that could affect different areas and when I was able to establish results that could have potential repercussions on them, I felt great emotions and gratifications.

I refer, for example, to the connections between asymptotic formulas for positive linear operators and the central limit theorems in Probability, or the equivalence between the Stone-Weierstrass density theorems and the Korovkin approximation theorems for positive linear operators, or the beautiful interrelationships between evolutionary differential equations, semigroups of positive operators, stochastic Markov processes, and positive linear approximation processes, all governed by a given positive linear operator.

If I could deliver a message to the younger generations, I would say to strive to continually enlarge one's knowledge, even outside one's own domain of specialization, not to chase routine results or insignificant generalizations, and to measure oneself with problems which are considered most interesting and important according to one's own knowledge, or one's own intuition or by the community in which one works.

A not marginal part of my professional activities and my time was spent in carrying out managing and organizing commitments towards the Graduate School in Mathematics, the Department of Mathematics, the PhD School of Mathematics, the recent establishment of the School of Sciences and Technologies, the Academic Senate of the University of Bari, and the Scientific Commission of the Italian Mathematical Union.

It is known that the academic community is mainly divided in two parts: those who have a rather elitist vision of their own role and, in principle, reject committing themselves towards their community, and those who, on the contrary, feel the duty of such commitments, not giving up on scientific activities, and thus redoubling the efforts in order to fulfill both commitments with self-denial and spirit of sacrifice.

I think that the prevalence or not of the former over the latter depends above all on the lesser or greater diffusion of the sense of community spread in various forms at departmental, university, and national level.

My hope is that this sense of community, in which I have believed and conformed all my academic activities, will be supported and strengthened at the Department and the University level with appropriate scientific and cultural initiatives. For these reasons, I hope that the recent actions adopted by our University do not lead to the impoverishment and fragmentation of the community of mathematicians who, in the past, have given so much prestige to our University.

In conclusion, let me end with a rather intimate remark. Let me express my deep thanks and affection to a special person, my wife Raffaella, who accompanies me and my life for more than 50 years with great love and spirit of self-sacrifice. Without her I would never have been able to achieve all the main goals of my scientific and academic life.

Bari, Italy
September 23, 2021

Francesco Altomare

Contents

List of Contributors

Ulrich Abel Technische Hochschule Mittelhessen, Fachbereich MND, Friedberg, Germany

Ana-Maria Acu Lucian Blaga University of Sibiu, Sibiu, Romania

Octavian Agratini Babeş-Bolyai University Faculty of Mathematics and Computer Science, Cluj-Napoca, Romania
Tiberiu Popoviciu Institute of Numerical Analysis Romanian Academy, Cluj-Napoca, Romania

Pietro Aiena Università di Palermo, Palermo, Italy

Angela A. Albanese Dipartimento di Matematica e Fisica "E. De Giorgi", Università del Salento, Lecce, Italy

Emilia Anna Alfano Department of Mathematics, University of Salerno, Salerno, Italy

Munirah Aljuaid Dept. of Mathematics, Northern Border University, Arar, Saudi Arabia

Francesco Altomare Dipartimento di Matematica, Università degli Studi di Bari Aldo Moro, Bari, Italy

Luciana Angiuli Dipartimento di Matematica e Fisica "E. De Giorgi", Università del Salento, Lecce, Italy

Rossella Bartolo Dipartimento di Meccanica, Matematica e Management, Politecnico di Bari, Bari, Italy

Mouffak Benchohra Djillali Liabes University, Sidi Bel-Abbes, Algeria

José Bonet IUMPA, Universitat Politècnica de València, Valencia, Spain

Stefano Bronzini Università degli Studi di Bari Aldo Moro, Piazza Umberto I, Bari, Italy

Alberto Cabada CITMAga, Santiago de Compostela, Spain
Departamento de Estatística, Análise Matemática e Optimización, Facultade de Matemáticas, Universidade de Santiago de Compostela, Santiago de Compostela, Spain

Alessandro Camasta Dipartimento di Matematica, Università degli Studi di Bari Aldo Moro, Bari, Italy

Michele Campiti Dipartimento di Matematica e Fisica "E. De Giorgi", Università del Salento, Lecce, Italy

Anna Maria Candela Dipartimento di Matematica, Università degli Studi di Bari Aldo Moro, Bari, Italy

Marco Cantarini Department of Medicine and Health Sciences, University of Molise, Campobasso, Italy

Mirella Cappelletti Montano Dipartimento di Matematica, Università degli Studi di Bari Aldo Moro, Bari, Italy

Simone Ciani Department of Mathematics, Technische Universität Darmstadt, Darmstadt, Germany

Silvia Cingolani Dipartimento di Matematica, Università degli Studi di Bari Aldo Moro, Bari, Italy

Flavia Colonna Dept. of Mathematical Sciences, George Mason University, Fairfax, VA, USA

Danilo Costarelli Department of Mathematics and Computer Science, University of Perugia, Perugia, Italy

Kristian P. Evans Department of Mathematics, Swansea University, Bay Campus, Swansea, UK

Genni Fragnelli Department of Ecology and Biology, Tuscia University, Largo dell'Università, Viterbo, Italy

Esther Gómez-Orts IUMPA, Universitat Politècnica de València, Valencia, Spain

Said R. Grace Department of Engineering Mathematics, Faculty of Engineering, Cairo University, Giza, Egypt

Umberto Guarnotta Dipartimento di Matematica e Informatica, Università degli Studi di Palermo, Palermo, Italy

Niels Jacob Department of Mathematics, Swansea University, Bay Campus, Swansea, UK

Jamal Eddine Lazreg Djillali Liabes University, Sidi Bel-Abbes, Algeria

Vita Leonessa Department of Mathematics, Computer Science and Economics, University of Basilicata, Potenza, Italy

Lucía López-Somoza CITMAga, Santiago de Compostela, Spain
Departamento de Estatística, Análise Matemática e Optimización, Facultade de Matemáticas, Universidade de Santiago de Compostela, Santiago de Compostela, Spain

Luca Lorenzi Dipartimento di Scienze Matematiche, Fisiche e Matematica, Plesso di Matematica, Università di Parma, Parma, Italy

Elisabetta Mangino Dipartimento di Matematica e Fisica "E. De Giorgi", Università del Salento, Lecce, Italy

Claudio Mele Dipartimento di Matematica e Fisica "E. De Giorgi", Università del Salento, Lecce, Italy

Federica Mennuni Dipartimento di Matematica, Università degli studi di Bari Aldo Moro, Bari, Italy

Giorgio Metafune Dipartimento di Matematica e Fisica "E. De Giorgi", Università del Salento, Lecce, Italy

Luigi Negro Dipartimento di Matematica e Fisica "E. De Giorgi", Università del Salento, Lecce, Italy

Dian K. Palagachev Dipartimento di Meccanica, Matematica e Management, Politecnico di Bari, Bari, Italy

Nikolaos S. Papageorgiou Department of Mathematics, National Technical University Zografou Campus, Athens, Greece

Ioan Raşa Technical University of Cluj-Napoca, Cluj-Napoca, Romania

Dušan D. Repovš Faculty of Education and Faculty of Mathematics and Physics, University of Ljubljana, Ljubljana, Slovenia
Institute of Mathematics, Physics and Mechanics, Ljubljana, Slovenia

Abdelkrim Salim Hassiba Benbouali University, Chlef, Algeria

Addolorata Salvatore Dipartimento di Matematica, Università degli Studi di Bari Aldo Moro, Bari, Italy

Lubomira Softova Department of Mathematics, University of Salerno, Salerno, Italy

Chiara Spina Dipartimento di Matematica e Fisica "E. De Giorgi", Università del Salento, Lecce, Italy

Kazunaga Tanaka Department of Mathematics, School of Science and Engineering, Waseda University, Tokyo, Japan

Vincenzo Vespri Dipartimento DIMAI, Università degli Studi di Firenze, Firenze, Italy

Calogero Vetro Department of Mathematics and Computer Science, University of Palermo, Palermo, Italy

Gianluca Vinti Department of Mathematics and Computer Science, University of Perugia, Perugia, Italy

Mouhcine Yousfi CITMAga, Santiago de Compostela, Spain
Departamento de Estatística, Análise Matemática e Optimización, Facultade de Matemáticas, Universidade de Santiago de Compostela, Santiago de Compostela, Spain

On Wachnicki's Generalization of the Gauss–Weierstrass Integral

Ulrich Abel and Octavian Agratini

Abstract The paper aims at a generalization of the Gauss–Weierstrass integral introduced by Eugeniusz Wachnicki two decades ago. It is intimately connected to a generalization of the heat equation. The main result is an asymptotic expansion for the operators when applied to a function belonging to a rather large class. An essential auxiliary result is a localization theorem which is interesting in itself.

Keywords Gauss–Weierstrass operator · Bessel function · Kummer function · Asymptotic expansion · Degree of approximation

1 Introduction

It is acknowledged that linear positive operators are a useful tool in approximating signals from various spaces. Referring to operators of either discrete or continuous type, a natural challenge is to highlight their properties. Two decades ago Eugeniusz Wachnicki [19] defined an integral operator representing a generalization of the classical Gauss–Weierstrass operators. One of the genuine Gauss–Weierstrass operator's write modes is as follows

$$W(f;x,t) = \frac{1}{2\sqrt{\pi t}} \int_{\mathbb{R}} \exp\left(-\frac{(x-y)^2}{4t}\right) f(y)dy, \tag{1}$$

U. Abel (✉)
Technische Hochschule Mittelhessen, Fachbereich MND, Friedberg, Germany
e-mail: ulrich.abel@mnd.thm.de

O. Agratini
Faculty of Mathematics and Computer Science, Babeş-Bolyai University, Cluj-Napoca, Romania

Tiberiu Popoviciu Institute of Numerical Analysis, Romanian Academy, Cluj-Napoca, Romania
e-mail: agratini@math.ubbcluj.ro

A. M. Candela et al. (eds.), *Recent Advances in Mathematical Analysis*,
Trends in Mathematics, https://doi.org/10.1007/978-3-031-20021-2_1

where $t > 0$ is a parameter, $x \in \mathbb{R}$ and $f : \mathbb{R} \to \mathbb{R}$ is chosen so that the integral exists and is finite. A concise but relevant presentation of these operators can be found, e.g., in the monograph [8, *Section 5.2.9*]. We mention that a generalization of the operators (1) was presented by Altomare and Milella, see [9]. Its complete asymptotic expansion was derived in [3, *Theorem 5.1*] and [4, *Corollary 3.2*].

Actually, the Wf transform is a smoothed out version of f obtained by averaging the values of f with a Gaussian signal centered at x. Specifically, it is the function G_t given by

$$G_t(x) = \frac{1}{2\sqrt{\pi t}} \int_{\mathbb{R}} f(y) \exp\left(-\frac{(x-y)^2}{4t}\right) dy = \frac{1}{2\sqrt{\pi t}} \int_{\mathbb{R}} f(x-y) \exp\left(-\frac{y^2}{4t}\right) dy$$

representing the convolution of f with the Gaussian function

$$x \mapsto \frac{1}{2\sqrt{\pi t}} \exp(-x^2/(4t)).$$

Weierstrass used a variant of this transform in his original proof of the famous approximation theorem which bears his name. Also, this transform is intimately related to the heat equation or, equivalently, the diffusion equation with constant diffusion coefficient. These are only a few reasons why it has been intensively studied over time. We note that an extension of this integral was performed even in q-calculus [10]. Among the most recent papers published in 2021, without intending to bring up them all, we mention [11, 20].

2 The Operators

First of all we recall the modified Bessel function of the first kind and fractional order $\alpha > -1$, see, e.g., [7, *Chapter 10*]. Denoting by I_α, it is described by the series

$$I_\alpha(z) = \left(\frac{z}{2}\right)^\alpha \sum_{k=0}^{\infty} \frac{\left(\frac{1}{4}z^2\right)^k}{k!\,\Gamma(\alpha+k+1)}, \tag{2}$$

where Γ is the Gamma function. An integral formula for this function (for $\mathrm{Re}(z) > 0$) will be read as follows

$$I_\alpha(z) = \frac{1}{\pi} \int_0^\pi e^{z\cos\theta} \cos(\alpha\theta) d\theta - \frac{\sin(\alpha\pi)}{\pi} \int_0^\infty e^{-z\cosh(t)-\alpha t} dt.$$

We indicate two particular cases useful throughout the paper

$$I_{-1/2}(z) = \sqrt{\frac{2}{\pi z}}\cosh(z), \quad I_{1/2}(z) = \sqrt{\frac{2}{\pi z}}\sinh(z), \tag{3}$$

see, e.g., [7, *p. 443*]. In the above, cosh, sinh represent the usual notation of hyperbolic functions.

Set $\mathbb{R}_+ = [0, \infty)$. For a fixed constant $K \geq 0$, we consider the space

$$\begin{aligned} E_K = \{ f : \mathbb{R}_+ \to \mathbb{R} \mid f \text{ is locally integrable and } (\exists)\ M_f \geq 0, \\ |f(s)| \leq M_f e^{Ks^2},\ s > 0 \}. \end{aligned}$$

The space is endowed with the norm $\|\cdot\|_K$,

$$\|f\|_K = \sup_{s \in \mathbb{R}_+} |f(s)| e^{-Ks^2}.$$

We consider the operator W_α defined on E_K by the following relation

$$W_\alpha(f; r, t) = \frac{1}{2t}\int_0^\infty r^{-\alpha} s^{\alpha+1} \exp\left(-\frac{r^2+s^2}{4t}\right) I_\alpha\left(\frac{rs}{2t}\right) f(s)ds, \tag{4}$$

where $\alpha \geq -1/2$, $(r, t) \in (0, \infty) \times (0, \infty)$ and I_α is given at (2).

W_α was introduced in [19, *Eq. (1)*] with a minor modification of the domain E_K in which the author inserted $f \in C(\mathbb{R}_+)$.

Since $W_\alpha f$, $f \in E_K$, is well defined for any $K > 0$, we can consider the domain of W_α as $E := \bigcup_{K>0} E_K$.

Wachnicki [19, *Theorem 4*] showed the convergence

$$\lim_{t \to 0^+} W_\alpha (f; r, t) = f(r),$$

for $f \in E \cap C(\mathbb{R}_+)$, uniformly on compact subintervals of $(0, +\infty)$.

For $\alpha \geq -1/2$, the function $W_\alpha (f; r, t)$ is an example of the solution of the generalized heat equation

$$\frac{\partial u}{\partial t} = \frac{\partial^2 u}{\partial r^2} + \frac{2\alpha+1}{r}\frac{\partial u}{\partial r}. \tag{5}$$

If $\alpha = n/2 - 1$, $n \in \mathbb{N}$, Eq. (5) is the heat equation in $\mathbb{R}^{n+1}$ in radial coordinates.

Recently, in [14] an extension of this operator was achieved for continuous functions defined on the domain $(0, \infty) \times \mathbb{R}$ and bounded by certain two-dimensional exponential functions.

Remark 2.1 For $\alpha = -1/2$ the operator defined by (4) becomes the authentic Gauss–Weierstrass operator, a specification that can be found in [19]. Because the statement was not accompanied by a proof, we insert it as a detail in our paper. More precisely we prove

$$W_{-1/2}f = W\widehat{f}, \quad f \in E_K, \tag{6}$$

see (1), where $\widehat{f}(s) = f((\operatorname{sgn} s)s)$, $s \in \mathbb{R}$.

Indeed, for any $(r, t) \in (0, \infty) \times (0, \infty)$, taking in view (3), we can write

$$\begin{aligned}
&W_{-1/2}(f; r, t)\\
&= \frac{1}{2t}\sqrt{\frac{4t}{\pi}} \int_0^\infty \exp\left(-\frac{r^2+s^2}{4t}\right) \cosh\left(\frac{rs}{2t}\right) f(s)ds\\
&= \frac{1}{\sqrt{\pi t}} \int_0^\infty \exp\left(-\frac{r^2+s^2}{4t}\right) \frac{1}{2}\left(\exp\left(\frac{rs}{2t}\right) + \exp\left(-\frac{rs}{2t}\right)\right) f(s)ds\\
&= \frac{1}{2\sqrt{\pi t}} \int_0^\infty \left(\exp\left(-\frac{(r-s)^2}{4t}\right) + \exp\left(-\frac{(r+s)^2}{4t}\right)\right) f(s)ds\\
&= \frac{1}{2\sqrt{\pi t}} \left(\int_0^\infty \exp\left(-\frac{(r-s)^2}{4t}\right) f(s)ds + \int_{-\infty}^0 \exp\left(-\frac{(r-s)^2}{4t}\right)\right) f(-s)ds\\
&= W(\widehat{f}; r, t).
\end{aligned}$$

3 The Asymptotic Expansion

To achieve our goal and to obtain a self contained exposure, we recall the following notions.

For factorial powers (falling respective rising factorial) we use the notations

$$u^{\underline{k}} = \prod_{\ell=0}^{k-1}(u-\ell), \quad u^{\overline{k}} = \prod_{\ell=0}^{k-1}(u+\ell), \quad k \in \mathbb{N}.$$

An empty product ($k = 0$) is taken to be 1.

Kummer's function of the first kind, also known as the confluent hypergeometric function of the first kind, is commonly denoted $M(a, b, z)$ or ${}_1F_1(a, b, z)$ and defined as follows

$$\begin{aligned}
M(a, b, z) \equiv {}_1F_1(a, b, z) &= 1 + \frac{a}{b}z + \frac{a(a+1)}{b(b+1)2!}z^2 + \dots\\
&= \sum_{k=0}^\infty \frac{a^{\overline{k}}}{b^{\overline{k}}k!} z^k, \; z \in \mathbb{C},
\end{aligned} \tag{7}$$

see [7, *Chapter 13*]. If $b \in \mathbb{Z} \setminus \mathbb{N}$, then M is undefined. Otherwise the series is convergent for all $z \in \mathbb{C}$.

Let us denote e_i the monomials $e_i(x) = x^i$, $i \in \mathbb{N}_0 = \mathbb{N} \cup \{0\}$. Clearly, $e_i \in E_K$, for each K. We determine all moments of the operator W_α.

Proposition 3.1 *Let W_α be defined by (4). For each $i \in \mathbb{N}_0$ the identity*

$$W_\alpha(e_i; r, t) = (4t)^{i/2} \exp\left(-\frac{r^2}{4t}\right) \frac{\Gamma(\alpha + 1 + i/2)}{\Gamma(\alpha + 1)} M\left(\alpha + 1 + \frac{i}{2}, \alpha + 1, \frac{r^2}{4t}\right) \tag{8}$$

takes place, where M stands for Kummer function (7). Moreover

$$W_\alpha(e_i; r, t) \sim r^i \sum_{n=0}^{\infty} \frac{1}{n!} \left(\frac{i}{2}\right)^{\underline{n}} \left(\alpha + \frac{i}{2}\right)^{\underline{n}} \left(\frac{4t}{r^2}\right)^n \quad (t \to 0^+). \tag{9}$$

Proof By a straightforward calculation we obtain

$$\begin{aligned} & W_\alpha(e_i; r, t) \\ &= \frac{1}{2t}\left(\frac{1}{4t}\right)^{\alpha} \exp\left(-\frac{r^2}{4t}\right) \sum_{k=0}^{\infty} \frac{\Gamma(k + \alpha + 1 + i/2)}{k!\,\Gamma(k + \alpha + 1)} \left(\frac{r}{4t}\right)^{2k} 2^{i+2\alpha+1+2k} t^{\alpha+1+k+i/2} \\ &= (4t)^{i/2} \exp\left(-\frac{r^2}{4t}\right) \sum_{k=0}^{\infty} \frac{\Gamma(k + \alpha + 1 + i/2)}{k!\,\Gamma(k + \alpha + 1)} \left(\frac{r^2}{4t}\right)^{k}. \end{aligned}$$

By using Kummer function defined by (7) and the well-known relation

$$\Gamma(u + k) = \Gamma(u) u^{\overline{k}}, \ u > 0, \ k \in \mathbb{N}_0,$$

we arrive at (8).

Further, we use the asymptotic expansion of Kummer function for large argument. To accomplish this, we use for example *Digital Library of Mathematical Functions* [16, *Formulas 13.2.4, 13.7.1*].

$$M(a, b, x) \sim \frac{\Gamma(b)}{\Gamma(a)} e^x x^{a-b} \sum_{n=0}^{\infty} \frac{(b-a)^{\overline{n}} (1-a)^{\overline{n}}}{n!} x^{-n} \quad (x \to \infty),\ a \notin \mathbb{Z} \setminus \mathbb{N}_0.$$

Returning at (8) with $x = \dfrac{r^2}{4t}$, we get

$$W_\alpha(e_i; r, t) \sim r^i \sum_{n=0}^{\infty} \frac{1}{n!} \left(-\frac{i}{2}\right)^{\overline{n}} \left(-\alpha - \frac{i}{2}\right)^{\overline{n}} \left(\frac{4t}{r^2}\right) \quad (t \to 0^+).$$

Knowing that $(-u)^{\overline{n}} = (-1)^n u^{\underline{n}}$, we obtain (9) and the proof is ended. □

Remark 3.1

Formula (9) means that

$$W_\alpha(e_i; r, t) = e_i(r) + \sum_{n=1}^{q} \frac{1}{n!} \left(\frac{i}{2}\right)^{\underline{n}} \left(\alpha + \frac{i}{2}\right)^{\underline{n}} \left(\frac{4t}{r^2}\right)^n + o(t^q) \quad (t \to 0^+)$$

for all $q \in \mathbb{N}$.

ii) If $i = 2k, k \in \mathbb{N}$, is an even integer, $W_\alpha\,(e_{2k}; r, t)$ is a polynomial as a function of r, since $M\left(\alpha + 1 + k, \alpha + 1, \frac{r^2}{4t}\right)$ is a finite sum. The polynomials determined in this way are called radial heat polynomials (see [12]). For $k = 0, 1, 2$ formula (8) easily leads us to the explicit formulas [19, *Eq. (13)*]

$$\begin{aligned} W_\alpha\,(e_0; r, t) &= 1, \\ W_\alpha\,(e_2; r, t) &= r^2 + 4\,(\alpha + 1)\,t, \\ W_\alpha\,(e_4; r, t) &= r^4 + 8\,(\alpha + 2)\,r^2 t + 16\,(\alpha + 1)\,(\alpha + 2)\,t^2. \end{aligned}$$

The first formula implies the fact that the operator reproduces the constants. In the special case $\alpha = -1/2$ we have the representation (see [17, *Eq. (1.2)*])

$$W_{-1/2}\,(e_{2k}; r, t) = k! \sum_{j=0}^{\lfloor k/2 \rfloor} \frac{x^{k-2j}}{(k-2j)!} \frac{t^j}{j!}.$$

Further we introduce the j-th central moments of W_α operator, $j \in \mathbb{N}_0$, i.e., $W_\alpha \psi_r^j$, where $\psi_r(s) = s - r$, $s > 0$, $r > 0$.

Lemma 3.1 *For each $j \in \mathbb{N}$, the operator W_α defined by (4) satisfies the relation*

$$W_\alpha(\psi_r^j; r, t) \sim C_\alpha(n, j) r^j \sum_{n=0}^{\infty} \frac{1}{n!} \left(\frac{4t}{r^2}\right)^n \quad (t \to 0^+), \tag{10}$$

where

$$C_\alpha(n, j) = \sum_{i=0}^{j} (-1)^{j-i} \binom{j}{i} \left(\frac{i}{2}\right)^{\underline{n}} \left(\alpha + \frac{i}{2}\right)^{\underline{n}}. \tag{11}$$

Proof Since $\psi_r^j(s) = \sum_{i=0}^{j} \binom{j}{i}(-r)^{j-i} e_i(s)$, based on relation (9) we can write

$$W_\alpha(\psi_r^j; r, t) \sim \sum_{i=0}^{j} \binom{j}{i}(-r)^{j-i} r^i \sum_{n=0}^{\infty} \frac{1}{n!}\left(\frac{i}{2}\right)^{\underline{n}}\left(\alpha + \frac{i}{2}\right)^{\underline{n}}\left(\frac{4t}{r^2}\right)^n \quad (t \to 0^+).$$

Using the coefficients defined in (11) we arrive at (10). □

Remark 3.2 Examining the coefficients $C_\alpha(n, j)$, $j \in \mathbb{N}_0$, we notice that they can be written as follows

$$\begin{aligned} C_\alpha(n, j) &= \sum_{i=0}^{j}(-1)^{j-i}\binom{j}{i}\left(\frac{\partial}{\partial w}\right)^n \left[w^{n+\alpha}\left(\frac{\partial}{\partial w}\right)^n w^{i/2}\right]\Big|_{w=1} \\ &= \left(\frac{\partial}{\partial w}\right)^n \left[w^{n+\alpha}\left(\frac{\partial}{\partial w}\right)^n (\sqrt{w}-1)^j\right]\Big|_{w=1}. \end{aligned} \tag{12}$$

The following theorem is our main result. It presents a complete asymptotic expansion of the operators $W_\alpha(f; r, t)$ as $t \to 0^+$, for functions $f \in E$ being sufficiently smooth at a point $r > 0$.

Theorem 3.1 *Let W_α be defined by (4) and $r > 0$, $q \in \mathbb{N}$ be given. If $f \in E$ is q times differentiable at r, then*

$$W_\alpha(f; r, t) = \sum_{n=0}^{q} \frac{1}{n!}\left(\frac{4t}{r^2}\right)^n \sum_{j=0}^{2q} \frac{f^{(j)}(r)}{j!} r^j C_\alpha(n, j) + o(t^q) \quad (t \to 0^+), \tag{13}$$

where the coefficients $C_\alpha(n, j)$, $j = 0, \ldots, 2q$, are defined at (11).

If f is a real analytic function, then

$$W_\alpha(f; r, t) \sim \sum_{n=0}^{\infty} c_n(\alpha, f, r) t^n \quad (t \to 0^+), \tag{14}$$

where

$$c_n(\alpha, f, r) = \frac{4^n}{n! r^{2n}}\left(\frac{\partial}{\partial w}\right)^n \left[w^{n+\alpha}\left(\frac{\partial}{\partial w}\right)^n f\left(r\sqrt{w}\right)\right]\Big|_{w=1}. \tag{15}$$

Proof Let $r > 0$ and put $U_\delta(r) = (r - \delta, r + \delta) \cap [0, +\infty)$, for $\delta > 0$. Let $\delta > 0$ be given. Choose a function $\varphi \in C^\infty([0, +\infty))$ with $\varphi(x) = 1$ on $U_\delta(r)$ and $\varphi(x) = 0$ on $[0, +\infty) \setminus U_{2\delta}(r)$. Put $\widetilde{f} = \varphi f$. Then we have $\widetilde{f} \equiv f$ on $U_\delta(r)$ and $\widetilde{f} \equiv 0$ on $[0, +\infty) \setminus U_{2\delta}(r)$. In particular $\widetilde{f}^{(i)}(r) = f^{(i)}(r)$, for $i = 0, \ldots, 2q$. By the localization theorem (Theorem 4.1), $W_\alpha\left(f - \widetilde{f}; r, t\right)$ deceases exponentially fast as $t \to 0^+$. Consequently, $\widetilde{f}$ and f possess the same asymptotic expansion of the form (13). Therefore, without loss of generality, we can assume that $f \equiv 0$ on $[0, +\infty) \setminus U_{2\delta}(r)$. By Lemma 3.1, we have $W_\alpha\left(\psi_r^{2s}; r, t\right) = O\left(t^s\right)$ as $t \to 0^+$. Under these conditions, a general approximation theorem due to Sikkema [18, *Theorem 3*] implies that

$$W_\alpha(f; r, t) = \sum_{j=0}^{2q} \frac{f^{(j)}(r)}{j!} W_\alpha(\psi_r^j; r, t) + o(t^q),$$

see [18, *Eq. (15)*]. Taking in view (10), identity (13) is proved.

If f is a real analytic function we have the possibility to write

$$f\left(r\sqrt{w}\right) = \sum_{j\geq 0} \frac{f^{(j)}(r)}{j!} \left(r\sqrt{w} - r\right)^j, \ w > 0.$$

Further, with the help of (12), formula (13) can be rewritten in the form

$$\begin{aligned} &W_\alpha(f; r, t) \\ &= \sum_{n=0}^{q} \frac{1}{n!} \left(\frac{4t}{r^2}\right)^n \left(\frac{\partial}{\partial w}\right)^n \left[w^{n+\alpha} \left(\frac{\partial}{\partial w}\right)^n f\left(r\sqrt{w}\right)\right]\Big|_{w=1} + o(t^q) \quad (t \to 0^+). \end{aligned}$$

Consequently we get

$$W_\alpha(f; r, t) \sim \sum_{n=0}^{\infty} \frac{1}{n!} \left(\frac{4t}{r^2}\right)^n \left(\frac{\partial}{\partial w}\right)^n \left[w^{n+\alpha} \left(\frac{\partial}{\partial w}\right)^n f\left(r\sqrt{w}\right)\right]\Big|_{w=1} \quad (t \to 0^+)$$

and (14) is substantiated. □

Applying formula (15) to find the first four coefficients of asymptotic expansion, we obtain the following values:

$$\begin{aligned} c_0(\alpha, f, r) &= f(r), \\ c_1(\alpha, f, r) &= \frac{2\alpha+1}{r} f'(r) + f''(r), \\ c_2(\alpha, f, r) &= \frac{1-4\alpha^2}{2r^3} f'(r) + \frac{4\alpha^2-1}{2r^2} f''(r) + \frac{2\alpha+1}{r} f'''(r) + \frac{1}{2} f^{(4)}(r), \\ c_3(\alpha, f, r) &= (4\alpha^2-1)\left(\frac{2\alpha-3}{2r^5} f'(r) - \frac{2\alpha-3}{2r^4} f''(r) + \frac{\alpha-3}{3r^3} f'''(r) + \frac{1}{2r^2} f^{(4)}(r)\right) \\ &\quad + \frac{2\alpha+1}{2r} f^{(5)}(r) + \frac{1}{6} f^{(6)}(r). \end{aligned}$$

For the special case $\alpha = -1/2$ (nearly Gauss–Weierstrass operator, see (6)) the respective coefficients become

$$c_0(-1/2, f, r) = f(r), \qquad c_1(-1/2, f, r) = f''(r),$$
$$c_2(-1/2, f, r) = \frac{1}{2} f^{(4)}(r),\ c_3(-1/2, f, r) = \frac{1}{6} f^{(6)}(r).$$

We have the following representation of the coefficients in the asymptotic expansion (cf. [6, *Eq. (1.5)*]).

Proposition 3.2 *For $\alpha = -1/2$ it holds*

$$c_n(-1/2, f, r) = \frac{1}{n!} f^{(2n)}(r)$$

Proof Since the formula is of algebraic form it is sufficient to prove it for polynomial functions f. For $f = e_i$, we have

$$\begin{aligned}
c_n(-1/2, e_i, r) &= \frac{4^n}{n! r^{2n}} \left(\frac{\partial}{\partial w}\right)^n \left[w^{n-1/2} \left(\frac{\partial}{\partial w}\right)^n \left(r\sqrt{w}\right)^i \right]\Bigg|_{w=1} \\
&= \frac{4^n r^i}{n! r^{2n}} \left(\frac{i}{2}\right)^{\underline{n}} \left(\frac{\partial}{\partial w}\right)^n w^{i/2-1/2}\Bigg|_{w=1} \\
&= \frac{4^n r^{i-2n}}{n!} \left(\frac{i}{2}\right)^{\underline{n}} \left(\frac{i-1}{2}\right)^{\underline{n}} \\
&= \frac{4^n r^{i-2n}}{n!} \frac{i}{2} \frac{i-1}{2} \frac{i-2}{2} \frac{i-3}{2} \cdots \frac{i-2n+2}{2} \frac{i-2n+1}{2} \\
&= \frac{1}{n!} i^{\underline{2n}} r^{i-2n} = \frac{1}{n!} e_i^{(2n)}(r).
\end{aligned}$$

□

At the end of this section it is worth mentioning that the first author obtained asymptotic expansions for various classes of approximation linear positive operators, see, e.g., [1, 2], as well as [5, 6].

4 Localization Result

The purpose of this paragraph is to characterize the function $W_\alpha f$ according to its growth rate, where $f \in E_K$ has a specific property which means that it vanishes in a neighborhood of a fixed value $r > 0$. For the modified Bessel function defined by (2), Luke [15, *Eq. (6.25)*] proved the estimate

$$1 < \Gamma(\alpha+1)\left(\frac{2}{z}\right)^\alpha I_\alpha(z) < \cosh(z),\ z > 0,\ \alpha > -\frac{1}{2}. \tag{16}$$

We emphasize that Ifantis and Siafarikas [13, *Eq. (2.1)*] gave more general bounds

$$\left(\frac{x}{y}\right)^\alpha \frac{\cosh(x)}{\cosh(y)} < \frac{I_\alpha(x)}{I_\alpha(y)} < \left(\frac{\cosh(x)}{\cosh(y)}\right)^{1/(2\alpha+2)},\ 0 < x < y,\ \alpha > -\frac{1}{2}. \tag{17}$$

From (2) is immediately deduced

$$\lim_{x\to 0^+} x^{-\alpha} I_\alpha(x) = \frac{2^{-\alpha}}{\Gamma(\alpha+1)}$$

and in this way, out of (17), inequalities (16) can be reobtained.

Knowing the particular value of $I_{-1/2}(z)$, see (3), from the previous relations we can finally write

$$I_\alpha(z) \le L(\alpha) z^\alpha e^z,\ z > 0,\ \alpha \ge -1/2, \tag{18}$$

where $L(\alpha) = 2^{-\alpha}/\Gamma(\alpha+1)$.

The following localization result has already been applied in the proof of Theorem 3.1. It is interesting in itself.

Theorem 4.1 *Let W_α be defined by (4). Let $r > 0$, $\delta > 0$ be fixed. We consider the function $f \in E_K$ satisfying the condition*

$$f(s) = 0,\ s \in (r-\delta, r+\delta) \cap [0, \infty).$$

Then, it holds

$$W_\alpha(f; r, t) = O\left(\exp\left(-\frac{\delta^2}{4t}\right)\right) \quad (t \to 0^+). \tag{19}$$

Proof By (18), we have

$$
\begin{aligned}
0 &\le \exp\left(-\frac{r^2+s^2}{4t}\right) I_\alpha\left(\frac{rs}{2t}\right) \\
&\le L(\alpha)\exp\left(-\frac{r^2+s^2}{4t}\right)\left(\frac{rs}{2t}\right)^\alpha \exp\left(\frac{rs}{2t}\right) \\
&= L(\alpha)\left(\frac{rs}{2t}\right)^\alpha \exp\left(-\frac{(s-t)^2}{4t}\right).
\end{aligned}
$$

Considering the conditions verified by the function f, we obtain

$$
|W_\alpha(f;r,t)| \le \frac{ML(\alpha)}{(2t)^{\alpha+1}}(I_1+I_2), \tag{20}
$$

where

$$
\begin{aligned}
I_1 &:= \int_0^{\max\{0,r-\delta\}} s^{2\alpha+1}\exp\left(-\frac{(s-r)^2}{4t}\right)e^{Ks^2}ds, \\
I_2 &:= \int_{r+\delta}^{\infty} s^{2\alpha+1}\exp\left(-\frac{(s-r)^2}{4t}\right)e^{Ks^2}ds.
\end{aligned}
$$

At first we estimate I_2. A change of variable leads to

$$
I_2 = \int_\delta^\infty (s+r)^{2\alpha+1}\exp\left(-\frac{s^2}{4t}+K(s+r)^2\right)ds.
$$

Next we consider $t>0$ small enough, for example $t<(8K)^{-1}$.
For each $m>1$, the derivative

$$
\begin{aligned}
&\frac{\partial}{\partial s}\left[(s+r)^m\exp\left(-\frac{s^2}{4t}+K(s+r)^2\right)\right] \\
&= \left[m+(s+r)\left(-\frac{2s}{4t}+2K(s+r)\right)\right](s+r)^{m-1}\exp\left(-\frac{s^2}{4t}+K(s+r)^2\right)
\end{aligned} \tag{21}
$$

has a unique positive zero, say $s(t)$, where

$$
s(t) = \frac{1}{2}(4Kt-1)^{-1}(r-8Krt-((8Krt-r)^2-8(4Kt-1)(2Kr^2+m)t)^{1/2}).
$$

As a function of t, $s(t)$ tends to 0 when $t \to 0^+$. Therefore, we can assume that $0 < s(t) < \delta$. Relation (21) and the above statements imply that the function h_r,

$$h_r(s) = (s+r)^{2\alpha+1} \exp\left(-\frac{s^2}{4t} + K(s+r)^2\right)$$

is monotonically decreasing. Consequently, for $s \geq \delta$ we have

$$\begin{aligned} I_2 &\leq (\delta+r)^{2\alpha+3} \exp\left(-\frac{\delta^2}{4t} + K(\delta+r)^2\right) \int_\delta^\infty (s+r)^{-2} ds \\ &= (\delta+r)^{2\alpha+2} \exp(K(\delta+r)^2) \exp\left(-\frac{\delta^2}{4t}\right) \\ &= O\left(\exp\left(-\frac{\delta^2}{4t}\right)\right) \qquad (t \to 0^+). \end{aligned}$$

We turn to the integral I_1. In the case $r \leq \delta$ there is nothing to prove. If $r > \delta$, we have

$$I_1 \leq (r-\delta)(r-\delta)^{2\alpha+1} \max_{0 \leq s \leq r-\delta} \exp\left(-\frac{(s-r)^2}{4t} + Ks^2\right).$$

Because $-(s-r)^2 \leq -\delta^2$, for $0 \leq s \leq r-\delta$, we infer that

$$I_1 \leq r^{2\alpha+2} e^{Kr^2} \exp\left(-\frac{\delta^2}{4t}\right).$$

Combining the estimates of the integrals I_1 and I_2, from (20) the statement (19) is proved. □

We note that Wachnicki [19, *Lemmas 2 and 3*] showed that

$$\lim_{t \to 0^+} \int_0^{r-\delta} K_\alpha(r,s,t)\, e^{Ks^2} ds = 0$$

uniformly on $r \in [\alpha, \beta]$, for $0 < \alpha < \delta < \beta$, and

$$\lim_{t \to 0^+} \int_{r+\delta}^{+\infty} K_\alpha(r,s,t)\, e^{Ks^2} ds = 0$$

uniformly on $r \in (0, \beta]$, for $\beta > 0$, which implies the weaker estimate

$$W_\alpha(f; r, t) = o(1) \qquad \left(t \to 0^+\right).$$

Acknowledgments The authors are grateful to the anonymous referee for valuable recommendations which led to several improvements of the manuscript. In particular, we thank for an additional reference.

References

1. Abel, U.: A Voronovskaya type result for simultaneous approximation by Bernstein–Chlodovsky polynomials. Results Math. **74**, Article number 117 (2019)
2. Abel, U.: Voronovskaja type theorems for positive linear operators related to squared fundamental functions. In: Draganov, B., Ivanov, K., Nikolov, G., Uluchev, R. (eds.) Constructive Theory of Functions, Sozopol 2019, pp. 1–21. Prof. Marin Drinov Publishing House of Bas, Sofia (2020)
3. Abel, U., Ivan, M.: Simultaneous approximation by Altomare operators. Suppl. Rend. Circ. Mat. Palermo **82**(2), 177–193 (2010)
4. Abel, U., Ivan, M.: Complete asymptotic expansions for Altomare operators. Mediterr. J. Math. **10**, 17–29 (2013)
5. Abel, U., Karsli, H.: A complete asymptotic expansion for Bernstein-Chodovsky polynomials for functions on $\mathbb{R}$. Mediterr. J. Math., **17**, Article number 201 (2020)
6. Abel, U., Agratini, O., Păltănea, R.: A complete asymptotic expansion for the quasi-interpolants of Gauss–Weierstrass operators. Mediterr. J. Math. **15**, Article number 156 (2018)
7. Abramowitz, M., Stegun, I.A. (eds.): Handbook of Mathematical Functions with Formulas, Graphs and Mathematical Tables. National Bureau of Standards Applied Mathematics Series, vol. 55. Tenth Printing (Issued, June 1964) (With corrections, December 1972)
8. Altomare, F., Campiti, M.: Korovkin-type Approximation Theory and its Applications. de Gruyter Series Studies in Mathematics, vol. 17. Walter de Gruyter, Berlin/New York (1994)
9. Altomare, F., Milella, S.: Integral-type operators on continuous function spaces on the real line. J. Approx. Theory **152**(2), 107–124 (2008)
10. Aral, A., Gal, S.G.: q-Generalizations of the Picard and Gauss–Weierstrass singular integrals. Taiwanese J. Math. **12**(9), 2051–2515 (2008)
11. Bardaro, C., Mantellini, I., Uysal, G., Yilmaz, B.: A class of integral operators that fix exponential functions. Mediterr. J. Math. **18**, Article number 179 (2021)
12. Bragg, L.R.: The radial heat polynomials and related functions. Trans. Am. Math. Soc. **119**, 270–290 (1965)
13. Ifantis, E.K., Siafarikas, P.D.: Bounds for modified Bessel functions. Rend. Circ. Mat. Palermo II **40**(3), 347–356 (1991)
14. Krech, G., Krech, I.: On some bivariate Gauss–Weierstrass operators. Constr. Math. Anal. **2**(2), 57–63 (2019)
15. Luke, Y.L.: Inequalities for generalized hypergeometric functions. J. Approx. Theory, **5**, 41–65 (1972)
16. NIST Digital Library of Mathematical Functions. F. W. J. Olver, A. B. Olde Daalhuis, D. W. Lozier, B. I. Schneider, R. F. Boisvert, C. W. Clark, B. R. Miller, B. V. Saunders, H. S. Cohl, and M. A. McClain, eds. https://dlmf.nist.gov
17. Rosenbloom, P., Widder, D. V.: Expansions in heat polynomials and associated functions. Trans. Am. Math. Soc. **92**, 220–266 (1959)
18. Sikkema, P.C.: On some linear positive operators. Ind. Math. **32**, 327–337 (1970)
19. Wachnicki, E.: On a Gauss–Weierstrass generalized integral. Rocznik Nauk.-Dydakt. Akad. Pedagogícznej W Krakow. Prace Mat. **17**(2000), 251–263
20. Yilmaz, B.: Approximation properties of modified Gauss–Weierstrass integral operators in exponential weighted L_p spaces. Facta Univ. (Niš) Ser. Math. Inform. **36**(1), 89–100 (2021)

Generalized Subharmonic and Weakly Convex Functions

Ana-Maria Acu and Ioan Raşa

Abstract This paper is devoted to a conjecture formulated by Francesco Altomare and Ioan Raşa in 1999. It is concerned with the relationship between some generalized subharmonic functions and generalized convex functions.

Keywords Generalized subharmonic functions · Generalized convex functions

1 Introduction

The main research area of Professor Francesco Altomare is the relationship among positive semigroups, initial-boundary value problems, Markov processes and approximation theory. The aim of this approach is to construct positive approximation processes whose iterates converge strongly to semigroups which furnish the solutions to the relevant initial-boundary value differential problems. This theory is presented in [3, 5, 6], and the references therein. The study involves in particular certain classes of generalized convex functions and generalized subharmonic functions.

Starting with a suitable positive linear projection T, F. Altomare and the second author defined in [4] the weakly T-convex functions. Using T, a C_0-semigroup of operators was constructed and the generalized A-subharmonic functions were defined, where A is the infinitesimal generator of the semigroup. It was proved that if a function is weakly T-convex, then it is generalized A-subharmonic. The authors of [4] conjectured that the converse is also true, but as far as we know this is still an open problem.

A.-M. Acu
Lucian Blaga University of Sibiu, Sibiu, Romania
e-mail: anamaria.acu@ulbsibiu.ro

I. Raşa (✉)
Technical University of Cluj-Napoca, Cluj-Napoca, Romania
e-mail: ioan.rasa@math.utcluj.ro

A. M. Candela et al. (eds.), *Recent Advances in Mathematical Analysis*,
Trends in Mathematics, https://doi.org/10.1007/978-3-031-20021-2_2

In this paper we present a result which is related to the conjecture. Namely, starting with the conjecture, we prove that a suitable stronger hypothesis entails a stronger conclusion. Other related results can be found in [9, Chapters 15–22].

2 A Conjecture

Let K be a convex compact subset of $\mathbb{R}^p$, $p \geq 1$, having nonempty interior. By $C(K)$ we denote the Banach lattice of all real-valued, continuous functions on K equipped with the supremum norm and the usual ordering. $C^2(K)$ stands for the subspace of all functions $f \in C(K)$ which are two times continuously differentiable on the interior $intK$ and whose partial derivatives of order ≤ 2 can be continuously extended to K. For every $u \in C^2(K)$ and $i, j \in \{1, \ldots, p\}$ we denote by $\frac{\partial u}{\partial x_i}$ and $\frac{\partial^2 u}{\partial x_i \partial x_j}$ the continuous extensions to K of the partial derivatives $\frac{\partial u}{\partial x_i}$ and $\frac{\partial^2 u}{\partial x_i \partial x_j}$ defined on $intK$.

For every $j = 1, \ldots, p$ we denote by $pr_j \in C(K)$ the function defined by

$$pr_j(x) = x_j, \ \ x = (x_1, \ldots, x_p) \in K.$$

Let $T : C(K) \to C(K)$ be a positive linear projection, i.e., a positive linear operator such that $T \circ T = T$. We assume that T is not the identity operator on $C(K)$, and

$$T\mathbf{1} = \mathbf{1}, \ \ T(pr_j) = pr_j, \ \ j = 1, \ldots, p, \tag{1}$$

where $\mathbf{1}$ is the function on K having constant value 1.

Let

$$H_T := T\,(C(K)) = \{h \in C(K) \mid Th = h\}.$$

For $z \in K$, $\alpha \in [0, 1]$ and $h \in H_T$ set

$$h_{z,\alpha}(x) := h(\alpha x + (1 - \alpha)z), \ \ x \in K.$$

We assume that

$$h_{z,\alpha} \in H_T, \ \text{for all}\ h \in H_T, \ z \in K, \ \alpha \in [0, 1].$$

For each $i, j = 1, \ldots, p$, denote

$$a_{ij} := T(pr_i pr_j) - pr_i pr_j.$$

Now we consider the differential operator $W_T : C^2(K) \to C(K)$ defined by

$$W_T u(x) := \frac{1}{2} \sum_{i,j=1}^{p} a_{ij}(x) \frac{\partial^2 u(x)}{\partial x_i \partial x_j},$$

for every $u \in C^2(K)$ and $x \in K$.

For each $m \geq 1$ let $A_m(K)$ be the subspace of the restrictions to K of all polynomial functions of degree $\leq m$. In particular, $A_1(K) = A(K)$ denotes the subspace of all continuous affine functions on K.

Now we are in the position to state a particular case of a result established in [2].

Theorem 2.1 (Altomare [2]) *Suppose that $T(A_2(K)) \subset A(K)$. Then the operator $(W_T, C^2(K))$ is closable and its closure $(A, D(A))$ generates a Feller semigroup on $C(K)$.*

Remark 2.1 The above considerations can be extended to the case where K is a compact convex subset of a locally convex space. In this case the set $\{\mathbf{1}, pr_1, \ldots, pr_p\}$ should be replaced by the subspace $A(K)$ of all real-valued affine continuous functions on K, the subspace $C^2(K)$ by the subspace

$$A_\infty(K) := \cup_{m=1}^{\infty} A_m(K),$$

where $A_m(K)$ is the subspace generated by

$$\left\{ \prod_{i=1}^{m} h_i \,\middle|\, h_i \in A(K),\ i = 1, \ldots, m \right\},$$

and the operator W_T by $V_T : A_\infty(K) \to C(K)$ defined by

$$V_T\left(\prod_{i=1}^{m} h_i\right) := \begin{cases} 0, \ m = 1, \\ T(h_1 h_2) - h_1 h_2, \ m = 2, \\ \displaystyle\sum_{1 \leq i < j \leq m} \left(T(h_i h_j) - h_i h_j\right) \prod_{\substack{r=1 \\ r \neq i,j}}^{m} h_r, \ m \geq 3. \end{cases}$$

Remark 2.2 Many examples, more general results, extensions, and many applications can be found in [2–5] and the references therein.

Definition 2.1 (Altomare and Raşa [4]) A function $f \in C(K)$ is said to be weakly T-convex if

$$f(z) \leq T(f_{z,\alpha})(z) \text{ for each } z \in K,\ \alpha \in [0, 1].$$

Definition 2.2 (Altomare and Raşa [4]) $f \in D(A)$ is called a generalized A-subharmonic function if $A(f) \geq 0$.

The following result was obtained in [4, Prop. 6.2].

Proposition 2.1 *Assume that $f \in A_\infty(K)$ or, respectively, $f \in C^2(K)$ if $K \subset \mathbb{R}^p$ and $\operatorname{int} K \neq \phi$. If f is weakly T-convex, then $Af \geq 0$.*

Conjecture 6.3 from the same paper reads as follows.

Conjecture 2.1 If $f \in D(A)$ and $Af \geq 0$, then f is weakly T-convex.

Examples and results supporting this conjecture can be found in [4] but as far as we know the problem is still open.

3 A Result Related to the Conjecture

In this section we present a result related to Conjecture 2.1: briefly speaking, we show that a stronger hypothesis entails a stronger conclusion.

First, we need

Lemma 3.1 *If $f \in D(A)$ is a convex function, then $Af \geq 0$.*

Proof If $f \in C(K)$ is convex, then $f_{z,\alpha}$ is convex for all $z \in K, \alpha \in [0, 1]$. Due to (1), $Tf_{z,\alpha} \geq f_{z,\alpha}$. In particular, $Tf_{x,\alpha}(x) \geq f_{x,\alpha}(x) = f(x),\ x \in K,\ \alpha \in [0, 1]$.

Now an application of Theorem 1.3 in [8] concludes the proof. □

For $z \in K$ and $0 \leq a \leq 1$ let $J_{z,a} : K \to K_{z,a}$ be defined by

$$J_{z,a}(x) := ax + (1 - a)z,\ \ x \in K,$$

where $K_{z,a} := J_{z,a}(K)$.

If $x \in K$, let

$$\mu_x(f) := Tf(x),\ \ f \in C(X).$$

With a slight abuse of notation we can consider $\mu_x \circ J_{x,a}^{-1}$ as a probability Borel measure on K, with support contained in $K_{x,a}$, defined by

$$\int_K f(t)d(\mu_x \circ J_{x,a}^{-1})(t) = \int_K (f \circ J_{x,a})(s)d\mu_x(s),\ \ f \in C(K). \tag{2}$$

The barycenter of a probability measure on K will be denoted by $r(\mu)$.

Lemma 3.2 *If $x \in X$, $0 \le a < b \le 1$, and $f \in C(K)$ is convex, then*

$$\int_K f(t)d(\mu_x \circ J_{x,b}^{-1})(t) \ge \int_K f(t)d(\mu_x \circ J_{x,a}^{-1})(t). \tag{3}$$

Proof According to [7, p. 107] or [1, p. 27] it suffices to construct a dilation which transforms $\mu_x \circ J_{x,a}^{-1}$ into $\mu_x \circ J_{x,b}^{-1}$. More precisely, we will associate to each $t \in K$ a probability Radon measure D_t such that

(a) $r(D_t) = t$,
(b) For each $f \in C(K)$ the function $t \to D_t(f)$ is Borel measurable on K and

$$\int_K f(t)d(\mu_x \circ J_{x,b}^{-1})(t) = \int_K D_t(f)d(\mu_x \circ J_{x,a}^{-1})(t). \tag{4}$$

By δ_x we denote the point evaluation at $x \in K$. Define

$$D_t := \begin{cases} \delta_t, \ t \in K \setminus K_{x,a}, \\ \dfrac{a}{b}\delta_{(bt+(a-b)x)/a} + \left(1 - \dfrac{a}{b}\right) \mu_x \circ J_{x,b}^{-1}, \ t \in K_{x,a}, \end{cases}$$

where if $a = 0$ we assign to $(bt + (a - b)x)/a$ an arbitrary value from K.

If $t \in K_{x,a}$, there exists $s \in K$ such that $t = as + (1 - a)x$. For $g \in C(K)$ we have

$$D_t(g) = \frac{a}{b}g(bs + (1 - b)x) + \left(1 - \frac{a}{b}\right)\left(\mu_x \circ J_{x,b}^{-1}\right)(g). \tag{5}$$

For $h \in A(K)$ (5) shows that

$$\begin{aligned} D_t(h) &= \frac{a}{b}\,(bh(s) + (1 - b)h(x)) + \left(1 - \frac{a}{b}\right)\int_K h(by + (1 - b)x)d\mu_x(y) \\ &= ah(s) + \left(\frac{a}{b} - a\right)h(x) + \left(1 - \frac{a}{b}\right)h(x) \\ &= ah(s) + (1 - a)h(x) = h(t). \end{aligned}$$

This shows that condition (a) is satisfied.

Let now $f \in C(K)$. Clearly the function $t \to D_t(f)$ is Borel measurable on K. Using (2) and (5) we can write

$$\begin{aligned} &\int_K D_t(f)d\left(\mu_x \circ J_{x,a}^{-1}\right)(t) = \int_K D_{as+(1-a)x}(f)d\mu_x(s) \\ &= \int_K \frac{a}{b} f(bs + (1 - b)x)d\mu_x(s) + \left(1 - \frac{a}{b}\right)\int_K (\mu_x \circ J_{x,b}^{-1})(f)d\mu_x(s) \end{aligned}$$

$$= \frac{a}{b}\int_K (f \circ J_{x,b})(s)d\mu_x(s) + \left(1 - \frac{a}{b}\right)\left(\mu_x \circ J_{x,b}^{-1}\right)(f)$$
$$= \frac{a}{b}\int_K f(t)d\left(\mu_x \circ J_{x,b}^{-1}\right)(t) + \left(1 - \frac{a}{b}\right)\int_K f(t)d\left(\mu_x \circ J_{x,b}^{-1}\right)(t)$$
$$= \int_K f(t)d(\mu_x \circ J_{x,b}^{-1})(t).$$

This proves (4) and so condition (b) is satisfied, too. This concludes the proof of Lemma 3.2. □

Theorem 3.1 *Given $f \in C(K)$, let $F : K \times [0, 1] \to \mathbb{R}$,*

$$F(x, a) = Tf_{x,a}(x) - f(x), \; x \in K, \; a \in [0, 1].$$

Then,

(i) $F(x, 0) = 0, \; x \in K.$
(ii) *If f is convex, then $F(x, \cdot)$ is increasing on $[0, 1]$ for each $x \in K$. In particular, $F \geq 0$ on $K \times [0, 1]$.*

Proof

(i) For a fixed $x \in K$, $f_{x,0}(t) = f(x)$, $t \in K$, so that $f_{x,0} = f(x)\mathbf{1}$. It follows that $F(x, 0) = 0$, $x \in K$.
(ii) Let f be convex and $0 \leq a < b \leq 1$. Using Lemma 3.2 we have

$$Tf_{x,a}(x) = \mu_x(f_{x,a}) = \int_K (f \circ J_{x,a})d\mu_x = \int_K fd\left(\mu_x \circ J_{x,a}^{-1}\right)$$
$$\leq \int_K fd\left(\mu_x \circ J_{x,b}^{-1}\right) = \int_K \left(f \circ J_{x,b}\right)d\mu_x = \mu_x(f_{x,b}) = Tf_{x,b}(x).$$

Now $F(x, a) = Tf_{x,a}(x) - f(x) \leq Tf_{x,b}(x) - f(x) = F(x, b)$, and the proof is finished.

□

Remark 3.1 Clearly $f \in C(K)$ is weakly T-convex if and only if $F \geq 0$ on $K \times [0, 1]$. Let $f \in D(A)$. In Conjecture 2.1 the hypothesis is $Af \geq 0$ and the conclusion is $F \geq 0$. According to Lemma 3.1 the hypothesis in Theorem 3.1, namely "f is convex", is stronger than that in Conjecture 2.1. On the other hand, the conclusion of Theorem 3.1, namely "$F(x, \cdot)$ is increasing on $[0, 1]$ and $F(x, 0) = 0, x \in K$" is stronger than the conclusion of Conjecture 2.1. The relationship between these hypotheses, conclusions, and other conditions/properties deserves to be further investigated.

References

1. Alfsen, E.M.: Compact Convex Sets and Boundary Integrals. Springer, Heidelberg (1971)
2. Altomare, F.: Lototsky-Schnabl operators on compact convex sets and their associated semigroups. Mh. Math. **114**, 1–13 (1992)
3. Altomare, F., Campiti, M.: Korovkin-type Approximation Theory and its Applications. De Gruyter Studies in Mathematics Series, vol. 17. Walter de Gruyter, Berlin (1994)
4. Altomare, F., Raşa, I.: Feller semigroups, Bernstein type operators and generalized convexity associated with positive projections. In: New Developments in Approximation Theory (Dortmund, 1998). International Series of Numerical Mathematics, vol. 132, pp. 9–32. Birkhäuser, Basel (1999)
5. Altomare, F., Cappelletti Montano, M., Leonessa, V., Raşa, I.: Markov Operators, Positive Semigroups and Approximation Processes. Walter de Gruyter, Berlin/Munich/Boston (2014)
6. Altomare, F., Cappelletti Montano, M., Leonessa, V., Raşa, I.: A generalization of Kantorovich operators for convex compact subsets. Banach J. Math. Anal. **11**(3), 591–614 (2017)
7. Phelps, R.R.: Lectures on Choquet's Theorem. Van Nostrand, Princeton (1966)
8. Raşa, I.: On some properties of Altomare projections. Conf. Sem. Mat. Univ. Bari **253**, 1–17 (1993)
9. Vladislav, T., Raşa, I.: Analiză Numerică. Aproximare, Problema lui Cauchy Abstractă, Proiectori Altomare, Editura Tehnică, Bucureşti (1999) (in Romanian)

A Strong Variant of Weyl's Theorem Under Functional Calculus and Perturbations

Pietro Aiena

Dedicated to F. Altomare on the occasion of his 70th birthday

Abstract The aim of this paper is the study of a new and strong variant (SW) of the classical Weyl's theorem, for operators defined on Banach spaces, under functional calculus. Furthermore, we give some results on the permanence of (SW) theorem under commuting perturbations, as algebraic or finite-dimensional commuting perturbations.

Keywords Property (gaz) · Localized SVEP · Weyl type theorems

1 Introduction

The classical Weyl's theorem, for bounded linear operators on an infinite-dimensional complex Banach space X, admits some stronger versions, as a-Weyl theorem and property (w), that have been studied in several papers by different authors, see [2, Chapters 6] for an extensive exposition. Property (w) and a-Weyl's theorem are independent one each other, and each one of them entails Weyl's theorem. A stronger version of Weyl type theorems has been recently introduced in [5] and the operators which satisfy this version are said to satisfy S-Weyl's theorem (*Strong Weyl's theorem*). This theorem entails both a-Weyl's theorem and property (w). Although S-Weyl's theorem requires rather strong spectral properties, for instance that the approximate point spectrum of T coincides with the whole spectrum $\sigma(T)$ there are important examples of operators which satisfy it, as the unilateral weighted left shift, the dual of non-invertible isometry which is not quasi-nilpotent, and the Toeplitz operators T_ϕ defined on the Hardy space $H^2(\mathbf{T})$ for

P. Aiena (✉)
Università di Palermo, Palermo, Italy
e-mail: pietro.aiena@unipa.it

A. M. Candela et al. (eds.), *Recent Advances in Mathematical Analysis*,
Trends in Mathematics, https://doi.org/10.1007/978-3-031-20021-2_3

which the symbol ϕ is a continuous function such that the orientation of the curve $\phi(\mathbf{T})$ traced out by ϕ is counterclockwise, see [5].

In this paper the permanence of S-Weyl's theorem is studied under functional calculus. In general, S-Weyl's theorem cannot be extended from an operator T to $f(T)$, where f is an analytic function defined on an open disc containing the spectrum $\sigma(T)$ and $f(T)$ is defined by the classical functional calculus. In this paper we shall prove that if T satisfies S-Weyl's theorem and T is isoloid then S-Weyl's theorem holds for $f(T)$ if and only if the spectral mapping theorem holds for the upper semi B-Weyl spectrum $\sigma_{\mathrm{ubw}}(T)$. In the last part of the paper we study the permanence of S-Weyl's theorem under suitable commuting perturbations, for instance Riesz commuting perturbations. or algebraic commuting perturbations.

2 Definitions and Preliminary Results

Let $T \in L(X)$ be a bounded linear operator defined on an infinite-dimensional complex Banach space X, and denote by $\alpha(T)$ the dimension of the kernel $\ker T$, while $\beta(T)$ denotes the codimension of the range $R(T) := T(X)$. Recall that $T \in L(X)$ is said to be *upper semi-Fredholm*, $T \in \Phi_+(X)$, if $\alpha(T) < \infty$ and $T(X)$ is closed, while $T \in L(X)$ is said to be *lower semi-Fredholm*, $T \in \Phi_-(X)$, if $\beta(T) < \infty$. The class of *Fredholm* operators is defined by $\Phi(X) := \Phi_+(X) \cap \Phi_-(X)$, while the class of *semi-Fredholm* operators is defined by $\Phi_\pm(X) := \Phi_+(X) \cup \Phi_-(X)$. If $T \in \Phi_\pm(X)$ then the index is defined by ind $(T) := \alpha(T) - \beta(T)$. The set of *Weyl operators* is defined by $W(X) := \{T \in \Phi(X) : \text{ind}\, T = 0\}$, the class of *upper semi-Weyl operators* is defined by $W_+(X) := \{T \in \Phi_+(X) : \text{ind}\, T \leq 0\}$, and class of *lower semi-Weyl operators* is defined by $W_-(X) := \{T \in \Phi_-(X) : \text{ind}\, T \geq 0\}$. Clearly, $W(X) = W_+(X) \cap W_-(X)$. These classes of operators generate the *Weyl spectrum*, defined by $\sigma_{\mathrm{w}}(T) := \{\lambda \in \mathbb{C} : \lambda I - T \notin W(X)\}$, and analogously are defined the *upper semi-Weyl spectrum* $\sigma_{\mathrm{uw}}(T)$ and the *lower semi-Weyl spectrum* $\sigma_{\mathrm{uw}}(T)$.

Let $p := p(T)$ be the *ascent* of an operator T; i.e. the smallest non-negative integer p for which $\ker T^p = \ker T^{p+1}$. If such integer does not exist we set $p(T) = \infty$. Analogously, let $q := q(T)$ be the *descent* of T; i.e the smallest non-negative integer q such that $T^q(X) = T^{q+1}(X)$, and if such integer does not exist we set $q(T) = \infty$. It is well known that if $p(T)$ and $q(T)$ are both finite then $p(T) = q(T)$, see [1, Theorem 3.3]. An important result in spectral theory shows that $0 < p(\lambda I - T) = q(\lambda I - T) < \infty$ if and only if λ is a pole of the resolvent, see [10, Proposition 50.2].

The following result, which will be needed in the sequel, is a particular case of Lemma 12 of [11].

Lemma 2.1 *If $T \in L(X)$ and $p = p(T) < \infty$ then the following statements are equivalent:*

(i) *There exists $n \geq p + 1$ such that $T^n(X)$ is closed;*
(ii) *$T^n(X)$ is closed for all $n \geq p$.*

The class of all *Browder operators* is defined as the set $B(X) := \{T \in \Phi(X) : p(T), q(T) < \infty\}$; the class of all *upper semi-Browder operators* is defined $B_+(X) := \{T \in \Phi_+(X) : p(T) < \infty\}$, and the class of all *lower semi-Browder operators* is defined $B_+(X) := \{T \in \Phi_-(X) : q(T) < \infty\}$. Obviously, $B(X) \subseteq W(X)$ and $B_+(X) \subseteq W_+(X)$ and $B_-(X) \subseteq W_-(X)$. The *upper semi-Browder spectrum* (respectively, *lower semi-Browder spectrum* and the *Browder spectrum*) is denoted by $\sigma_{\mathrm{ub}}(T)$ (respectively, $\sigma_{\mathrm{lb}}(T)$, $\sigma_{\mathrm{b}}(T)$). In the sequel we denote by $\sigma_{\mathrm{ap}}(T)$ the classical *approximate point spectrum*, defined by $\sigma_{\mathrm{ap}}(T) := \{\lambda \in \mathbb{C} : \lambda I - T \text{ is not bounded below}\}$, where an operator is said to be *bounded below* if it is injective and has closed range. The *surjective spectrum* of T is denoted by $\sigma_{\mathrm{s}}(T)$.

An operator $T \in L(X)$ is said to satisfy *Browder's theorem* if $\sigma_{\mathrm{w}}(T) = \sigma_{\mathrm{b}}(T)$, or equivalently $\Delta(T) = p_{00}(T)$, where

$$\Delta(T) := \sigma(T) \setminus \sigma_{\mathrm{w}}(T) \quad \text{and} \quad p_{00}(T) = \sigma(T) \setminus \sigma_{\mathrm{b}}(T).$$

The operator $T \in L(X)$ is said to satisfy *a-Browder's theorem* if $\sigma_{\mathrm{uw}}(T) = \sigma_{\mathrm{ub}}(T)$, or equivalently $\Delta_a(T) = p_{00}^a(T)$, where

$$\Delta_a(T) := \sigma_{\mathrm{ap}}(T) \setminus \sigma_{\mathrm{uw}}(T) \quad \text{and} \quad p_{00}^a(T) := \sigma_{\mathrm{ap}}(T) \setminus \sigma_{\mathrm{ub}}(T).$$

It is known that a-Browder's theorem entails Browder's theorem, [2, Corollary 5.30].

Semi-Fredholm operators have been generalized by Berkani [6–8] in the following way: for every $T \in L(X)$ and a nonnegative integer n let us denote by $T_{[n]}$ the restriction of T to $T^n(X)$, viewed as a map from the space $T^n(X)$ into itself (we set $T_{[0]} = T$). $T \in L(X)$ is said to be *semi B-Fredholm*, (resp. *B-Fredholm, upper semi B-Fredholm, lower semi B-Fredholm,*) if for some integer $n \geq 0$ the range $T^n(X)$ is closed and $T_{[n]}$ is a semi-Fredholm operator (resp. Fredholm, upper semi-Fredholm, lower semi-Fredholm). In this case $T_{[m]}$ is a semi-Fredholm operator for all $m \geq n$ with the same index of $T_{[n]}$. This enables one to define the index of a semi B-Fredholm as $\text{ind}\, T = \text{ind}\, T_{[n]}$.

A bounded operator $T \in L(X)$ is said to be *B-Weyl* (respectively, *upper semi B-Weyl, lower semi B-Weyl*) if for some integer $n \geq 0$ the range $T^n(X)$ is closed and $T_{[n]}$ is Weyl (respectively, upper semi-Weyl. lower semi-Weyl). The *B-Weyl spectrum* is defined by $\sigma_{\mathrm{bw}}(T) := \{\lambda \in \mathbb{C} : \lambda I - T \text{ is not B-Weyl}\}$, and the *upper semi B-Weyl spectrum* of T is defined by $\sigma_{\mathrm{ubw}}(T) := \{\lambda \in \mathbb{C} : \lambda I - T \text{ is not upper semi B-Weyl}\}$. Analogously, the *lower semi B-Weyl spectrum* of T is defined by $\sigma_{\mathrm{lbw}}(T) := \{\lambda \in \mathbb{C} : \lambda I - T \text{ is not lower semi B-Weyl}\}$. Note that if $T \in L(X)$ is upper semi B-Weyl and $\alpha(T) < \infty$ then T is upper semi-Weyl, see [1].

The concept of Drazin invertibility has been introduced in a more abstract setting than operator theory. In the case of the Banach algebra $L(X)$, $T \in L(X)$ is said to be *Drazin invertible* (with a finite index), $T \in D(X)$, if $p(T) = q(T) < \infty$. Clearly, $\lambda I - T \in L(X)$ is Drazin invertible if and only if $\lambda I - T$ is invertible or λ is a pole of the resolvent. The concept of Drazin invertibility for bounded operators suggests the following definition:

Definition 2.1 An operator $T \in L(X)$, is said to be *left Drazin invertible*, $T \in LD(X)$, if $p := p(T) < \infty$ and $T^{p+1}(X)$ is closed. $T \in L(X)$, is said to be *right Drazin invertible*, $T \in RD(X)$, if $q := q(T) < \infty$ and $T^q(X)$ is closed. If $\lambda I - T$ is left Drazin invertible and $\lambda \in \sigma_{\rm ap}(T)$ then λ is said to be a *left pole*. A left pole λ is said to have finite rank if $\alpha(\lambda I - T) < \infty$. If $\lambda I - T$ is right Drazin invertible and $\lambda \in \sigma_{\rm s}(T)$ then λ is said to be a *right pole*. A right pole λ is said to have finite rank if $\beta(\lambda I - T) < \infty$.

It should be noted that there is a perfect duality, i.e., T is left Drazin invertible (respectively, right Drazin invertible) if and only if T^* is right Drazin invertible (respectively, left Drazin invertible). Furthermore, $T \in L(X)$ is Drazin invertible if and only if T is both left Drazin invertible and right Drazin invertible, see [2, Chapter 1].

The *Drazin spectrum* is defined as $\sigma_{\rm d}(T) := \{\lambda \in \mathbb{C} : \lambda I - T \notin D(X)\}$. The *left Drazin spectrum* is defined as $\sigma_{\rm ld}(T) := \{\lambda \in \mathbb{C} : \lambda I - T \notin LD(X)\}$, while the *right Drazin spectrum* is defined as $\sigma_{\rm rd}(T) := \{\lambda \in \mathbb{C} : \lambda I - T \notin RD(X)\}$. Evidently, $\sigma_{\rm d}(T) = \sigma_{\rm ld}(T) \cup \sigma_{\rm rd}(T)$. Every left Drazin invertible operator is upper semi B-Weyl (respectively, every right Drazin invertible operator is lower semi B-Weyl, every Drazin invertible operator is B-Weyl), so we have:

$$\sigma_{\rm ubw}(T) \subseteq \sigma_{\rm ld}(T) \quad \text{and}\, \sigma_{\rm bw}(T) \subseteq \sigma_{\rm d}(T).$$

Denote by $\Pi(T)$, $\Pi_a(T)$ and $\Pi_s(T)$ the set of all poles, the set of left poles of T, and the set of right poles respectively. Clearly,

$$\Pi(T) = \sigma(T) \setminus \sigma_{\rm d}(T) \quad \text{and} \quad \Pi_a(T) = \sigma_{\rm ap}(T) \setminus \sigma_{\rm ld}(T)$$

and, analogously, $\Pi_s(T) = \sigma_{\rm s}(T) \setminus \sigma_{\rm rd}(T)$.

Lemma 2.2 *If $T \in L(X)$ then*

$$\Pi_a(T) \subseteq iso\, \sigma_{\rm ap}(T) \quad and \quad \Pi_s(T) \subseteq iso\, \sigma_{\rm s}(T). \tag{1}$$

Proof In fact, if $\lambda_0 \in \Pi_a(T)$ then $\lambda I - T$ is left Drazin invertible and hence $p(\lambda_0 I - T) < \infty$. Since $\lambda I - T$ has topological uniform descent (see [9], for definition and details), it then follows, from [9, Corollary 4.8], that $\lambda I - T$ is bounded below in a punctured disc centered at λ_0. An analogous reasoning shows that $\Pi_s(T) \subseteq \text{iso}\, \sigma_{\rm s}(T)$ for all $T \in L(X)$. □

The single-valued extension property has a basic importance in local spectral theory since it is satisfied by a wide variety of linear bounded operators in the spectral decomposition problem.

Definition 2.2 $T \in L(X)$ is said to have *the single valued extension property* at $\lambda_0 \in \mathbb{C}$ (abbreviated SVEP at λ_0), if for every open disc U of λ_0, the only analytic function $f : U \to X$ which satisfies the equation

$$(\lambda I - T)f(\lambda) = 0 \quad \text{for all } \lambda \in U$$

is the function $f \equiv 0$. An operator $T \in L(X)$ is said to have SVEP if T has SVEP at every point $\lambda \in \mathbb{C}$.

Evidently, both T and the dual T^* have SVEP at the points λ which belong to the boundary $\partial\sigma(T)$ of the spectrum. In particular both T and T^* have SVEP at the isolated points of the spectrum. Furthermore, if $\sigma_{\mathrm{ap}}(T) \subseteq \partial\sigma(T)$ then T has SVEP, and dually the inclusion $\sigma_{\mathrm{su}}(T) \subseteq \partial\sigma(T)$ entails that T' has SVEP. Furthermore, it is not difficult to see that

$$\sigma_{\mathrm{ap}}(T) \text{ does not cluster at } \lambda \Rightarrow T \text{ has SVEP at } \lambda,$$

and dually,

$$\sigma_{\mathrm{su}}(T) \text{ does not cluster at } \lambda \Rightarrow T^* \text{ has SVEP at } \lambda.$$

Define

$$\Delta_a^g(T) := \sigma_{\mathrm{ap}}(T) \setminus \sigma_{\mathrm{ubw}}(T) \quad \text{and} \quad \Delta_1^g(T) := \sigma(T) \setminus \sigma_{\mathrm{ubw}}(T).$$

Since $\sigma_{\mathrm{ubw}}(T) \subseteq \sigma_{\mathrm{ld}}(T)$, we have

$$\Pi_a(T) \subseteq \Delta_a^g(T) \subseteq \Delta_1^g(T).$$

Define $p_{00}(T) := \sigma(T) \setminus \sigma_{\mathrm{b}}(T)$ and $p_{00}^a(T) := \sigma_{\mathrm{ap}}(T) \setminus \sigma_{\mathrm{ub}}(T)$. It is easily seen that $p_{00}(T) \subseteq p_{00}^a(T)$ and obviously every point of $p_{00}(T)$ is an isolated point of $\sigma(T)$, as well as of $\sigma_{\mathrm{ap}}(T)$. An operator $T \in L(X)$ is said to satisfies *generalized a-Browder's theorem* if $\sigma_{\mathrm{ubw}}(T) = \sigma_{\mathrm{ld}}(T)$. Generalized a-Browder's theorem and a-Browder's theorem are equivalent, see [2, Theorem 5.38]

Definition 2.3 An operator $T \in L(X)$ is said to satisfy *property* (gaz) if $\Delta_1^g(T) = \Pi_a(T)$, or equivalently $\sigma(T) \setminus \sigma_{\mathrm{uw}}(T) = p_{00}^a(T)$, see [4, Theorem 3.5], or [12].

Note that

$$\sigma_{\mathrm{rd}}(T) = \sigma_{\mathrm{d}}(T) \Leftrightarrow \sigma_{\mathrm{ap}}(T) = \sigma(T), \tag{2}$$

see [5, Lemma 2.4]. Analogously,

$$\sigma_{\mathrm{rd}}(T) = \sigma_{\mathrm{d}}(T) \Leftrightarrow \sigma_{\mathrm{s}}(T) = \sigma(T), \tag{3}$$

The proof of the following result may be found in [4].

Theorem 2.1 *Let $T \in L(X)$. Then T has property (gaz) if and only if a-Browder's theorem holds for T and $\sigma_{\mathrm{ap}}(T) = \sigma(T)$.*

The next result shows that property (gaz) for T, or T^*, entails that several other spectra coincide.

Theorem 2.2 *Let $T \in L(X)$. Then we have:*

(i) *If T has property (gaz) then*

$$\sigma_{\mathrm{ubw}}(T) = \sigma_{\mathrm{bw}}(T) = \sigma_{\mathrm{ld}}(T) = \sigma_{\mathrm{d}}(T) \tag{4}$$

Consequently, $\Pi(T) = \Pi_a(T)$.

(ii) *If T^* has property (gaz) then*

$$\sigma_{\mathrm{ubw}}(T^*) = \sigma_{\mathrm{bw}}(T^*) = \sigma_{\mathrm{rd}}(T) = \sigma_{\mathrm{d}}(T) \tag{5}$$

Consequently, $\Pi(T) = \Pi_s(T)$.

Proof

(i) By Theorem 2.1 we have $\sigma_{\mathrm{ap}}(T) = \sigma(T)$ and hence, from the equivalence (2), we have $\sigma_{\mathrm{ld}}(T) = \sigma_{\mathrm{d}}(T)$. By Theorem 2.1 T also satisfies generalized a-Browder's theorem, so $\sigma_{\mathrm{ubw}}(T) = \sigma_{\mathrm{ld}}(T)$. Moreover, since generalized a-Browder's theorem entails generalized Browder's theorem, we have $\sigma_{\mathrm{bw}}(T) = \sigma_{\mathrm{d}}(T)$. Therefore, the equalities (4) hold. Since $\sigma_{\mathrm{ap}}(T) = \sigma(T)$ we also have

$$\Pi_a(T) = \sigma_{\mathrm{ap}}(T) \setminus \sigma_{\mathrm{ld}}(T) = \sigma(T) \setminus \sigma_{\mathrm{d}}(T) = \Pi(T).$$

(ii) By Theorem 2.1 we have

$$\sigma_{\mathrm{s}}(T) = \sigma_{\mathrm{ap}}(T^*) = \sigma(T^*) = \sigma(T),$$

and hence, from the equivalence (3), $\sigma_{\mathrm{rd}}(T) = \sigma_{\mathrm{d}}(T)$. By Theorem 2.1 T^* also satisfies generalized a-Browder's theorem, so $\sigma_{\mathrm{ubw}}(T^*) = \sigma_{\mathrm{ld}}(T^*)$, or equivalently, $\sigma_{\mathrm{ubw}}(T^*) = \sigma_{\mathrm{rd}}(T)$. Moreover, since generalized a-Browder's theorem entails generalized Browder's theorem, we have $\sigma_{\mathrm{bw}}(T^*) = \sigma_{\mathrm{d}}(T^*) = \sigma_{\mathrm{d}}(T)$. Therefore, the equalities (5) hold. Since $\sigma_{\mathrm{s}}(T) = \sigma_{\mathrm{ap}}(T^*) = \sigma(T^*) = \sigma(T)$, we have

$$\Pi_s(T) = \sigma_{\mathrm{s}}(T) \setminus \sigma_{\mathrm{rd}}(T) = \sigma(T) \setminus \sigma_{\mathrm{d}}(T) = \Pi(T),$$

as desired.

□

Let $\mathcal{H}(\sigma(T))$ be the set of all analytic functions f defined on an open disc containing the spectrum $\sigma(T)$, and let $f(T)$ be defined by the classical functional calculus. Recall that the spectral mapping theorem holds for $\sigma_{\rm ap}(T)$ and $\sigma_{\rm s}(T)$. The spectral mapping theorem does not hold for the semi B-Weyl spectra, we only have the inclusions

$$f(\sigma_{\rm ubw}(T)) \subseteq \sigma_{\rm ubw}(f(T)) \quad \text{and} \quad f(\sigma_{\rm lbw}(T)) \subseteq \sigma_{\rm lbw}(f(T)), \tag{6}$$

see [2, Chapter 3].

Remark 2.1 Let $U \in L(X)$ be invertible, and $UT = TU$. It is easily seen that if UT is upper semi B-Weyl then T is upper semi B-Weyl with ind $(UT) =$ ind (T).

In the sequel we shall need the following simple result.

Lemma 2.3 *Let $T \in L(X)$ and suppose that $f \in \mathcal{H}(\sigma(T))$ is injective on $\sigma_{\rm ubw}(T)$. Then $\sigma_{\rm ubw}(f(T) = f(\sigma_{\rm ubw}(T))$, and $\sigma_{\rm lbw}(f(T) = f(\sigma_{\rm lbw}(T))$.*

Proof To show the first assertion, we need only to prove the inclusion $\sigma_{\rm ubw}(f(T) \subseteq f(\sigma_{\rm ubw}(T)$. Suppose that $\mu_0 \in f(\sigma_{\rm ubw}(T))$ and let $\lambda_0 \in \sigma_{\rm ubw}(T)$ be such that $\mu_0 = f(\lambda_0)$, so $\lambda_0 I - T$ is not upper semi B-Weyl. Define an analytic function g in the same domain of f as

$$g(\lambda) := \begin{cases} \dfrac{f(\lambda) - f(\lambda_0)}{\lambda - \lambda_0} & \text{if } \lambda \neq \lambda_0, \\ f'(\lambda_0) & \text{if } \lambda = \lambda_0. \end{cases}$$

Since f is injective on $\sigma_{\rm ubw}(T)$, the function g does not vanish on $\sigma_{\rm ubw}(T)$, hence $0 \notin g(\sigma(T)) = \sigma(g(T))$, thus $g(T)$ is invertible. Evidently,

$$g(T)(\lambda_0 I - T) = (\lambda_0 I - T)g(T) = \mu_0 I - f(T).$$

Suppose now that $\mu_0 \notin \sigma_{\rm ubw}(f(T))$. Then $\mu_0 I - f(T)$ is upper semi B-Weyl and from Remark 2.1 we have $\lambda_0 I - T$ is upper semi B-Weyl, i.e., $\lambda_0 \notin \sigma_{\rm ubw}(T)$, a contradiction. Therefore, $\mu_0 \in \sigma_{\rm ubw}(f(T))$, as desired.
The proof of the second assertion is similar. □

Theorem 2.3 *Suppose that $T \in L(X)$ satisfies property (gaz) and $f \in \mathcal{H}(\sigma(T))$. Then the following are equivalent:*

(i) *$f(T)$ satisfies property (gaz);*
(ii) *$f(\sigma_{\rm ubw}(T)) = \sigma_{\rm ubw}(f(T))$.*

Proof (i) $\Rightarrow$ (ii). Since T and $f(T)$ have property (gaz) we have, by Theorem 2.2, $\sigma_{\rm ubw}(T) = \sigma_{\rm ld}(T)$ and $\sigma_{\rm ubw}(f(T)) = \sigma_{\rm ld}(f(T))$. The set of left Drazin invertible operators is a regularity, see [2, Chapter 3], so the spectral mapping theorem holds for the left Drazin spectrum. Therefore,

$$\sigma_{\mathrm{ubw}}(f(T)) = \sigma_{\mathrm{ld}}(f(T)) = f(\sigma_{\mathrm{ld}}(T)) = f(\sigma_{\mathrm{ubw}}(T)).$$

(ii) $\Rightarrow$ (i) Since T has property (gaz) we have $\sigma(T) = \sigma_{\mathrm{ap}}(T)$. The spectral mapping theorem holds for the approximate point spectrum, so

$$\sigma_{\mathrm{ap}}(f(T)) = f(\sigma_{\mathrm{ap}}(T)) = f(\sigma(T)) = \sigma_{\mathrm{ap}}(f(T)).$$

Now, T satisfies Browder's theorem, or equivalently generalized a-Browder's theorem, so $\sigma_{\mathrm{ubw}}(T) = \sigma_{\mathrm{ld}}(T)$, from which we obtain

$$\sigma_{\mathrm{ubw}}(f(T)) = f(\sigma_{\mathrm{ubw}}(T)) = f(\sigma_{\mathrm{ld}}(T)) = \sigma_{\mathrm{ld}}(f(T)),$$

i.e., $f(T)$ satisfies generalized Browder's theorem. By Theorem 2.1 it then follows that $f(T)$ satisfies property (gaz). □

Corollary 2.1 *Let $T \in L(X)$ and suppose that either T or T^* have SVEP. If T satisfies property (gaz) then $f(T)$ satisfies property (gaz) for every $f \in \mathcal{H}(\sigma(T))$.*

Proof If either T or T^* have SVEP then the spectral mapping theorem holds for $\sigma_{\mathrm{ubw}}(T)$, see [2, Corollary 3.123]. □

From Lemma 2.3 we also have:

Corollary 2.2 *Let $T \in L(X)$ and suppose that $f \in \mathcal{H}(\sigma(T))$ is injective. If T satisfies property (gaz) then $f(T)$ satisfies property (gaz).*

3 S-Weyl's Theorem

Define

$$\pi_{00}(T) := \{\lambda \in \operatorname{iso}\sigma(T) : 0 < \alpha(\lambda I - T) < \infty\}$$

and

$$\pi_{00}^{a}(T) := \{\lambda \in \operatorname{iso}\sigma_{\mathrm{ap}}(T) : 0 < \alpha(\lambda I - T) < \infty\}.$$

Evidently,

$$p_{00}(T) \subseteq \pi_{00}(T) \subseteq \pi_{00}^{a}(T) \quad \text{and} \quad p_{00}^{a}(T) \subseteq \pi_{00}^{a}(T).$$

Weyl type theorems are defined as follows:

Definition 3.1 A bounded operator $T \in L(X)$ is said to satisfy *Weyl's theorem* if $\sigma(T) \setminus \sigma_{\mathrm{w}}(T) = \pi_{00}(T)$, $T \in L(X)$ is said to satisfy *a-Weyl's theorem* if $\sigma_{\mathrm{ap}}(T) \setminus \sigma_{\mathrm{uw}}(T) = \pi_{00}^{a}(T)$, $T \in L(X)$ is said to satisfy property (w) if $\sigma_{\mathrm{ap}}(T) \setminus \sigma_{\mathrm{uw}}(T) = \pi_{00}(T)$.

The following diagram resume the relationships between Weyl's theorems, a-Browder's theorem and property (w).

$$\begin{array}{ccc} \text{Property } (w) & \Rightarrow & a\text{-Browder's theorem} \\ \Downarrow & & \Uparrow \\ \text{Weyl's theorem} & \Leftarrow & a\text{-Weyl's theorem} \end{array}$$

see [2, Chapter 6]. Property (w) is not intermediate between Weyl's theorem and a-Weyl's theorem.

We now introduce a strong variant of Weyl's theorem.

Definition 3.2 A bounded operator $T \in L(X)$ is said to satisfy the *S-Weyl's theorem*, abbreviated (SW), if $\sigma(T) \setminus \sigma_{\mathrm{ubw}}(T) = \pi_{00}(T)$.

The following result has been proved in [5] .

Theorem 3.1 *An operator $T \in L(X)$ satisfies S-Weyl's theorem if and only if T satisfies property (gaz) and $\Pi_a(T) = \pi_{00}(T)$.*

Actually, S-Weyl's theorem is a stronger version of Weyl type theorems. In fact, see [5, Theorem 3.11], we have:

$$\text{S-Weyl's theorem} \Rightarrow \text{a-Weyl's theorem and property } (w).$$

Recall that $T \in L(X)$ is said to be *isoloid* if every $\lambda \in \operatorname{iso} \sigma(T)$ is an eigenvalue of T. $T \in L(X)$ is said to be *a-isoloid* if every $\lambda \in \operatorname{iso} \sigma_{\mathrm{ap}}(T)$ is an eigenvalue of T.

Theorem 3.2 *Let $\{\lambda_1, \lambda_2, \ldots, \lambda_n\}$ be a finite subset of $\mathbb{C}$ such that $\lambda_i \neq \lambda_j$ for $i \neq j$. Assume that $\{\nu_1, \nu_2, \ldots, \nu_n\} \subset \mathbb{N}$ and set $h(\lambda) = \prod_{i=1}^{k} (\lambda_i - \lambda)^{\nu_i}$. If $T \in L(X)$ set $h(T) := \prod_{i=1}^{k} (\lambda_i I - T)^{\nu_i}$.*

(i) *We have*

$$0 \in \Pi_a(h(T)) \Leftrightarrow \lambda_i \in \Pi_a(T) \quad \textit{for all } i = 1, \ldots, n.$$

(ii) *If T is a-isoloid then*

$$\sigma_{\mathrm{ap}}(f(T)) \setminus \pi_{00}^{a}(f(T)) = f(\sigma(T) \setminus \pi_{00}^{a}(T)).$$

Proof

(i) It is known, see [2, Lemma 1.76] that

$$\ker h(T) = \bigoplus_{i=1}^{n} \ker (\lambda_i I - T)^{\nu_i}, \tag{7}$$

and

$$h(T)(X) = \bigcap_{i=1}^{n} (\lambda_i I - T)^{\nu_i}(X). \tag{8}$$

To show the implication $\Rightarrow$), suppose that $\lambda_i \in \Pi_a(T)$ for all $i = 1, \ldots, n$. Then $\lambda_i \in \sigma_{\mathrm{ap}}(T)$ and $\lambda_i I - T$ is left Drazin invertible for all $i = 1, \ldots, n$. so the ascent $p(\lambda_i I - T) < \infty$ and $(\lambda_i I - T)^{p_i+1}(X)$ is closed for every i. Choose $m \in \mathbb{N}$ such that

$$\nu_i m > p_i \quad \text{for all } i = 1, \ldots, n.$$

Then $\ker (\lambda_i I - T)^{\nu_i m} = \ker (\lambda_i I - T)^{\nu_i m+1}$ and from (7) we obtain $\ker h(T)^{\nu_i m} = \ker h(T)^{\nu_i m+1}$. By Lemma 2.1 we know that $(\lambda_i - T)^{\nu_i m+1}(X)$ is closed for all $i = 1, \ldots, n$, hence from (8) we deduce that $h(T)^{\nu_i m+1}(X)$ is closed, thus $h(T)$ is left Drazin invertible. Since $\lambda_i \in \sigma_{\mathrm{ap}}(T)$, from the equality $\sigma_{\mathrm{ap}}(h(T)) = h(\sigma_{\mathrm{ap}}(T)$, we obtain $0 = h(\lambda_i) \in \sigma_{\mathrm{ap}}(h(T)$, so $0 \in \Pi_a(h(T))$.

Conversely, to show the implication $\Leftarrow$), assume that $\lambda_j \notin \Pi_a(T)$ for some j.
We have either $\lambda_j \notin \sigma_{\mathrm{ap}}(T)$ or $\lambda_j \in \sigma_{\mathrm{ap}}(T)$. If $\lambda_j \notin \sigma_{\mathrm{ap}}(T)$ then $\lambda_j I - T$ is bounded below and again from the equality $\sigma_{\mathrm{ap}}(h(T)) = h(\sigma_{\mathrm{ap}}(T)$ we see that $0 = h(\lambda_j) \notin \sigma_{\mathrm{ap}}(h(T))$, thus $0 \notin \Pi_a(h(T)) = \sigma_{\mathrm{ap}}(h(T)) \setminus \sigma_{\mathrm{ld}}(h(T))$. Assume the other case, $\lambda_j \in \sigma_{\mathrm{ap}}(T)$. Since $\lambda_j \notin \Pi_a(T)$ then $\lambda_j I - T$ is not left Drazin invertible, so either $p_j := p(\lambda_j - T) = \infty$ or $p_j < \infty$ but $(\lambda_j I - T)^{p_j+1}(X)$ is not closed.

If $p_j = \infty$ then $\ker (\lambda_j I - T)^k$ is properly contained in $\ker (\lambda_j I - T)^{k+1}$ for all $k \in \mathbb{N}$. Being $\ker (\lambda_j I - T)^{\nu_j} \cap \ker (\lambda_i I - T)^{\nu_i} = \emptyset$ for $i \neq j$ we then deduce that $\ker h(T)^k$ is property contained in $\ker h(T)^{k+1}$ for all $k \in \mathbb{N}$, so $h(T)$ has infinite ascent. In the other case where $(\lambda_j I - T)^{p_j+1}(X)$ is not closed, from (8) we see that $h(T)^{p_j+1}(X)$ is not closed, so $0 \in \sigma_{\mathrm{ld}}(h(T))$, and hence $0 \notin \Pi_a(h(T))$, as desired.

(ii) This equality has been known, see [2, Lemma 6.78].

□

Theorem 3.3 *Let $T \in L(X)$ be isoloid and $f \in \mathcal{H}(\sigma(T))$. If T satisfies S-Weyl's theorem then the following assertions are equivalent:*

(i) *$f(T)$ satisfies S-Weyl's theorem;*
(ii) $f(\sigma_{\mathrm{ubw}}(T)) = \sigma_{\mathrm{ubw}}(f(T))$.

Proof Note first that since T satisfies T satisfies S-Weyl's theorem and hence property (gaz) then $\sigma(T) = \sigma_{\text{ap}}(T)$ and $\sigma_{\text{ubw}}(T) = \sigma_{\text{ld}}(T)$, by Theorem 2.2, in particular T is a-isoloid and $\pi_{00}(T) = \pi_{00}^a(T)$. Furthermore, by Theorem 3.1, $\Pi_a(T) = \pi_{00}(T)$.

(i) $\Rightarrow$ (ii) Since $f(T)$satisfies S-Weyl's theorem and hence property (gaz) we have $\sigma_{\text{ubw}}(f((T)) = \sigma_{\text{ld}}(f(T))$, and taking into account that the spectral mapping theorem holds for $\sigma_{\text{ld}}(T)$ we have

$$f(\sigma_{\text{ubw}}(f(T)) = f(\sigma_{\text{ld}}(T)) = \sigma_{\text{ld}}(f(T)) = \sigma_{\text{ubw}}(f(T)).$$

(ii) $\Rightarrow$ (i) T satisfies (gaz) and by Theorem 2.3, the condition (ii) entails property (gaz) for $f(T)$, hence $\sigma_{\text{ubw}}(f(T)) = \sigma_{\text{ld}}(f(T))$ and $\sigma(f(T)) = \sigma_{\text{ap}}(f(T))$, by Theorem 2.2, and $\pi_{00}^a(f(T)) = \pi_{00}(f(T))$. To show that $f(T)$ satisfies S-Weyl's theorem it suffices, by Theorem 3.1 to prove $\Pi_a(f(T)) = \pi_{00}(f(T))$. By Theorem 3.1 we have

$$\Pi_a(T) = \sigma_{\text{ap}}(T) \setminus \sigma_{\text{ld}}(T) = \pi_{00}(T) = \pi_{00}^a(T),$$

so

$$\sigma_{\text{ubw}}(f(T)) = \sigma_{\text{ld}}(f(T)) = f(\sigma_{\text{ld}}(T)) = f(\sigma_{\text{ap}}(T) \setminus \pi_{00}(T) = f(\sigma_{\text{ap}}(T) \setminus \pi_{00}^a(T)$$

From Theorem 3.2 we then have

$$\sigma_{\text{ubw}}(f(T)) = \sigma_{\text{ap}}(f(T)) \setminus \pi_{00}^a(f(T)). \tag{9}$$

From the equality (9) we then obtain

$$\begin{aligned}\pi_{00}(f(T)) &= \pi_{00}^a(f(T)) = \sigma_{\text{ap}}(f(T)) \setminus \sigma_{\text{ubw}}(f(T)) \\ &= \sigma_{\text{ap}}(f(T)) \setminus \sigma_{\text{ld}}(f(T)) = \Pi_a(f(T)),\end{aligned}$$

as desired. □

Since the spectral mapping theorem holds for $\sigma_{\text{ubw}}(T)$ if T or T^* has SVEP, or if f is injective, by Lemma 2.3, we have:

Corollary 3.1 *Let $T \in L(X)$ satisfies S-Weyl's theorem.*

(i) *If either T or T^* have SVEP then S-Weyl's theorem holds for $f(T)$ for every $f \in \mathcal{H}(\sigma(T))$.*
(ii) *If $f \in \mathcal{H}(\sigma(T))$ is injective then S-Weyl's theorem holds for $f(T)$.*

4 S-Weyl's Theorem Under Commuting Perturbations

In this section we give some results concerning property (gaz) and S-Weyl theorem under commuting perturbations.

Theorem 4.1 *Suppose that T^* has SVEP and $\text{iso}\,\sigma_{\text{ap}}(T) = \emptyset$. Then $f(T + K)$ satisfies S-Weyl's theorem for every finite dimensional operator $K \in L(X)$ for which $TK = KT$ and for every $f \in \mathcal{H}(\sigma(T))$. Analogously, if T has SVEP and $\text{iso}\,\sigma_{\text{s}}(T) = \emptyset$ then $f(T^* + K^*)$ satisfies S-Weyl's theorem for every $f \in \mathcal{H}(\sigma(T))$.*

Proof Suppose first that $\text{iso}\,\sigma_{\text{ap}}(T) = \emptyset$ and that T^* has SVEP. By Aiena [2, Theorem 3.29] then

$$\text{iso}\,\sigma_{\text{ap}}(T + K) = \text{iso}\,\sigma_{\text{ap}}(T) = \emptyset,$$

and from [2, Lemma 2.6] we obtain that

$$\text{iso}\,\sigma_{\text{ap}}(f(T + K)) \subseteq f(\text{iso}\,\sigma_{\text{ap}}(T + K)) = \emptyset.$$

Since a finite-dimensional operator is algebraic, then $T^* + K^*$ has SVEP, see [2, Theorem 2.145], and consequently, by Aiena [2, Theorem 2.86], $f(T^* + K^*) = [f(T + K)]^*$ has SVEP for every $f \in \mathcal{H}(\sigma(T + K))$. By Aiena et al. [5, Theorem 3.13] we conclude that $f(T + K)$ satisfies S-Weyl's theorem.

Analogously, suppose that

$$\text{iso}\,\sigma_{\text{ap}}(T^*) = \text{iso}\,\sigma_{\text{ap}}(T) = \emptyset$$

and that T has SVEP. Since K^* is finite-dimensional and $T^*K^* = K^*T^*$, we have, always by Aiena [2, Theorem 3.29], that

$$\text{iso}\,\sigma_{\text{s}}(T + K) = \text{iso}\,\sigma_{\text{ap}}(T^* + K^*) = \sigma_{\text{ap}}(T^*) = \emptyset.$$

Moreover, the SVEP for T is transmitted to $T + K$ and hence to $f(T + K)$. Again by Aiena et al. [5, Theorem 3.13] we conclude that $f(T^* + K^*)$ satisfies S-Weyl's theorem. □

It is known that $\sigma(T) = \sigma(T + Q)$ and $\sigma_{\text{ap}}(T) = \sigma_{\text{ap}}(T + Q)$ for every quasi-nilpotent commuting perturbation Q, see [2, Corollary 3.24]. Since a-Browder's theorem (or equivalently, generalized a-Browder's theorem) is preserved under quasi-nilpotent commuting perturbations, see [2, Corollary 5.5], it then follows that property (gaz) is preserved under commuting quasi-nilpotent perturbations.

Consider the case that K^n is a finite-dimensional operator for some $n \in \mathbb{N}$. It is known that $\sigma_{\text{ap}}(T)$ and $\sigma_{\text{ap}}(T + K)$ may differ at the isolated points, see [2, Theorem 3.26].

An operator $K \in L(X)$ is said to be a *Riesz operator* if $\lambda I - T \in \Phi(X)$ for all $\lambda \neq 0$. Recall that if K^n is finite-dimensional operator for some $n \in \mathbb{N}$ then K is

Riesz, see [2, Theorem 3.4] and that operator $T \in L(X)$ is said to have *property* (R) if $p_{00}^{a}(T) = \pi_{00}(T)$. By Aiena [2, Theorem 6.89] we have

$$\text{property } (w) \Rightarrow \text{property } (R).$$

Theorem 4.2 *Suppose that* $T, K \in L(X)$, $TK = KT$, K^n *finite-dimensional for some* $n \in \mathbb{N}$ *and* iso $\sigma_{\rm ap}(T) =$ iso $\sigma_{\rm ap}(T + K)$. *Then we have:*

(i) *If* T *has property* (gaz) *then* $T + K$ *has property* (gaz).
(ii) *If* T *is isoloid and satisfies* S*-Weyl's theorem then also* $T + K$ *satisfies* S*-Weyl's theorem.*

Proof Suppose that T has property (gaz). By Theorem 2.1 it suffices to prove that $T + K$ satisfies generalized a-Browder's theorem, or equivalently, a-Browder's theorem, and $\sigma_{\rm ap}(T) = \sigma_{\rm ap}(T + K)$. Since T satisfies a-Browder's then $\sigma_{\rm uw}(T) = \sigma_{\rm ub}(T)$. But these spectra are invariant Riesz commuting perturbation, see Chapter 3 of [2] so a-Browder's theorem (and hence generalized a-Browder's theorem) holds for $T + K$, since K is a Riesz operator.

Now, by Aiena [2, Theorems 3.27 and 3.28], the assumption iso $\sigma_{\rm ap}(T) =$ iso $\sigma_{\rm ap}(T + K)$ entails $\sigma_{\rm ap}(T) = \sigma_{\rm ap}(T + K)$. Indeed, these two spectra have the same accumulation points, see [2, Theorem 3.26], so

$$\sigma_{\rm ap}(T+K) = \operatorname{iso}\sigma_{\rm ap}(T+K) \cup \operatorname{acc}\sigma_{\rm ap}(T+K) = \operatorname{iso}\sigma_{\rm ap}(T) \cup \operatorname{acc}\sigma_{\rm ap}(T) = \sigma_{\rm ap}(T).$$

Similarly, by Aiena [2, Theorem 3.20] we have $\sigma(T + K) = \sigma(T)$. Since $\sigma_{\rm ap}(T) = \sigma(T)$, by Theorem 2.1, it then follows that $\sigma_{\rm ap}(T + K) = \sigma(T + K)$, so, again by Theorem 2.1, we conclude that $T + K$ has property (gaz).

(ii) From part (i) we know that $T + K$ satisfies property (gaz), so it suffices, by Theorem 3.1, to prove the equality $\Pi_a(T + K) = \pi_{00}(T + K)$. As observed before S-Weyl' s theorem entails property (w), and hence property (R). Since T is isoloid then, from [2, Theorem 3.78] we obtain $\pi_{00}(T) = \pi_{00}(T + K)$, hence, taking into account that $\sigma_{\rm ld}(T + K) = \sigma_{\rm ld}(T)$, see [2, Theorem 3.78], we obtain that

$$\begin{aligned}\Pi_a(T + K) &= \sigma_{\rm ap}(T + K) \setminus \sigma_{\rm ld}(T + K) = \sigma_{\rm ap}(T) \setminus \sigma_{\rm ld}(T) = \Pi_a(T) \\ &= \pi_{00}(T) = \pi_{00}(T + K),\end{aligned}$$

as desired. □

In the special case where iso $\sigma_{\rm ap}(T) = \emptyset$ we have iso $\sigma_{\rm ap}(T + K) = \emptyset$, see [2, Theorem 3.29].

Corollary 4.1 *Suppose that* $T, K \in L(X)$, $TK = KT$, *and that* K^n *is finite-dimensional for some* $n \in \mathbb{N}$. *If* iso $\sigma_{\rm ap}(T) = \emptyset$ *and* T *has property* (gaz) *then* $T + K$ *has property* (gaz). *If* T *satisfies* S*-Weyl theorem then also* $T + K$ *satisfies* S*-Weyl's theorem.*

Proof The first part is clear from Theorem 4.2. Since $\text{iso}\,\sigma_{\text{ap}}(T+K) = \emptyset$ then $\Pi_a(T+K) = \emptyset$, since $\Pi_a(T+K) \subseteq \text{iso}\,\sigma_{\text{ap}}(T+K)$, and also $\pi_{00}(T+K) = \emptyset$, since every isolated point of the spectrum $\sigma(T+K)$ belongs to $\sigma_{\text{ap}}(T+K)$, and hence is isolated in $\sigma_{\text{ap}}(T+K)$. Theorem 3.1 then applies. □

An immediate consequence of Corollary 4.1 is that if L is the classical left shift operator in $\ell^2(\mathbb{N})$ then $T+K$ has property (gaz) for all commuting $K \in \ell^2(\mathbb{N})$, for which K^n is finite-dimensional for some $n \in \mathbb{N}$.

An operator $T \in L(X)$ is said to be *hereditarily polaroid*, if every isolated point of the spectrum of the restriction $T|M$, M a closed T-invariant subspace, is a pole of the resolvent of $T|M$, (see [2, Chapter 4]). An operator $K \in L(X)$ is said to be *algebraic* if there exists a complex nontrivial polynomial h such that $h(T) = 0$. Examples of algebraic operators are the operators $K \in L(X)$ for which K^n is finite-dimensional for some $n \in \mathbb{N}$, nilpotent and idempotent operators.

Theorem 4.3 *If $T \in L(X)$ is hereditarily polaroid then $T^* + K^*$ satisfies property (gaz) for every algebraic operator K which commutes with T.*

Proof If T is hereditarily polaroid then T has SVEP, see [2, Theorem 4.31]. The SVEP of T is preserved under commuting algebraic perturbations, see [3], so Corollary 3.7 of [4] applies. □

References

1. Aiena, P.: Fredholm and Local Spectral Theory II, with Application to Multipliers. Kluwer Academic Publishers, Dordrecht (2004)
2. Aiena, P.: Fredholm and Local Spectral Theory II, with Application to Weyl-Type Theorems. Springer Lecture Notes of Math, vol. 2235. Springer, Cham (2018)
3. Aiena, P., Neumann, M.M.: On the stability of the localized single-valued extension property under commuting perturbations. Proc. Am. Math. Soc. **141**(6), 2039–2050 (2013)
4. Aiena, P., Aponte, E., Guillén, J.: Property (gaz) through localized SVEP. Mat. Vesn. **72**(4), 314–326 (2020)
5. Aiena, P., Aponte, E., Guillén, J.: A strong variant of Weyl's theorem. Syphax J. Math. Nonlinear Anal. Oper. Syst. **1**, 1–16 (2021)
6. Berkani, M.: On a class of quasi-Fredholm operators. Integr. Equ. Oper. Theory **34**(1), 244–249 (1999)
7. Berkani, M.: Index of B-Fredholm operators and generalization of a Weyl's theorem. Proc. Am. Math. Soc. **130**(6), 1717–1723 (2001)
8. Berkani, M., Sarih, M.: On semi B-Fredholm operators. Glasg. Math. J. **43**, 457–465 (2001)
9. Grabiner, S.: Uniform ascent and descent of bounded operators. J. Math. Soc. Jpn. **34**, 317–337 (1982)
10. Heuser, H.: Functional Analysis. Marcel Dekker, New York (1982)
11. Mbekhta, M., Müller, V.: On the axiomatic theory of the spectrum II. Stud. Math. **119**, 129–147 (1996)
12. Zariouh, H: Property (gz) for bounded linear operators. Mat. Vesn. **65**, 94–103 (2013)

Multiplication and Convolution Topological Algebras in Spaces of ω-Ultradifferentiable Functions of Beurling Type

Angela A. Albanese and Claudio Mele

Abstract We determine multiplication and convolution topological algebras for classes of ω-ultradifferentiable functions of Beurling type. Hypocontinuity and discontinuity of the multiplication and convolution mappings are also investigated.

Keywords Multipliers · Convolutors · Topological algebras · Weight functions · Ultradifferentiable rapidly decreasing function spaces of Beurling type

1 Introduction

Schwartz started in 1966 the study of multipliers and convolutors of the space $\mathcal{S}(\mathbb{R}^N)$ of rapidly decreasing functions. The interest lies in the importance of their application to the study of partial differential equations. Since then many authors introduced and studied particular aspects of the spaces of multipliers and of convolutors for ultradifferentiable classes of rapidly decreasing functions of Beurling or Roumieu type in the sense of Komatsu [15] (see [8–12] for recent results in this setting). In the last years the attention has focused on the study of the space $\mathcal{S}_\omega(\mathbb{R}^N)$ of the ultradifferentiable rapidly decreasing functions of Beurling type, as introduced by Björck [3] (see [4–6, 11], for instance, and the references therein). Inspired by this line of research and by the previous work, in [1, 2] the authors introduced and studied the space $\mathcal{O}_{M,\omega}(\mathbb{R}^N)$ of the slowly increasing functions of Beurling type in the setting of ultradifferentiable function spaces of Beurling type, showing that it is the space of the multipliers of $\mathcal{S}_\omega(\mathbb{R}^N)$ and of its dual $\mathcal{S}'_\omega(\mathbb{R}^N)$, and the space $\mathcal{O}_{C,\omega}(\mathbb{R}^N)$ of the very slowly increasing functions of Beurling type, whose strong dual $\mathcal{O}'_{C,\omega}(\mathbb{R}^N)$ is the space of the convolutors of $\mathcal{S}_\omega(\mathbb{R}^N)$ and of its dual $\mathcal{S}'_\omega(\mathbb{R}^N)$. Their rich topological structure led us to determine deeply results concerning regularity, equivalent systems of seminorms, representations of

A. A. Albanese (✉) · C. Mele
Dipartimento di Matematica e Fisica "E. De Giorgi", Università del Salento, Lecce, Italy
e-mail: angela.albanese@unisalento.it; claudio.mele1@unisalento.it

A. M. Candela et al. (eds.), *Recent Advances in Mathematical Analysis*,
Trends in Mathematics, https://doi.org/10.1007/978-3-031-20021-2_4

the ultradistributions and the action of the Fourier transform, that is a topological isomorphism from $\mathcal{O}_{M,\omega}(\mathbb{R}^N)$ to $\mathcal{O}'_{C,\omega}(\mathbb{R}^N)$.

In this paper the authors continue the study of these spaces. The aim is to establish that $(\mathcal{O}_{M,\omega}(\mathbb{R}^N),\cdot)$, $(\mathcal{S}_\omega(\mathbb{R}^N),\cdot)$ are multiplication topological algebras, while $(\mathcal{S}_\omega(\mathbb{R}^N),\star)$ and $(\mathcal{O}'_{C,\omega}(\mathbb{R}^N),\star)$ are convolution topological algebras. We also determine the spaces of multipliers and of convolutors of the spaces $\mathcal{O}_{M,\omega}(\mathbb{R}^N)$, $\mathcal{O}_{C,\omega}(\mathbb{R}^N)$ and their duals. Furthermore, we analyze the continuity of the multiplication and convolution bilinear mappings on some pairs between the spaces $\mathcal{O}_{M,\omega}(\mathbb{R}^N)$, $\mathcal{O}_{C,\omega}(\mathbb{R}^N)$, $\mathcal{S}_\omega(\mathbb{R}^N)$ and their duals, studied classically by Schwartz [17] (see also Larcher [16] and the references therein). This approach is done not only treating the hypocontinuity, but also describing and investigating the continuity properties of these mappings.

The paper is organized as follows. Section 2 is devoted to recall some definitions and properties of the weights ω and of the ω-ultradifferentiable functions, that we use in the following. In Sect. 3 we prove the results about the multiplication and convolution (topological) algebras and determine the spaces of multipliers and of convolutors of the spaces under consideration. Finally, in the last section, we analyze the multiplication and convolution bilinear mappings on some pairs between the spaces $\mathcal{O}_{M,\omega}(\mathbb{R}^N)$, $\mathcal{O}_{C,\omega}(\mathbb{R}^N)$, $\mathcal{S}_\omega(\mathbb{R}^N)$ and their duals, proving when are hypocontinuous or discontinuous.

2 Definitions and Preliminary Results

We first give the definition of non-quasianalytic weight function in the sense of Braun et al. [7] suitable for the Beurling case, i.e., we also consider the logarithm as a weight function.

Definition 2.1 A non-quasianalytic weight function is a continuous increasing function $\omega : [0,\infty) \to [0,\infty)$ satisfying the following properties:

(α) there exists $K \geq 1$ such that $\omega(2t) \leq K(1+\omega(t))$ for every $t \geq 0$;
(β) $\int_1^\infty \frac{\omega(t)}{1+t^2}\,dt < \infty$;
(γ) there exist $a \in \mathbb{R}$, $b > 0$ such that $\omega(t) \geq a + b\log(1+t)$, for every $t \geq 0$;
(δ) $\varphi_\omega(t) = \omega \circ \exp(t)$ is a convex function.

We recall some known properties of the weight functions that shall be useful in the following (the proofs can be found in the literature):

(1) Condition (α) implies that

$$\omega(t_1+t_2) \leq K(1+\omega(t_1)+\omega(t_2)), \;\; \forall t_1, t_2 \geq 0. \tag{1}$$

Observe that this condition is weaker than subadditivity (i.e., $\omega(t_1 + t_2) \le \omega(t_1) + \omega(t_2)$). The weight functions satisfying (α) are not necessarily subadditive in general.

(2) Condition (α) implies that there exists $L \ge 1$ such that

$$\omega(et) \le L(1 + \omega(t)), \ \forall t \ge 0. \tag{2}$$

(3) By condition (γ) we have that

$$e^{-\lambda\omega(t)} \in L^p(\mathbb{R}^N), \ \forall \lambda \ge \frac{N+1}{bp}. \tag{3}$$

Given a non-quasianalytic weight function ω, we define the Young conjugate φ^*_ω of φ_ω as the function $\varphi^*_\omega : [0, \infty) \to [0, \infty)$ by

$$\varphi^*_\omega(s) := \sup_{t \ge 0}\{st - \varphi_\omega(t)\}, \quad s \ge 0.$$

There is no loss of generality to assume that ω vanishes on $[0, 1]$. Therefore, φ^*_ω is convex and increasing, $\varphi^*_\omega(0) = 0$ and $(\varphi^*_\omega)^* = \varphi_\omega$.

Further useful properties of φ^*_ω are listed below (see [7]):

(1) $\frac{\varphi^*_\omega(t)}{t}$ is an increasing function in $(0, \infty)$.

(2) For every $s, t \ge 0$ and $\lambda > 0$

$$2\lambda\varphi^*_\omega\left(\frac{s+t}{2\lambda}\right) \le \lambda\varphi^*_\omega\left(\frac{s}{\lambda}\right) + \lambda\varphi^*_\omega\left(\frac{t}{\lambda}\right) \le \lambda\varphi^*_\omega\left(\frac{s+t}{\lambda}\right). \tag{4}$$

(3) For every $t \ge 0$ and $\lambda > 0$

$$\lambda L\varphi^*_\omega\left(\frac{t}{\lambda L}\right) + t \le \lambda\varphi^*_\omega\left(\frac{t}{\lambda}\right) + \lambda L,$$

where $L \ge 1$ is the constant appearing in formula (2).

(4) For all $m, M \in \mathbb{N}$ with $M \ge mL$, where L is the constant appearing in formula (2), and for every $t \ge 0$

$$2^t \exp\left(M\varphi^*_\omega\left(\frac{t}{M}\right)\right) \le C \exp\left(m\varphi^*_\omega\left(\frac{t}{m}\right)\right), \tag{5}$$

with $C := e^{mL}$.

We now introduce the ultradifferentiable function space $\mathcal{S}_\omega(\mathbb{R}^N)$ in the sense of Björk [3].

Definition 2.2 Let ω be a non-quasianalytic weight function. We denote by $\mathcal{S}_\omega(\mathbb{R}^N)$ the set of all functions $f \in L^1(\mathbb{R}^N)$ such that $f, \hat{f} \in C^\infty(\mathbb{R}^N)$ and for all $\lambda > 0$ and $\alpha \in \mathbb{N}_0^N$ we have

$$\|\exp(\lambda\omega)\partial^\alpha f\|_\infty < \infty \text{ and } \|\exp(\lambda\omega)\partial^\alpha \hat{f}\|_\infty < \infty ,$$

where $\hat{f}$ denotes the Fourier transform of f. The elements of $\mathcal{S}_\omega(\mathbb{R}^N)$ are called *ω-ultradifferentiable rapidly decreasing functions of Beurling type*. We denote by $\mathcal{S}'_\omega(\mathbb{R}^N)$ the dual of $\mathcal{S}_\omega(\mathbb{R}^N)$ endowed with its strong topology.

The space $\mathcal{S}_\omega(\mathbb{R}^N)$ is a Fréchet space with different equivalent systems of seminorms (see [4, Theorem 4.8] and [6, Theorem 2.6]). In the following, we will use the following system of norms generating the Fréchet topology of $\mathcal{S}_\omega(\mathbb{R}^N)$:

$$q_{\lambda,\mu}(f) := \sup_{\alpha\in\mathbb{N}_0^N} \|\exp(\mu\omega)\partial^\alpha f\|_\infty \exp\left(-\lambda\varphi^*_\omega\left(\frac{|\alpha|}{\lambda}\right)\right), \quad \lambda, \mu > 0, \; f \in \mathcal{S}_\omega(\mathbb{R}^N),$$

or equivalently, the sequence of norms $\{q_{m,n}\}_{m,n\in\mathbb{N}}$.

We point out that the space $\mathcal{S}_\omega(\mathbb{R}^N)$ is a nuclear Fréchet space, see, f.i., [5, Theorem 3.3] or [11, Theorem 1.1].

We refer to [7] for the definition and the main properties of the ultradifferentiable function spaces $\mathcal{E}_\omega(\Omega)$, $\mathcal{D}_\omega(\Omega)$ and their duals of Beurling type in the sense of Braun, Meise and Taylor. We only recall that for an open subset Ω of $\mathbb{R}^N$, the space $\mathcal{E}_\omega(\Omega)$ is defined as

$$\mathcal{E}_\omega(\Omega) := \left\{f \in C^\infty(\Omega) \colon p_{K,m}(f) < \infty \; \forall K \Subset \Omega, \; m \in \mathbb{N}\right\},$$

where

$$p_{K,m}(f) := \sup_{x\in K} \sup_{\alpha\in\mathbb{N}_0^N} |\partial^\alpha f(x)| \exp\left(-m\varphi^*_\omega\left(\frac{|\alpha|}{m}\right)\right).$$

$\mathcal{E}_\omega(\Omega)$ is a nuclear Fréchet space with respect to the lc-topology generated by the system of seminorms $\{p_{K,m}\}_{K\Subset\Omega, m\in\mathbb{N}}$ (see [7, Proposition 4.9]). The elements of $\mathcal{E}_\omega(\Omega)$ are called *ω-ultradifferentiable functions of Beurling type* on Ω.

The spaces $\mathcal{O}_{M,\omega}(\mathbb{R}^N)$ and $\mathcal{O}_{C,\omega}(\mathbb{R}^N)$ have been introduced in [2] and the definition has been given there in terms of weighted L^∞-norms as it follows.

Definition 2.3 Let ω be a non-quasianalytic weight function.

(a) The space $\mathcal{O}_{M,\omega}(\mathbb{R}^N)$ of *slowly increasing functions of Beurling type* on $\mathbb{R}^N$ is defined by

$$\mathcal{O}_{M,\omega}(\mathbb{R}^N) := \bigcap_{m=1}^{\infty}\bigcup_{n=1}^{\infty} \mathcal{O}^m_{n,\omega}(\mathbb{R}^N),$$

where

$$\mathcal{O}^m_{n,\omega}(\mathbb{R}^N) :=$$

$$\left\{ f \in C^\infty(\mathbb{R}^N) \colon r_{m,n}(f) := \sup_{\alpha\in\mathbb{N}_0^N} \sup_{x\in\mathbb{R}^N} |\partial^\alpha f(x)| \exp\left(-n\omega(x) - m\varphi^*_\omega\left(\frac{|\alpha|}{m}\right)\right) < \infty \right\}$$

endowed with the norm $r_{m,n}$ is a Banach space for any $m, n \in \mathbb{N}$. $\mathcal{O}_{M,\omega}(\mathbb{R}^N)$ is endowed with its natural lc-topology t, i.e., $\mathcal{O}_{M,\omega}(\mathbb{R}^N) = \mathrm{proj}_{\overleftarrow{m}}\, \mathrm{ind}_{\overrightarrow{n}}\, \mathcal{O}^m_{n,\omega}(\mathbb{R}^N)$ is the projective limit of the (LB)-spaces $\mathcal{O}^m_\omega(\mathbb{R}^N) := \mathrm{ind}_{\overrightarrow{n}}\, \mathcal{O}^m_{n,\omega}(\mathbb{R}^N)$.

(b) The space $\mathcal{O}_{C,\omega}(\mathbb{R}^N)$ of *very slowly increasing functions of Beurling type* on $\mathbb{R}^N$ is defined by

$$\mathcal{O}_{C,\omega}(\mathbb{R}^N) := \bigcup_{n=1}^{\infty} \bigcap_{m=1}^{\infty} \mathcal{O}^m_{n,\omega}(\mathbb{R}^N).$$

The space $\mathcal{O}_{C,\omega}(\mathbb{R}^N)$ endowed with its natural lc-topology is an (LF)-space, i.e., $\mathcal{O}_{C,\omega}(\mathbb{R}^N) = \mathrm{ind}_{\overrightarrow{n}}\, \mathcal{O}_{n,\omega}(\mathbb{R}^N)$, where $\mathcal{O}_{n,\omega}(\mathbb{R}^N) := \mathrm{proj}_{\overleftarrow{m}}\, \mathcal{O}^m_{n,\omega}(\mathbb{R}^N)$.

Let $\mathcal{O}'_{M,\omega}(\mathbb{R}^N)$ ($\mathcal{O}'_{C,\omega}(\mathbb{R}^N)$, resp.) denote the strong dual of $\mathcal{O}_{M,\omega}(\mathbb{R}^N)$ (of $\mathcal{O}_{C,\omega}(\mathbb{R}^N)$, resp.) endowed with its strong topology.

Recall that the inclusions $\mathcal{D}_\omega(\mathbb{R}^N) \hookrightarrow \mathcal{S}_\omega(\mathbb{R}^N) \hookrightarrow \mathcal{O}_{C,\omega}(\mathbb{R}^N) \hookrightarrow \mathcal{O}_{M,\omega}(\mathbb{R}^N) \hookrightarrow \mathcal{E}_\omega(\mathbb{R}^N)$ are well-defined, continuous with dense range, see [2, Theorems 3.8, 3.9 and 5.2(1)]. Hence, the inclusions $\mathcal{E}'_\omega(\mathbb{R}^N) \hookrightarrow \mathcal{O}'_{M,\omega}(\mathbb{R}^N) \hookrightarrow \mathcal{O}'_{C,\omega}(\mathbb{R}^N) \hookrightarrow \mathcal{S}'_\omega(\mathbb{R}^N) \hookrightarrow \mathcal{D}'_\omega(\mathbb{R}^N)$ are also well-defined and continuous. On the other hand, $\mathcal{O}_{M,\omega}(\mathbb{R}^N) \hookrightarrow \mathcal{S}'_\omega(\mathbb{R}^N)$ and $\mathcal{O}_{C,\omega}(\mathbb{R}^N) \hookrightarrow \mathcal{S}'_\omega(\mathbb{R}^N)$ continuously, as it is easy to see.

The space $\mathcal{O}_{M,\omega}(\mathbb{R}^N)$ is the space of multipliers of $\mathcal{S}_\omega(\mathbb{R}^N)$ and of its dual space $\mathcal{S}'_\omega(\mathbb{R}^N)$ as proved in [2, 8], i.e., $f \in \mathcal{O}_{M,\omega}(\mathbb{R}^N)$ if, and only if, $fg \in \mathcal{S}_\omega(\mathbb{R}^N)$ for all $g \in \mathcal{S}_\omega(\mathbb{R}^N)$ if, and only if, $fT \in \mathcal{S}'_\omega(\mathbb{R}^N)$ for all $T \in \mathcal{S}'_\omega(\mathbb{R}^N)$. Moreover, if $f \in \mathcal{O}_{M,\omega}(\mathbb{R}^N)$, then the linear operators $M_f : \mathcal{S}_\omega(\mathbb{R}^N) \to \mathcal{S}_\omega(\mathbb{R}^N)$ defined by $M_f(g) := fg$, for $g \in \mathcal{S}_\omega(\mathbb{R}^N)$, and $M_f \colon \mathcal{S}'_\omega(\mathbb{R}^N) \to \mathcal{S}'_\omega(\mathbb{R}^N)$ defined by $M_f(T) := fT$, for $T \in \mathcal{S}'_\omega(\mathbb{R}^N)$, are continuous. Furthermore, $\mathcal{O}_{M,\omega}(\mathbb{R}^N)$ is an ultrabornological space as shown by Debrouwere and Neyt [8, Theorem 5.3]. Due to [8, Theorem 5.3] and the results in [2], we have that $\mathcal{O}_{M,\omega}(\mathbb{R}^N)$ is a closed subspace of $\mathcal{L}_b(\mathcal{S}_\omega(\mathbb{R}^N))$ ($\mathcal{L}_b(\mathcal{S}_\omega(\mathbb{R}^N))$ denotes the space of all continuous linear operators from $\mathcal{S}_\omega(\mathbb{R}^N)$ into itself endowed with the topology of the uniform convergence on bounded subsets of $\mathcal{S}_\omega(\mathbb{R}^N)$). In particular, a fundamental system of continuous norms on $\mathcal{O}_{M,\omega}(\mathbb{R}^N)$ is given by

$$q_{m,g}(f) := \sup_{\alpha\in\mathbb{N}_0^N} \sup_{x\in\mathbb{R}^N} |g(x)||\partial^\alpha f(x)| \exp\left(-m\varphi_\omega^*\left(\frac{|\alpha|}{m}\right)\right), \quad g \in \mathcal{S}_\omega(\mathbb{R}^N),\ m \in \mathbb{N}, \tag{6}$$

see [2, Proposition 5.6 and Theorem 5.9].

For any non-quasianalytic weight function ω satisfying the condition $\log(1+t) = o(\omega(t))$ as $t \to \infty$, the space $\mathcal{O}'_{C,\omega}(\mathbb{R}^N)$ is the space of convolutors of $\mathcal{S}_\omega(\mathbb{R}^N)$ and of its dual space $\mathcal{S}'_\omega(\mathbb{R}^N)$ as proved in [1], i.e., $T \in \mathcal{O}'_{C,\omega}(\mathbb{R}^N)$ if, and only if, $T\star f \in \mathcal{S}_\omega(\mathbb{R}^N)$ for all $f \in \mathcal{S}_\omega(\mathbb{R}^N)$ if, and only if, $T \star S \in \mathcal{S}'_\omega(\mathbb{R}^N)$ for all $S \in \mathcal{S}'_\omega(\mathbb{R}^N)$. Moreover, if $T \in \mathcal{O}'_{C,\omega}(\mathbb{R}^N)$, then the linear operators $C_T : \mathcal{S}_\omega(\mathbb{R}^N) \to \mathcal{S}_\omega(\mathbb{R}^N)$ defined by $C_T(f) := T \star f$, for $f \in \mathcal{S}_\omega(\mathbb{R}^N)$, and $C_T\colon \mathcal{S}'_\omega(\mathbb{R}^N) \to \mathcal{S}'_\omega(\mathbb{R}^N)$ defined by $C_T(S) := T \star S$, for $S \in \mathcal{S}'_\omega(\mathbb{R}^N)$, are continuous.

In [1] the following spaces have been introduced.

Definition 2.4 Let ω be a non-quasianalytic weight function and $1 \le p < \infty$.

(a) The space $\mathcal{O}_{M,\omega,p}(\mathbb{R}^N)$ is defined by $\mathcal{O}_{M,\omega,p}(\mathbb{R}^N) := \bigcap_{m=1}^\infty \bigcup_{n=1}^\infty \mathcal{O}^m_{n,\omega,p}(\mathbb{R}^N)$, where

$$\mathcal{O}^m_{n,\omega,p}(\mathbb{R}^N) :=$$

$$\left\{ f \in C^\infty(\mathbb{R}^N) \colon r^p_{m,n,p}(f) := \sum_{\alpha\in\mathbb{N}_0^N} \|\exp(-n\omega)\partial^\alpha f\|_p^p \exp\left(-mp\varphi_\omega^*\left(\frac{|\alpha|}{m}\right)\right) < \infty \right\}$$

endowed with the norm $r_{m,n,p}$ is a Banach space for any $m, n \in \mathbb{N}$. The space is endowed with its natural lc-topology, i.e., $\mathcal{O}_{M,\omega,p}(\mathbb{R}^N) = \operatorname{proj}_{\underset{m}{\leftarrow}} \operatorname{ind}_{\underset{n}{\rightarrow}} \mathcal{O}^m_{n,\omega,p}(\mathbb{R}^N)$ is the projective limit of the (LB)-spaces $\mathcal{O}^m_{\omega,p}(\mathbb{R}^N) := \operatorname{ind}_{\underset{n}{\rightarrow}} \mathcal{O}^m_{n,\omega,p}(\mathbb{R}^N)$.

(b) The space $\mathcal{O}_{C,\omega,p}(\mathbb{R}^N)$ is defined by $\mathcal{O}_{C,\omega,p}(\mathbb{R}^N) := \bigcup_{n=1}^\infty \bigcap_{m=1}^\infty \mathcal{O}^m_{n,\omega,p}(\mathbb{R}^N)$. $\mathcal{O}_{C,\omega,p}(\mathbb{R}^N)$ endowed with its natural lc-topology is an (LF)-space, i.e., $\mathcal{O}_{C,\omega,p}(\mathbb{R}^N) = \operatorname{ind}_{\underset{n}{\rightarrow}} \mathcal{O}_{n,\omega,p}(\mathbb{R}^N)$, where $\mathcal{O}_{n,\omega,p}(\mathbb{R}^N) := \operatorname{proj}_{\underset{m}{\leftarrow}} \mathcal{O}^m_{n,\omega,p}(\mathbb{R}^N)$.

As shown in [1, Proposition 3.8] it holds true that $\mathcal{O}^m_\omega(\mathbb{R}^N) = \mathcal{O}^m_{\omega,p}(\mathbb{R}^N)$ and $\mathcal{O}_{n,\omega}(\mathbb{R}^N) = \mathcal{O}_{n,\omega,p}(\mathbb{R}^N)$ algebraically and topologically for any $1 \le p < \infty$, thereby obtaining that $\mathcal{O}_{M,\omega}(\mathbb{R}^N) = \mathcal{O}_{M,\omega,p}(\mathbb{R}^N)$ and $\mathcal{O}_{C,\omega}(\mathbb{R}^N) = \mathcal{O}_{C,\omega,p}(\mathbb{R}^N)$ algebraically and topologically for any $1 \le p < \infty$. Since the Banach spaces $\mathcal{O}^m_{n,\omega,p}(\mathbb{R}^N)$ are reflexive whenever $1 < p < \infty$, by a result of Vogt [19, Proposition 4.4] we can conclude that $\mathcal{O}^m_\omega(\mathbb{R}^N) = \mathcal{O}^m_{\omega,p}(\mathbb{R}^N)$ is a reflexive and complete (LB)-space, hence a *regular (LB)-space*, being it an inductive limit of reflexive Banach spaces. Recall that an (LF)-space $E = \operatorname{ind}_{\underset{n}{\rightarrow}} E_n$ is said to be *regular* if every bounded subset in E is contained and bounded in E_n for some $n \in \mathbb{N}$ and that complete (LF)-spaces are always regular. An (LF)-space $E = \operatorname{ind}_{\underset{n}{\to}} E_n$ is said to be *sequentially retractive* if for every null sequence in E there exists $n \in \mathbb{N}$ such that the sequence is contained and converges to zero in E_n. We observe that sequential retractivity implies completeness, see [20, Corollary 2.8].

We end this section by showing that the (LB)-spaces $\mathcal{O}_\omega^m(\mathbb{R}^N)$, with $m \in \mathbb{N}$, are sequentially retractive and Montel. For this, we need the following lemma.

Lemma 2.1 *Let ω be a non-quasianalytic weight function and let $m \in \mathbb{N}$. Then the spaces $\mathcal{O}_{n+1,\omega}^m(\mathbb{R}^N)$ and $\mathcal{E}_\omega(\mathbb{R}^N)$ induce the same topology on the closed unit ball B_n^m of $\mathcal{O}_{n,\omega}^m(\mathbb{R}^N)$ for all $n \in \mathbb{N}$.*

Proof Fix $n \in \mathbb{N}$ and $\varepsilon > 0$. The set $U := \{f \in \mathcal{O}_{n+1,\omega}^m(\mathbb{R}^N) : r_{m,n+1}(f) < \varepsilon\}$ is a 0-neighborhood of $\mathcal{O}_{n+1,\omega}^m(\mathbb{R}^N)$. Now, let $M > 0$ such that $\exp(-\omega(x)) < \varepsilon$ for every $|x| \geq M$ and $V := \{f \in \mathcal{E}_\omega(\mathbb{R}^N) : p_{K,m}(f) < \varepsilon\}$, where $K := \{x \in \mathbb{R}^N : |x| \leq M\}$. Then $V \cap B_n^m \subseteq U \cap B_n^m$. Indeed, if $f \in V \cap B_n^m$, then

$$|\partial^\alpha f(x)| \leq \exp\left(n\omega(x) + m\varphi_\omega^*\left(\frac{|\alpha|}{m}\right)\right) < \varepsilon \exp\left((n+1)\omega(x) + m\varphi_\omega^*\left(\frac{|\alpha|}{m}\right)\right)$$

for every $\alpha \in \mathbb{N}_0^N$ and $|x| \geq M$. Moreover, $f \in V \cap B_n^m$ also implies that

$$|\partial^\alpha f(x)| < \varepsilon \exp\left(m\varphi_\omega^*\left(\frac{|\alpha|}{m}\right)\right) \leq \varepsilon \exp\left((n+1)\omega(x) + m\varphi_\omega^*\left(\frac{|\alpha|}{m}\right)\right)$$

for every $\alpha \in \mathbb{N}_0^N$ and $|x| \leq M$. It follows that $r_{m,n+1}(f) < \varepsilon$ and so, $f \in U \cap B_n^m$. Since $\varepsilon > 0$ is arbitrary, we get the thesis. □

We can show that the (LB)-spaces $\mathcal{O}_\omega^m(\mathbb{R}^N)$, with $m \in \mathbb{N}$, are sequentially retractive.

Theorem 2.1 *Let ω be a non-quasianalytic weight function. For every $m \in \mathbb{N}$ the (LB)-space $\mathcal{O}_\omega^m(\mathbb{R}^N)$ is sequentially retractive.*

Proof Let $\{f_j\}_{j\in\mathbb{N}}$ be a null sequence of $\mathcal{O}_\omega^m(\mathbb{R}^N)$. Then $B := \{f_j : j \in \mathbb{N}\}$ is a bounded subset of $\mathcal{O}_\omega^m(\mathbb{R}^N)$. Since $\mathcal{O}_\omega^m(\mathbb{R}^N)$ is a regular (LB)-space, B is contained and bounded in $\mathcal{O}_{n,\omega}^m(\mathbb{R}^N)$ for some $n \in \mathbb{N}$. On the other hand, $\{f_j\}_{j\in\mathbb{N}}$ converges to 0 in $\mathcal{E}_\omega(\mathbb{R}^N)$, as $\mathcal{O}_\omega^m(\mathbb{R}^N)$ is continuously included in $\mathcal{E}_\omega(\mathbb{R}^N)$. Since $\mathcal{O}_{n+1,\omega}^m(\mathbb{R}^N)$ and $\mathcal{E}_\omega(\mathbb{R}^N)$ induce the same topology on B by Lemma 2.1, it follows that $\{f_j\}_{j\in\mathbb{N}}$ converges to 0 in $\mathcal{O}_{n+1,\omega}^m(\mathbb{R}^N)$. □

Theorem 2.2 *Let ω be a non-quasianalytic weight function. Then the spaces $\mathcal{O}_\omega^m(\mathbb{R}^N)$, for $m \in \mathbb{N}$, and $\mathcal{O}_{M,\omega}(\mathbb{R}^N)$ are complete Montel spaces.*

Proof The spaces $\mathcal{O}_\omega^m(\mathbb{R}^N)$, for $m \in \mathbb{N}$, are clearly barrelled as inductive limits of barrelled spaces. On the other hand, $\mathcal{O}_{M,\omega}(\mathbb{R}^N)$ is ultrabornological (see [8]), hence barrelled. Hence, it remains to show that each bounded set in such spaces is relatively compact. This follows from Lemma 2.1. Indeed, fixed $m \in \mathbb{N}$ and a bounded subset B of $\mathcal{O}_\omega^m(\mathbb{R}^N)$, from the regularity of $\mathcal{O}_\omega^m(\mathbb{R}^N)$ there exists $n \in \mathbb{N}$ such that B is contained and bounded in $\mathcal{O}_{n,\omega}^m(\mathbb{R}^N)$. By Lemma 2.1 the spaces $\mathcal{O}_{n+1,\omega}^m(\mathbb{R}^N)$ and $\mathcal{E}_\omega(\mathbb{R}^N)$ induce the same topology on B and hence, B is relatively compact in $\mathcal{O}_{n+1,\omega}^m(\mathbb{R}^N)$, after having observed that $\mathcal{E}_\omega(\mathbb{R}^N)$ is Montel. Therefore, B is also relatively compact in $\mathcal{O}_\omega^m(\mathbb{R}^N)$.

Since $\mathcal{O}_{M,\omega}(\mathbb{R}^N)$ is the projective limit of the Montel spaces $\mathcal{O}_\omega^m(\mathbb{R}^N)$, each bounded subset of $\mathcal{O}_{M,\omega}(\mathbb{R}^N)$ is clearly relatively compact. The completeness of the space $\mathcal{O}_{M,\omega}(\mathbb{R}^N)$ follows from the fact that it is the projective limit of the complete spaces $\mathcal{O}_\omega^m(\mathbb{R}^N)$. □

Remark 2.1 We remark that the completeness of the space $\mathcal{O}_{M,\omega}(\mathbb{R}^N)$ follows also from [2, Theorem 5.2(ii)] combined with [8, Theorem 5.3].

3 The Algebras $(\mathcal{O}_{M,\omega}(\mathbb{R}^N), \cdot)$, $(\mathcal{S}_\omega(\mathbb{R}^N), \cdot)$, $(\mathcal{S}_\omega(\mathbb{R}^N), \star)$ and $(\mathcal{O}'_{C,\omega}(\mathbb{R}^N), \star)$

Let us recall that a bilinear mapping $b\colon E \times F \to G$ between locally convex Hausdorff spaces E, F, G is continuous if, and only if, for every continuous seminorm p_1 on G there exist continuous seminorms p_2 and p_3 on E and F, respectively, such that the inequality

$$p_1(b(v, w)) \leq p_2(v)p_3(w)$$

holds for every pair $(v, w) \in E \times F$. A locally convex algebra (topological algebra, briefly) over the field $\mathbb{K} = \mathbb{R}$ or $\mathbb{K} = \mathbb{C}$ is a lcHs E together with a bilinear map $b\colon E \times E \to E$ that turns E into an algebra over $\mathbb{K}$ with b continuous.

The aim of this section is to establish that $(\mathcal{O}_{M,\omega}(\mathbb{R}^N), \cdot)$, $(\mathcal{S}_\omega(\mathbb{R}^N), \cdot)$ are multiplication topological algebras, while $(\mathcal{S}_\omega(\mathbb{R}^N), \star)$ and $(\mathcal{O}'_{C,\omega}(\mathbb{R}^N), \star)$ are convolution topological algebras.

We first consider the case $\mathcal{O}_{M,\omega}(\mathbb{R}^N)$ and establish the following fact.

Lemma 3.1 *Let ω be a non-quasianalytic weight function. Then for all $k \in \mathcal{S}_\omega(\mathbb{R}^N)$ there exists $l \in \mathcal{S}_\omega(\mathbb{R}^N)$ such that $|k(x)| \leq l^2(x)$ for every $x \in \mathbb{R}^N$.*

Proof Let $k \in \mathcal{S}_\omega(\mathbb{R}^N)$ be fixed. The function $h(x) := \sqrt{|k(x)|}$ for $x \in \mathbb{R}^N$ is a non-negative function such that

$$\lim_{|x|\to\infty} \exp(\lambda\omega(x))h(x) = \lim_{|x|\to\infty} \sqrt{\exp\left(\frac{\lambda}{2}\omega(x)\right)|k(x)|} = 0$$

for all $\lambda > 0$. By [2, Lemma 5.8] there exists $l \in \mathcal{S}_\omega(\mathbb{R}^N)$ for which $h(x) \leq l(x)$ for every $x \in \mathbb{R}^N$. Accordingly, $|k(x)| \leq l^2(x)$ for every $x \in \mathbb{R}^N$. □

Theorem 3.1 *Let ω be a non-quasianalytic weight function. Then $(\mathcal{O}_{M,\omega}(\mathbb{R}^N), \cdot)$ is a multiplication topological algebra.*

Proof To see this, we fix $m \in \mathbb{N}$ and $k \in \mathcal{S}_\omega(\mathbb{R}^N)$. We choose $m' \in \mathbb{N}$ such that $m' \geq Lm$, where L is the constant appearing in (2), and $l \in \mathcal{S}_\omega(\mathbb{R}^N)$ as in Lemma 3.1. Now, let $f, g \in \mathcal{O}_{M,\omega}(\mathbb{R}^N)$. Then we have for every $\alpha \in \mathbb{N}_0^N$ and $x \in \mathbb{R}^N$ that

$$\begin{aligned}|k(x)||\partial^\alpha(fg)(x)| &\leq \sum_{\gamma\leq\alpha}\binom{\alpha}{\gamma} l^2(x)|\partial^\gamma f(x)||\partial^{\alpha-\gamma}g(x)| \\ &\leq q_{m',l}(f)q_{m',l}(g)\sum_{\gamma\leq\alpha}\binom{\alpha}{\gamma}\exp\left(m'\varphi_\omega^*\left(\frac{|\gamma|}{m'}\right)\right)\exp\left(m'\varphi_\omega^*\left(\frac{|\alpha-\gamma|}{m'}\right)\right).\end{aligned}$$

Applying (4) and (5), we obtain for every $\alpha \in \mathbb{N}_0^N$ and $x \in \mathbb{R}^N$ that

$$\begin{aligned}|k(x)||\partial^\alpha(fg)(x)| &\leq q_{m',l}(f)q_{m',l}(g)2^{|\alpha|}\exp\left(m'\varphi_\omega^*\left(\frac{|\alpha|}{m'}\right)\right) \\ &\leq Cq_{m',l}(f)q_{m',l}(g)\exp\left(m\varphi_\omega^*\left(\frac{|\alpha|}{m}\right)\right),\end{aligned}$$

where $C := e^{mL}$. Accordingly,

$$q_{m,k}(fg) \leq Cq_{m',l}(f)q_{m',l}(g). \tag{7}$$

Since $f, g \in \mathcal{O}_{M,\omega}(\mathbb{R}^N)$, $m \in \mathbb{N}$ and $k \in \mathcal{S}_\omega(\mathbb{R}^N)$ are arbitrary, we can conclude from (7) that the multiplication operator $M\colon \mathcal{O}_{M,\omega}(\mathbb{R}^N) \times \mathcal{O}_{M,\omega}(\mathbb{R}^N) \to \mathcal{O}_{M,\omega}(\mathbb{R}^N)$, $(f, g) \mapsto fg$, is well-defined and continuous, i.e., $\mathcal{O}_{M,\omega}(\mathbb{R}^N)$ is a multiplication topological algebra. □

Arguing in a similar way with simple changes (indeed, it suffices to consider $\exp(n\omega)$ instead of the function k and to take $\exp\left(\frac{n}{2}\omega\right)$ instead of the function l), one shows the same result for $\mathcal{S}_\omega(\mathbb{R}^N)$.

Theorem 3.2 *Let ω be a non-quasianalytic weight function. Then $(\mathcal{S}_\omega(\mathbb{R}^N), \cdot)$ is a multiplication topological algebra.*

We now introduce the following definition.

Definition 3.1 Let ω be a non-quasianalytic weight function. Let E be a lcHs of ω-ultradifferentiable functions on $\mathbb{R}^N$ continuously included in $\mathcal{E}_\omega(\mathbb{R}^N)$ with dense range. We denote by $M(E)$ the space of all multipliers of E, i.e., the largest space of ω-ultradifferentiable functions on $\mathbb{R}^N$ satisfying the following conditions:

1. the multiplication operator on $E \times M(E) \to \mathcal{E}_\omega(\mathbb{R}^N)$, $(f, g) \mapsto fg$, is well-defined and takes values in E;
2. for all $f \in M(E)$, the operator $M_f : E \to E$, $g \mapsto fg$ is continuous.

If E' is the strong dual of E, we denote by $M(E')$ the space of all multipliers of E', i.e., the largest space of ω-ultradifferentiable functions on $\mathbb{R}^N$ for which the following conditions are satisfied:

1. for all $T \in E'$ and $f \in M(E')$ we have that fT is well-defined on E and belongs to E';
2. for all $f \in M(E')$, the operator $M_f : E' \to E'$, $T \mapsto fT$ is continuous.

Remark 3.1 We first recall that $M(\mathcal{S}_\omega(\mathbb{R}^N)) = M(\mathcal{S}'_\omega(\mathbb{R}^N)) = \mathcal{O}_{M,\omega}(\mathbb{R}^N)$.

We now observe that if E is a lcHs of ω-ultradifferentiable functions on $\mathbb{R}^N$ continuously included in $\mathcal{E}_\omega(\mathbb{R}^N)$ with dense range and such that the constant functions belong to E, then

$$M(E) = \{1\} \cdot M(E) \subseteq E \cdot M(E) \subseteq E.$$

Since $\mathcal{O}_{M,\omega}(\mathbb{R}^N)$ and $\mathcal{O}_{C,\omega}(\mathbb{R}^N)$ contain the constant functions, we get

$$M(\mathcal{O}_{M,\omega}(\mathbb{R}^N)) \subseteq \mathcal{O}_{M,\omega}(\mathbb{R}^N), \quad M(\mathcal{O}_{C,\omega}(\mathbb{R}^N)) \subseteq \mathcal{O}_{C,\omega}(\mathbb{R}^N).$$

But $\mathcal{O}_{M,\omega}(\mathbb{R}^N)$ is a multiplication algebra as shown in Theorem 3.1. Therefore, we clearly have that $\mathcal{O}_{M,\omega}(\mathbb{R}^N) \subseteq M(\mathcal{O}_{M,\omega}(\mathbb{R}^N))$. Thus, $M(\mathcal{O}_{M,\omega}(\mathbb{R}^N)) = \mathcal{O}_{M,\omega}(\mathbb{R}^N)$.

It is also true that $M(\mathcal{O}_{C,\omega}(\mathbb{R}^N)) = \mathcal{O}_{C,\omega}(\mathbb{R}^N)$, as the following result shows.

Proposition 3.1 *Let ω be a non-quasianalytic weight function.*

Then $M(\mathcal{O}_{C,\omega}(\mathbb{R}^N)) = \mathcal{O}_{C,\omega}(\mathbb{R}^N)$. Hence, $(\mathcal{O}_{C,\omega}(\mathbb{R}^N), \cdot)$ is a multiplication algebra.

Proof By Remark 3.1 it suffices to show that $\mathcal{O}_{C,\omega}(\mathbb{R}^N) \subseteq M(\mathcal{O}_{C,\omega}(\mathbb{R}^N))$. In order to do this, we fix $n, m \in \mathbb{N}$ with $n \geq 2$ and choose $m' \geq Lm$, with L the constant appearing in (2). Let $n_1, n_2 \in \mathbb{N}$ be such that $n_1 + n_2 = n$. If $f \in \mathcal{O}_{n_1,\omega}(\mathbb{R}^N) = \cap_{r=1}^{\infty} \mathcal{O}^r_{n_1,\omega}(\mathbb{R}^N)$ and $g \in \mathcal{O}_{n_2,\omega}(\mathbb{R}^N) = \cap_{r=1}^{\infty} \mathcal{O}^r_{n_2,\omega}(\mathbb{R}^N)$, then via (4) and (5) we have for every $\alpha \in \mathbb{N}_0^N$ and $x \in \mathbb{R}^N$ that

$$\begin{aligned}
|\partial^\alpha (fg)(x)| &\leq \sum_{\gamma \leq \alpha} \binom{\alpha}{\gamma} |\partial^\gamma f(x)| |\partial^{\alpha-\gamma} g(x)| \\
&\leq r_{m',n_1}(f) r_{m',n_2}(g) \sum_{\gamma \leq \alpha} \binom{\alpha}{\gamma} \exp\left(n_1 \omega(x) + m' \varphi^*_\omega\left(\frac{|\gamma|}{m'}\right)\right) \times
\end{aligned}$$

$$\times \exp\left(n_2\omega(x) + m'\varphi^*_\omega\left(\frac{|\alpha-\gamma|}{m'}\right)\right)$$

$$\leq Cr_{m',n_1}(f)r_{m',n_2}(g)\exp(n\omega(x))\exp\left(m\varphi^*_\omega\left(\frac{|\alpha|}{m}\right)\right),$$

where $C := e^{mL}$. Accordingly, we have

$$r_{m,n}(fg) \leq Cr_{m',n_1}(f)r_{m',n_2}(g). \tag{8}$$

Since $m \in \mathbb{N}$, $f \in \mathcal{O}_{n_1,\omega}(\mathbb{R}^N)$ and $g \in \mathcal{O}_{n_2,\omega}(\mathbb{R}^N)$ are arbitrary, from (8) it follows that the multiplication operator $M : \mathcal{O}_{n_1,\omega}(\mathbb{R}^N) \times \mathcal{O}_{n_2,\omega}(\mathbb{R}^N) \to \mathcal{O}_{n,\omega}(\mathbb{R}^N)$ is continuous. But also $n \in \mathbb{N}$ is arbitrary and $\mathcal{O}_{C,\omega}(\mathbb{R}^N)$ is the inductive limit of the Fréchet spaces $\mathcal{O}_{n,\omega}(\mathbb{R}^N)$. So, we can conclude that the multiplication operator $M : \mathcal{O}_{C,\omega}(\mathbb{R}^N) \times \mathcal{O}_{C,\omega}(\mathbb{R}^N) \to \mathcal{O}_{C,\omega}(\mathbb{R}^N)$ is well-defined. Moreover, (8) also implies that the operator $M_f : \mathcal{O}_{C,\omega}(\mathbb{R}^N) \to \mathcal{O}_{C,\omega}(\mathbb{R}^N)$, $g \mapsto fg$, is continuous for all $f \in \mathcal{O}_{C,\omega}(\mathbb{R}^N)$. Therefore, $\mathcal{O}_{C,\omega}(\mathbb{R}^N) \subseteq M(\mathcal{O}_{C,\omega}(\mathbb{R}^N))$. This completes the proof. □

Remark 3.2 For any non-quasianalytic weight function ω such that $\log(1+t) = o(\omega(t))$ as $t \to \infty$., the space $(\mathcal{O}_{C,\omega}(\mathbb{R}^N), \cdot)$ is not a multiplication topological algebra (see Sect. 4).

Corollary 3.1 *Let ω be a non-quasianalytic weight function. Then $M(\mathcal{O}'_{M,\omega}(\mathbb{R}^N)) = \mathcal{O}_{M,\omega}(\mathbb{R}^N)$ and $M(\mathcal{O}'_{C,\omega}(\mathbb{R}^N)) = \mathcal{O}_{C,\omega}(\mathbb{R}^N)$.*

Proof The result follows from Theorem 3.1 and Proposition 3.1, taking into account of the fact that the multiplication of ω-ultradistributions with ω-ultradifferentiable functions is defined by transposition. □

We now pass to the case $(\mathcal{S}_\omega(\mathbb{R}^N), \star)$, for which the following result is true.

Theorem 3.3 *Let ω be a non-quasianalytic weight function. Then $(\mathcal{S}_\omega(\mathbb{R}^N), \star)$ is a convolution topological algebra.*

Proof To see this we have to show that the bilinear map $\star : \mathcal{S}_\omega(\mathbb{R}^N) \times \mathcal{S}_\omega(\mathbb{R}^N) \to \mathcal{S}_\omega(\mathbb{R}^N)$ is continuous. So, let $n_0 \in \mathbb{N}$ be fixed such that $n_0 \geq \frac{N+1}{b}$. Hence, $\exp(-n_0\omega) \in L^1(\mathbb{R}^N)$ by (3). Fixed $n \in \mathbb{N}$ and $f, g \in \mathcal{S}_\omega(\mathbb{R}^N)$, we have

$$\partial^\alpha(f \star g)(x) = \int_{\mathbb{R}^N} f(y)\partial^\alpha g(x-y)\,dy, \ x \in \mathbb{R}^N, \ \alpha \in \mathbb{N}_0^N.$$

By (1) $\omega(x) = \omega(x - y + y) \leq K(1 + \omega(x - y) + \omega(y))$. Hence, we get for every $x \in \mathbb{R}^N$ and $\alpha \in \mathbb{N}_0^N$ that

$$
\begin{aligned}
&|\partial^\alpha (f \star g)(x)| \exp(n\omega(x)) \\
&\quad \leq e^{Kn} \int_{\mathbb{R}^N} |f(y)| \exp(Kn\omega(y)) |\partial^\alpha g(x - y)| \exp(Kn\omega(x - y))\, dy \\
&\quad \leq e^{Kn} \| \exp(-n_0\omega)\|_1 \| f \exp((Kn + n_0)\omega)\|_\infty \|\partial^\alpha g \exp(Kn\omega)\|_\infty .
\end{aligned}
$$

Therefore, for all $m, n \in \mathbb{N}$

$$
\begin{aligned}
q_{m,n}(f \star g) &\leq e^{Kn} \| \exp(-n_0\omega)\|_1 \| f \exp((Kn + n_0)\omega)\|_\infty q_{m,Kn}(g) \\
&\leq e^{Kn} \| \exp(-n_0\omega)\|_1 q_{m,Kn+n_0}(f) q_{m,Kn}(g) \\
&\leq e^{Kn} \| \exp(-n_0\omega)\|_1 q_{m,n'}(f) q_{m,n'}(g),
\end{aligned}
$$

with $n' \in \mathbb{N}$ such that $n' \geq Kn + n_0$. This completes the proof. □

Remark 3.3 We point out that the result in Theorem 3.3 can be achieved by applying Theorem 3.2 combined with the use of the Fourier transform, which is a topological isomorphism from $\mathcal{S}_\omega(\mathbb{R}^N)$ onto itself such that $\widehat{f \star g} = \hat{f}\hat{g}$, for all $f, g \in \mathcal{S}_\omega(\mathbb{R}^N)$, see [3].

The next aim is to show that $(\mathcal{O}'_{C,\omega}(\mathbb{R}^N), \star)$ is a convolution topological algebra. Hence, let us recall that in case $T \in \mathcal{O}'_{C,\omega}(\mathbb{R}^N)$ and $S \in \mathcal{S}'_\omega(\mathbb{R}^N)$ the convolution $T \star S$ is well defined on $\mathcal{S}_\omega(\mathbb{R}^N)$ and belongs to $\mathcal{S}'_\omega(\mathbb{R}^N)$. Indeed, for all $f \in \mathcal{S}_\omega(\mathbb{R}^N)$ we have $\check{T} \star f \in \mathcal{S}_\omega(\mathbb{R}^N)$ and the operator $C_T : \mathcal{S}_\omega(\mathbb{R}^N) \to \mathcal{S}_\omega(\mathbb{R}^N)$, $f \mapsto \check{T} \star f$, is continuous, see, [1, Theorem 5.3] (the distribution $\check{T}$ is defined by $f \mapsto \check{T}(f) := T(\check{f})$, with $\check{f}(x) := f(-x)$ for all $x \in \mathbb{R}^N$. Furthermore, for R a distribution and f a function the convolution $R \star f$ is defined by $(R \star f)(x) := \langle R_y, \tau_x \check{f}\rangle$, where the notation R_y means that the distribution R acts on a function $\phi(x - y)$, when the latter is regarded as a function of the variable y). Hence, the convolution $T \star S$ defined by $\langle T \star S, f\rangle = \langle S, \check{T} \star f\rangle$, for $f \in \mathcal{S}_\omega(\mathbb{R}^N)$, is clearly a well-defined element of $\mathcal{S}'_\omega(\mathbb{R}^N)$. Since C_T is a continuous operator from $\mathcal{S}_\omega(\mathbb{R}^N)$ into $\mathcal{O}_{C,\omega}(\mathbb{R}^N)$ and the space $\mathcal{S}_\omega(\mathbb{R}^N)$ is dense in $\mathcal{O}_{C,\omega}(\mathbb{R}^N)$, we also have that $T \star S \in \mathcal{O}'_{C,\omega}(\mathbb{R}^N)$ (in the sense that $T \star S$ extends continuously on whole $\mathcal{O}_{C,\omega}(\mathbb{R}^N)$).

Moreover, for any ω non-quasianalytic weight function satisfying the condition $\log(1 + t) = o(\omega(t))$ as $t \to \infty$, the Fourier transform $\mathcal{F}$ is a topological isomorphism from the space $\mathcal{O}'_{C,\omega}(\mathbb{R}^N)$ onto the space $\mathcal{O}_{M,\omega}(\mathbb{R}^N)$ (see [1, Theorem 6.1]). Accordingly, we have $\mathcal{F}(\mathcal{O}'_{C,\omega}(\mathbb{R}^N)) = \mathcal{O}_{M,\omega}(\mathbb{R}^N)$ and $\mathcal{F}(\mathcal{O}_{M,\omega}(\mathbb{R}^N)) = \mathcal{O}'_{C,\omega}(\mathbb{R}^N)$. In particular, for all $T \in \mathcal{O}'_{C,\omega}(\mathbb{R}^N)$ and $S \in \mathcal{S}'_\omega(\mathbb{R}^N)$ the convolution $T \star S$ satisfies the following property:

$$
\mathcal{F}(T \star S) = \mathcal{F}(T)\mathcal{F}(S). \tag{9}
$$

We can now state the following result.

Theorem 3.4 *Let ω be a non-quasianalytic weight function such that* $\log(1+t) = o(\omega(t))$ *for* $t \to \infty$. *Then* $(\mathcal{O}'_{C,\omega}(\mathbb{R}^N), \star)$ *is a convolution topological algebra.*

Proof We first observe that if $S, T \in \mathcal{O}'_{C,\omega}(\mathbb{R}^N) \subset \mathcal{S}'_\omega(\mathbb{R}^N)$, then the convolutions $T \star S$ and $S \star T$ are well defined and belong to $\mathcal{S}'_\omega(\mathbb{R}^N)$. Since $\mathcal{F}(S), \mathcal{F}(T) \in \mathcal{O}_{M,\omega}(\mathbb{R}^N)$ and so, $\mathcal{F}(S)\mathcal{F}(T) = \mathcal{F}(T)\mathcal{F}(S)$, by (9) it follows that $S \star T = T \star S \in \mathcal{O}'_{C,\omega}(\mathbb{R}^N)$. This means that $(\mathcal{O}'_{C,\omega}(\mathbb{R}^N), \star)$ is a convolution algebra. Finally, since the Fourier transform is a topological isomorphim from $\mathcal{O}'_{C,\omega}(\mathbb{R}^N)$ onto $\mathcal{O}_{M,\omega}(\mathbb{R}^N)$ satisfying Eq. (9), we can apply Theorem 3.1 to obtain that $(\mathcal{O}'_{C,\omega}(\mathbb{R}^N), \star)$ is a topological algebra. □

As done in the case of multipliers of lcHs' of ω-ultradifferentiable functions on $\mathbb{R}^N$, we now introduce the space of convolutors.

Definition 3.2 Let E be a lcHs of ω-ultradifferentiable functions on $\mathbb{R}^N$. We denote by $C(E)$ the space of all convolutors of E, i.e., the largest space of ω-ultradistributions on $\mathbb{R}^N$ satisfying the following conditions:

1. the convolution operator on $E \times C(E) \to \mathcal{E}_\omega(\mathbb{R})$, $(f, T) \mapsto T \star f$, is well-defined and takes value in E;
2. for all $T \in C(E)$ the operator $C_T : E \to E$, $f \mapsto T \star f$, is continuous.

We denote by $C(E')$ the space of all convolutors of E', i.e., the largest space of ω-ultradistributions on $\mathbb{R}^N$ satisfying the following conditions:

1. for all $S \in E'$ and $T \in C(E')$ we have that $T \star S$ is well-defined on E and belongs to E';
2. for all $T \in C(E')$ the operator $C_T : E' \to E'$, $S \mapsto T \star S$, is continuous.

Remark 3.4 Let ω be a non-quasianalytic weight function ω such that $\log(1+t) = o(\omega(t))$ for $t \to \infty$. Then $C(\mathcal{S}_\omega(\mathbb{R}^N)) = C(\mathcal{S}'_\omega(\mathbb{R}^N)) = \mathcal{O}'_{C,\omega}(\mathbb{R}^N)$.

We point out that the Eq. (9) is also satisfied for any pairs (T, S) of ω-ultradistributions in $\mathcal{O}'_{C,\omega}(\mathbb{R}^N) \times \mathcal{O}_{C,\omega}(\mathbb{R}^N)$, $\mathcal{O}'_{M,\omega}(\mathbb{R}^N) \times \mathcal{O}_{M,\omega}(\mathbb{R}^N)$, $\mathcal{O}'_{C,\omega}(\mathbb{R}^N) \times \mathcal{O}'_{C,\omega}(\mathbb{R}^N)$ and $\mathcal{O}'_{M,\omega}(\mathbb{R}^N) \times \mathcal{O}'_{M,\omega}(\mathbb{R}^N)$, because these spaces are continuously included in $\mathcal{O}'_{C,\omega}(\mathbb{R}^N) \times \mathcal{S}'_\omega(\mathbb{R}^N)$. On the other hand, by Remark 3.1 and Proposition 3.1 we have that $M(\mathcal{O}_{C,\omega}(\mathbb{R}^N)) = \mathcal{O}_{C,\omega}(\mathbb{R}^N)$ and $M(\mathcal{O}_{M,\omega}(\mathbb{R}^N)) = \mathcal{O}_{M,\omega}(\mathbb{R}^N)$. Recalling that the Fourier transform is a topological isomorphism from $\mathcal{O}'_{C,\omega}(\mathbb{R}^N)$ onto $\mathcal{O}_{M,\omega}(\mathbb{R}^N)$, these facts yield that $C(\mathcal{O}'_{M,\omega}(\mathbb{R}^N)) = \mathcal{O}'_{M,\omega}(\mathbb{R}^N)$ and $C(\mathcal{O}'_{C,\omega}(\mathbb{R}^N)) = \mathcal{O}'_{C,\omega}(\mathbb{R}^N)$. Furthemore, by Corollary 3.1 we have that $M(\mathcal{O}'_{C,\omega}(\mathbb{R}^N)) = \mathcal{O}_{C,\omega}(\mathbb{R}^N)$ and $M(\mathcal{O}'_{M,\omega}(\mathbb{R}^N)) = \mathcal{O}_{M,\omega}(\mathbb{R}^N)$. Thus, by the same arguments we obtain that $C(\mathcal{O}_{M,\omega}(\mathbb{R}^N)) = \mathcal{O}'_{M,\omega}(\mathbb{R}^N)$ and $C(\mathcal{O}_{C,\omega}(\mathbb{R}^N)) = \mathcal{O}'_{C,\omega}(\mathbb{R}^N)$. We summarize our results in this simple table.

E	M(E)	C(E)
$\mathcal{S}_\omega(\mathbb{R}^N)$	$\mathcal{O}_{M,\omega}(\mathbb{R}^N)$	$\mathcal{O}'_{C,\omega}(\mathbb{R}^N)$
$\mathcal{S}'_\omega(\mathbb{R}^N)$	$\mathcal{O}_{M,\omega}(\mathbb{R}^N)$	$\mathcal{O}'_{C,\omega}(\mathbb{R}^N)$
$\mathcal{O}_{M,\omega}(\mathbb{R}^N)$	$\mathcal{O}_{M,\omega}(\mathbb{R}^N)$	$\mathcal{O}'_{M,\omega}(\mathbb{R}^N)$
$\mathcal{O}'_{M,\omega}(\mathbb{R}^N)$	$\mathcal{O}_{M,\omega}(\mathbb{R}^N)$	$\mathcal{O}'_{M,\omega}(\mathbb{R}^N)$
$\mathcal{O}_{C,\omega}(\mathbb{R}^N)$	$\mathcal{O}_{C,\omega}(\mathbb{R}^N)$	$\mathcal{O}'_{C,\omega}(\mathbb{R}^N)$
$\mathcal{O}'_{C,\omega}(\mathbb{R}^N)$	$\mathcal{O}_{C,\omega}(\mathbb{R}^N)$	$\mathcal{O}'_{C,\omega}(\mathbb{R}^N)$

From Remark 3.4, the next result follows.

Corollary 3.2 *Let ω be a non-quasianalytic weight function such that* $\log(1+t) = o(\omega(t))$ *for $t \to \infty$. Then $\mathcal{O}'_{M,\omega}(\mathbb{R}^N)$ is a convolution algebra.*

Remark 3.5 For any non-quasianalytic weight function ω such that $\log(1+t) = o(\omega(t))$ for $t \to \infty$, the space $(\mathcal{O}'_{M,\omega}(\mathbb{R}^N), \star)$ is not a convolution topological algebra (see Sect. 4).

4 Hypocontinuity and Discontinuity

4.1 Hypocontinuity

In this final section we discuss the hypocontinuity of the multiplication mapping on some pairs between the spaces $\mathcal{O}_{M,\omega}(\mathbb{R}^N)$, $\mathcal{O}_{C,\omega}(\mathbb{R}^N)$, $\mathcal{S}_\omega(\mathbb{R}^N)$ and their duals.

Let us recall that if E, F and G are topological vector spaces and $b\colon E \times F \to G$ is bilinear map, then b is called *hypocontinuous* if the following holds:

(1) For every bounded subset A of E, the set $\{b(x,\cdot)\colon x \in A\}$ is equicontinuous in $\mathcal{L}(F, G)$;
(2) For every bounded subset B of F, the set $\{b(\cdot, y)\colon y \in B\}$ is equicontinuous in $\mathcal{L}(E, G)$.

Proposition 4.1 *Let ω be a non-quasianalytic weight function. Then the multiplication operator $M\colon \mathcal{O}_{M,\omega}(\mathbb{R}^N) \times \mathcal{S}_\omega(\mathbb{R}^N) \to \mathcal{S}_\omega(\mathbb{R}^N)$, $(f, g) \mapsto fg$, is separately continuous and hence, a hypocontinuous bilinear mapping.*

Proof By [2, Theorem 4.4] the operator $M_f := M(f,\cdot)\colon \mathcal{S}_\omega(\mathbb{R}^N) \to \mathcal{S}_\omega(\mathbb{R}^N)$ is continuous for all $f \in \mathcal{O}_{M,\omega}(\mathbb{R}^N)$.

Let $g \in \mathcal{S}_\omega(\mathbb{R}^N)$ be fixed. We claim that $M_g := M(\cdot, g)\colon \mathcal{O}_{M,\omega}(\mathbb{R}^N) \to \mathcal{S}_\omega(\mathbb{R}^N)$ is a continuous operator. To show the claim, we suppose that $\{f_i\}_i \subset \mathcal{O}_{M,\omega}(\mathbb{R}^N)$ is any net such that $f_i \to f$ in $\mathcal{O}_{M,\omega}(\mathbb{R}^N)$ and $M_g(f_i) = f_i g \to h$ in $\mathcal{S}_\omega(\mathbb{R}^N)$. Then by [2, Theorem 5.2(1) and Proposition 5.6] we have that $f_i \to f$ in $\mathcal{E}_\omega(\mathbb{R}^N)$ and hence, $M_g(f_i) = f_i g \to fg$ in $\mathcal{E}_\omega(\mathbb{R}^N)$ too. But $M_g(f_i) = f_i g \to h$ also in $\mathcal{E}_\omega(\mathbb{R}^N)$. Therefore, $h = fg = M_g(f)$. Since $\{f_i\}_i \subset \mathcal{O}_{M,\omega}(\mathbb{R}^N)$ is arbitrary, this shows that the graph of the operator M_g is closed. But, the space

$\mathcal{O}_{M,\omega}(\mathbb{R}^N)$ is ultrabornological (see [8]), hence barrelled, and $\mathcal{S}_\omega(\mathbb{R}^N)$ is a Fréchet space, and so M_g is necessarily continuous.

Finally, as the operator M is separately continuous and $\mathcal{O}_{M,\omega}(\mathbb{R}^N)$ and $\mathcal{S}_\omega(\mathbb{R}^N)$ are barrelled spaces, applying [18, Theorem 41.2] we can conclude that M is a hypocontinuous bilinear mapping. □

Corollary 4.1 *Let ω be a non-quasianalytic weight function. Then the multiplication operator $M\colon \mathcal{O}_{C,\omega}(\mathbb{R}^N) \times \mathcal{S}_\omega(\mathbb{R}^N) \to \mathcal{S}_\omega(\mathbb{R}^N)$, $(f, g) \mapsto fg$, is separately continuous and hence, a hypocontinuous bilinear mapping.*

Proof By [2, Theorem 3.8(1)] the space $\mathcal{O}_{C,\omega}(\mathbb{R}^N)$ is continuously included in $\mathcal{O}_{M,\omega}(\mathbb{R}^N)$. So, via Proposition 4.1 it follows that the multiplication operator M is separately continuous and so hypocontinuous. □

Proposition 4.2 *Let ω be a non-quasianalytic weight function. Then the multiplication operator $M\colon \mathcal{O}_{M,\omega}(\mathbb{R}^N)\times\mathcal{S}'_\omega(\mathbb{R}^N) \to \mathcal{S}'_\omega(\mathbb{R}^N)$, $(f, T) \mapsto fT$, is separately continuous and hence, a hypocontinuous bilinear mapping.*

Proof By [2, Theorem 4.6] the multiplication operator $M_f := M(f, \cdot)\colon \mathcal{S}'_\omega(\mathbb{R}^N) \to \mathcal{S}'_\omega(\mathbb{R}^N)$ is continuous for all $f \in \mathcal{O}_{M,\omega}(\mathbb{R}^N)$.

Let $T \in \mathcal{S}'_\omega(\mathbb{R}^N)$ be fixed. We claim that $M_T := M(\cdot, T)\colon \mathcal{O}_{M,\omega}(\mathbb{R}^N) \to \mathcal{S}'_\omega(\mathbb{R}^N)$ is a continuous operator. To show the claim, we fix a 0-neighborhood V in $\mathcal{S}'_\omega(\mathbb{R}^N)$. We can suppose that $V = \{S \in \mathcal{S}'_\omega(\mathbb{R}^N)\colon \sup_{g\in B} |\langle S, g\rangle| \leq \varepsilon\}$ for some $\varepsilon > 0$ and a bounded subset B of $\mathcal{S}_\omega(\mathbb{R}^N)$. On the other hand, as $T \in \mathcal{S}'_\omega(\mathbb{R}^N)$, there exist $m, n \in \mathbb{N}$ and $c > 0$ such that $|\langle T, g\rangle| \leq c\, q_{m,n}(g)$ for all $g \in \mathcal{S}_\omega(\mathbb{R}^N)$. We now set $U := \{g \in \mathcal{S}_\omega(\mathbb{R}^N)\colon q_{m,n}(g) \leq c^{-1}\varepsilon\}$ which is a 0-neighborhood in $\mathcal{S}_\omega(\mathbb{R}^N)$, and define the set

$$W := \{T \in \mathcal{L}(\mathcal{S}_\omega(\mathbb{R}^N))\colon T(B) \subseteq U\}.$$

Then W is a 0-neighborhood in $\mathcal{L}_b(\mathcal{S}_\omega(\mathbb{R}^N))$. Since $\mathcal{O}_{M,\omega}(\mathbb{R}^N)$ is a subspace of $\mathcal{L}_b(\mathcal{S}_\omega(\mathbb{R}^N))$, there exists a 0-neighborhood W_0 in $\mathcal{O}_{M,\omega}(\mathbb{R}^N)$ for which $fg \in U$ for all $f \in W_0$ and $g \in B$ (i.e., $M_f(B) \subseteq U$ for all $f \in W_0$). Accordingly, $M_T(W_0) \subseteq V$. Indeed, for a fixed $f \in W_0$, $fg \in U$ for all $g \in B$ and hence, $q_{m,n}(fg) \leq c^{-1}\varepsilon$ for all $g \in B$. This yields for all $g \in B$ that

$$|\langle M_T(f), g\rangle| = |\langle fT, g\rangle| = |\langle T, fg\rangle| \leq c\, q_{m,n}(fg) \leq \varepsilon.$$

This means that $M_T(f) \in V$. Since $M_T(W_0) \subseteq V$, it is obvious that $W_0 \subseteq M_T^{-1}(V)$. Finally, since V is an arbitrary 0-neighborhood in $\mathcal{S}'_\omega(\mathbb{R}^N)$, we can conclude that the operator M_T is continuous.

We now observe that the space $\mathcal{S}_\omega(\mathbb{R}^N)$ is nuclear and hence distinguished, i.e., its strong dual $\mathcal{S}'_\omega(\mathbb{R}^N)$ is a barrelled lcHs. So, since M is separately continuous and $\mathcal{O}_{M,\omega}(\mathbb{R}^N)$ and $\mathcal{S}'_\omega(\mathbb{R}^N)$ are barrelled lcHs, applying [18, Theorem 41.2] we get that M is a hypocontinuous bilinear mapping. □

Corollary 4.2 *Let ω be a non-quasianalytic weight function satisfying the condition $\log(1+t) = o(\omega(t))$ as $t \to \infty$. Then the multiplication operator $M\colon \mathcal{O}_{M,\omega}(\mathbb{R}^N) \times \mathcal{O}'_{C,\omega}(\mathbb{R}^N) \to \mathcal{S}'_\omega(\mathbb{R}^N)$, $(f,T) \mapsto fT$, is separately continuous and hence, a hypocontinuous bilinear mapping.*

Proof By [2, Theorems 3.8(2) and Theorem 3.9] the space $\mathcal{O}'_{C,\omega}(\mathbb{R}^N)$ is continuously included in $\mathcal{S}'_\omega(\mathbb{R}^N)$. So, via Proposition 4.2 it follows that the multiplication operator M is separately continuous and so hypocontinuous, being $\mathcal{O}'_{C,\omega}(\mathbb{R}^N)$ topologically isomorphic to $\mathcal{O}_{M,\omega}(\mathbb{R}^N)$ via the Fourier transform and hence barrelled. □

Proposition 4.3 *Let ω be a non-quasianalytic weight function. Then the multiplication operator $M\colon \mathcal{O}_{M,\omega}(\mathbb{R}^N) \times \mathcal{O}'_{M,\omega}(\mathbb{R}^N) \to \mathcal{O}'_{M,\omega}(\mathbb{R}^N)$, $(f,T) \mapsto fT$, is separately continuous and hence, a hypocontinuous bilinear mapping.*

Proof By Corollary 3.1 the multiplication operator $M_f := M(f,\cdot)\colon \mathcal{O}'_{M,\omega}(\mathbb{R}^N) \to \mathcal{O}'_{M,\omega}(\mathbb{R}^N)$ is continuous for all $f \in \mathcal{O}_{M,\omega}(\mathbb{R}^N)$.

Let $T \in \mathcal{O}'_{M,\omega}(\mathbb{R}^N)$ be fixed. We claim that $M_T := M(\cdot,T)\colon \mathcal{O}_{M,\omega}(\mathbb{R}^N) \to \mathcal{O}'_{M,\omega}(\mathbb{R}^N)$ is a continuous operator. To this end, we observe that there exist a function $k \in \mathcal{S}_\omega(\mathbb{R}^N)$, $m \in \mathbb{N}$ and $c > 0$ such that for all $g \in \mathcal{O}_{M,\omega}(\mathbb{R}^N)$ we have

$$|\langle T, g\rangle| \le c\, q_{m,k}(g). \tag{10}$$

If B is any bounded subset of $\mathcal{O}_{M,\omega}(\mathbb{R}^N)$, then from (10) it follows for all $f \in \mathcal{O}_{M,\omega}(\mathbb{R}^N)$ that

$$\sup_{g\in B} |M_T(f)(g)| = \sup_{g\in B} |\langle T, fg\rangle| \le c \sup_{g\in B} q_{m,k}(fg). \tag{11}$$

Now, as it is shown in Theorem 3.1 there exist $l \in \mathcal{S}_\omega(\mathbb{R}^N)$ and $m' \in N$ such that $q_{m,k}(uv) \le C q_{m',l}(u) q_{m',l}(v)$ whenever $u, v \in \mathcal{S}_\omega(\mathbb{R}^N)$ and for $C = e^{mL}$. Accordingly, we obtain via (11) that

$$\sup_{g\in B} |M_T(f)(g)| \le Cc \sup_{g\in B} q_{m',l}(g) q_{m',l}(f)$$

for all $f \in \mathcal{O}_{M,\omega}(\mathbb{R}^N)$, where $D := \sup_{g\in B} q_{m',l}(g) < \infty$, being B a bounded subset of $\mathcal{O}_{M,\omega}(\mathbb{R}^N)$. Since B is an arbitrary bounded subset of $\mathcal{O}_{M,\omega}(\mathbb{R}^N)$, this yields that the multiplication operator $M_T\colon \mathcal{O}_{M,\omega}(\mathbb{R}^N) \to \mathcal{O}'_{M,\omega}(\mathbb{R}^N)$ is continuous.

Finally, since M is separately continuous and $\mathcal{O}_{M,\omega}(\mathbb{R}^N)$ and $\mathcal{O}'_{M,\omega}(\mathbb{R}^N)$ are barrelled lcHs (the latter space is barrelled because it is reflexive as the strong dual of a Montel space), applying [18, Theorem 41.2] we get that M is a hypocontinuous bilinear mapping. □

Proposition 4.4 *Let ω be a non-quasianalytic weight function satisfying the condition $\log(1+t) = o(\omega(t))$ as $t \to \infty$. Then the multiplication operator $M: \mathcal{O}_{C,\omega}(\mathbb{R}^N) \times \mathcal{O}'_{C,\omega}(\mathbb{R}^N) \to \mathcal{O}'_{C,\omega}(\mathbb{R}^N)$, $(f, T) \mapsto fT$, is separately continuous and hence, a hypocontinuous bilinear mapping.*

Proof By Corollary 3.1 the multiplication operator $M_f := M(f, \cdot): \mathcal{O}'_{C,\omega}(\mathbb{R}^N) \to \mathcal{O}'_{C,\omega}(\mathbb{R}^N)$ is continuous for all $f \in \mathcal{O}_{C,\omega}(\mathbb{R}^N)$.

Let $T \in \mathcal{O}'_{C,\omega}(\mathbb{R}^N)$ be fixed. We show that $M_T := M(\cdot, T): \mathcal{O}_{C,\omega}(\mathbb{R}^N) \to \mathcal{O}'_{C,\omega}(\mathbb{R}^N)$ is a continuous operator. To see this, let $\{f_i\}_i \subset \mathcal{O}_{C,\omega}(\mathbb{R}^N)$ be any net such that $f_i \to f$ in $\mathcal{O}_{C,\omega}(\mathbb{R}^N)$ and $M_T(f_i) = f_i T \to S$ in $\mathcal{O}'_{C,\omega}(\mathbb{R}^N)$. Then $(f_i T)(g) = T(f_i g) \to S(g)$ for all $g \in \mathcal{O}_{C,\omega}(\mathbb{R}^N)$. On the other hand, $f_i g \to fg$ in $\mathcal{O}_{C,\omega}(\mathbb{R}^N)$ for all $g \in \mathcal{O}_{C,\omega}(\mathbb{R}^N)$ by Proposition 3.1, thereby implying that $T(f_i g) \to T(fg)$ for all $g \in \mathcal{O}_{C,\omega}(\mathbb{R}^N)$. Therefore, $T(fg) = S(g)$ for all $g \in \mathcal{O}_{C,\omega}(\mathbb{R}^N)$. This means that $S = fT = M_T(f)$. Since $\{f_i\}_i \subset \mathcal{O}_{C,\omega}(\mathbb{R}^N)$ is arbitrary, this shows that the graph of the operator M_T is closed. But, the space $\mathcal{O}_{C,\omega}(\mathbb{R}^N)$ is an (LF)-space and $\mathcal{O}'_{C,\omega}(\mathbb{R}^N)$ is a webbed space, and so M_T is necessarily continuous (see [14]).

Finally, since M is separately continuous and $\mathcal{O}_{C,\omega}(\mathbb{R}^N)$ and $\mathcal{O}'_{C,\omega}(\mathbb{R}^N)$ are barrelled lcHs, applying [18, Theorem 41.2] we get that M is a hypocontinuous bilinear mapping. □

4.2 Discontinuity

In this last part we show examples of multiplication and convolution mapping on some pairs between the spaces $\mathcal{O}_{M,\omega}(\mathbb{R}^N)$, $\mathcal{O}_{C,\omega}(\mathbb{R}^N)$, $\mathcal{S}_\omega(\mathbb{R}^N)$ and their duals that are not continuous.

Proposition 4.5 *Let ω be a non-quasianalytic weight function. Then the following assertions hold true:*

(i) The multiplication operator $M: \mathcal{O}_{M,\omega}(\mathbb{R}^N) \times \mathcal{O}'_{M,\omega}(\mathbb{R}^N) \to \mathcal{O}'_{M,\omega}(\mathbb{R}^N)$, $(f, T) \mapsto fT$, is discontinuous;
(ii) The multiplication operator $M: \mathcal{O}_{C,\omega}(\mathbb{R}^N) \times \mathcal{O}'_{C,\omega}(\mathbb{R}^N) \to \mathcal{O}'_{C,\omega}(\mathbb{R}^N)$, $(f, T) \mapsto fT$, is discontinuous.

Proof Since the proof is analogous in both the cases, we show only assertion (i).

Assume by contradiction that $M: \mathcal{O}_{M,\omega}(\mathbb{R}^N) \times \mathcal{O}'_{M,\omega}(\mathbb{R}^N) \to \mathcal{O}'_{M,\omega}(\mathbb{R}^N)$, $(f, T) \mapsto fT$, is continuous. Thus, the canonical bilinear form $\langle \cdot, \cdot \rangle: \mathcal{O}_{M,\omega}(\mathbb{R}^N) \times \mathcal{O}'_{M,\omega}(\mathbb{R}^N) \to \mathbb{C}$, $(f, T) \mapsto \langle f, T \rangle := T(f)$ is continuous, being it equals to the composition map $\mathbf{1} \circ M$ of the continuous mapping $\mathbf{1}: \mathcal{O}'_{M,\omega}(\mathbb{R}^N) \to \mathbb{C}$, $T \mapsto T(\mathbf{1})$ (note that the function $\mathbf{1} \in \mathcal{O}_{M,\omega}(\mathbb{R}^N)$) and the multiplication M. But the canonical bilinear form of a lcHs E is continuous if, and only if, the space E is normed (see [13, Page 359]), which is not the case for $\mathcal{O}_{M,\omega}(\mathbb{R}^N)$. □

Proposition 4.6 *Let ω be a non-quasianalytic weight function satisfying the condition* $\log(1+t) = o(\omega(t))$ *as* $t \to \infty$. *Then the following properties hold true:*

(i) *The convolution operator* $C\colon \mathcal{O}_{M,\omega}(\mathbb{R}^N) \times \mathcal{O}'_{M,\omega}(\mathbb{R}^N) \to \mathcal{O}_{M,\omega}(\mathbb{R}^N)$, $(f, T) \mapsto T \star f$, *is discontinuous.*

(ii) *The convolution operator* $C\colon \mathcal{O}_{C,\omega}(\mathbb{R}^N) \times \mathcal{O}'_{C,\omega}(\mathbb{R}^N) \to \mathcal{O}_{C,\omega}(\mathbb{R}^N)$, $(f, T) \mapsto T \star f$, *is discontinuous.*

Proof We first observe that the mapping are well defined by Remark 3.4.

Since the proof is analogous in both the cases, we show only property (i).

Assume by contradiction that $C\colon \mathcal{O}_{M,\omega}(\mathbb{R}^N) \times \mathcal{O}'_{M,\omega}(\mathbb{R}^N) \to \mathcal{O}_{M,\omega}(\mathbb{R}^N)$, $(f, T) \mapsto T \star f$, is continuous. Thus, the canonical bilinear form $\langle\cdot,\cdot\rangle\colon \mathcal{O}_{M,\omega}(\mathbb{R}^N) \times \mathcal{O}'_{M,\omega}(\mathbb{R}^N) \to \mathbb{C}$, $(f, T) \mapsto \langle f, T\rangle := T(f)$, is continuous, being it equal to the composition map $\delta \circ R \circ C$ of the continuous mappings $\delta\colon \mathcal{O}_{M,\omega}(\mathbb{R}^N) \to \mathbb{C}$, $f \mapsto f(0)$, $R\colon \mathcal{O}_{M,\omega}(\mathbb{R}^N) \to \mathcal{O}_{M,\omega}(\mathbb{R}^N)$, $f \mapsto \check{f}$, and the convolution C. But the canonical bilinear form of a lcHs E is continuous if, and only if, the space E is normed (see [13, Page 359]), which is not the case for $\mathcal{O}_{M,\omega}(\mathbb{R}^N)$. □

Remark 4.1 The proof of Proposition 4.6 can be also achieved applying Proposition 4.5 and using the properties of the Fourier transform.

Thanks to Propositions 4.3, 4.4 and 4.5, we get that the multiplication operators $M\colon \mathcal{O}_{M,\omega}(\mathbb{R}^N) \times \mathcal{O}'_{M,\omega}(\mathbb{R}^N) \to \mathcal{O}'_{M,\omega}(\mathbb{R}^N)$ and $M\colon \mathcal{O}_{C,\omega}(\mathbb{R}^N) \times \mathcal{O}'_{C,\omega}(\mathbb{R}^N) \to \mathcal{O}'_{C,\omega}(\mathbb{R}^N)$ are hypocontinuous but not continuous. The same also holds true for the hypocontinuous multiplication operator $M\colon \mathcal{O}_{M,\omega}(\mathbb{R}^N) \times \mathcal{S}'_{\omega}(\mathbb{R}^N) \to \mathcal{S}'_{\omega}(\mathbb{R}^N)$, as we show in the following result.

Proposition 4.7 *Let ω be a non-quasianalytic weight function. Then the multiplication operator* $M\colon \mathcal{O}_{M,\omega}(\mathbb{R}^N) \times \mathcal{S}'_{\omega}(\mathbb{R}^N) \to \mathcal{S}'_{\omega}(\mathbb{R}^N)$, $(f, T) \mapsto fT$, *is discontinuous.*

Proof Assume by contradiction that $M\colon \mathcal{O}_{M,\omega}(\mathbb{R}^N) \times \mathcal{S}'_{\omega}(\mathbb{R}^N) \to \mathcal{S}'_{\omega}(\mathbb{R}^N)$ is continuous. Since $\mathcal{S}_{\omega}(\mathbb{R}^N)$ is continuously included in $\mathcal{O}_{M,\omega}(\mathbb{R}^N)$, it follows that $M\colon \mathcal{S}_{\omega}(\mathbb{R}^N) \times \mathcal{S}'_{\omega}(\mathbb{R}^N) \to \mathcal{S}'_{\omega}(\mathbb{R}^N)$ is also continuous. Arguing as in the proof of Proposition 4.5, we get that the canonical bilinear form of $\mathcal{S}_{\omega}(\mathbb{R}^N)$ is continuous. This is a contradiction. □

Using the properties of the Fourier transform, we obtain the following result as a consequence.

Corollary 4.3 *Let ω be a non-quasianalytic weight function satisfying the condition* $\log(1+t) = o(\omega(t))$ *as* $t \to \infty$. *Then the convolution operator* $C\colon \mathcal{O}'_{C,\omega}(\mathbb{R}^N) \times \mathcal{S}'_{\omega}(\mathbb{R}^N) \to \mathcal{S}'_{\omega}(\mathbb{R}^N)$, $(T, S) \mapsto T \star S$, *is discontinuous.*

Now, we prove that $(\mathcal{O}'_{M,\omega}(\mathbb{R}^N), \star)$ is not a topological algebra.

Proposition 4.8 *Let ω be a non-quasianalytic weight function satisfying the condition* $\log(1+t) = o(\omega(t))$ *as* $t \to \infty$. *Then the convolution operator* $C\colon \mathcal{O}'_{M,\omega}(\mathbb{R}^N) \times \mathcal{O}'_{M,\omega}(\mathbb{R}^N) \to \mathcal{O}'_{M,\omega}(\mathbb{R}^N)$, $(T, S) \mapsto T \star S$, *is discontinuous.*

Proof We first observe that by Remark 3.4 the convolution C is well-defined. Now, assume by contradiction that $C\colon \mathcal{O}'_{M,\omega}(\mathbb{R}^N) \times \mathcal{O}'_{M,\omega}(\mathbb{R}^N) \to \mathcal{O}'_{M,\omega}(\mathbb{R}^N)$ is continuous. Since $\mathcal{O}_{M,\omega}(\mathbb{R}^N)$ is continuously included in the space $\mathcal{O}_M(\mathbb{R}^N)$ of multipliers of $\mathcal{S}(\mathbb{R}^N)$ with dense range, $\mathcal{O}'_M(\mathbb{R}^N)$ is continuously included in $\mathcal{O}'_{M,\omega}(\mathbb{R}^N)$. Therefore, we get that the map $C\colon \mathcal{O}'_M(\mathbb{R}^N) \times \mathcal{O}'_M(\mathbb{R}^N) \to \mathcal{O}'_{M,\omega}(\mathbb{R}^N)$ is continuous with range a subset of $\mathcal{O}'_M(\mathbb{R}^N)$. This yields that the map $\mathcal{O}'_M(\mathbb{R}^N) \times \mathcal{O}'_M(\mathbb{R}^N) \to \mathcal{O}'_M(\mathbb{R}^N)$ has closed graph, as it is easy to verify. By applying the closed graph theorem for (LF)-spaces (see, f.i., [14, Chap. 5, 5.4.1]) we obtain that $\mathcal{O}'_M(\mathbb{R}^N) \times \mathcal{O}'_M(\mathbb{R}^N) \to \mathcal{O}'_M(\mathbb{R}^N)$ is continuous, after having observed that $\mathcal{O}'_M(\mathbb{R}^N)$ in an (LF)-space. This is a contradiction with [16, Proposition 4]. □

Thanks to Proposition 4.8 and the properties of the Fourier transform we get that $(\mathcal{O}_{C,\omega}(\mathbb{R}^N), \cdot)$ is not a topological algebra.

Corollary 4.4 *Let ω be a non-quasianalytic weight function satisfying the condition $\log(1+t) = o(\omega(t))$ as $t \to \infty$. Then the multiplication operator $M\colon \mathcal{O}_{C,\omega}(\mathbb{R}^N) \times \mathcal{O}_{C,\omega}(\mathbb{R}^N) \to \mathcal{O}_{C,\omega}(\mathbb{R}^N)$, $(f,g) \mapsto fg$, is discontinuous.*

Acknowledgments The authors would like to thank Andrew Debrouwere for finding a gap in the earlier version of Theorem 2.1.

References

1. Albanese, A.A., Mele, C.: Convolutors on $\mathcal{S}_\omega(\mathbb{R}^N)$. RACSAM **115**, Article 157 (2021)
2. Albanese, A.A., Mele, C.: Multipliers on $\mathcal{S}_\omega(\mathbb{R}^N)$. J. Pseudo-Differ. Oper. Appl. **12**, Article 35 (2021)
3. Björck, G.: Linear partial differential operators and generalized distributions. Ark. Mat. **6**, 351–407 (1965)
4. Boiti, C., Jornet, D., Oliaro, A.: Regularity of partial differential operators in ultradifferentiable spaces and Wigner type transforms. J. Math. Anal. Appl. **446**, 920–944 (2017)
5. Boiti, C., Jornet, D., Oliaro, A., Schindl, G.: Nuclearity of rapidly decreasing ultradifferentiable functions and time-frequency analysis. Collect. Math. **72**, 423–442 (2021)
6. Boiti, C, Jornet, D., Oliaro, A.: Real Paley-Wiener theorems in spaces of ultradifferentiable functions. J. Funct. Anal. **278**, 1–45 (2020)
7. Braun, R.W., Meise, R., Taylor, B.A.: Ultradifferentiable functions and Fourier analysis. Result. Math. **17**, 206–237 (1990)
8. Debrouwere, A., Neyt, L.: Weighted (PLB)-spaces of ultradifferentiable functions and multiplier spaces. Monatsh Math **198**, 31–60 (2022)
9. Debrouwere, A., Vindas, J.: On weighted inductive limits of spaces of ultradifferentiable functions and their duals. Math. Nachr. **292**, 573–602 (2019)
10. Debrouwere, A., Vindas, J.: Topological properties of convolutor spaces via the short-time Fourier transform. Trans. Am. Math. Soc. **374**, 829–861 (2021)
11. Debrouwere, A., Neyt, L., Vindas, J.: Characterization of nuclearity for Beurling-Björk spaces. Proc. Am. Math. Soc. **148**, 5171–5180 (2020)
12. Dimovski, P., Pilipovíc, S., Prangoski, B., Vindas, J.: Convolution of ultradistributions and ultradistribution spaces associated to translation-invariant Banach spaces. Kyoto J. Math. **56**, 401–440 (2016)

13. Horvath, J.: Topological Vector Spaces and Distributions, vol. 1. Addison-Wesley Publishing Company, Boston (1966)
14. Jarchow, H.: Locally Convex Spaces. B.G. Teubner, Stuttgart (1981)
15. Komatsu, H.: Ultradistributions 1. Structure theorems and a characterization. J. Fac. Sci. Tokyo Sec. IA **20**, 25–105 (1973)
16. Larcher, J.: Multiplications and convolutions in L. Schwartz spaces of test functions and distributions and their continuity. Analysis (Int. J. Anal. Appl.) **33**, 319–332 (2013)
17. Schwartz, L.: Théorie des Distributions. Hermann, Paris (1966)
18. Treves, F.: Topological Vector Spaces, Distributions and Kernels. Academic Press, New York (1967)
19. Vogt, D.: Regularity properties of (LF)-spaces. In: Progress in Functional Analysis. North-Holland Mathematics Studies, vol. 170, pp. 57–84. Royal Irish Academy, Dublin (1992)
20. Wengenroth, J.: Acyclic inductive spectra of Fréchet spaces. Stud. Math. **120**, 247–258 (1996)

Higher Order Elliptic Equations in Generalized Morrey Spaces

Emilia Anna Alfano, Dian K. Palagachev, and Lubomira Softova

To Francesco Altomare with all the best wishes on the occasion of his 70th anniversary

Abstract We study the generalized Morrey regularity of the strong solutions to higher-order uniformly elliptic equations with VMO principal coefficients.

Keywords Higher-order elliptic equations · Strong solutions · VMO coefficients · A priori estimate · Generalized Morrey spaces

1 Introduction

We deal with regularity theory in generalized Morrey spaces of higher-order linear elliptic operators with discontinuous coefficients. Precisely, we obtain local regularity results and derive interior a priori estimates for the strong solutions of the uniformly elliptic equation

$$\mathcal{L}(x, D)u := \sum_{|\alpha|=2b} a_\alpha(x) D^\alpha u(x) + \sum_{|\beta|<2b} b_\beta(x) D^\beta u(x) = f(x),$$

E. A. Alfano · L. Softova
Department of Mathematics, University of Salerno, Salerno, Italy
e-mail: ealfano@unisa.it; lsoftova@unisa.it

D. K. Palagachev (✉)
Department of Mechanics, Mathematics and Management, Polytechnic University of Bari, Bari, Italy
e-mail: dian.palagachev@poliba.it

A. M. Candela et al. (eds.), *Recent Advances in Mathematical Analysis*,
Trends in Mathematics, https://doi.org/10.1007/978-3-031-20021-2_5

where the principal coefficients a_α are assumed to be functions with *vanishing mean oscillation (VMO),* while b_β are measurable and essentially bounded.

In the previous papers [11, 14, 15] a *Calderón–Zygmund* type theory in the framework of the classical Morrey spaces $L^{p,\lambda}$ has been developed for linear and quasi-linear elliptic and parabolic systems with VMO principal coefficients. On the other hand, in the recent years an exhaustive Calderón–Zygmund theory has been elaborated both for elliptic and parabolic equations/systems in *divergence form* with VMO-coefficients in the framework of the *generalized Morrey spaces* $L^{p,\omega}$ (cf. [4, 5] and the survey [6]). This last generalization of the classical Morrey spaces allows finer control on the local oscillation properties of a function near its singular points and that is why regularity results in $L^{p,\omega}$ of solutions to PDEs with discontinuous coefficients are of great importance in the applications to differential geometry, stochastic control, nonlinear optimization, adaptive discontinuous Galerkin FEMs, etc. As it concerns regularity results in other function spaces, we can mention also the recent results [1, 2] that consider linear higher order elliptic equations in Grand Lebesgue spaces.

In the present work we are going to extend the results obtained in [15–17, 20] to higher-order uniformly elliptic equations with discontinuous coefficients in the framework of generalized Morrey spaces.

In what follows we use the standard notations:

- $x = (x_1, \dots, x_n) \in \mathbb{R}^n$, $r > 0$ and $\mathcal{B}_r(x) = \{y \in \mathbb{R}^n : |x - y| < r\}$;
- $\Omega \subset \mathbb{R}^n, n \geq 2$, is a bounded domain, $|\Omega|$ is the Lebesgue measure of Ω, $\Omega_r(x) = \Omega \cap \mathcal{B}_r(x)$;
- $\mathbb{S}^{n-1} = \{\xi \in \mathbb{R}^n : |\xi| = 1\}$ is the unit sphere in $\mathbb{R}^n$;
- For any function f and any domain D with $f : D \to \mathbb{R}$ we write

$$f_D = \fint_D f(y)dy = \frac{1}{|D|} \int_D f(y)\, dy,$$

$$\|f\|^p_{p,D} = \|f\|^p_{L^p(D)} = \int_D |f(y)|^p\, dy.$$

Throughout this paper, the standard summation convention on repeated upper and lower indexes is adopted. The letter C is used for various constants and may change from one occurrence to another.

2 Definitions and Preliminary Results

We are interested in operators with *discontinuous coefficients* a_α belonging to the Sarason function class *VMO*.

Definition 2.1 For $a \in L^1_{\text{loc}}(\mathbb{R}^n)$ and any $R > 0$ set

$$\gamma_a(R) := \sup_{\mathcal{B}_r, r \le R} \frac{1}{|\mathcal{B}_r|} \int_{\mathcal{B}_r} |a(y) - a_{\mathcal{B}_r}|\, dy,$$

where $\mathcal{B}_r$ is any ball in $\mathbb{R}^n$. We say that

- $a \in BMO$ if

$$\|a\|_* = \sup_{R>0} \gamma_a(R) < \infty;$$

- $a \in VMO$ with VMO-modulus γ_a if $a \in BMO$ and

$$\lim_{R \to 0} \gamma_a(R) = 0.$$

We call *weight* a measurable function $\omega\colon\ \mathbb{R}^n \times \mathbb{R}_+ \to \mathbb{R}_+$ and for any ball $\mathcal{B}_r(x)$ we write $\omega(x, r)$ instead of $\omega(\mathcal{B}_r(x))$. In addition we assume that there exist positive constants κ_1, κ_2 and κ_3 such that

$$\begin{cases} \kappa_1 < \dfrac{\omega(x_0, s)}{\omega(x_0, r)} < \kappa_2 \quad \forall\, 0 < r \le s \le 2r, \quad \forall\, x_0 \in \mathbb{R}^n; \\ \displaystyle\int_r^\infty \frac{\omega(x_0, s)}{s^{n+1}}\, ds \le \kappa_3 \frac{\omega(x_0, r)}{r^n}. \end{cases} \tag{1}$$

Definition 2.2 (Nakai [13]) A function $f \in L^p(\Omega)$ with $1 \le p < \infty$ belongs to the generalized Morrey space $L^{p,\omega}(\Omega)$ if the following norm is finite:

$$\|f\|_{p,\omega;\Omega} = \left(\sup_{\mathcal{B}_r(x)} \frac{1}{\omega(x, r)} \int_{\Omega_r(x)} |f(y)|^p\, dy \right)^{1/p},$$

where the supremum is taken over all balls centered at any $x \in \Omega$ and of radius $r \in (0, \operatorname{diam} \Omega]$.

The generalized Sobolev–Morrey space $W^{2b,p,\omega}(\Omega)$ consists of all functions $u \in L^p(\Omega)$ with generalized derivatives $D^\alpha u$, $|\alpha| \le 2b$, belonging to $L^{p,\omega}(\Omega)$ and endowed with the norm

$$\|u\|_{W^{2b,p,\omega}(\Omega)} = \sum_{s=0}^{2b} \sum_{|\alpha|=s} \|D^\alpha u\|_{p,\omega;\Omega}.$$

Remark 2.1 It is clear that if $\omega(x, r) = r^\lambda$ with $\lambda \in (0, n)$, then $L^{p,\omega}$ gives rise to the classical Morrey space $L^{p,\lambda}$, while $L^{p,1} \equiv L^p$ and $W^{2b,p,1}$ reduces to the classical Sobolev space $W^{2b,p}$ when $\omega \equiv 1$.

In the sequel, we will use also a localized version $W^{2b,p,\omega}_{\mathrm{loc}}(\Omega)$ of $W^{2b,p,\omega}(\Omega)$, consisting of all functions that belong to $W^{2b,p,\omega}(\Omega')$ for each $\Omega' \Subset \Omega$.

Definition 2.3 Let $\mathcal{K}(x;\xi)\colon \mathbb{R}^n \times (\mathbb{R}^n \setminus \{0\}) \to \mathbb{R}$ be a *variable Calderón–Zygmund kernel,* i.e.

1. for each fixed $x \in \mathbb{R}^n$, $\mathcal{K}(x;\cdot)$ is a Calderón–Zygmund kernel:

$$\begin{array}{ll} \text{(a)} & \mathcal{K}(x;\cdot) \in C^\infty(\mathbb{R}^n \setminus \{0\}) \\ \text{(b)} & \mathcal{K}(x;\mu\xi) = \mu^{-n}\mathcal{K}(x,\xi) \quad \forall \mu > 0 \\ \text{(c)} & \displaystyle\int_{\mathbb{S}^{n-1}} \mathcal{K}(x;\xi)\, d\sigma_\xi = 0 \qquad \int_{\mathbb{S}^{n-1}} |\mathcal{K}(x;\xi)|\, d\sigma_\xi < \infty; \end{array}$$

2. for every multi-index β : $\sup\limits_{\xi \in \mathbb{S}^{n-1}} |D^\beta_\xi \mathcal{K}(x;\xi)| \le C(\beta)$ independently of x, where $\mathbb{S}^{n-1}$ is the unit sphere in $\mathbb{R}^n$.

Given a function $f \in L^1(\Omega)$, define the singular integral operator

$$\mathfrak{K}f(x) := P.V. \int_{\mathbb{R}^n} \mathcal{K}(x; x-y) f(y)\, dy$$

and its commutator with multiplication by a function $a \in L^\infty(\mathbb{R}^n)$ as

$$\begin{aligned} \mathfrak{C}[a,f](x) &:= P.V. \int_{\mathbb{R}^n} \mathcal{K}(x;x-y)[a(y)-a(x)]f(y)\, dy \\ &= \mathfrak{K}(af)(x) - a(x)\mathfrak{K}f(x). \end{aligned}$$

The L^p and $L^{p,\omega}$-boundedness of the operators $\mathfrak{K}$ and $\mathfrak{C}$ have been obtained in [3, 10] and [18, 19], respectively. For the sake of completeness, we summarize these results here.

Proposition 2.1 *Let ω be a weight satisfying* (1) *and $f \in L^{p,\omega}(\Omega)$ with $p \in (1,\infty)$. Then there exists a positive constant $C = C(p,\omega,\mathcal{K})$ such that*

$$\|\mathfrak{K}f\|_{p,\omega;\Omega} \le C\|f\|_{p,\omega;\Omega}, \qquad \|\mathfrak{C}[a,f]\|_{p,\omega;\Omega} \le C\|a\|_*\|f\|_{p,\omega;\Omega}.$$

In addition, if $a \in VMO$, then for each $\varepsilon > 0$ there exists $r_0 = r_0(\varepsilon,\gamma_a) > 0$ such that for any $r \in (0,r_0)$ and any ball $\mathcal{B}_r$ the following inequality holds:

$$\|\mathfrak{C}[a,f]\|_{p,\omega;\mathcal{B}_r} \le C\varepsilon\|f\|_{p,\omega;\mathcal{B}_r}.$$

3 Statement of the Problem and the Main Result

Let Ω be a domain in $\mathbb{R}^n$ with $n \geq 2$. Fixed an integer $b \geq 1$ and given a measurable function $f: \Omega \to \mathbb{R}$, we deal with the $2b$-order linear equation

$$\begin{aligned}\mathcal{L}(x, D)u &:= \sum_{|\alpha|=2b} a_\alpha(x) D^\alpha u(x) + \sum_{|\beta|<2b} b_\beta D^\beta u(x) \\ &=: L(x, D)u + Q(x, D)u \\ &= f(x) \quad \text{almost everywhere in } \Omega.\end{aligned} \tag{2}$$

The operator $\mathcal{L}(x, D)$ is supposed to be *uniformly elliptic* with generally *discontinuous* coefficients. We assume precisely, that there exists a positive constant Λ such that

$$\Lambda^{-1}|\xi|^{2b} \leq \sum_{|\alpha|=2b} a_\alpha(x)\xi^\alpha \leq \Lambda|\xi|^{2b} \tag{3}$$

for almost all $x \in \Omega$ and all $\xi \in \mathbb{R}^n$. The principal coefficients a_α of $\mathcal{L}$ are supposed to be VMO-functions, while the lower-order coefficients b_α are only essentially bounded, $b_\beta \in L^\infty(\Omega)$. That is,

$$a_\alpha \in VMO \cap L^\infty(\Omega), \qquad b_\beta \in L^\infty(\Omega), \tag{4}$$

and we set $\gamma_a = \sum_{|\alpha|=2b} \gamma_{a_\alpha}$ for the cumulative VMO-modulus of a_α's.

We consider strong solutions of (2), that is, functions $u: \Omega \to \mathbb{R}$ belonging to the Sobolev space $u \in W^{2b,p}(\Omega)$ with $p \in (1, \infty)$ and satisfying (2) for almost all $x \in \Omega$.

Our main result asserts *Calderón–Zygmund* property of the operator $\mathcal{L}$ in the framework of the generalized Sobolev–Morrey spaces. Namely

Theorem 3.1 *Suppose* (3) *and* (4)*, let* $f \in L^{p,\omega}_{\text{loc}}(\Omega)$ *with* ω *satisfying* (1) *and* $p \in (1, \infty)$, *and assume* $u \in W^{2b,p}_{\text{loc}}(\Omega)$ *is a strong solution to the Eq.* (2)*, such that* $D^\beta u \in L^{p,\omega}_{\text{loc}}(\Omega)$ *for* $|\beta| \leq 2b - 1$.

Then $D^{2b}u \in L^{p,\omega}_{\text{loc}}(\Omega)$ *and*

$$\|u\|_{W^{2b,p,\omega}(\Omega')} \leq C\Big(\|f\|_{p,\omega;\Omega''} + \|u\|_{p,\omega;\Omega''}\Big) \tag{5}$$

for all $\Omega' \Subset \Omega'' \subset \Omega$, *where the constant* C *depends on* $n, p, b, \omega, \|a_\alpha\|_{\infty;\Omega}$, *the* VMO*-moduli* γ_a *of the coefficients* a_α, $\|b_\alpha\|_{\infty;\Omega}$, *and on* $\operatorname{dist}(\Omega', \partial\Omega'')$.

The proof of Theorem 3.1 relies on some real analysis results regarding boundedness of Calderón–Zygmund type singular integral operators and their commutators, obtained in [14, 15, 18, 19].

To begin with, we fix the principal coefficients a_α at a point $x_0 \in \Omega$ and consider the constant coefficients operator

$$L(x_0, D) := \sum_{|\alpha|=2b} a_\alpha(x_0) D^\alpha.$$

The method of the decomposition into plane waves leads to *John's fundamental solution* of $L(x_0, D)$ with pole at the origin, that is given as

$$\Gamma(x_0; \xi) = |\xi|^{2b-n} P(x_0, \overline{\xi}), \qquad \overline{\xi} = \frac{\xi}{|\xi|},$$

with $P(x_0, \overline{\xi})$ being a real analytic function with respect to the second variable $\overline{\xi}$.

Let us recall the main properties of the function $\Gamma(x_0; \xi)$:

1. $D_\xi^{2b}\Gamma(x_0; \xi) \in C^\infty\big(\mathbb{R}^n \setminus \{0\}\big)$;
2. $D_\xi^{2b}\Gamma(x_0; \mu\xi) = \mu^{-n} D_\xi^{2b}\Gamma(x_0; \xi) \qquad$ for all $\mu > 0$;
3. $\displaystyle\int_{\mathbb{S}^{n-1}} D_\xi^{2b}\Gamma(x_0; \xi)\, d\sigma_\xi = 0, \qquad \int_{\mathbb{S}^{n-1}} |D_\xi^{2b}\Gamma(x_0; \xi)|\, d\sigma_\xi < \infty.$

These mean that the $2b$-order ξ-derivatives of Γ are classical Calderón-Zygmund kernels.

Take now a ball $\mathcal{B}_r(x_0) \Subset \Omega$ and a cut-off function $v \in C_0^\infty(\mathcal{B}_r)$ and write

$$L(x_0, D)v(y) = \big(L(x_0, D) - L(y, D)\big)v(y) + L(y, D)v(y). \tag{6}$$

A standard approach (cf. [7–9]) leads to an explicit representation formula for $v(x)$ in terms of *Newtonian potentials,*

$$\begin{aligned} v(x) = &\int_{\mathcal{B}_r} \boldsymbol{\Gamma}(x_0; x-y) L(y, D) v(y)\, dy \\ &+ \int_{\mathcal{B}_r} \boldsymbol{\Gamma}(x_0; x-y)\big(L(x_0, D) - L(y, D)\big)v(y)\, dy. \end{aligned}$$

We take now the α-derivatives with $|\alpha| = 2b$ and then unfreeze the coefficients by putting $x_0 = x$ in order to get

$$\begin{aligned} D^\alpha v(x) = P.V. &\int_{\mathcal{B}_r} D^\alpha \boldsymbol{\Gamma}(x; x-y) L(y, D) v(y)\, dy \\ &+ \sum_{|\alpha'|=2b} P.V. \int_{\mathcal{B}_r} D^\alpha \boldsymbol{\Gamma}(x; x-y)\Big(a_{\alpha'}(x) - a_{\alpha'}(y)\Big) D^{\alpha'} v(y)\, dy \end{aligned} \tag{7}$$

$$+ \sum_{|\beta^s|=2b-1} \int_{\mathbb{S}^{n-1}} D^{\beta^s}\mathbf{\Gamma}(x;y)\nu_s\, d\sigma_y\, L(x,D)v(x)$$

$$=: \mathfrak{K}_\alpha(Lv)(x) + \sum_{|\alpha'|=2b} \mathfrak{C}_\alpha[a_{\alpha'}, D^{\alpha'}v](x) + Q_\alpha(x)L(x,D)v(x).$$

Here the derivatives $D^\alpha\mathbf{\Gamma}(\cdot;\cdot)$ and $D^{\beta^s}\mathbf{\Gamma}(\cdot;\cdot)$ are taken with respect to the second variable, the multi-indices β^s depend on α and are such that

$$\beta^s = (\alpha_1,\dots,\alpha_{s-1},\alpha_s - 1,\alpha_{s+1},\dots,\alpha_n), \quad |\beta^s| = 2b-1,$$

and $\boldsymbol{\nu} = (\nu_1,\dots,\nu_n)$ is the outer normal to $\mathbb{S}^{n-1}$. In view of the properties of $\mathbf{\Gamma}$ mentioned above, it is clear that $\mathfrak{K}_\alpha$ are Calderón–Zygmund type singular integral operators, $\mathfrak{C}_\alpha$ are commutators of $\mathfrak{K}_\alpha$ with multiplication by VMO functions $a_{\alpha'}$, while Q_α are bounded integrals (cf. [7, 8, 15]).

Fix now an arbitrary $x_0 \in \operatorname{supp} u$ and take a small enough $r > 0$ to ensure that the ball $\mathcal{B}_r \equiv \mathcal{B}_r(x_0)$ is compactly imbedded in Ω. The representation formula (7) remains valid for functions $v \in W^{2b,p,\omega}(\mathcal{B}_r)$ belonging to the closure $W_0^{2b,p}(\mathcal{B}_r)$ of $C_0^\infty(\mathcal{B}_r)$ with respect to the norm in $W^{2b,p}(\mathcal{B}_r)$. Taking the $L^{p,\omega}(\mathcal{B}_r)$-norm of the both sides of (7) and making use of the regularity assumptions (4) on the coefficients of $\mathcal{L}$, Proposition 2.1 ensures that for each $\varepsilon > 0$ there exists $r_0 = r_0(\varepsilon,\gamma_a)$ such that

$$\|D^{2b}v\|_{p,\omega;\mathcal{B}_r} \le C\big(\|L(\cdot,D)v\|_{p,\omega;\mathcal{B}_r} + \varepsilon\|D^{2b}v\|_{p,\omega;\mathcal{B}_r}\big)$$

whenever $r < r_0$. Choosing ε small enough, we obtain

$$\|D^{2b}v\|_{p,\omega;\mathcal{B}_r} \le C\|L(\cdot,D)v\|_{p,\omega;\mathcal{B}_r} \tag{8}$$

for all $v \in W^{2b,p,\omega}(\mathcal{B}_r) \cap W_0^{2b,p}(\mathcal{B}_r)$ with a constant C independent of v.

Let $\theta \in (0,1)$, $\theta' = (\theta+1)/2 > 0$ and define the cut-off function $\varphi(x) \in C_0^\infty(\mathcal{B}_r)$, $0 \le \varphi \le 1$, such that

$$\varphi(x) = \begin{cases} 1 & x \in \mathcal{B}_{\theta r}(x_0) \\ 0 & x \notin \mathcal{B}_{\theta' r}(x_0). \end{cases}$$

Since $\theta' - \theta = (1-\theta)/2$, direct calculations give

$$|D^s\varphi| \le C(s)(1-\theta)^{-s}r^{-s}, \qquad \forall\, s = 1,2,\dots,2b.$$

We will prove first the a priori estimate (5), assuming that the strong solution u of (2) belongs to $W_{\text{loc}}^{2b,p,\omega}(\Omega)$. For, setting $v = \varphi u$, it is clear that $v \in W_0^{2b,p}(\mathcal{B}_r) \cap$

$W^{2b,p,\omega}(\mathcal{B}_r)$ and therefore (8) yields

$$\begin{aligned}
\|D^{2b}u\|_{p,\omega;\mathcal{B}_{\theta r}} &\leq \|D^{2b}v\|_{p,\omega;\mathcal{B}_{\theta' r}} \leq C\|L(\cdot,D)v\|_{p,\omega;\mathcal{B}_{\theta' r}} \\
&= C\|\mathcal{L}(\cdot,D)v - \mathcal{Q}(\cdot,D)v\|_{p,\omega;\mathcal{B}_{\theta' r}} \\
&\leq C\left(\|\varphi\mathcal{L}u(\cdot,D)u\|_{p,\omega;\mathcal{B}_{\theta' r}} + \sum_{s=0}^{2b}\|D^s\varphi\|_{\infty,\mathcal{B}_r}\|D^{2b-s}u\|_{p,\omega;\mathcal{B}_{\theta' r}}\right) \\
&\leq C\left(\|\mathcal{L}u(\cdot,D)u\|_{p,\omega;\mathcal{B}_r} + \sum_{s=1}^{2b-1}\frac{\|D^{2b-s}u\|_{p,\omega;\mathcal{B}_{\theta' r}}}{(1-\theta)^s r^s} + \frac{\|u\|_{p,\omega;\mathcal{B}_{\theta' r}}}{(1-\theta)^{2b}r^{2b}}\right),
\end{aligned}$$

where the last constant depends also on $\|a_\alpha\|_{\infty,\Omega}$ and $\|b_\beta\|_{\infty,\Omega}$. We multiply the both sides of the last inequality by $(1-\theta)^{2b}r^{2b}$ and use $(1-\theta) = 2(1-\theta') < 1$ in order to conclude

$$\begin{aligned}
&(1-\theta)^{2b}r^{2b}\|D^{2b}u\|_{p,\omega;\mathcal{B}_{\theta r}} \\
&\quad\leq C\Bigg(r^{2b}\|f\|_{p,\omega;\mathcal{B}_r} + \sum_{s=1}^{2b-1}(1-\theta)^{2b-s}r^{2b-s}\|D^{2b-s}u\|_{p,\omega;\mathcal{B}_{\theta' r}} \\
&\qquad + \|u\|_{p,\omega;\mathcal{B}_{\theta' r}}\Bigg),
\end{aligned}$$

that becomes

$$\Theta_{2b} \leq C\left(r^{2b}\|f\|_{p,\omega;\mathcal{B}_r} + \sum_{s=1}^{2b-1}\Theta_s + \Theta_0\right) \tag{9}$$

after taking the supremum in $\theta \in (0,1)$ and where

$$\Theta_s := \sup_{0<\theta<1}(1-\theta)^s r^s\|D^s u\|_{p,\omega;\mathcal{B}_{\theta r}}.$$

In order to estimate the seminorms Θ_s we need the following *interpolation inequality* which follows as in [14, 21].

Lemma 3.1 *For any* $s = 1,\ldots,2b-1$, *there is a constant* C, *depending on* n, b, p, s, *such that*

$$\Theta_s \leq \varepsilon\Theta_{2b} + \frac{C}{\varepsilon^{s/(2b-s)}}\Theta_0 \qquad \text{for each } \varepsilon > 0. \tag{10}$$

Proof It follows from the definition of the seminorms Θ_s that, fixed an s there exists $\theta_0 \in (0,1)$ with the property

$$\Theta_s \le 2(1-\theta_0)^s r^s \|D^s u\|_{p,\omega;\mathcal{B}_{\theta_0 r}}.$$

Employing the standard L^p-interpolation inequality from [12, Theorem 7.28], simple scaling arguments lead to

$$\|D^s u\|_{p,\omega;\mathcal{B}_{\theta_0 r}} \le \delta \|D^{2b} u\|_{p,\omega;\mathcal{B}_{\theta_0 r}} + \frac{C_1}{\delta^{s/(2b-s)}} \|u\|_{p,\omega;\mathcal{B}_{\theta_0 r}} \qquad \forall \delta > 0,$$

and the desired estimate (10) follows by choosing $\delta = \frac{\varepsilon}{2}(1-\theta_0)^{2b-s} r^{2b-s}$. □

Turning back to (9) and interpolating the intermediate seminorms by the aid of (10) with suitable choice of ε, we get

$$\Theta_{2b} \le C\Big(r^{2b}\|f\|_{p,\omega;\mathcal{B}_r} + \Theta_0\Big).$$

Therefore,

$$\|D^{2b}u\|_{p,\omega;\mathcal{B}_{\theta r}} \le \frac{C}{(1-\theta)^{2b} r^{2b}} \Big(r^{2b}\|f\|_{p,\omega;\mathcal{B}_r} + \|u\|_{p,\omega;\mathcal{B}_r}\Big) \qquad \forall \theta \in (0,1).$$

The desired estimate (5) follows now by fixing $\theta = 1/2$ above, covering Ω' with a finite number of balls $\mathcal{B}_{r/2}$ with $r < \operatorname{dist}(\Omega', \partial\Omega'')$ and using a partition of unity subordinated to this covering.

We are in a position now to show also the *Calderón–Zygmund* regularizing property of the operator $\mathcal{L}$ in generalized Morrey spaces. Namely, we will prove that

$$u \in W^{2b,p}_{\mathrm{loc}}(\Omega), \quad \mathcal{L}u \in L^{p,\omega}_{\mathrm{loc}}(\Omega), \quad D^\beta u \in L^{p,\omega}_{\mathrm{loc}}(\Omega) \text{ for } |\beta| \le 2b-1 \tag{11}$$

yield

$$D^{2b}u \in L^{p,\omega}_{\mathrm{loc}}(\Omega). \tag{12}$$

For, arguing as in [7, 14], we take a ball $\mathcal{B}_r \Subset \Omega$ with $r > r_0$, where r_0 is the number used in getting (8), and for each couple of multi-indices α and α', $|\alpha| = |\alpha'| = 2b$, define the operators

$$C_{\alpha\alpha'}g(x) := P.V. \int_{\mathcal{B}_r} D^\alpha \boldsymbol{\Gamma}(x; x-y)\Big(a_{\alpha'}(x) - a_{\alpha'}(y)\Big) g(y)\, dy$$

where, as in (7), the derivatives of $\mathbf{\Gamma}$ are taken with respect to the second variable. Proposition 2.1 ensures that the operators $C_{\alpha\alpha'}$ act from $L^{p,\omega}(\mathcal{B}_r)$ into itself and

$$\sum_{\alpha,\alpha'} \|C_{\alpha\alpha'}\| < 1 \tag{13}$$

if r is small enough.

Take now a cut-off function $\varphi \in C_0^\infty(\mathcal{B}_r)$ with $\varphi \equiv 1$ in $\mathcal{B}_{r/2}$ and consider $v := \varphi u$. Remembering (2), we have $L(\cdot, D)v = \varphi L(\cdot, D)u + L'(\cdot, D)u$ with linear differential operator $L'(\cdot, D)$ of order $2b - 1$, whence

$$L(\cdot, D)v = \varphi\big(\mathcal{L}(\cdot, D)u - Q(\cdot, D)u\big) + L'(\cdot, D)u \in L^{p,\omega}(\mathcal{B}_r)$$

as consequence of (11). Therefore, for each multi-index α, $|\alpha| = 2b$, and with Q_α as in (7), the functions

$$\mathcal{D}_\alpha(x) := P.V. \int_{\mathcal{B}_r} D^\alpha \mathbf{\Gamma}(x; x-y) L(y, D)v(y)\, dy + Q_\alpha(x) L(x, D)v(x)$$

are well-defined and belong to $L^{p,\omega}(\mathcal{B}_r)$.

Set $\mathcal{A}$ for the collection of all multi-indices α with $|\alpha| = 2b$, and let m be its cardinality. Define the linear operator

$$\mathcal{T}\colon \big(L^{p,\omega}(\mathcal{B}_r)\big)^m \to \big(L^{p,\omega}(\mathcal{B}_r)\big)^m$$

by the setting

$$\mathcal{T}(\mathbf{w}) = \left(\mathcal{D}_\alpha + \sum_{\alpha' \in \mathcal{A}} C_{\alpha\alpha'}(w_{\alpha'})\right)_{\alpha\in\mathcal{A}}, \qquad \mathbf{w} = \big(w_{\alpha'}\big)_{\alpha'\in\mathcal{A}}.$$

If r is small enough, (13) ensures that $\mathcal{T}$ is a contraction mapping and therefore admits a unique fixed point $\overline{\mathbf{w}} \in \big(L^{p,\omega}(\mathcal{B}_r)\big)^m$. Since also $\mathbf{v} = \big(D^\alpha v\big)_{\alpha\in\mathcal{A}}$ is a *fixed* point of $\mathcal{T}$ as shows (7), we conclude $D^\alpha v \in L^{p,\omega}(\mathcal{B}_r)$ for each α with $|\alpha| = 2b$ and thus the claim (12).

This completes the proof of Theorem 3.1.

Remark 3.1 For the sake of simplicity we assumed essential boundedness of the lower-order coefficients b_β of the operator $\mathcal{L}(x, D)$ (cf. (4)). It is not hard to weaken this hypothesis, taking b_β's in suitable Lebesgue spaces with exponents depending on n, p and $|\beta|$ and such that $b_\beta D^\beta u \in L^{p,\omega}_{\text{loc}}(\Omega)$ as $u \in W^{2b,p,\omega}_{\text{loc}}(\Omega)$.

Acknowledgments All the authors are members of *INdAM—GNAMPA.*
The research of E.A. Alfano is partially supported by the project *ALPHA-MENTE,* Lotto1/Ambito AV01.

The work of D.K. Palagachev was supported by the Italian Ministry of Education, University and Research under the Program “Department of Excellence” L. 232/2016 (Grant No. CUP - D94I18000260001).

References

1. Bilalov, B.T., Sadigova, S.R.: On solvability in the small of higher order elliptic equations in grand-Sobolev spaces. Complex Var. Elliptic Equ. **66**(12), 2117–2130 (2021). https://doi.org/10.1080/17476933.2020.1807965
2. Bilalov, B.T., Sadigova, S.R.: Interior Schauder-type estimates for higher-order elliptic operators in Grand-Sobolev spaces. Sahand Commun. Math. Anal. **18**(2), 129–148 (2021)
3. Bramanti, M., Cerutti, M.C.: Commutators of singular integrals on homogeneous spaces. Boll. Un. Mat. Ital. B (VII) **10**(4), 843–883 (1996)
4. Byun, S.-S., Softova, L.: Gradient estimates in generalized Morrey spaces for parabolic operators. Math. Nachr. **288**(14–15), 1602–1614 (2015). https://doi.org/10.1002/mana.201400113
5. Byun, S.-S., Softova, L.: Asymptotically regular operators in generalized Morrey spaces. Bull. Lond. Math. Soc. **52**(1), 64–76 (2020). https://doi.org/10.1112/blms.12306
6. Byun, S.-S., Palagachev, D.K., Softova, L.: Survey on gradient estimates for nonlinear elliptic equations in various function spaces. Algebra i Anal. **31**(3), 10–35 (2019). Reprinted in St. Petersburg Math. J. **31**(3), 401–419 (2020). https://doi.org/10.1090/SPMJ/1605
7. Chiarenza, F., Frasca, M., Longo, P.: Interior $W^{2,p}$-estimates for non divergence elliptic equations with discontinuous coefficients. Ricerche Mat. **60**(1), 149–168 (1991)
8. Chiarenza, F., Franciosi, M., Frasca, M.: L^p-estimates for linear elliptic systems with discontinuous coefficients. Atti Accad. Naz. Lincei Cl. Sci. Fis. Mat. Natur. Rend. Lincei (9) Mat. Appl. **5**(1), 27–32 (1994)
9. Douglis, A., Nirenberg, L.: Interior estimates for elliptic systems of partial differential equations. Commun. Pure Appl. Math. **8**, 503–538 (1955). https://doi.org/10.1002/cpa.3160080406
10. Fabes, E.B., Rivière, N.: Singular integrals with mixed homogeneity. Stud. Math. **27**, 19–38 (1966)
11. Fattorusso, L., Softova, L.: Precise Morrey regularity of the weak solutions to a kind of quasilinear systems with discontinuous data. Electron. J. Qual. Theory Differ. Equ. **36**, 13 pp. (2020). https://doi.org/10.14232/ejqtde.2020.1.36
12. Gilbarg, D., Trudinger, N.S.: Elliptic Partial Differential Equations of Second Order, Reprint of the 1998 edition. Classics in Mathematics. Springer-Verlag, Berlin (2001). https://doi.org/10.1007/978-3-642-61798-0
13. Nakai, F.: Hardy-Littlewood maximal operator, singular integral operators and the Riesz potentials on generalized Morrey spaces. Math. Nachr. **166**, 95–103 (1994). https://doi.org/10.1002/mana.19941660108
14. Palagachev, D.K., Softova, L.: Singular integral operators, Morrey spaces and fine regularity of solutions to PDE's. Potential Anal. **20**(3), 237–263 (2004). https://doi.org/10.1023/B:POTA.0000010664.71807.f6
15. Palagachev, D.K., Softova, L.: Fine regularity for elliptic systems with discontinuous ingredients. Arch. Math. **86**(2), 145–153 (2006). https://doi.org/10.1007/s00013-005-1336-8
16. Palagachev, D.K., Softova, L.: Generalized Morrey regularity of $2b$-parabolic systems. Appl. Math. Lett. **112**, Paper No. 106838, 6 pp. (2021). https://doi.org/10.1016/j.aml.2020.106838
17. Palagachev, D.K., Softova, L.: Elliptic systems in generalized Morrey spaces. Azerb. J. Math. **11**(2), 153–162 (2021)
18. Softova, L.: Singular integrals and commutators in generalized Morrey spaces. Acta Math. Sin. (Engl. Ser.) **22**(3), 757–766 (2006). https://doi.org/10.1007/s10114-005-0628-z
19. Softova, L.: Singular integral operators in functional spaces of Morrey type. In: N.M. Chuong et al. (eds.) Advances in Deterministic and Stochastic Analysis, pp. 33–42. World Scientific Publishing, Hackensack (2007). https://doi.org/10.1142/9789812770493_0003

20. Softova, L.: The Dirichlet problem for elliptic equations with VMO coefficients in generalized Morrey spaces. In: Advances in Harmonic Analysis and Operator Theory, vol. 229, pp. 371–386. Operator Theory: Advances and Applications. Birkhäuser/Springer Basel AG, Basel (2013). https://doi.org/10.1007/978-3-0348-0516-2_21
21. Solonnikov, V.A.: On boundary value problems for linear parabolic systems of differential equations of general form. Trudy Mat. Inst. Steklov **83**, 3–163 (1965)

Norm and Essential Norm of Composition Operators Mapping into Weighted Banach Spaces of Harmonic Mappings

Munirah Aljuaid and Flavia Colonna

In honor of the 70th birthday and retirement of Prof. Franco Altomare, who has worked tirelessly to inspire many generations of mathematicians. Through his editorial initiatives he has helped strengthen and disseminate mathematical research in many countries around the world.

Abstract Let X be a Banach space of harmonic mappings on the unit disk $\mathbb{D}$ in the complex plane whose point-evaluation functionals are bounded. In this work, we study the boundedness and compactness of composition operators from X into the harmonic growth space $\mathcal{H}_\mu^\infty$, where μ is an arbitrary positive continuous function on $\mathbb{D}$. We give a formula of the operator norm, and under some restrictions on the space X, we obtain an approximation of the essential norm. We apply our results to the case when the composition operator acts on the members of a class of harmonic Hilbert spaces. We obtain an exact formula of the essential norm, which holds in particular for the harmonic Hardy space, the harmonic Bergman space, the harmonic Dirichlet space, and the harmonic Bloch space.

Keywords Harmonic mapping · Weighted Hardy space · Bloch space · Composition operator

M. Aljuaid
Department of Mathematics, Northern Border University, Arar, Saudi Arabia
e-mail: maljuaid@gmu.edu

F. Colonna (✉)
Department of Mathematical Sciences, George Mason University, Fairfax, VA, USA
e-mail: fcolonna@gmu.edu

A. M. Candela et al. (eds.), *Recent Advances in Mathematical Analysis*,
Trends in Mathematics, https://doi.org/10.1007/978-3-031-20021-2_6

1 Introduction

Given a region Ω in the complex plane $\mathbb{C}$, a **harmonic mapping** with domain Ω is a complex-valued function h defined on Ω satisfying the Laplace equation

$$\Delta h := 4h_{z\overline{z}} \equiv 0 \quad \text{on } \Omega,$$

having denoted by $h_{z\overline{z}}$ the mixed complex second partial derivative of h.

It is well known that a harmonic mapping h admits a representation of the form $f + \overline{g}$, where f and g are analytic functions. This representation is unique if, fixing a base point z_0 in Ω, the function g is chosen so that $g(z_0) = 0$.

Due to this representation, we see that the composition operator C_φ induced by an analytic or a conjugate analytic self-map φ of Ω, defined as

$$C_\varphi h = h \circ \varphi,$$

is a linear transformation over $\mathbb{C}$ on the class of harmonic mappings on Ω.

Let $\mathbb{D}$ denote the open unit disk in $\mathbb{C}$ and denote by $H(\mathbb{D})$ and $\mathcal{H}(\mathbb{D})$ the class of analytic functions and harmonic mappings on $\mathbb{D}$, respectively. In this work, we shall be considering harmonic mappings with domain $\mathbb{D}$ and choose as base point $z_0 = 0$. Thus, the canonical representation of a mapping $h \in \mathcal{H}(\mathbb{D})$ is $h = f + \overline{g}$, where $f, g \in H(\mathbb{D})$ and $g(0) = 0$.

For a general reference on the theory of harmonic functions, see [5].

The operator theory of Banach spaces of analytic functions on $\mathbb{D}$ has been thoroughly studied in the last few decades, and a massive amount of papers on this topic have appeared in the literature. We were surprised that, until very recently, in our research we could not find a similarly extensive coverage in the setting of the harmonic extensions of classical spaces of analytic functions. We are very grateful to the referee for having brought to our attention numerous important references and hope to help make better known the development of the operator theory of harmonic mappings, by highlighting the references below.

In [25], Shields and Williams studied Banach spaces of harmonic functions and in [24] they gave characterizations of duality for weighted spaces of harmonic functions on the unit disk. In [11], the second author introduced and studied Bloch harmonic mappings on $\mathbb{D}$ as Lipschitz maps from the hyperbolic disk into $\mathbb{C}$. In [20], Lusky investigated weighted spaces of harmonic functions on $\mathbb{D}$ and, in [21], isomorphism classes of weighted spaces of holomorphic and harmonic functions with a radial weight on $\mathbb{C}$ and on $\mathbb{D}$. In [26], Yoneda studied harmonic Bloch spaces and harmonic Besov spaces. In [19], Laitila and Tylli characterized the weak compactness of the composition operators on vector-valued harmonic Hardy spaces and on the spaces of vector-valued Cauchy transforms for reflexive Banach spaces. Characterizations of the isometries between weighted spaces of harmonic functions were provided by Boyd and Rueda in [6]. In [18], Jordá and Zarco studied Banach spaces of harmonic functions and composition operators between weighted Banach spaces of pluriharmonic functions. Isomorphisms on weighted Banach spaces of harmonic and holomorphic functions were treated in [17].

Other references in the unit disk setting include [2] for characterizations of Bloch-type spaces of harmonic mappings, [4] for the study harmonic Zygmund spaces, [8] for the study of harmonic ν-Bloch mappings and [9, 10] for the study of harmonic Lipschitz-type spaces and harmonic Hardy classes. In the setting of the unit ball in $\mathbb{C}^n$, see [7] for the study of the harmonic Bloch spaces, and [16] for extensions of the main results in [26].

In [3], we studied composition operators with analytic symbol on several Banach spaces of harmonic mappings. We showed that such operators on the harmonic extensions of the α-Bloch spaces, the growth spaces, the Zygmund space, the Besov spaces, are bounded, bounded below, closed range, compact, respectively, if and only if they are bounded, bounded below, closed range, compact, respectively, as operators on the analytic counterparts. We also showed that the isometries are the same, except possibly for the cases of the minimal Möbius invariant space and BMOA, for which the determination of the isometries on the harmonic extensions is still incomplete. In addition, we studied the eigenfunctions of the composition operators.

The results obtained in [3] lead to the question of whether the operator theoretic properties of composition operators between Banach spaces of harmonic mappings are the same as those between the analytic counterparts.

For general references on composition operators on Banach spaces of analytic functions, we refer the interested reader to [15, 23].

By a **functional Banach space of harmonic mappings** we mean a Banach space $X \subset \mathcal{H}(\mathbb{D})$ with norm $\|\cdot\|$ whose point-evaluation functionals $\delta_z : h \mapsto h(z)$ (for $z \in \mathbb{D}$) are bounded. Then

$$\|\delta_z\| = \sup\{|h(z)| : h \in X, \|h\| \leq 1\}. \tag{1}$$

Thus, for each $z \in \mathbb{D}$ and $h \in X$,

$$|h(z)| \leq \|h\| \, \|\delta_z\|. \tag{2}$$

A **reproducing kernel Hilbert space of harmonic mappings** with domain $\mathbb{D}$ is a Hilbert space $\mathcal{H}$ contained in $\mathcal{H}(\mathbb{D})$ with inner product $\langle \cdot, \cdot \rangle$ whose point-evaluation functionals are bounded. By the Riesz Representation Theorem, for each $z \in \mathbb{D}$, there exists $K_z \in \mathcal{H}$ such that

$$h(z) = \langle h, K_z \rangle \quad \text{for each } h \in \mathcal{H}.$$

Thus, the role of δ_z is taken over by K_z and its norm in (2) is replaced by $\|K_z\| = K_z(z)^{1/2}$.

Let μ be a positive continuous function on $\mathbb{D}$ and let $\mathcal{H}_\mu^\infty$ denote the collection of harmonic mappings h on $\mathbb{D}$ such that

$$\|h\|_\mu := \sup_{z\in\mathbb{D}} \mu(z)|h(z)| < \infty.$$

Equipped with the norm $\|\cdot\|_\mu$, $\mathcal{H}_\mu^\infty$ is a functional Banach space of harmonic mappings we call the *harmonic growth space with weight* μ. Clearly, $\mathcal{H}_\mu^\infty \cap H(\mathbb{D})$ is the weighted Banach space of analytic functions H_μ^∞.

Let $\mathcal{H}_{\mu,0}^\infty$ denote the subspace of $\mathcal{H}_\mu^\infty$ whose elements h satisfy the condition

$$\lim_{|z|\to 1} \mu(z)|h(z)| = 0.$$

Let φ be a non-constant analytic self-map of $\mathbb{D}$ and μ a weight. In this work, our aim is to provide characterizations of the bounded and the compact composition operators C_φ from a general functional Banach space X of harmonic mappings into the harmonic growth space $\mathcal{H}_\mu^\infty$ as well as determine the operator norm and an approximation of the essential norm of C_φ.

The principal motivation of this study is to identify conditions on the space X that allow us to obtain formulas for the norm and the essential norm of the operators $C_\varphi : X \to \mathcal{H}_\mu^\infty$ that are valid independently of the choice of the domain space X. This project is an extension to the harmonic setting of the work done by the second author and Tjani in [14] for the weighted composition operators (i.e. the composition product of the multiplication operator and the composition operator) acting on a reproducing kernel Hilbert space of analytic functions and in [13] for operators acting on a functional Banach space of analytic functions. In this work, we limit our study to the composition operators because a multiplication operator acting on a non-trivial space of harmonic mappings preserves harmonicity only when the multiplier is constant.

Below is the list of axiomatic conditions inspired by Colonna and Tjani [13], which include those that will be used in this work.

(I) There is a positive constant C such that

$$\|\delta_{rz}\| \le C\|\delta_z\|,$$

for all $z \in \mathbb{D}$ and $0 < r < 1$.

(II) The unit ball of X is relatively compact with respect to the topology of uniform convergence on compact subsets of $\mathbb{D}$.

(III) $\lim_{|z|\to 1} \|\delta_z\| = \infty$.

(IV) The map $z \mapsto \|\delta_z\|$ is bounded below by a positive constant on compact subsets of $\mathbb{D}$.

(V) For $0 < r < 1$, the linear operator $T_r : X \to X$ defined as $h_r(z) = h(rz)$ is compact.

(VI) There is $C > 0$ such that for all $\mathcal{S} \in Aut(\mathbb{D})$ and all $f \in X$, $\mathcal{S}f \in X$ and $\|\mathcal{S}f\| \leq C\|f\|$.

(VII) There is a constant $C > 0$ such that for all $h \in X$ and all $z \in \mathbb{D}$,

$$(1-|z|^2)\big(|h_z(z)|+|h_{\overline{z}}(z)|\big) \leq C\|h\|\|\delta_z\|.$$

(VIII) With T_r as in (V),

$$\sup_{0<r<1} \|T_r\| < \infty.$$

As noted above, the multiplication operator fails to preserve harmonic mappings unless the symbol is constant. Thus, condition (VI), used in [13] to construct test functions in order to characterize compactness of the weighted composition operator, is not applicable in the harmonic settings.

As in the analytic case (see Remarks 2.2 and 2.3 of [13]), the following properties hold.

Proposition 1.1 *Let X be a non-trivial Banach space of harmonic mappings.*

(a) If X contains the constant functions, then condition (IV) *holds.*
(b) If X satisfies condition (I) *and contains mappings h not vanishing at* 0*, then* $\|\delta_z\| > 0$, *for all* $z \in \mathbb{D}$.
(c) The mapping $z \mapsto \|\delta_z\|$ *is bounded on compact subsets of* $\mathbb{D}$.
(d) If X is reflexive, then condition (II) *holds.*

Proof

(a) If X contains the constant functions, then by (1), for each $z \in \mathbb{D}$, $\|\delta_z\| \geq \|1\|^{-1}$. Hence, condition (IV) holds.

(b) Suppose X satisfies condition (I) and contains mappings non-vanishing at 0. If X contains the constants, the result is an obvious consequence of part (a). So assume there exists $h \in X$ non-constant such that $h(0) \neq 0$. Scaling h, we may assume $\|h\| \leq 1$. Then $\|\delta_0\| \geq |h(0)| > 0$. Moreover, for any fixed $z \neq 0$, there is an $r \in (0, 1)$ such that $h(z) \neq h(rz)$. Then by condition (I), there is a constant $C > 0$ independent of z and r such that

$$0 < |h(z) - h(rz)| \leq \|h\|\ \|\delta_z\| + \|h\|\ \|\delta_{rz}\| \leq C\|h\|\ \|\delta_z\|.$$

Therefore, $\|\delta_z\| > 0$, for all $z \in \mathbb{D}$.

(c) Let E be a compact subset of $\mathbb{D}$. Since each $h \in X$ is continuous,

$$\sup_{z\in E} |h(z)| = \sup_{z\in E} |\delta_z(h)|$$

is finite. The closed unit ball of X is defined as $B_X = \{h \in X : \|h\| \leq 1\}$. By the Uniform Boundedness Principle, we have

$$\sup_{z\in E} \|\delta_z\| = \sup_{z\in E} \sup_{h\in B_X} |\delta_z(h)| < \infty.$$

(d) Assume X is reflexive. Let $\{h_n\}$ be a sequence in B_X. Then by (1) and part (c), $\{h_n\}$ is uniformly bounded on compact sets. Hence, so are the sequences of its real and imaginary parts. Theorem 2.6 in [5] states that any locally uniformly bounded sequence of real-valued harmonic functions defined on an open subset of $\mathbb{R}^n$ has a subsequence converging uniformly on compact subsets. Thus, letting $u_n = \operatorname{Re} h_n$ and $v_n = \operatorname{Im} h_n$, and applying this theorem to the sequence $\{u_n\}$, some subsequence $\{u_{n_k}\}_{k\in\mathbb{N}}$ converges uniformly on compact subsets of $\mathbb{D}$. Reapplying the theorem to the sequence $\{v_{n_k}\}_{k\in\mathbb{N}}$, we can find a subsequence $\{v_{n_{k_j}}\}_{j\in\mathbb{N}}$ converging uniformly on compact subsets of $\mathbb{D}$. Then the sequence $\{h_{n_{k_j}}\}$ does also. Therefore, B_X is relatively compact with respect to the topology of uniform convergence on compact subset of $\mathbb{D}$.

□

Organization of the Paper In Sect. 2, we characterize the bounded operators $C_\varphi : X \to \mathcal{H}_\mu^\infty$ and determine the operator norm for any functional Banach space X.

In Sect. 3, we give an approximation of the essential norm under some restrictions on the domain space X, which in particular yields a characterization of the compact operator C_φ.

In Sect. 4, we apply our results to the case when X is a harmonic weighted Hardy space. In fact, we obtain an exact formula for the essential norm.

In Sect. 5, we apply the results in Sects. 2 and 3 to the case when X is the harmonic Bloch space $\mathcal{B}_H$. We show that Theorem 3.1, which requires the reflexivity of the domain space X, also holds for $\mathcal{B}_H$, which is not reflexive. In addition, we obtain an exact formula of the essential norm.

We note that since the operator C_ψ, where ψ is the conjugate of an analytic self-map of $\mathbb{D}$, preserves harmonic mappings, a similar study with this choice of symbol would give results equivalent (under the action of the conjugation operator) to those obtained in this work.

Throughout this paper, we shall assume that X is a functional Banach space of harmonic mappings, φ is a non-constant analytic self-map of $\mathbb{D}$ and μ is a weight. We shall use the notation C for a positive constant independent of the variables involved, whose value may change at each occurrence. By the notation $A \asymp B$ we mean $C_1 A \leq B \leq C_2 A$ for some positive constants C_1 and C_2.

2 Boundedness, Compactness, and Operator Norm

The first result of the section is a characterization of the bounded composition operator $C_\varphi : X \to \mathcal{H}_\mu^\infty$ in terms of the weight μ and the norm of the point-evaluation functional with no restriction imposed on the space X.

Theorem 2.1 *The operator $C_\varphi : X \to \mathcal{H}_\mu^\infty$ is bounded if and only if*

$$\tau_\varphi := \sup_{z\in\mathbb{D}} \mu(z)\|\delta_{\varphi(z)}\| < \infty$$

in which case $\|C_\varphi\| = \tau_\varphi$.

Proof Assume C_φ is bounded. Fix $z \in \mathbb{D}$ and let $h \in X$ with $\|h\| \leq 1$. Then

$$\mu(z)|h(\varphi(z))| \leq \|C_\varphi h\|_\mu \leq \|C_\varphi\|\|h\| \leq \|C_\varphi\|.$$

Taking the supremum over all such mappings h, we have $\mu(z)\|\delta_{\varphi(z)}\| \leq \|C_\varphi\|$. Hence taking the supremum over all $z \in \mathbb{D}$, we obtain

$$\tau_\varphi \leq \|C_\varphi\|, \tag{3}$$

proving that τ_φ is finite.

Conversely, suppose $\tau_\varphi < \infty$. Let $h \in X$ with $\|h\| \leq 1$. Then, for each $z \in \mathbb{D}$, we have

$$\mu(z)|h(\varphi(z))| \leq \mu(z)\|\delta_{\varphi(z)}\| \leq \tau_\varphi.$$

Taking the supremum over $\mathbb{D}$, we obtain $\|C_\varphi h\|_\mu \leq \tau_\varphi$. Hence C_φ is bounded and

$$\|C_\varphi\| \leq \tau_\varphi,$$

which combined with (3) proves that $\|C_\varphi\| = \tau_\varphi$. □

We next provide a sufficient condition for compactness.

Theorem 2.2 *Let X be a functional Banach space of harmonic mappings satisfying conditions* (II) *and* (IV)*, and φ an analytic self map of $\mathbb{D}$ such that $C_\varphi : X \to \mathcal{H}_\mu^\infty$ is bounded. If*

$$\lim_{s\to 1} \sup_{|\varphi(z)|>s} \mu(z)\|\delta_{\varphi(z)}\| = 0, \tag{4}$$

then the operator $C_\varphi : X \to \mathcal{H}_\mu^\infty$ is compact.

Proof Suppose (4) holds. Let $\{h_n\}$ is a sequence in B_X converging to 0 uniformly on compact subsets of $\mathbb{D}$. We wish to show that $\|C_\varphi h_n\|_\mu \to 0$.

If $\|\varphi\|_\infty < 1$, then $\mathbb{C}_\varphi$ is clearly compact. So assume $\|\varphi\|_\infty = 1$. Fix $\varepsilon > 0$ and choose a number $s \in (0, 1)$ such that $\mu(z)\|\delta_{\varphi(z)}\| < \varepsilon$ whenever $|\varphi(z)| > s$. Since for all $w \in \mathbb{D}$, $|h_n(w)| \leq \|\delta_w\|$, if $|\varphi(z)| > s$, then

$$\mu(z)|h_n(\varphi(z))| < \varepsilon. \tag{5}$$

On the other hand, since $h_n \to 0$ uniformly on the disk $\{w : |w| \leq s\}$, there exists a natural number N such that $|h_n(w)| < \varepsilon$ for all $n \geq N$ whenever $|w| \leq s$. By the boundedness of C_φ, Theorem 2.1, and since by condition (IV), $z \mapsto \|\delta_{\varphi(z)}\|$ is bounded away from 0 on compact sets, there is a positive constant c only dependent on s such that

$$\|C_\varphi\| \geq \mu(z)\|\delta_{\varphi(z)}\| \geq c\,\mu(z) \tag{6}$$

whenever $|\varphi(z)| \leq s$. Hence, from (5) and (6), we see that

$$\begin{aligned}\|C_\varphi h_n\|_\mu &= \sup_{z\in\mathbb{D}} \mu(z)|h_n(\varphi(z))| \leq \sup_{|\varphi(z)|>s} \mu(z)|h_n(\varphi(z))| \\ &+ \sup_{|\varphi(z)|\leq s} \mu(z)|h_n(\varphi(z))| \leq \varepsilon + \|C_\varphi\|c^{-1}\varepsilon.\end{aligned}$$

Since ε is arbitrary, the result follows. □

The converse of Theorem 2.2 requires the use of a specific test function. In the analytic case shown in [13], this required the use of condition (VI), which fails in the harmonic setting.

In Sect. 3, we shall obtain an approximation of the essential norm under some additional assumptions on the space X, which will show that under such assumptions, (4) does characterize compactness.

We end the section by showing that all bounded composition operators from a functional Banach space of harmonic mappings into $\mathcal{H}^\infty_{\mu,0}$ are compact.

Theorem 2.3 *Let X be a functional Banach space of harmonic mappings containing the constants and satisfying condition* (II), *φ an analytic self map of $\mathbb{D}$, and μ a weight. Then the following statements are equivalent:*

(a) $C_\varphi : X \to \mathcal{H}^\infty_{\mu,0}$ *is compact;*
(b) $C_\varphi : X \to \mathcal{H}^\infty_{\mu,0}$ *is bounded;*
(c) $\lim_{|z|\to 1} \mu(z)\|\delta_{\varphi(z)}\| = 0$.

In addition, under any of the above equivalent assumptions,

$$\|C_\varphi\|_{X\to\mathcal{H}^\infty_{\mu,0}} = \|C_\varphi\|_{X\to\mathcal{H}^\infty_\mu} = \sup_{z\in\mathbb{D}} \mu(z)\|\delta_{\varphi(z)}\|.$$

Proof $(a) \implies (b)$ is obvious. To prove $(b) \implies (c)$, assume $C_\varphi : X \to \mathcal{H}^\infty_{\mu,0}$ is bounded and that there exist a positive number δ and a sequence $\{z_n\}$ in $\mathbb{D}$, $|z_n| \to 1$, such that $\mu(z_n)\|\delta_{\varphi(z_n)}\| > \delta$ for each $n \in \mathbb{N}$. Then for every $n \in \mathbb{N}$, there exists $h_n \in B_X$ such that $|h_n(\varphi(z_n))| > \|\delta_{\varphi(z_n)}\| - \delta/2$. Thus

$$\mu(z_n)|h_n(\varphi(z_n))| > \delta - \delta/2\, \mu(z_n). \tag{7}$$

On the other hand, since X contains the constant functions and the operator C_φ has target space $\mathcal{H}^\infty_{\mu,0}$, then $\mu(z_n) \to 0$ as $n \to \infty$. Therefore, from (7) it follows that $C_\varphi h_n \notin \mathcal{H}^\infty_{\mu,0}$, which contradicts the boundedness of $C_\varphi : X \to \mathcal{H}^\infty_{\mu,0}$.

Lastly, to prove that $(c) \implies (a)$, suppose (c) holds. Then for each $h \in X$,

$$\mu(z)|h(\varphi(z))| \leq \|h\|\, \mu(z)\|\delta_{\varphi(z)}\| \to 0$$

as $|z| \to 1$, so C_φ maps X boundedly into $\mathcal{H}^\infty_{\mu,0}$.

Recalling part (a) of Proposition 1.1, we see that X satisfies conditions (II) and (IV). The compactness of the operator $C_\varphi : X \to \mathcal{H}^\infty_{\mu,0}$ now follows by arguing as in the proof of Theorem 2.2 to prove the compactness of the operator $C_\varphi : X \to \mathcal{H}^\infty_{\mu}$, noting that if $|\varphi(z)| \to 1$, then $|z| \to 1$ also.

The formula of the operator norm when the target space is $\mathcal{H}^\infty_{\mu,0}$ follows at once from Theorem 2.1. □

3 Essential Norm

Our next objective is to characterize the compact composition operators C_φ acting on a large class of Banach spaces of harmonic mappings X mapping into $\mathcal{H}^\infty_{\mu}$. A standard argument aimed at characterizing the compactness in terms of the "little-oh" condition corresponding to $\tau_\varphi < \infty$, valid in the analytic setting, fails in the harmonic setting as the test functions needed are build in terms of a product of a disk automorphism and a harmonic mapping in the space, which is why condition (VI) is needed for the argument to work. Since, as noted in the Introduction, non-trivial functional Banach spaces of harmonic mappings do not satisfy condition (VI), we determine suitable conditions that allow us to obtain estimates of the essential norm. These then yield a characterization of compactness as a special case.

The following result extends to Banach spaces of harmonic mappings Lemma 3.1 of [13].

Lemma 3.1 *Let X be a non-trivial Banach space of harmonic mappings.*

(a) If X contains harmonic mappings non-vanishing at 0 *and satisfying either* (I) *or* (VIII), *then for each $r \in (0, 1)$,*

$$\sup_{\|h\|\leq 1} \sup_{z\in\mathbb{D}} \|\delta_z\|^{-1}|((I - T_r)h)(z)| < \infty. \tag{8}$$

(b) *If* X *satisfies either condition* (VII) *or conditions* (II) *and* (VIII), *then for each* $s \in (0, 1)$ *and each* $\varepsilon > 0$, *there exists* $r \in (0, 1)$ *such that*

$$\sup_{\|h\|\leq 1} \sup_{|z|\leq s} |((I - T_r)h)(z)| < \varepsilon. \tag{9}$$

If, in addition, X *satisfies condition* (IV), *then for each* $s \in (0, 1)$ *and each* $\varepsilon > 0$, *there exists an* $r \in (0, 1)$ *such that*

$$\sup_{\|h\|\leq 1} \sup_{|z|\leq s} \|\delta_z\|^{-1}\Big(\big|((I - T_r)h)_z(z)\big| + \big|((I - T_r)h)_{\overline{z}}(z)\big|\Big) < \varepsilon. \tag{10}$$

Proof Let $r \in (0, 1), h \in X$, and $z \in \mathbb{D}$. Then by (2), there is a constant $C > 0$ such that

$$|h(z) - h(rz)| \leq \|h\|\, \|\delta_z\| + \|h\|\, \|\delta_{rz}\| \leq C\|h\|\|\delta_z\|. \tag{11}$$

To prove (a), suppose X contains harmonic mappings non-vanishing at 0. Note that if (I) or (VIII) holds, then $\|\delta_z\| > 0$ for all $z \in \mathbb{D}$. Indeed, under the assumption (I), by part (b) of Proposition 1.1, $\|\delta_z\| > 0$ for all $z \in \mathbb{D}$.

Next, suppose condition (VIII) holds. Then, by (2), for each $r \in (0, 1)$, $z \in \mathbb{D}$, there is $h \in X$ non-constant, such that

$$0 < |h(z) - h(rz)| \leq \|(I - T_r)h\|\, \|\delta_z\| \leq C\|h\|\|\delta_z\|,$$

where the constant C does not depend on r. Therefore, $\|\delta_z\| > 0$.

Thus, under either assumption (I) or (VIII), dividing (11) by $\|\delta_z\|$, and taking the supremum over all $z \in \mathbb{D}$, we obtain

$$\sup_{z\in\mathbb{D}} \|\delta_z\|^{-1}\big|((I - T_r)h)(z)\big| \leq C.$$

Then (a) follows by taking the supremum over all h in the unit ball of X.

To prove (b), first assume that condition (VII) holds. Let $h \in X$ with $\|h\| \leq 1$ and fix $r, s \in (0, 1)$ and $z \in \mathbb{D}$, with $|z| \leq s$. Then parametrizing the line segment from rz to z by $\gamma(t) = (1 - t)rz + tz$, for $0 \leq t \leq 1$, we have

$$\begin{aligned} |h(z) - h(rz)| &= \left|\int_\gamma \Big(h_z(\zeta) + \overline{h_{\overline{z}}(\zeta)}\Big)\, d\zeta\right| \\ &\leq (1 - r)|z| \int_0^1 \Big(|h_z(\gamma(t))| + |h_{\overline{z}}(\gamma(t))|\Big)\, dt. \end{aligned} \tag{12}$$

Thus, by (VII), there is a positive constant C such that

$$(1-|\gamma(t)|^2)\Big(|h_z(\gamma(t))|+|h_{\overline{z}}(\gamma(t))|\Big) \le C\|h\|\,\|\delta_{\gamma(t)}\| \le C\,\|\delta_{\gamma(t)}\|.$$

Moreover, by part (c) of Proposition 1.1, since $|\gamma(t)| \le s$ for all $t \in [0,1]$, $\|\delta_{\gamma(t)}\|$ is bounded above by some constant ℓ_s only dependent on s. Thus, from (12), we obtain

$$|h(z)-h(rz)| \le (1-r)s\int_0^1 \frac{C\|\delta_{\gamma(t)}\|}{1-|\gamma(t)|^2}dt \le (1-r)\frac{Cs\ell_s}{1-s^2} \to 0 \quad \text{as } r \to 1,$$

and the convergence is uniform on $\{z : |z| \le s\}$, proving (9) in this case.

Next, suppose conditions (II) and (VIII) hold. Fix $s \in (0,1)$, $z \in \mathbb{D}$, $r_n \in (0,1)$ with $r_n \to 1$, and let $h \in X$ with $\|h\| \le 1$. Then, for each $z \in \mathbb{D}$, $(I - T_{r_n})h(z) \to 0$ as $n \to \infty$. By condition (VIII), there exists a constant $C > 0$ such that the sequence $\{C(I - T_{r_n})h\}$ is in the unit ball of X. Thus, by condition (II), there exists a subsequence $\{r_{n_j}\}_{j\in\mathbb{N}}$ in $(0,1)$ such that $(I - T_{r_{n_j}})h \to 0$ as $j \to \infty$ uniformly on compact sets. Hence, (9) holds.

Finally, assume condition (IV) also holds. With $\{r_{n_j}\}$ as above, we also have that $((I - T_{r_{n_j}})h)_z \to 0$ and $((I - T_{r_{n_j}})h)_{\overline{z}} \to 0$ uniformly on compact subsets of $\mathbb{D}$. Since $z \mapsto \|\delta_z\|$ is bounded away from zero on compact subsets, (10) follows. □

We next provide an approximation of the essential norm.

Theorem 3.1 *Let X be a non-trivial reflexive Banach space of harmonic mappings and X contains harmonic mappings non-vanishing at 0 and φ an analytic self-map of $\mathbb{D}$ such that $\|\delta_{\varphi(z)}\| \to \infty$ whenever $|\varphi(z)| \to 1$,*

$$\inf_{z\in\mathbb{D}} \|\delta_{\varphi(z)}\| > 0, \tag{13}$$

and satisfying condition (V) along with either both (I) and (VII) or condition (VIII).

Let μ be a weight such that $C_\varphi : X \to \mathcal{H}_\mu^\infty$ is bounded. Then

$$\|C_\varphi\|_e \asymp \lim_{s\to 1}\ \sup_{|\varphi(z)|>s}\ \mu(z)\|\delta_{\varphi(z)}\|.$$

In particular, C_φ is a compact operator if and only if

$$\lim_{s\to 1}\ \sup_{|\varphi(z)|>s}\ \mu(z)\|\delta_{\varphi(z)}\| = 0.$$

Proof Define the quantity

$$A(\varphi) := \lim_{s\to 1}\ \sup_{|\varphi(z)|>s}\ \mu(z)\|\delta_{\varphi(z)}\|.$$

We begin by proving that

$$\|C_\varphi\|_e \geq A(\varphi). \tag{14}$$

Since $\|\varphi\|_\infty < 1$ the operator is clearly compact, we shall assume $\|\varphi\|_\infty = 1$. Let $\{z_n\}$ be a sequence in $\mathbb{D}$ such that $|\varphi(z_n)| \to 1$ and

$$A(\varphi) := \lim_{n\to\infty} \mu(z_n)\|\delta_{\varphi(z)}\|.$$

Since C_φ is bounded, by Theorem 2.1, $\tau_\varphi < \infty$, so by our assumption (13), the sequence $\mu(z_n)$ is bounded.

Fix a positive number ε. For each $n \in \mathbb{N}$, there exists $h_n \in X$ with $\|h_n\| \leq 1$ such that

$$|h_n(\varphi(z_n))| > \|\delta_{\varphi(z_n)}\| - \varepsilon. \tag{15}$$

Without loss of generality, we may assume $\{h_n\}$ converges uniformly (and hence uniformly bounded) on compact sets. Therefore by since by assumption $\|\delta_{\varphi(z)}\| \to \infty$ whenever $|\varphi(z)| \to 1$, the sequence $\{H_n\}$ defined as $H_n(z) = \|\delta_{\varphi(z_n)}\|^{-1}h_n(z)$ is bounded in X and converges to 0 uniformly on compact subsets of $\mathbb{D}$. Since the space X is reflexive, by Lemma 3.1, if T is any compact operator from X into $\mathcal{H}_\mu^\infty$, then $\|TH_n\|_{\mathcal{H}_\mu^\infty} \to 0$ as $n \to \infty$. Therefore, by (15),

$$\begin{aligned}
\|C_\varphi - T\| &\geq \limsup_{n\to\infty} \|(C_\varphi - T)H_n\|_{\mathcal{H}_\mu^\infty} \\
&\geq \limsup_{n\to\infty} \frac{\|\delta_{\varphi(z_n)}\|\,\|C_\varphi H_n\|_{\mathcal{H}_\mu^\infty}}{\|h_n\|} \\
&\geq \limsup_{n\to\infty} \|\delta_{\varphi(z_n)}\|\mu(z_n)\|H_n(\varphi(z_n))\| \\
&= \limsup_{n\to\infty} \mu(z_n)\|h_n(\varphi(z_n))\| \\
&\geq \limsup_{n\to\infty} \mu(z_n)\big(\|\delta_{\varphi(z_n)}\| - \varepsilon\big) \\
&\geq \limsup_{n\to\infty} \mu(z_n)\|\delta_{\varphi(z_n)}\| - \varepsilon \sup_{n\in\mathbb{N}} \mu(z_n).
\end{aligned}$$

Since ε is arbitrary, we obtain $\|C_\varphi - T\| \geq A(\varphi)$. Taking the infimum over all compact operators $T : X \to \mathcal{H}_\mu^\infty$, the lower estimate (14) follows.

Next, fix $\varepsilon > 0$ and $s \in (0, 1)$. For $0 < r < 1$, since by condition *(V)*, T_r is compact as an operator on X, and the operator C_φ is bounded, the composition product $C_\varphi T_r : X \to \mathcal{H}_\mu^\infty$ is compact as well. Thus,

$$\|C_\varphi\|_e \leq \|C_\varphi - C_\varphi T_r\| = \sup_{\|h\|\leq 1} \|C_\varphi(I - T_r)h\|_{\mathcal{H}_\mu^\infty} \leq I + II, \tag{16}$$

where the terms I and II are defined as

$$I := \sup_{\|h\|\le 1} \sup_{|\varphi(z)|\le s} \mu(z)\big|\big((I - T_r)h\big)\varphi(z)\big| \text{ and } II := \sup_{\|h\|\le 1} \sup_{|\varphi(z)|>s} \mu(z)|((I - T_r)h)\varphi(z)|.$$

Observe that since by assumption, $d := \inf_{z\in\mathbb{D}} \|\delta_{\varphi(z)}\| > 0$, choosing $r \in (0, 1)$ as in Lemma 3.1(b), we have

$$\begin{aligned} I &\le \sup_{\|h\|\le 1} \sup_{|\varphi(z)|\le s} \mu(z)\|\delta_{\varphi(z)}\| \times \|\delta_{\varphi(z)}\|^{-1}\big|\big((I - T_r)h\big)\varphi(z)\big| \\ &\le \|C_\varphi\| d^{-1}\varepsilon. \end{aligned}$$

Furthermore, by Lemma 3.1(a),

$$\begin{aligned} II &\le \sup_{|\varphi(z)|>s} \mu(z)\|\delta_{\varphi(z)}\| \times \sup_{\|h\|\le 1} \sup_{|\varphi(z)|>s} \|\delta_{\varphi(z)}\|^{-1}\big|\big((I - T_r)h\big)\varphi(z)\big| \\ &\le C \sup_{|\varphi(z)|>s} \mu(z)\|\delta_{\varphi(z)}\|. \end{aligned}$$

Therefore, from (16) it follows that

$$\|C_\varphi\|_e \le \|C_\varphi\| d^{-1}\varepsilon + C \sup_{|\varphi(z)|>s} \mu(z)\|\delta_{\varphi(z)}\|.$$

Since ε is arbitrary, letting $s \to 1$, we conclude that $\|C_\varphi\|_e \le C\,A(\varphi)$, where C is the constant in part (a) of Lemma 3.1. □

4 Applications to Hilbert Spaces of Harmonic Mappings

In this section, we apply our results to the case when the domain space belongs to a class of spaces we call harmonic weighted Hardy spaces. Such spaces are the natural extensions to harmonic mappings of the weighted Hardy spaces treated in [15].

Definition 4.1 A Hilbert space $\mathcal{H}$ with norm $\|\cdot\|$ whose vectors are harmonic mappings on $\mathbb{D}$ is called a **harmonic weighted Hardy space** if $1, z, \overline{z}, z^2, \overline{z}^2, \ldots$ form a complete orthogonal set of non-zero vectors in $\mathcal{H}$.

For $j \in \mathbb{Z}$, set

$$\beta(j) = \begin{cases} \|z^j\| & \text{for } j \ge 0, \\ \|\overline{z}^{-j}\| & \text{for } j < 0. \end{cases} \tag{17}$$

Then each $h \in \mathcal{H}$ admits the representation

$$h(z) = \sum_{j=0}^{\infty} a_j z^j + \sum_{j=1}^{\infty} a_{-j}\overline{z}^j, \quad \text{for } z \in \mathbb{D}, \tag{18}$$

so its squared norm can be expressed as

$$\|h\|^2 = \sum_{j\in\mathbb{Z}} |a_j|^2 \beta(j)^2,$$

and the associated inner product is given by

$$\left\langle \sum_{j=0}^{\infty} a_j z^j + \sum_{j=1}^{\infty} a_{-j}\overline{z}^j, \sum_{j=0}^{\infty} b_j z^j + \sum_{j=1}^{\infty} b_{-j}\overline{z}^j \right\rangle = \sum_{j\in\mathbb{Z}} a_j \overline{b_j} \beta(j)^2.$$

Conforming to the notation adopted in [15], such a space will be denoted by $\mathcal{H}^2(\beta)$.

It is evident that the space $\mathcal{H}^2(\beta) \cap H(\mathbb{D})$ is the *analytic* weighted Hardy space $H^2_+(\beta)$ generated by the sequence $\{\beta(j) : j \geq 0\}$. Similarly, the space formed by taking the complex conjugates of the conjugate-analytic parts of the mappings in $\mathcal{H}^2(\beta)$ is the analytic weighted Hardy space $H^2_-(\beta)$ generated by the sequence $\{\beta(j) : j < 0\}$. The completeness of both these spaces guarantees that the power series $\sum_{j=0}^{\infty} a_j z^j$ defines an analytic function in $H^2_+(\beta)$ if and only if the series $\sum_{j=0}^{\infty} |a_j|^2 \beta(j)^2$ converges. Similarly, $\sum_{j=1}^{\infty} \overline{a_{-j}} z^j$ defines an analytic function in $H^2_-(\beta)$ if and only if $\sum_{j=1}^{\infty} |a_{-j}|^2 \beta(-j)^2 < \infty$.

As noted in the Introduction, in a reproducing kernel Hilbert space the role of the point-evaluation functional is taken by the reproducing kernel. A useful tool to obtain a formula for the reproducing kernel of a weighted Hardy space is the generating function.

Definition 4.2 The **generating function** for $\mathcal{H}^2(\beta)$ is the function k defined on $\mathbb{D}$ by

$$k(z) = \sum_{j=0}^{\infty} \frac{1}{\beta(j)^2} z^j + \sum_{j=1}^{\infty} \frac{1}{\beta(-j)^2} \overline{z}^j. \tag{19}$$

Then the functions $f(z) = \sum\limits_{j=0}^{\infty} \frac{1}{\beta(j)^2} z^j$ and $g(z) = \sum\limits_{j=1}^{\infty} \frac{1}{\beta(-j)^2} z^j$ are analytic in $\mathbb{D}$ and $k = f + \overline{g}$. Thus, k is a harmonic mapping on $\mathbb{D}$ and for $w \in \mathbb{D}$, taking the inner product of $h \in \mathcal{H}^2(\beta)$ represented as in (18) with the mapping K_w defined by

$K_w(z) = k(\overline{w}z)$, we see that

$$\langle h, K_w \rangle = h(w).$$

Moreover,

$$\|K_w\|^2 = K_w(w) = k(|w|^2) = \sum_{j\in\mathbb{Z}} \frac{|w|^{2|j|}}{\beta(j)^2}. \tag{20}$$

Proposition 4.1 *Given a sequence $\beta = \{\beta(j) : j \in \mathbb{Z}\}$ of positive numbers such that $\beta(0) = 1$ and $\liminf_{|j|\to\infty} \beta(j)^{1/|j|} \geq 1$, then the space*

$$\mathcal{H} := \left\{ \sum_{j=0}^{\infty} a_j z^j + \sum_{j=1}^{\infty} a_{-j}\overline{z}^j : \sum_{j\in\mathbb{Z}} |a_j|^2 \beta(j)^2 < \infty \right\}$$

with inner product

$$\left\langle \sum_{j=0}^{\infty} a_j z^j + \sum_{j=1}^{\infty} a_{-j}\overline{z}^j, \sum_{j=0}^{\infty} b_j z^j + \sum_{j=1}^{\infty} b_{-j}\overline{z}^j \right\rangle = \sum_{j\in\mathbb{Z}} a_j \overline{b_j} \beta(j)^2 \tag{21}$$

defines a weighted Hardy space of harmonic mappings.

Proof The assumption $\liminf_{|j|\to\infty} \beta(j)^{1/|j|} \geq 1$ says that the power series

$$\sum_{j=0}^{\infty} \frac{1}{\beta(j)^2} z^j \quad \text{and} \quad \sum_{j=1}^{\infty} \frac{1}{\beta(-j)^2} z^j$$

have radius of convergence at least 1 and thus they define respectively analytic functions f and g on $\mathbb{D}$. Then $f + \overline{g}$ is the mapping k defined in (19), which is therefore harmonic.

Letting $\|\cdot\|$ denote the norm induced by (21), $1, z, \overline{z}, z^2, \overline{z}^2, \ldots$ form a complete orthogonal set of non-zero vectors in $\mathcal{H}$ and (17) holds. Thus $\mathcal{H} = \mathcal{H}^2(\beta)$ with generating function k. □

From now on, we shall assume that $\beta = \{\beta(j) : j \in \mathbb{Z}\}$ is a doubly infinite sequence of positive numbers such that $\beta(0) = 1$ and $\liminf_{|j|\to\infty} \beta(j)^{1/|j|} \geq 1$.

Remark 4.1 By formula (20), if the series $\sum_{j\in\mathbb{Z}} \beta(j)^{-2}$ diverges, then $\|K_w\| \to \infty$ as $|w| \to 1$. Thus, under this assumption, condition (III) holds for the weighted Hardy space $\mathcal{H}^2(\beta)$.

For $j \in \mathbb{N}$, let R_j be the operator on $\mathcal{H}^2(\beta)$ defined by

$$(R_j h)(z) = \sum_{k=j}^{\infty} \left(a_k z^k + a_{-k}\overline{z}^k\right) \quad \text{for } z \in \mathbb{D}, \tag{22}$$

where $h \in \mathcal{H}^2(\beta)$ is represented as in (18).

Then the operator $Q_j = I - R_j$ is given by

$$(Q_j h)(z) = a_0 + \sum_{k=1}^{j-1} \left(a_k z^k + a_{-k}\overline{z}^k\right).$$

Being a finite-rank operator, Q_j is compact.

The following result is an extension to harmonic mappings of Lemma 3.16 in [15]. Its proof is similar.

Lemma 4.1 *If $C_\varphi : \mathcal{H}^2(\beta) \to \mathcal{H}^2(\beta)$ is bounded, then*

$$\|C_\varphi\|_e = \lim_{j\to\infty} \|C_\varphi R_j\|.$$

We next provide a characterization of the bounded composition operator from a general weighted Hardy space into $\mathcal{H}_\mu^\infty$ and give precise formulas of the operator norm and the essential norm (the latter, under the assumption that the sequence $\{1/\beta(j)\}_{j\in\mathbb{Z}}$ is not square summable).

Theorem 4.1 *Let φ be an analytic self-map of $\mathbb{D}$ and μ a weight. Then the operator $C_\varphi : \mathcal{H}^2(\beta) \to \mathcal{H}_\mu^\infty$ is bounded if and only if*

$$\sup_{z\in\mathbb{D}} \mu(z) \left(\sum_{j\in\mathbb{Z}} \frac{|\varphi(z)|^{2|j|}}{\beta(j)^2}\right)^{1/2} < \infty.$$

If $C_\varphi : \mathcal{H}^2(\beta) \to \mathcal{H}_\mu^\infty$ is bounded, then

$$\|C_\varphi\| = \sup_{z\in\mathbb{D}} \mu(z) \left(\sum_{j\in\mathbb{Z}} \frac{|\varphi(z)|^{2|j|}}{\beta(j)^2}\right)^{1/2}.$$

If, in addition, $\sum_{j\in\mathbb{Z}} \beta(j)^{-2} = \infty$, then

$$\|C_\varphi\|_e = \lim_{s\to 1} \sup_{|\varphi(z)|>s} \mu(z) \left(\sum_{j\in\mathbb{Z}} \frac{|\varphi(z)|^{2|j|}}{\beta(j)^2}\right)^{1/2}. \tag{23}$$

Proof The characterization of boundedness and the formula of the operator norm are an immediate consequence of Theorem 2.1 and formula (20).

To prove the formula of the essential norm, first observe that $\mathcal{H}^2(\beta)$ is clearly reflexive and contains the constants, so (13) holds. In addition, since the series $\sum_{j\in\mathbb{Z}} \beta(j)^{-2}$ diverges, by Remark 4.1, it follows that $\|K_{\varphi(z)}\| \to \infty$ as $|\varphi(z)| \to 1$. We next prove that conditions (V) and (VIII) hold.

For $0 < r < 1$, the operator T_r is bounded on $\mathcal{H}^2(\beta)$ and, with h represented as in (18),

$$\|T_r h\|^2 = \sum_{j\in\mathbb{Z}} |a_j|^2 r^{2|j|}\beta(j)^2 \leq \sum_{j\in\mathbb{Z}} |a_j|^2\beta(j)^2 = \|h\|^2.$$

Thus, $\|T_r\| \leq 1$, proving the validity of condition (VIII).

To prove condition (V), we need to show that for $0 < r < 1$, the operator $T_r : \mathcal{H}^2(\beta) \to \mathcal{H}^2(\beta)$ is compact. Observe that $T_r h(z) = h(rz)$ is the composition operator C_{ϕ_r} induced by the dilation function $\phi_r(z) = rz$. Then for $h \in \mathcal{H}^2(\beta)$ represented as in (18), $j \in \mathbb{N}$, R_j defined in (22), and $z \in \mathbb{D}$,

$$(T_r R_j h)(z) = \sum_{k=j}^{\infty} \left(a_k r^k z^k + a_{-k} r^k \bar{z}^k\right),$$

so that $T_r R_j$ is a bounded operator on $\mathcal{H}^2(\beta)$. By Lemma 4.1, it follows that

$$\|T_r\|_e = \lim_{j\to\infty} \|T_r R_j\|.$$

For $h \in \mathcal{H}^2(\beta)$ represented as above,

$$\|T_r R_j h\|^2 = \sum_{|k|=j}^{\infty} |a_k|^2 r^{2|k|}\beta(k)^2 \leq r^{2j} \sum_{k\in\mathbb{Z}} |a_k|^2\beta(k)^2 = r^{2j}\|h\|^2 \to 0$$

as $j \to \infty$, since $0 < r < 1$. Therefore, $\|T_r R_j\| \to 0$, and so $\|T_r\|_e = 0$, proving the compactness of T_r.

We may now apply Theorem 3.1. Letting $A(\varphi)$ be the right side of (23), we see that $\|C_\varphi\|_e \asymp A(\varphi)$. In the proof of Theorem 3.1, it was shown that

$$A(\varphi) \leq \|C_\varphi\|_e \leq C\, A(\varphi),$$

where C is the constant in part (a) of Lemma 3.1. Thus, to prove that $\|C_\varphi\|_e = A(\varphi)$, it suffices to show that $C = 1$.

For $h \in \mathcal{H}^2(\beta)$ represented as in (18), $0 < r < 1$, and $z \in \mathbb{D}$,

$$h(z) - h(rz) = \sum_{j=1}^{\infty} (1 - r^j)\left(a_j z^j + a_{-j}\bar{z}^j\right).$$

Therefore,

$$\|h - T_r h\| = \left(\sum_{0 \neq j \in \mathbb{Z}} (1 - r^{|j|})^2 |a_j|^2 \beta(j)^2 \right)^{1/2} \leq \left(\sum_{j \in \mathbb{Z}} |a_j|^2 \beta(j)^2 \right)^{1/2} = \|h\|.$$

In particular, if $\|h\| \leq 1$, then

$$|h(z) - h(rz)| \leq \|h - T_r h\| \|K_z\| \leq \|K_z\| \quad \text{for each } z \in \mathbb{D}.$$

Dividing by $\|K_z\|$, and taking the supremum over all $z \in \mathbb{D}$ and over all h in the closed unit ball of $\mathcal{H}^2(\beta)$, we obtain

$$\sup_{\|h\| \leq 1} \sup_{z \in \mathbb{D}} \|K_z\|^{-1} |h(z) - h(rz)| \leq 1.$$

Therefore, the constant C in part (a) of Lemma 3.1 is 1. The proof is now complete. □

We next give examples of harmonic weighted Hardy spaces that are extensions of the classical Hardy Hilbert space, the Bergman Hilbert space and the Dirichlet space.

4.1 *The Harmonic Hardy Space*

It is the space $\mathcal{H}^2$ defined as

$$\left\{ h \in \mathcal{H}(\mathbb{D}) \middle| h(z) = \sum_{j=0}^{\infty} a_j z^j + \sum_{j=1}^{\infty} a_{-j} \overline{z}^j, \ \|h\|^2_{\mathcal{H}^2} := \sum_{j \in \mathbb{Z}} |a_j|^2 < \infty \right\}.$$

Thus, $\mathcal{H}^2$ is the weighted Hardy space $\mathcal{H}^2(\beta)$ where for $j \in \mathbb{Z}$, $\beta(j) = 1$.

The generating function for $\mathcal{H}^2$ is given by

$$k(z) = \sum_{j=0}^{\infty} z^j + \sum_{j=1}^{\infty} \overline{z}^j.$$

Thus, the reproducing kernel for $\mathcal{H}^2$ is

$$K_w(z) = k(\overline{w}z) = \frac{1}{1 - \overline{w}z} + \frac{1}{1 - w\overline{z}} - 1$$

and

$$\|K_w\|^2_{\mathcal{H}^2} = k(|w|^2) = \frac{2}{1-|w|^2} - 1 = \frac{1+|w|^2}{1-|w|^2}. \tag{24}$$

In particular, $\|K_w\| \to \infty$ as $|w| \to 1$.

Applying Theorem 4.1 and using (24), we deduce the following result.

Corollary 4.1 *The operator* $C_\varphi : \mathcal{H}^2 \to \mathcal{H}^\infty_\mu$ *is bounded if and only if*

$$\sup_{z\in\mathbb{D}} \mu(z)\Big(\frac{1+|\varphi(z)|^2}{1-|\varphi(z)|^2}\Big)^{1/2} < \infty.$$

If $C_\varphi : \mathcal{H}^2 \to \mathcal{H}^\infty_\mu$ *is bounded, then*

$$\|C_\varphi\| = \sup_{z\in\mathbb{D}} \mu(z)\Big(\frac{1+|\varphi(z)|^2}{1-|\varphi(z)|^2}\Big)^{1/2} \quad \text{and}$$

$$\|C_\varphi\|_e = \lim_{s\to 1} \sup_{|\varphi(z)|>s} \mu(z)\Big(\frac{1+|\varphi(z)|^2}{1-|\varphi(z)|^2}\Big)^{1/2}.$$

4.2 The Harmonic Bergman Space

The harmonic Bergman space $\mathcal{A}^2$ is defined as

$$\Big\{h \in \mathcal{H}(\mathbb{D}) \Big| \|h\|^2_{\mathcal{A}^2} := \int_{\mathbb{D}} |h(z)|^2 \, dA(z) < \infty\Big\},$$

where dA denotes the normalized Lebesgue area measure.

Using orthogonality, the harmonic Bergman norm can be expressed in terms of the coefficients of the powers of z and $\overline{z}$. Indeed, for $h \in \mathcal{A}^2$, $h(z) = \sum_{j=0}^{\infty} a_j z^j + \sum_{j=1}^{\infty} a_{-j}\overline{z}^j$, we have

$$\|h\|^2_{\mathcal{A}^2} = \sum_{j\in\mathbb{Z}} \frac{1}{|j|+1} |a_j|^2.$$

So $\mathcal{A}^2$ is the harmonic weighted Bergman space $\mathcal{H}^2(\beta)$, where for $j \in \mathbb{Z}$, $\beta(j) = \dfrac{1}{(|j|+1)^{1/2}}$.

The generating function k for $\mathcal{A}^2$ is given by

$$k(z) = \sum_{j=0}^{\infty}(j+1)z^j + \sum_{j=1}^{\infty}(j+1)\overline{z}^j.$$

Thus, the reproducing kernel for $\mathcal{A}^2$ and the corresponding norm are

$$K_w(z) = \frac{1}{(1-\overline{w}z)^2} + \frac{1}{(1-w\overline{z})^2} - 1 \text{ and } \|K_w\|^2 = K_w(w) = \frac{2}{(1-|w|^2)^2} - 1. \quad (25)$$

Thus, $\|K_w\| \to \infty$ as $|w| \to 1$.

The following result is an immediate corollary of Theorems 2.1 and 4.1 and formula (25).

Corollary 4.2 *The operator $C_\varphi : \mathcal{A}^2 \to \mathcal{H}_\mu^\infty$ is bounded if and only if*

$$\sup_{z\in\mathbb{D}} \mu(z) \left[\frac{2}{(1-|\varphi(z)|^2)^2} - 1\right]^{1/2} < \infty.$$

If $C_\varphi : \mathcal{A}^2 \to \mathcal{H}_\mu^\infty$ is bounded, then

$$\|C_\varphi\| = \sup_{z\in\mathbb{D}} \mu(z) \left[\frac{2}{(1-|\varphi(z)|^2)^2} - 1\right]^{1/2} \quad \text{and}$$

$$\|C_\varphi\|_e = \lim_{s\to 1} \sup_{|\varphi(z)|>s} \mu(z) \left[\frac{2}{(1-|\varphi(z)|^2)^2} - 1\right]^{1/2}.$$

4.3 *The Harmonic Dirichlet Space*

The harmonic Dirichlet space $\mathcal{D}_H$ is the space of harmonic mappings h on $\mathbb{D}$ satisfying the condition

$$\int_{\mathbb{D}} \left(|h_z(z)|^2 + |h_{\overline{z}}(z)|^2\right) dA(z) < \infty.$$

As in the case of the harmonic Bergman space, the Dirichlet norm $\|\cdot\|_{\mathcal{D}_H}$, defined by

$$\|h\|_{\mathcal{D}_H} := \left(|h(0)|^2 + \int_{\mathbb{D}} \left(|h_z(z)|^2 + |h_{\overline{z}}(z)|^2\right) dA(z)\right)^{1/2},$$

can be expressed in terms of the coefficients of the powers of z and $\overline{z}$.

For $h \in \mathcal{D}_H$, with series expansion $h(z) = \sum_{j=0}^{\infty} a_j z^j + \sum_{j=1}^{\infty} a_{-j}\overline{z}^j$, differentiating the series with respect to z and $\overline{z}$ and using orthogonality, we see that

$$|h(0)|^2 + \int_{\mathbb{D}} \left(|h_z(z)|^2 + |h_{\overline{z}}(z)|^2\right) dA(z) = \sum_{j=-\infty}^{\infty} (|j|+1)|a_j|^2.$$

Thus, the harmonic Dirichlet space is the weighted Hardy space $\mathcal{H}^2(\beta)$, where for $j \in \mathbb{Z}$,

$$\beta(j) = (|j|+1)^{1/2}.$$

The generating function k for $\mathcal{D}_H$ is given by

$$k(z) = \sum_{j=0}^{\infty} \frac{1}{j+1} z^j + \sum_{j=1}^{\infty} \frac{1}{j+1} \overline{z}^j.$$

A straightforward calculation shows that the reproducing kernel K_w for the harmonic Dirichlet space and its norm are

$$K_w(z) = \frac{1}{\overline{w}z} \log \frac{1}{1-\overline{w}z} + \frac{1}{w\overline{z}} \log \frac{1}{1-w\overline{z}} - 1 \quad \text{and}$$

$$\|K_w\|^2 = K_w(w) = \frac{2}{|w|^2} \log \frac{1}{1-|w|^2} - 1. \tag{26}$$

In particular, $\|K_0\| = 1$ and $\|K_w\| \to \infty$ as $|w| \to 1$.

From (26) and Theorems 2.1 and 4.1, we obtain the following result.

Corollary 4.3 *The operator $C_\varphi : \mathcal{D}_H \to \mathcal{H}_\mu^\infty$ is bounded if and only if*

$$\sup_{z\in\mathbb{D}} \mu(z)\left[\frac{1}{|\varphi(z)|^2} \log \frac{1}{1-|\varphi(z)|^2} - 1\right]^{1/2} < \infty.$$

Moreover, if $C_\varphi : \mathcal{D}_H \to \mathcal{H}_\mu^\infty$ is bounded, then

$$\|C_\varphi\| = \sup_{z\in\mathbb{D}} \mu(z)\left[\frac{1}{|\varphi(z)|^2} \log \frac{1}{1-|\varphi(z)|^2} - 1\right]^{1/2} \quad \text{and}$$

$$\|C_\varphi\|_e = \lim_{s\to 1} \sup_{|\varphi(z)|>s} \mu(z)\left[\frac{2}{|\varphi(z)|^2} \log \frac{1}{1-|\varphi(z)|^2} - 1\right]^{1/2}.$$

Remark 4.2 Boundedness (respectively, compactness) of the composition operator $C_\varphi : X \to \mathcal{H}_\mu^\infty$ where X is any of the spaces $\mathcal{H}^2$, $\mathcal{A}^2$, and $\mathcal{D}_H$ is equivalent to boundedness (respectively, compactness) of the corresponding operator acting between the analytic counterparts, although the formulas of the operator norm and essential norm are different (see Sections 4.1 and 4.2 of [13], where formulas of the operator norm and essential norm are provided for the Hardy spaces and the weighted Bergman spaces; see also Corollary 1 and Theorem 5 of [12] for the Dirichlet space case).

We may generate new examples by considering sequences that differ for the positive and negative indices. An example is the space $\mathcal{H}^2(\beta)$, where the sequence $\beta = \{\beta(j)\}_{j\in\mathbb{Z}}$ is defined by

$$\beta(j) = \begin{cases} 1 & \text{if } j \geq 0, \\ (|j|+1)^{-1/2} & \text{if } j < 0. \end{cases}$$

Then $\mathcal{H}^2(\beta)$ is the space of harmonic mappings h on $\mathbb{D}$, $h(z) = \sum_{j=0}^\infty a_j z^j + \sum_{j=1}^\infty a_{-j}\overline{z}^j$ such that

$$\|h\|^2 := \sum_{j=0}^\infty |a_j|^2 + \sum_{j=1}^\infty \frac{1}{j+1}|a_{-j}|^2 < \infty.$$

From this definition we see that the analytic parts of the elements of $\mathcal{H}^2(\beta)$ are the functions in the classical Hardy Hilbert space H^2 and the conjugate analytic parts (after taking the conjugate of a_{-j}) are functions in the Bergman space A^2. From the formulas derived above we see that the reproducing kernel at $w \in \mathbb{D}$ and the corresponding squared norm are given by

$$K_w(z) = \frac{1}{1-\overline{w}z} + \frac{1}{(1-w\overline{z})^2} - 1 \quad \text{and} \quad \|K_w\|^2 = \frac{1}{1-|w|^2} + \frac{1}{(1-|w|^2)^2} - 1.$$

5 An Application to the Harmonic Bloch Space

In the section, we give an application to a functional Banach space that is not reflexive: the harmonic Bloch space.

In [11], the second author defined the harmonic Bloch functions as the harmonic mappings on $\mathbb{D}$ that are Lipschitz functions when regarded as maps between the hyperbolic disk and $\mathbb{C}$ endowed with the Euclidean metric. Such harmonic mappings h are also characterized by the condition

$$b_h := \sup_{z\in\mathbb{D}} (1-|z|^2)\big(|h_z(z)| + |h_{\overline{z}}(z)|\big) < \infty. \tag{27}$$

Indeed, the following equality holds

$$b_h = \sup_{z,w\in\mathbb{D},\ z\neq w} \frac{|h(z)-h(w)|}{\rho(z,w)}, \tag{28}$$

where ρ denotes the hyperbolic metric on $\mathbb{D}$. This metric is Möbius invariant and

$$\rho(z,0) = \frac{1}{2}\log\frac{1+|z|}{1-|z|}.$$

The collection $\mathcal{B}_H$ of such mappings h, called the *harmonic Bloch space* was further studied by the first author in [1], where in particular it was shown that $\mathcal{B}_H$ is a functional Banach space with norm

$$\|h\|_{\mathcal{B}_H} = |h(0)| + b_h.$$

If h is analytic, then b_h is precisely the well-known Bloch semi-norm $\sup_{z\in\mathbb{D}}(1-|z|^2)|h'(z)|$. The *harmonic little Bloch space* is the subspace $\mathcal{B}_{H,0}$ of $\mathcal{B}_H$ whose elements h satisfy the condition

$$\lim_{|z|\to 1}(1-|z|^2)\big(|h_z(z)|+|h_{\bar{z}}(z)|\big) = 0.$$

Of course, $\mathcal{B}_{H,0} \cap H(\mathbb{D})$ is the little Bloch space $\mathcal{B}_0$.

It is straightforward to verify that a harmonic mapping $h = f + \overline{g}$, with f, g analytic, is in $\mathcal{B}_H$ (respectively, in $\mathcal{B}_{H,0}$) if and only if $f, g \in \mathcal{B}$ (respectively, in $\mathcal{B}_0$).

Then $\mathcal{B}_{H,0}$ is the closure in $\mathcal{B}_H$ of the harmonic polynomials (i.e. harmonic mappings of the form $p + \overline{q}$, where $p, q \in \mathbb{C}[z]$).

As shown in Theorem 3.6 of [1], for $h \in \mathcal{B}_H$ and $z \in \mathbb{D}$,

$$|h(z)| \leq \|h\|_{\mathcal{B}_H} \max\left\{1, \frac{1}{2}\log\frac{1+|z|}{1-|z|}\right\}, \tag{29}$$

where the maximum on the right side of (29) is precisely the norm of the point-evaluation functional at z for the analytic Bloch space $\mathcal{B}$ (see Section 4.3 of [13]). Since the $\mathcal{B}$ is a subspace of the harmonic Bloch space, it follows that the functional δ_z for the harmonic Bloch space has the same norm:

$$\|\delta_z\| = \max\left\{1, \frac{1}{2}\log\frac{1+|z|}{1-|z|}\right\}. \tag{30}$$

Moreover, in [13] it was shown that the norms of δ_z for the spaces $\mathcal{B}$ and $\mathcal{B}_0$ coincide. Thus, the same holds for the harmonic counterparts.

The operator norm and the essential norm of the bounded composition operators from $\mathcal{B}_H$ into the weighted Banach space $\mathcal{H}_\mu^\infty$ are given in the following theorem.

Theorem 5.1 *The operator $C_\varphi : \mathcal{B}_H \to \mathcal{H}_\mu^\infty$ is bounded if and only if*

$$\sup_{z\in\mathbb{D}} \mu(z) \max\left\{1, \frac{1}{2}\log\frac{1+|\varphi(z)|}{1-|\varphi(z)|}\right\} < \infty.$$

If $C_\varphi : \mathcal{B}_H \to \mathcal{H}_\mu^\infty$ is bounded, then

$$\|C_\varphi\| = \sup_{z\in\mathbb{D}} \mu(z) \max\left\{1, \frac{1}{2}\log\frac{1+|\varphi(z)|}{1-|\varphi(z)|}\right\} \quad \text{and}$$

$$\|C_\varphi\|_e = \lim_{s\to 1} \sup_{|\varphi(z)|>s} \mu(z)\frac{1}{2}\log\frac{1+|\varphi(z)|}{1-|\varphi(z)|}. \tag{31}$$

Proof The characterization of boundedness and the formula of the operator norm follows immediately from Theorem 2.1 and formula (30). To prove the formula of the essential norm, set $A(\varphi)$ equal to the right-hand side of (31). We wish to show that, save the reflexivity of the domain space, all other conditions in Theorem 3.1 hold for $\mathcal{B}_H$.

In $\mathcal{B}_H$ condition (V) holds. Indeed, in [3] it was shown that a composition operator C_ϕ on $\mathcal{B}_H$ is compact if and only if it is compact as an operator on $\mathcal{B}$, i.e. [22]

$$\lim_{s\to 1} \sup_{|\phi(z)|>s} \frac{(1-|z|^2)|\phi'(z)|}{1-|\phi(z)|^2} = 0.$$

In the special case of $\phi(z) = rz$, for a fixed $r \in (0, 1)$, the above condition clearly holds.

Next note that $\rho(z, 0) \le 1$ if and only if $|z| \le (e^2-1)/(e^2+1)$. Thus, from (30) it follows that $\|\delta_z\| \to \infty$ as $|z| \to 1$.

Moreover, since the function $x \mapsto \log\frac{1+x}{1-x}$ is increasing on $(0, 1)$, for $z \in \mathbb{D}$ and $0 < r < 1$, we have $\|\delta_{rz}\| \le \|\delta_z\|$, proving that condition (I) holds for $\mathcal{B}_H$.

Next observe that for each $z \in \mathbb{D}$ and each $h \in \mathcal{B}_H$, by the expression in (27) of the Bloch semi-norm and since by (30) $\|\delta_z\| \ge 1$, we have

$$(1-|z|^2)\big(|h_z(z)| + |h_{\bar{z}}(z)\big) \le \|h\|_{\mathcal{B}_H} \le \|h\|_{\mathcal{B}_H}\|\delta_z\|.$$

Thus, condition (VII) holds.

The space $\mathcal{B}_H$ is not reflexive. However, the estimate $\|C_\varphi\| \ge A(\varphi)$, that in the proof of Theorem 3.1 made use of the reflexivity of the space X, holds for the space $\mathcal{B}_H$. The proof is based on the fact that if $\{h_n\}$ is a bounded sequence in $\mathcal{B}_{H,0}$ converging to 0 uniformly on compact subsets of $\mathbb{D}$, then it converges to 0 weakly, so if $T : \mathcal{B}_{H,0} \to \mathcal{H}_\mu^\infty$ is a compact operator, then $\|Th_n\|_\mu \to 0$ as $n \to \infty$. Therefore, arguing as in the proof of Theorem 3.1 and since the norms of the point-evaluation functionals at $z \in \mathbb{D}$ for $\mathcal{B}_H$ and $\mathcal{B}_{H,0}$ are equal, we see that viewing C_φ as an operator from $\mathcal{B}_{H,0} \to \mathcal{H}_\mu^\infty$, $\|C_\varphi\|_e \geq A(\varphi)$.

Applying Theorem 3.1, we obtain $\|C_\varphi\|_e \leq C\,A(\varphi)$, where C is the constant in part (a) of Lemma 3.1. Thus, to prove the estimate $\|C_\varphi\|_e \leq A(\varphi)$, it suffices to show that for $0 < r < 1$ and $h \in \mathcal{B}_H$ with $\|h\|_{\mathcal{B}_H} \leq 1$,

$$\sup_{z\in\mathbb{D}} \|\delta_z\|^{-1}|h(z) - h(rz)| \leq 1. \tag{32}$$

For $z \in \mathbb{D}$, using (28), and the Möbius invariance of ρ, we have

$$\begin{aligned} |h(z) - h(rz)| &\leq b_h\,\rho(z, rz) \leq \rho(z, rz) \\ &= \rho(0, \phi_z(rz)) = \frac{1}{2}\log\frac{1 + |\phi_z(rz)|}{1 - |\phi_z(rz)|}, \end{aligned} \tag{33}$$

where ϕ_z denotes the disk automorphism that interchanges 0 and z, that is,

$$\phi_z(w) = \frac{z - w}{1 - \bar{z}w}.$$

A straightforward calculation shows that

$$\frac{1 + |\phi_z(rz)|}{1 - |\phi_z(rz)|} = \frac{1 - r|z|^2 + (1 - r)|z|}{1 - r|z|^2 - (1 - r)|z|} \leq \frac{1 + |z|}{1 - |z|}.$$

Hence, from (33), we obtain

$$|h(z) - h(rz)| \leq \frac{1}{2}\log\frac{1 + |z|}{1 - |z|} \leq \|\delta_z\|,$$

which yields (32). □

From Theorem 5.1 and Theorem 4.3 of [13], we see that $C_\varphi : \mathcal{B}_H \to \mathcal{H}_\mu^\infty$ is bounded (respectively, compact) if and only if $C_\varphi : \mathcal{B} \to H_\mu^\infty$ is bounded (respectively, compact). Moreover, the operator norms and essential norms are the same.

References

1. Aljuaid, M.: The operator theory on some spaces of harmonic mappings. Doctoral Dissertation, George Mason University (2019)
2. Aljuaid, M., Colonna, F.: Characterizations of Bloch-type spaces of harmonic mappings. J. Function Spaces Appl. **2019**, Article ID 5687343, 11 pp. (2019). https://doi.org/10.1155/2019/5687343

3. Aljuaid, M., Colonna, F.: Composition operators on some Banach spaces of harmonic mappings. J. Funct. Spaces **2020**, Article ID 9034387, 11 pp. (2020). https://doi.org/10.1155/2020/9034387
4. Aljuaid, M., Colonna, F.: On the harmonic Zygmund spaces. Bull. Aust. Math. Soc. **101**(3), 466–476 (2020)
5. Axler, S., Bourdon, P, Ramey, W.: Harmonic Function Theory, vol. 137, 2nd edn. Graduate Texts in Mathematics. Springer, New York (2001)
6. Boyd, C., Rueda, P.: Isometries of weighted spaces of harmonic functions. Potential Anal. **29**(1), 37–48 (2008)
7. Chen, S., Wang, X.: On harmonic Bloch spaces in the unit ball of $\mathbb{C}^n$. Bull. Aust. Math. Soc. **84**, 67–78 (2011)
8. Chen, S., Ponnusamy, S., Wang, X.: Landau's theorem and Marden constant for harmonic ν-Bloch mappings. Bull. Aust. Math. Soc. **84**, 19–32 (2011)
9. Chen, S., Ponnusamy, S., Wang, X.: On planar harmonic Lipschitz and planar harmonic Hardy classes. Ann. Acad. Sci. Fen. Math. **36**, 567–576 (2011)
10. Chen, S., Ponnusamy, S., Rasila, A.: Lengths, areas and Lipschitz-type spaces of planar harmonic mappings. Nonlinear Anal. **115**, 62–70 (2015)
11. Colonna, F.: The Bloch constant of bounded harmonic mappings. Indiana U. Math. J. **38**(4), 829–840 (1989)
12. Colonna, F., Tjani, M.: Weighted composition operators from the Besov spaces into the weighted-type space H^∞_μ,. J. Math. Anal. Appl. **402**, 594–611 (2013)
13. Colonna, F., Tjani, M.: Operator norms and essential norms of weighted composition operators between Banach spaces of analytic functions. J. Math. Anal. Appl. **434**, 93–124 (2016)
14. Colonna, F., Tjani, M.: Essential norms of weighted composition operators from reproducing kernel Hilbert spaces into weighted-type spaces. Houst. J. Math. **42**(3), 877–903 (2016)
15. Cowen, C., MacCluer, B.: Composition Operators on Spaces of Analytic Functions. Studies in Advanced Mathematics. CRC Press, Boca Raton (1995)
16. Fu, X., Liu, X.: On characterizations of Bloch spaces and Besov spaces of pluriharmonic mappings. J. Inequal. Appl. **2015**, 360 (2015). https://doi.org/10.1186/s13660-015-0884-0
17. Jordá, E., Zarco, A.-M.: Isomorphisms on weighted Banach spaces of harmonic and holomorphic functions. J. Funct. Spaces Appl. **2013**, Article ID 178460, 6 pp. (2013)
18. Jordá, E., Zarco, A.-M.: Weighted Banach spaces of harmonic functions. Rev. R. Acad. Cienc. Exactas Fis. Nat. Ser. A Mat. **108**(2), 405–418 (2014)
19. Laitila, J., Tylli, H.O.: Composition operators on vector-valued harmonic functions and Cauchy transforms. Indiana Univ. Math. J. **55**(2), 719–746 (2006)
20. Lusky, W.: On weighted spaces of harmonic and holomorphic functions. J. Lond. Math. Soc. **51**, 309–320 (1995)
21. Lusky, W.: On the isomorphism classes of weighted spaces of harmonic and holomorphic functions. Stud. Math. **175**(1), 19–45 (2006)
22. Madigan, K., Matheson, A.: Compact composition operators on the Bloch space. Trans. Am. Math. Soc. **347**, 2679–2687 (1995)
23. Shapiro, J.H.: Composition Operators and Classical Function Theory. Universitext: Tracts in Mathematics. Springer-Verlag, New York (1993)
24. Shields, A.L., Williams, D.L.: Bounded projections, duality, and multipliers in spaces of harmonic functions. J. Reine Angew. Math. **299**(300), 256–279 (1978)
25. Shields, A.L., Williams, D.L.: Bounded projections and the growth of harmonic conjugates in the unit disc. Mich. Math. J. **29**, 3–25 (1982)
26. Yoneda, R.: A characterization of the harmonic Bloch space and the harmonic Besov spaces by an oscillation. Proc. Edinb. Math. Soc. **45**, 229–239 (2002)

Generalized Gaussian Estimates for Elliptic Operators with Unbounded Coefficients on Domains

Luciana Angiuli, Luca Lorenzi, and Elisabetta Mangino

Dedicated to Prof. Francesco Altomare for his 70th birthday

Abstract We consider second-order elliptic operators $\mathcal{A}$ in divergence form with coefficients belonging to $L^\infty_{\mathrm{loc}}(\Omega)$, when $\Omega \subseteq \mathbb{R}^d$ is a sufficiently smooth (unbounded) domain. We prove that the realization of $\mathcal{A}$ in $L^2(\Omega)$, with Neumann-type boundary conditions, generates a contractive, strongly continuous and analytic semigroup $(T(t))$ which has a kernel k satisfying generalized Gaussian estimates, written in terms of a distance function induced by the diffusion matrix and the potential term. Examples of operators where such a distance function is equivalent to the Euclidean one are also provided.

Keywords Second-order elliptic operators · Neumann-type boundary conditions · Semigroups of operators · Generalized Gaussian estimates

1 Introduction

Gaussian estimates for heat kernels of semigroups associated to strictly elliptic operators with bounded coefficients and their relevant consequences on the regularity of the semigroups, are nowadays a classical and well investigated topic, see for instance [4, 5, 9] and the monograph [27]. On the other hand, after the pioneering researches by Davies [10] and Davies and Simon [12], which were concerned with Schrödinger operators, due to the relevant interplay with other subjects such

L. Angiuli · E. Mangino (✉)
Dipartimento di Matematica e Fisica "E. De Giorgi", Università del Salento, Lecce, Italy
e-mail: luciana.angiuli@unisalento.it; elisabetta.mangino@unisalento.it

L. Lorenzi
Dipartimento di Scienze Matematiche, Fisiche e Informatiche, Plesso di Matematica, Università di Parma, Parma, Italy
e-mail: luca.lorenzi@unipr.it

A. M. Candela et al. (eds.), *Recent Advances in Mathematical Analysis*, Trends in Mathematics, https://doi.org/10.1007/978-3-031-20021-2_7

as stochastic differential equations and their applications (where no assumptions of boundedness on the coefficients are required), the interest towards the study of semigroups associated with elliptic operators with unbounded coefficients has recently grown considerably. The fact that the coefficients of the operators are unbounded does not represent only a mere technical difficulty but has significant consequences for the solutions. For instance, in general, uniqueness does not hold in the space of bounded and continuous functions and, whenever they exist, the semigroups associated to such operators usually are not strongly continuous and could not preserve the L^p spaces. Some properties of such semigroups such as compactness, invariance properties of function spaces and gradient estimates have been recently studied both in the whole space and in unbounded domains, where Dirichlet and Neumann type boundary conditions are considered, see [1, 2, 14]. Much less is known about kernel estimates for such semigroups. These estimates can be split into two main classes: Gaussian (or generalized Gaussian) estimates and upper bounds estimates. The main difference is that, in the (generalized) Gaussian estimates, the kernel $k(t, x, y)$ is estimated in terms of a function which, besides on t, depends on the norm of the difference $x - y$. In particular, these estimates are relevant outside the main diagonal and lead to relevant consequences such as the analyticity of the semigroup on the L^p-scale on a sector which is independent of p. On the other hand, upper bounds estimates provide estimates on $k(t, x, y)$, which depend on t and on $|x|$ and $|y|$. The major efforts and the main results in this direction have covered so far, mainly the case of upper bounds estimates for uniformly elliptic operators, with coefficients defined in the whole $\mathbb{R}^d$, having unbounded potentials (see e.g., [21]), also unbounded drift coefficients (see e.g., [25, 28]) and in some cases, unbounded second-order coefficients are allowed (see e.g., [7, 8, 13, 17, 18, 20, 22]). See also [19] for further references. To the best of our knowledge, only few results are available in the literature related to Gaussian (or generalized Gaussian) estimates for semigroups associated with elliptic operators with unbounded coefficients. More precisely, we are aware of [6], where the operators considered may have unbounded drift and potential terms over $\mathbb{R}^d$, and [3, 26], where also the diffusion coefficients may be unbounded over $\mathbb{R}^d$. Different techniques have been used and developed to prove kernel estimates in the previous quoted papers (and some of them are contained in the monographs [11, 27]) and the estimates obtained are not always comparable since most the times the scopes are different: one may be interested just in the behaviour of the kernel with respect to large values of the spatial variables or with respect to time for large or small times.

In this paper we prove generalized Gaussian estimates for the heat kernel $k(t, \cdot, \cdot)$ of the semigroup generated by the realization of a second-order uniformly elliptic operator in $L^2(\Omega)$ subject to Neumann-type boundary conditions, when all the coefficients of the operator, even the diffusion coefficients are allowed to be unbounded. More specifically, we consider a smooth enough domain $\Omega \subseteq \mathbb{R}^d$ and non-self-adjoint operators in divergence form, defined on smooth enough functions u by

$$\mathcal{A}u = \mathrm{div}(Q\nabla u) - (B, \nabla u) + \mathrm{div}(Cu) - Vu,$$

and having real-valued measurable coefficients $Q = (q_{ij})$, $B = (B_i)_i$, $C = (C_i)_i$, $(i, j = 1, \ldots, d)$ and V which belong to $L^\infty_{loc}(\Omega)$. The main assumptions on the coefficients are that $\frac{Q+Q^t}{2}$ is uniformly definite positive and controls $(Q - Q^t)$, V is uniformly bounded from below and the unboundedness of the drift coefficients B and C is controlled by Q and V. The assumptions (A), (B), (C) below will detail these hypotheses. We point out in particular that no assumptions on the symmetry of the diffusion matrix Q is required. This less restrictive hypothesis on Q is useful since it allows to consider problems with more general boundary conditions (see the examples in the last section).

We will use the approach with sesquilinear forms to define the realization of such operators in $L^2(\Omega; \mathbb{C})$ and to associate with them analytic semigroups on $L^2(\Omega; \mathbb{C})$. Inspired by the so-called "Davies trick" we prove upper bounds like

$$k(x, y, t) \leq c_1(t)e^{-c_2(t)(\psi(x)-\psi(y))^a} \tag{1}$$

for any $t > 0$ and almost every $x, y \in \Omega$. Here a is a positive real number, while the positive functions $c_1(t)$ and $c_2(t)$ have quite an explicit behaviour near 0. Concerning the real-valued function ψ, it belongs to a set of smooth functions satisfying the condition $\|(Q\nabla\psi, \nabla\psi)\|_\infty \leq h$, where h is a function having a prescribed growth with respect to V. This point represents the novelty key of the paper: indeed we recall that $h \equiv 1$ in the original Davies trick. Taking the supremum on ψ in (1) yields an estimate of $k(t, \cdot, \cdot)$ in terms of $\exp(-c_2(t)(d_{Q,V}(x, y))^a)$, see (16), where $d_{Q,V}$ is a distance depending on Q and V. This distance in general is not equivalent to the Euclidean one if Q is unbounded, but the introduction of the function h will allow to get the equivalence of the two distances in several interesting cases and thus getting an explicit estimate for the kernel k. We point out that, differently from the existing literature where kernel estimates are obtained essentially for operators with coefficients having a prescribed growth at infinity, mainly of polynomial type, our sets of hypotheses are satisfied in a greater generality and allow to consider a large class of operators.

The paper is organized as follows. Section 2 is a review of classical results that we decided to collect for easiness of reading of the subsequent sections. In Sect. 3, we introduce some of the assumptions on the coefficients of the operator $\mathcal{A}$ and a sesquilinear form in $L^2(\Omega; \mathbb{C})$, which is associated with the realization of operator $\mathcal{A}$ with Neumann-type boundary conditions. The core of the paper is the fourth section, where by combining L^∞-contractivity, $L^p \to L^p$-estimates, Sobolev inequalities and a bootstrap argument, we prove the generalized Gaussian estimate. In Sect. 5, we address the L^p-analyticity of the semigroup $(T(t))$ associated with our operator. As it is well known, the analyticity of a semigroup in $L^2(\Omega)$ together with its boundedness in $L^q(\Omega; \mathbb{C})$ for some $q > 2$ carry over the analyticity from $L^2(\Omega; \mathbb{C})$ to $L^p(\Omega; \mathbb{C})$ for any $p \in [2, q[$. However, in general the sector of analyticity depends on p and q. Thanks to the kernel estimates we have proved, we provide sufficient conditions for the semigroup $(T(t))$ to be extended to a bounded analytic semigroup in $L^p(\Omega, \mathbb{C})$ in a sector uniformly with respect to

$p \in [1, +\infty[$. Finally, we consider some cases where the distance $d_{Q,V}$ is equivalent to the Euclidean one and, in the last section, we collect some examples of operators to which our results can be applied.

Notation Throughout the paper, we use the following set of notation. Given $\xi, \eta \in \mathbb{C}^d$, we denote by (ξ, η) the Euclidean inner product in $\mathbb{C}^d$. Moreover, we find it useful to use also the notation $\xi \cdot \eta = \sum_{i=1}^d \xi_i \eta_i$. Given a complex-valued function u, its signum is the function $\mathrm{sign}(u) = |u|^{-1} u \chi_{\{u \neq 0\}}$, where χ_A denotes the characteristic function of a set A. If u is a real valued function, then u^+ and u^- denote, respectively, the positive and the negative parts of u, i.e., $u^+ = \max\{u, 0\}$, $u^- = \max\{-u, 0\}$. Next, for every $p \in [1, +\infty[\cup\{+\infty\}$, we denote by $\|\cdot\|_p$ the usual norm of the space $L^p(\mathbb{R}^d; \mathbb{C})$ with respect to the Lebesgue measure. Given an open set $\Omega \subseteq \mathbb{R}^d$ and a bounded operator $S : L^p(\Omega; \mathbb{C}) \to L^q(\Omega; \mathbb{C})$ $p, q \in [1, +\infty[\cup\{+\infty\}$, we denote by $\|S\|_{p \to q}$ its operator norm. When we use a function space X of real-valued functions defined on $\mathbb{R}^d$, we simply write $X(\Omega)$ instead of $X(\Omega; \mathbb{R})$. Finally, the subscript "b" means bounded, whereas the subscript "c" means compactly supported.

2 Preliminaries

In this section, we review some classical results about sesquilinear forms in infinite- and finite-dimensional spaces, as they are needed throughout the whole paper.

2.1 Sesquilinear Forms and Properties of the Associated Semigroups

We start by recalling the following result that will be used at several stages in the paper.

Theorem 2.1 *Let $\mathcal{H}$, $\mathcal{V}$ be Hilbert spaces with $\mathcal{V}$ dense and continuously embedded in $\mathcal{H}$ and let $\mathfrak{a} : \mathcal{V} \times \mathcal{V} \to \mathbb{C}$ be a sesquilinear form. Assume that $\mathfrak{a}$ is continuous with respect to the norm $\|\cdot\|_{\mathcal{V}}$ of $\mathcal{V}$ and coercive, namely there exist $\omega \in \mathbb{R}$, $\mu > 0$ and a dense subspace D of $\mathcal{V}$ such that*

$$\mathrm{Re}\, \mathfrak{a}(u, u) + \omega \|u\|_{\mathcal{H}}^2 \geq \mu \|u\|_{\mathcal{V}}^2, \qquad u \in D.$$

Then, we can associate to $\mathfrak{a}$ an operator $(A, D(A))$ defined by

$$\begin{cases} D(A) = \{u \in \mathcal{V} \mid \exists w \in \mathcal{H} \text{ s.t. } (w, \varphi) = \mathfrak{a}(u, \varphi)\ \forall \varphi \in \mathcal{V}\}, \\ Au = w, \quad u \in D(A), \end{cases}$$

such that $-A$ *generates a strongly continuous analytic semigroup* $(T(t))$ *on* $\mathcal{H}$. *Moreover, if* $\omega_0 = \inf\{\omega \in \mathbb{R} \mid \operatorname{Re} a(u,u) + \omega\|u\|_{\mathcal{H}}^2 \geq 0 \; \forall u \in \mathcal{V}\}$, *then*

$$\|T(t)\| \leq e^{\omega_0 t}, \qquad t \geq 0.$$

In particular, if the form $\mathfrak{a}$ is accretive (i.e. $\operatorname{Re}\mathfrak{a}(u,u) \geq 0$ for every $u \in \mathcal{V}$), densely defined, continuous and coercive, then the opposite of the associated operator generates a contractive semigroup.

The adjoint form of a sesquilinear form $\mathfrak{a} : \mathcal{V} \times \mathcal{V} \to \mathbb{C}$ is defined by

$$\mathfrak{a}^*(u, w) = \overline{\mathfrak{a}(w, u)}, \qquad u, w \in \mathcal{V}.$$

If $\mathfrak{a}$ is densely defined, continuous and coercive, then the same holds for $\mathfrak{a}^*$ and the operator associated with $\mathfrak{a}^*$ is exactly the adjoint operator A^* of the operator A associated with $\mathfrak{a}$.

Recall that if $\mathcal{H} = L^2(\Omega; \mathbb{C})$, with Ω open subset of $\mathbb{R}^d$, endowed with the Lebesgue measure, then a semigroup $(T(t))$ on $\mathcal{H}$ is said to be L^∞-contractive if

$$\|T(t)f\|_\infty \leq \|f\|_\infty, \qquad f \in L^2(\Omega; \mathbb{C}) \cap L^\infty(\Omega; \mathbb{C}).$$

In this case, if $(T(t))$ is a contraction semigroup on $L^2(\Omega; \mathbb{C})$, then, by interpolation, we can extend each operator $T(t)$ as a contraction operator on $L^p(\Omega; \mathbb{C})$ for every $p \in [2, +\infty[\cup\{+\infty\}$ and $(T(t))$ becomes a contraction semigroup on every $L^p(\Omega; \mathbb{C})$ with $p \in [2, +\infty[$.

In case of semigroups associated with sesquilinear forms, the following characterizations hold. We refer to [27, Theorems 2.6, 2.15, Corollary 2.17] for their proofs.

Proposition 2.1 *Let* $\mathfrak{a}$ *be a sesquilinear form satisfying the assumptions of Theorem 2.1, with* $\mathcal{H} = L^2(\Omega; \mathbb{C})$ *and let* $(T(t))$ *be the semigroup associated with* $\mathfrak{a}$. *Then, the following properties are satisfied.*

(a) $(T(t))$ *is* L^∞*-contractive if and only if for every* $u \in D$, *it holds that* $(|u| \wedge 1)\operatorname{sign}(u) \in \mathcal{V}$ *and* $\operatorname{Re}\mathfrak{a}((|u| \wedge 1)\operatorname{sign}(u), (|u| - 1)^+\operatorname{sign}(u)) \geq 0$.
(b) *If* $\mathfrak{a}$ *is accretive, then* $(T(t))$ *is positive if and only if for every* $u \in D$ *it holds that* $\operatorname{Re} u^+ \in \mathcal{V}$, $\mathfrak{a}(\operatorname{Re} u, \operatorname{Im} u) \in \mathbb{R}$ *and* $\mathfrak{a}(\operatorname{Re} u^+, \operatorname{Re} u^-) \leq 0$.
(c) *If* $(T(t))$ *is positive, then* $(T(t))$ *is* L^∞*-contractive if and only if for every* $u \in D$, $u \geq 0$ *almost everywhere in* Ω, *it holds that* $u \wedge 1 \in \mathcal{V}$ *and* $\mathfrak{a}(u \wedge 1, (u - 1)^+) \geq 0$.

2.2 Finite-Dimensional Sesquilinear Forms with Real Coefficients

Let Q be a real-valued matrix. To begin with, we observe that for every $\xi, \eta \in \mathbb{C}^d$

$$\begin{aligned} \mathrm{Re}\,(Q\xi,\eta) &= (Q\mathrm{Re}\,\xi, \mathrm{Re}\,\eta) + (Q\mathrm{Im}\,\xi, \mathrm{Im}\,\eta), \\ \mathrm{Im}\,(Q\xi,\eta) &= (Q\mathrm{Im}\,\xi, \mathrm{Re}\,\eta) - (Q\mathrm{Re}\,\xi, \mathrm{Im}\,\eta). \end{aligned} \tag{2}$$

Moreover, since $\overline{(Q\eta,\xi)} = (Q^t\xi,\eta)$, we get that

$$\begin{aligned} \mathrm{Re}\,(Q\xi,\xi) &= \frac{1}{2}\big((Q+Q^t)\xi,\xi\big), \\ \mathrm{Im}\,(Q\xi,\xi) &= \frac{1}{2i}\big((Q-Q^t)\xi,\xi\big). \end{aligned}$$

If $\mathrm{Re}\,(Q\xi,\xi) \geq 0$ for every $\xi \in \mathbb{C}^d$ or, equivalently, $(Q\xi,\xi) \geq 0$ for every $\xi \in \mathbb{R}^d$, then by (2)

$$(Q\mathrm{Re}\,\xi, \mathrm{Re}\,\xi) \leq \mathrm{Re}\,(Q\xi,\xi), \qquad \xi \in \mathbb{C}^d, \tag{3}$$

and, moreover,

$$\frac{1}{2}|((Q+Q^t)\xi,\eta)| \leq \Big(\mathrm{Re}\,(Q\xi,\xi)\Big)^{\frac{1}{2}}\Big(\mathrm{Re}\,(Q\eta,\eta)\Big)^{\frac{1}{2}}, \qquad \xi,\eta \in \mathbb{C}^d. \tag{4}$$

Further, if there exists a positive constant q_0 such that

$$|\mathrm{Im}\,(Q\xi,\xi)| = \frac{1}{2}\big|((Q-Q^t)\xi,\xi)\big| \leq q_0\mathrm{Re}\,(Q\xi,\xi), \qquad \xi \in \mathbb{C}^d,$$

then

$$\frac{1}{2}|((Q-Q^t)\xi,\eta)| \leq q_0\Big(\mathrm{Re}\,(Q\xi,\xi)\Big)^{\frac{1}{2}}\Big(\mathrm{Re}\,(Q\eta,\eta)\Big)^{\frac{1}{2}}, \qquad \xi,\eta \in \mathbb{C}^d \tag{5}$$

and

$$|(Q\xi,\eta)| \leq (1+q_0)\Big(\mathrm{Re}\,(Q\xi,\xi)\Big)^{\frac{1}{2}}\Big(\mathrm{Re}\,(Q\eta,\eta)\Big)^{\frac{1}{2}}, \qquad \xi,\eta \in \mathbb{C}^d, \tag{6}$$

(we refer for the last two inequalities, e.g., to [27, Proposition 1.8] and its proof).

3 A Second Order Elliptic Operator on Ω with Unbounded Coefficients

Let Ω be an open subset of $\mathbb{R}^d$. We are going to introduce a sesquilinear form on $L^2(\Omega;\mathbb{C})$ and to study its properties. To this aim, we first consider a Hilbert space that will be the domain of the form.

Hypothesis (A) Let $Q=(q_{ij})_{i,j=1}^d\in L^\infty_{\text{loc}}(\Omega;\mathbb{R}^d\times\mathbb{R}^d)$ and $V\in L^\infty_{\text{loc}}(\Omega)$. Assume that there exist constants $q_0\ge 0$, $V_0>0$ and a locally uniformly strictly positive function $\nu:\Omega\to]0,+\infty[$ such that for every $\xi\in\mathbb{C}^d$ and almost every $x\in\Omega$

$$|\text{Im}(Q(x)\xi,\xi)|\le q_0\text{Re}\,(Q(x)\xi,\xi), \tag{7}$$

$$\text{Re}\,(Q(x)\xi,\xi)\ge \nu(x)|\xi|^2, \tag{8}$$

$$V(x)\ge V_0. \tag{9}$$

Define

$$D_{Q,V}:=\{u\in W^{1,2}_{\text{loc}}(\Omega;\mathbb{C})\cap L^2(\Omega;\mathbb{C})\ |\ (\text{Re}\,(Q\nabla u,\nabla u))^{\frac{1}{2}},\ V^{\frac{1}{2}}u\in L^2(\Omega;\mathbb{C})\}$$

and endow it with the inner product

$$\langle u,w\rangle_{Q,V}=\int_\Omega\left(\left(\frac{Q+Q^t}{2}\nabla u,\nabla w\right)+Vu\overline{w}\right)dx,\qquad u,v\in D_{Q,V}, \tag{10}$$

and the associated norm

$$\|u\|_{Q,V}=\langle u,u\rangle^{\frac{1}{2}}_{Q,V}=\left(\int_\Omega\left(\text{Re}\,(Q\nabla u,\nabla u)+V|u|^2\right)dx\right)^{\frac{1}{2}},\qquad u\in D_{Q,V}.$$

Proposition 3.1 *Under Hypothesis (A), $D_{Q,V}$, endowed with the inner product* (10), *is a Hilbert space and it is continuously and densely embedded in $L^2(\Omega;\mathbb{C})$.*

Proof If (u_n) is a Cauchy sequence in $D_{Q,V}$, then (u_n) converges to a function $u\in W^{1,2}_{\text{loc}}(\Omega;\mathbb{C})\cap L^2(\Omega;\mathbb{C})$, due to conditions (8) and (9), and, up to a subsequence, ∇u_n converges to ∇u and u_n converges to u almost everywhere in Ω. Taking into account that (u_n) is bounded in $D_{Q,V}$, a straightforward application of Fatou's lemma shows that $u\in D_{Q,V}$ and $\lim_{n\to+\infty}\|u_n-u\|_{Q,V}=0$. Since $\|u\|_{Q,V}\ge V_0^{\frac{1}{2}}\|u\|_2$ for every $u\in D_{Q,V}$, we conclude that $D_{Q,V}$ is continuously embedded in $L^2(\Omega;\mathbb{C})$ and it is therein dense since it contains the space of test functions. □

Next, we introduce the space

$$\widetilde{D}_{Q,V}=\overline{\{u_{|\Omega}\ |\ u\in C_c^\infty(\mathbb{R}^d,\mathbb{C})\}}^{D_{Q,V}}.$$

Remark 3.1 We observe that if $u \in W^{1,2}_{\mathrm{loc}}(\Omega;\mathbb{C})$, and in particular if $u \in D_{Q,V}$, then $\nabla|u|^2 = 2\mathrm{Re}\,(u\nabla\overline{u})$, and $\nabla|u| = \mathrm{Re}\,(\mathrm{sign}\,(u)\nabla u)$. Thus, by (3),

$$(Q\nabla|u|^2, \nabla|u|^2) = 4(Q\mathrm{Re}\,(u\nabla\overline{u}), \mathrm{Re}\,(u\nabla\overline{u})) \le 4\mathrm{Re}\,(Q\nabla u, \nabla u)|u|^2.$$

Remark 3.2 A straightforward modification of [27, Lemmas 4.4, 4.9] (see also [5, Lemma 2.5]) shows that $\mathrm{Re}\,u^+$, $\mathrm{Re}\,u^-$, $|u|$, $|u| \vee 1$ belong to $\widetilde{D}_{Q,V}$ for every $u \in \widetilde{D}_{Q,V}$.

At this point we can introduce a sesquilinear form on $\widetilde{D}_{Q,V}$, adding another hypothesis.

Hypothesis (B) Let $B = (b_1, \dots, b_d)$, $C = (c_1, \dots, c_d) \in L^\infty_{\mathrm{loc}}(\Omega;\mathbb{R}^d)$ and assume that there exists a positive constant $\kappa \in (0, 1/2)$ such that

$$|B(x)\cdot\xi| \le \kappa\Big(\mathrm{Re}\,(Q(x)\xi,\xi)\Big)^{\frac12} V(x)^{\frac12},$$

$$|C(x)\cdot\xi| \le \kappa\Big(\mathrm{Re}\,(Q(x)\xi,\xi)\Big)^{\frac12} V(x)^{\frac12}$$

for every $\xi \in \mathbb{C}^d$ and almost every $x \in \Omega$.

For every $u, w \in \widetilde{D}_{Q,V}$, set

$$\begin{aligned}\mathfrak{a}(u,w) &= \int_\Omega \Bigg(\sum_{i,j=1}^d q_{ij} D_j u \overline{D_i w} + \overline{w}\sum_{i=1}^d b_i D_i u + u\sum_{i=1}^d c_i D_i\overline{w} + Vu\overline{w}\Bigg)dx\\ &= \int_\Omega \Big((Q\nabla u, \nabla w) + \overline{w}B\cdot\nabla u + uC\cdot\nabla\overline{w} + Vu\overline{w}\Big)dx. \qquad (11)\end{aligned}$$

The form $\mathfrak{a}$ is well defined and continuous on $\widetilde{D}_{Q,V}$ since for every $u, w \in \widetilde{D}_{Q,V}$ it holds that:

$$|(Q\nabla u, \nabla w)| \le (1+q_0)\,(\mathrm{Re}\,(Q\nabla u, \nabla u))^{\frac12}\,(\mathrm{Re}\,(Q\nabla w, \nabla w))^{\frac12}, \qquad (12)$$

$$|\overline{w}B\cdot\nabla u| \le \kappa V^{\frac12}\,(\mathrm{Re}\,(Q\nabla u, \nabla u))^{\frac12}\,|w| \le \frac{\varepsilon}{2}\mathrm{Re}\,(Q\nabla u, \nabla u) + \frac{\kappa^2}{2\varepsilon}V|w|^2, \qquad (13)$$

$$|uC\cdot\nabla\overline{w}| \le \kappa V^{\frac12}\,(\mathrm{Re}\,(Q\nabla w, \nabla w))^{\frac12}\,|u| \le \frac{\varepsilon}{2}\mathrm{Re}\,(Q\nabla w, \nabla w) + \frac{\kappa^2}{2\varepsilon}V|u|^2, \qquad (14)$$

for every $\varepsilon > 0$, where we used (6) and (7) to deduce (12), whereas (13) and (14) follow straightforwardly from Hypothesis (B).

Moreover, using (8) and the first inequalities in (13) and (14), together with Young inequality, we get that

$$\operatorname{Re} \mathfrak{a}(u,u) \geq (1-\varepsilon)\int_\Omega \operatorname{Re}(Q\nabla u, \nabla u)dx + \left(1-\frac{\kappa^2}{\varepsilon}\right)\int_\Omega V|u|^2 dx \tag{15}$$

for every $u \in C_c^\infty(\mathbb{R}^d;\mathbb{C})$. Taking $\varepsilon = \kappa$, we deduce that $\mathfrak{a}$ is coercive on $\widetilde{D}_{Q,V}$, and therefore, by Theorem 2.1, the opposite of the operator $(A, D(A))$ associated with $\mathfrak{a}$ generates a strongly continuous analytic semigroup. Since $C_c^\infty(\Omega;\mathbb{C}) \subseteq \widetilde{D}_{Q,V}$, if Ω is smooth enough, then we get that for every u smooth,

$$Au = -\sum_{i=1}^d D_i(q_{ij}D_j u) + \sum_{i=1}^d B_i D_i u - \sum_{i=1}^d D_i(C_i u) + Vu$$

distributionally. Moreover, $\dfrac{\partial u}{\partial \eta} + (C,\nu)u = 0$ on $\partial\Omega$, where $\eta = Q^t\nu$ and $\nu(x)$ is the exterior unit normal vector to $\partial\Omega$ at $x \in \partial\Omega$.

Since $\operatorname{Re}(Q\xi,\xi) = \operatorname{Re}(Q^t\xi,\xi)$ and $|\operatorname{Im}(Q\xi,\xi)| = |\operatorname{Im}(Q^t\xi,\xi)|$ for every $\xi \in \mathbb{C}^d$, the previous considerations hold also for the adjoint form

$$\mathfrak{a}^*(u,w) = \overline{\mathfrak{a}(w,u)} = \int_\Omega \left((Q^t\cdot\nabla u, \nabla w) + \overline{w}C\cdot\nabla u + uB\cdot\nabla\overline{w} + Vu\overline{w}\right)dx$$

and the operator associated to $\mathfrak{a}^*$ is the adjoint operator of A, which distributionally is given by

$$A^*u = -\sum_{i=1}^d D_i(q_{ji}D_j u) + \sum_{i=1}^d C_i D_i u - \sum_{i=1}^d D_i(B_i u) + Vu.$$

The semigroup $(T(t))$ and its dual $(T^*(t))$ are positive. Indeed, if $v \in C_c^\infty(\mathbb{R}^d,\mathbb{C})$ and u is its restriction to Ω, since $\operatorname{Re} u^+ \in \widetilde{D}_{Q,V}$ by Remark 3.2 and

$$\nabla(\operatorname{Re} u^+) = (\nabla \operatorname{Re} u)\chi_{\{\operatorname{Re} u>0\}}, \qquad \nabla(\operatorname{Re} u^-) = -(\nabla \operatorname{Re} u)\chi_{\{\operatorname{Re} u<0\}},$$

it follows that $\mathfrak{a}(\operatorname{Re} u^+, \operatorname{Re} u^-) = 0$. Moreover, clearly $\mathfrak{a}(\operatorname{Re} u, \operatorname{Im} u) \in \mathbb{R}$. Hence, $(T(t))$ is positive by Proposition 2.1(b).

Remark 3.3 We point out that condition (9) can be weakened. Indeed, we can actually assume that the essential infimum of V is real. Also Hypothesis (B) can be weakened by assuming that

$$|B(x)\cdot\xi| \le \kappa\Big(\mathrm{Re}\,(Q(x)\xi,\xi)\Big)^{\frac{1}{2}}(V(x)+\gamma)^{\frac{1}{2}}$$

$$|C(x)\cdot\xi| \le \kappa\Big(\mathrm{Re}\,(Q(x)\xi,\xi)\Big)^{\frac{1}{2}}(V(x)+\gamma)^{\frac{1}{2}}$$

for every $\xi\in\mathbb{C}^d$, almost every $x\in\Omega$ and some constants $\kappa\in(0,1/2)$ and $\gamma>0$. For ease of presentation, we prove our results under condition (9) and Hypotheses (B).

4 Gaussian Estimates

Along this section, we will need stronger conditions on the coefficients of the form $\mathfrak{a}$ and on Ω, as specified below.

Hypothesis (C)

(a) There exists a positive constant ν_0 such that

$$\mathrm{Re}\,(Q(x)\xi,\xi)\ge\nu_0|\xi|^2 \qquad \xi\in\mathbb{C}^d,\ \text{a.e. } x\in\Omega.$$

(b) The set Ω enjoys the extension property.

Remark 4.1 Thanks to Hypothesis (C), it follows that $\widetilde{D}_{Q,V}\subseteq W^{1,2}(\Omega)$, thus $\widetilde{D}_{Q,V}$ embeds continuously in $L^r(\Omega)$ with $r=2^*=\frac{2d}{d-2}$ if $d\ge 3$ and r is any real number strictly greater than 2 if $d\le 2$.

Under Hypotheses (A), (B) and (C), we are going to prove that the semigroup associated with the form $\mathfrak{a}$ is defined by a kernel satisfying generalized Gaussian estimates. Due to the presence of a diffusion matrix and a potential term which can be unbounded, the Gaussian estimate will be expressed in terms of a distance depending on them, namely

$$d_{Q,V}(x,y)=\sup\{|\psi(x)-\psi(y)|\mid \psi\in C_b(\Omega)\cap C^1(\Omega),\ \|(Q\nabla\psi,\nabla\psi)\|_{L^\infty(\Omega)}\le h\} \tag{16}$$

for every $x,y\in\Omega$, where h is a measurable function, locally bounded, with positive essential infimum over Ω, for which there exist $\alpha, C>0$ such that

$$h(x)\le\gamma V(x)+C\gamma^{-\alpha} \tag{17}$$

for almost every $x \in \Omega$ and for every $\gamma > 0$. This assumption is satisfied for example if $h = V^a$ with some $a \in]0, 1[$, since in this case

$$V^a \le \gamma V + \left(\frac{\gamma}{a}\right)^{\frac{a}{a-1}} (1-a).$$

The distance $d_{Q,V}$ is equivalent to the Euclidean metric if there exist two positive constants μ_0 and μ_1 such that $\mu_0 h(x)|\xi|^2 \le \mathrm{Re}\,(Q(x)\xi, \xi) \le \mu_1 h(x)|\xi|^2$ for every $\xi \in \mathbb{C}^d$ and almost every $x \in \Omega$ (see e.g., [10, Theorem 7]). In our case, in general it holds true only that for every compact set $K \subset \Omega$, there exists a positive constant C_K such that

$$d_{Q,V}(x, y) \le C_K|x - y|, \qquad x, y \in K, \tag{18}$$

and the previous estimate holds true in the whole Ω if $\mu_0 h(x)|\xi|^2 \le \mathrm{Re}\,(Q(x)\xi, \xi)$ for every $\xi \in \mathbb{R}^d$ and almost every $x \in \Omega$.

Theorem 4.1 *Assume that Hypotheses (A), (B), (C) hold true and let A be the operator associated with the form* $\mathfrak{a}$ *introduced in* (11)*. Then the semigroup* $(T(t))$ *generated by* $-A$ *is given by a kernel k satisfying the estimate*

$$\begin{aligned} k(t, x, y) \le & H_1[(1-\kappa)\nu_0]^{-\frac{d}{2}} e^{-\frac{1-\kappa}{4} V_0 t} t^{-\frac{d}{2}} e^{-H_2 (d_{Q,V}(x,y))^{\frac{2\alpha+2}{2\alpha+1}} t^{-\frac{1}{2\alpha+1}}} \\ & \times \left(1 + \frac{1-\kappa}{8} V_0 t + H_3 \frac{\left(d_{Q,V}(x, y)\right)^{\frac{2\alpha+2}{2\alpha+1}}}{t^{\frac{1}{\alpha+1}}}\right)^{\frac{d}{2}} \end{aligned} \tag{19}$$

for every $t > 0$ *and almost every* $x, y \in \Omega$*, where* H_i $(i = 1, 2, 3)$ *are strictly positive constants depending on* d, α, κ, q_0, C *and* Ω*, explicitly computed in the proof.*

Remark 4.2 In [3], kernel estimates have been proved for operators defined in the whole $\mathbb{R}^d$, with a similar technique but using a distance which depends only on Q, i.e., in that paper, h is taken identically equal to one. The latter distance turns out to be a bounded distance over $\mathbb{R}^d$ if Q grows too fast at infinity. Consider for instance the one dimensional case where $q(x) = (1 + |x|^2)^2$ for every $x \in \mathbb{R}$. Then,

$$d_Q(x, y) = \sup\{|\psi(x) - \psi(y)| : \psi \in C_b^1(\mathbb{R}),\ (1 + |x|^2)|\psi'(x)| \le 1,\ x \in \mathbb{R}\}.$$

Clearly,

$$|\psi(x) - \psi(y)| \le \left|\int_x^y \frac{1}{1+s^2} ds\right| = |\arctan(x) - \arctan(y)| \le \pi, \qquad x, y \in \mathbb{R}.$$

Hence, $d_Q(x, y) \le \pi$ for every $x, y \in \mathbb{R}$ and it follows that, in this situation, any estimate like (19) does not give the expected information on the kernel.

Adding the function h in the definition of the distance $d_{Q,V}$ allows us to prevent this situation and yields spatial exponential decaying kernels also when Q grows quickly at infinity provided its growth is suitably balanced by the growth of the potential term V.

Remark 4.3 The proof is inspired by the so-called Davies trick, where the assertion is usually obtained by studying the L^∞-contractivity properties of the semigroup $(T^{\sigma,\psi}(t))$ associated with the sesquilinear form $\mathfrak{a}(e^{-\sigma\psi}\cdot, e^{\sigma\psi}\cdot)$, where $\sigma \in \mathbb{R}$ and ψ belongs to a suitable class of bounded functions with bounded derivative (see e.g., [5, 6]).

In our situation, due to the presence of the (possibly) unbounded diffusion coefficients, $(T^{\sigma,\psi}(t))$ is not L^∞-contractive in general. Hence, we have to follow a different path.

We prove first that each operator $T^{\sigma,\psi}(t)$ is a bounded operator on $L^p(\Omega;\mathbb{C})$ for every $p \in [2,+\infty[$, with an explicit estimate of the norm. Here, we use the idea of the proof of [27, Theorem 4.28]. Next, we gain in summability proving that $T^{\sigma,\psi}(t) : L^2(\Omega;\mathbb{C}) \to L^q(\Omega;\mathbb{C})$ for some $q > 2$, by using Sobolev's inequality and taking into account that $\widetilde{D}_{Q,V} \subseteq W^{1,2}(\Omega;\mathbb{C})$.

At this point, combining a bootstrap argument with interpolation and duality arguments, we conclude that

$$\begin{aligned}\|T^{\sigma,\psi}(t)\|_{1\to\infty} \leq& H_d[(1-\kappa)\nu_0]^{-\frac{d}{2}} t^{-\frac{d}{2}} e^{|\sigma|^{2+2\alpha}K_1 t} e^{-\frac{1-\kappa}{4}V_0 t}\\ &\times\left(1+\frac{1-\kappa}{8}V_0 t + |\sigma|^{2+2\alpha}\frac{c_{d,\alpha,q_0,\kappa}}{2}t\right)^{\frac{d}{2}}\end{aligned}$$

for every $t > 0$, $\sigma \in \mathbb{R}$ and positive constants H_d and $c_{d,\alpha,q_0,\kappa}$, explicitly computed in the proof.

The existence of the kernel will follow by considering $\sigma = 0$, while the Gaussian estimate will be a consequence of a minimum argument.

Proof of Theorem 4.1 Set

$$W = \{\psi \in C_b(\Omega) \cap C^1(\Omega) \mid (Q\nabla\psi, \nabla\psi) \leq h \text{ a.e. in } \Omega\},$$

where h is a measurable positive function for which there exist $\alpha, C > 0$ such that

$$h(x) \leq \gamma V(x) + C\gamma^{-\alpha}$$

for almost every $x \in \Omega$ and for every $\gamma > 0$.

Clearly, $e^{\sigma\psi}\widetilde{D}_{Q,V} = \widetilde{D}_{Q,V}$ for every $\psi \in W$ and $\sigma \in \mathbb{R}$. Indeed, for every $u \in \widetilde{D}_{Q,V}$, $h^{\frac{1}{2}}u$ belongs to $L^2(\Omega)$ and the function $e^{\sigma\psi}u$ belongs to $\widetilde{D}_{Q,V}$ as it follows from the following formula

$$\begin{aligned}
&\big(Q\nabla(e^{\sigma\psi}u), \nabla(e^{\sigma\psi}u)\big)\\
&= e^{2\sigma\psi}\Big[(Q\nabla u, \nabla u) + \sigma^2|u|^2(Q\nabla\psi, \nabla\psi) + \sigma u(Q\nabla\psi, \nabla u) + \sigma\overline{u}\,(Q\nabla u, \nabla\psi)\Big]\\
&\le e^{2\sigma\psi}\Big[(Q\nabla u, \nabla u) + \sigma^2|u|^2 h + \sigma u(Q\nabla\psi, \nabla u) + \sigma\overline{u}\,(Q\nabla u, \nabla\psi)\Big],
\end{aligned}$$

which, taking also (5) into account, gives

$$\begin{aligned}
&\int_\Omega \Big(\mathrm{Re}\,\big(Q\nabla(ue^{\sigma\psi}), \nabla(ue^{\sigma\psi})\big) + Ve^{2\sigma\psi}|u|^2\Big)\,dx\\
&\le e^{2|\sigma|\|\psi\|_\infty}\bigg(\int_\Omega \mathrm{Re}\,(Q\nabla u, \nabla u)dx + \sigma^2\int_\Omega h|u|^2dx\\
&\quad +2(q_0+1)|\sigma|\int_\Omega |u|(Q\nabla\psi, \nabla\psi)^{\frac{1}{2}}\,(\mathrm{Re}\,(Q\nabla u, \nabla\, u))^{\frac{1}{2}}\,dx + \int_\Omega V|u|^2dx\bigg)\\
&\le e^{2|\sigma|\|\psi\|_\infty}\bigg(\int_\Omega \mathrm{Re}\,(Q\nabla u, \nabla u)dx + \sigma^2\int_\Omega h|u|^2dx\\
&\quad +2(q_0+1)|\sigma|\int_\Omega |u|h^{\frac{1}{2}}\,(\mathrm{Re}\,(Q\nabla u, \nabla\, u))^{\frac{1}{2}}\,dx + \int_\Omega V|u|^2dx\bigg)\\
&\le e^{2|\sigma|\|\psi\|_\infty}\bigg(((q_0+1)|\sigma|+1)\int_\Omega \mathrm{Re}\,(Q\nabla u, \nabla u)dx\\
&\qquad\qquad +((q_0+1)|\sigma|+\sigma^2)\int_\Omega h|u|^2dx + \int_\Omega V|u|^2dx\bigg).
\end{aligned}$$

As a consequence, fixing $\sigma \in \mathbb{R}$ and $\psi \in W$, the sesquilinear form, defined for every $u, w \in \widetilde{D}_{Q,V}$ by

$$\begin{aligned}
\mathfrak{a}_{\sigma,\psi}(u,w) :=& \mathfrak{a}(e^{-\sigma\psi}u, e^{\sigma\psi}w)\\
=& \mathfrak{a}(u,w) - \sigma\int_\Omega (Q\nabla\psi, \nabla w)u\,dx + \sigma\int_\Omega (Q\nabla u, \nabla\psi)\overline{w}\,dx\\
&- \sigma^2\int_\Omega (Q\nabla\psi, \nabla\psi)u\overline{w}\,dx + \sigma\int_\Omega u\overline{w}(C-B)\cdot\nabla\psi\,dx\\
=& \int_\Omega \Big[(Q\nabla u\cdot\nabla w) + \overline{w}(B+\sigma Q^t\nabla\psi)\cdot\nabla u + u(C-\sigma Q\nabla\psi)\cdot\nabla\overline{w}\\
&\qquad + \Big(V - \sigma^2(Q\nabla\psi, \nabla\psi) + \sigma(C-B)\nabla\psi\Big)u\overline{w}\Big]dx,
\end{aligned}$$

is well defined and continuous on $\widetilde{D}_{Q,V}$.

Moreover, taking (5), (15) and Remark 3.1 into account, we can show that, for every $\varepsilon, \varepsilon_i, i = 1, 2, 3$ and every $u \in C_c^\infty(\mathbb{R}^d, \mathbb{C})$, it holds that

$$
\begin{aligned}
\operatorname{Re} \mathfrak{a}_{\sigma,\psi}(u,u) =& \operatorname{Re} \mathfrak{a}(u,u) - \sigma \int_\Omega \operatorname{Re}[(Q\nabla\psi, \nabla u)u]dx \\
& + \sigma \int_\Omega \operatorname{Re}[(Q\nabla u, \nabla\psi)\overline{u}]dx - \sigma^2 \int_\Omega (Q\nabla\psi, \nabla\psi)|u|^2 dx \\
& + \sigma \int_\Omega (C - B)\cdot\nabla\psi |u|^2 dx \\
=& \operatorname{Re} \mathfrak{a}(u,u) - \frac{\sigma}{2}\int_\Omega (Q\nabla\psi, \nabla|u|^2)dx + \frac{\sigma}{2}\int_\Omega (Q\nabla|u|^2, \nabla\psi)dx \\
& - \sigma^2 \int_\Omega (Q\nabla\psi, \nabla\psi)|u|^2 dx + \sigma\int_\Omega (C - B)\cdot\nabla\psi|u|^2 dx \\
=& \operatorname{Re} \mathfrak{a}(u,u) + \frac{\sigma}{2}\int_\Omega ((Q - Q^t)\nabla|u|^2, \nabla\psi)dx \\
& - \sigma^2 \int_\Omega (Q\nabla\psi, \nabla\psi)|u|^2 dx + \sigma\int_\Omega (C - B)\cdot\nabla\psi|u|^2 dx \\
\geq& \operatorname{Re} \mathfrak{a}(u,u) - |\sigma| q_0 \int_\Omega (Q\nabla|u|^2, \nabla|u|^2)^{\frac{1}{2}}(Q\nabla\psi, \nabla\psi)^{\frac{1}{2}} dx \\
& - \sigma^2 \int_\Omega h|u|^2 dx - |\sigma|\int_\Omega |(C - B)\cdot\nabla\psi||u|^2 dx \\
\geq& \operatorname{Re} \mathfrak{a}(u,u) - 2|\sigma| q_0 \int_\Omega (\operatorname{Re}(Q\nabla u, \nabla u))^{\frac{1}{2}} h^{\frac{1}{2}} |u| dx \\
& - \sigma^2 \int_\Omega h|u|^2 dx - 2|\sigma|\kappa \int_\Omega |V|^{\frac{1}{2}} h^{\frac{1}{2}} |u|^2 dx \\
\geq& \operatorname{Re} \mathfrak{a}(u,u) - 2|\sigma| q_0 \int_\Omega (\operatorname{Re}(Q\nabla u, \nabla u))^{\frac{1}{2}} h^{\frac{1}{2}} |u| dx \\
& - \sigma^2 \int_\Omega h|u|^2 dx - \kappa^2\varepsilon_1 \int_\Omega V|u|^2 dx - \frac{\sigma^2}{\varepsilon_1}\int_\Omega h|u|^2 dx \\
\geq& (1 - \varepsilon - \varepsilon_2)\int_\Omega \operatorname{Re}(Q\nabla u, \nabla u)dx \\
& + \left(1 - \frac{\kappa^2}{\varepsilon} - \varepsilon_1\kappa^2\right)\int_\Omega V|u|^2 dx - \sigma^2\left(\frac{q_0^2}{\varepsilon_2} + 1 + \frac{1}{\varepsilon_1}\right)\int_\Omega h|u|^2 dx \\
\geq& (1 - \varepsilon - \varepsilon_2)\int_\Omega \operatorname{Re}(Q\nabla u, \nabla u)dx \\
& + \left(1 - \frac{\kappa^2}{\varepsilon} - \varepsilon_1\kappa^2\right)\int_\Omega V|u|^2 dx - \varepsilon_3\sigma^2\left(\frac{q_0^2}{\varepsilon_2} + 1 + \frac{1}{\varepsilon_1}\right)\int_\Omega V|u|^2 dx \\
& - C\varepsilon_3^{-\alpha}\sigma^2\left(\frac{q_0^2}{\varepsilon_2} + 1 + \frac{1}{\varepsilon_1}\right)\int_\Omega |u|^2 dx.
\end{aligned}
$$

Choosing $\varepsilon = \kappa$, $\varepsilon_1 = \frac{1-\kappa}{2\kappa^2}$, $\varepsilon_2 = \frac{1-\kappa}{2}$, and, if $\sigma \neq 0$,

$$\varepsilon_3 = \sigma^{-2}\left(\frac{q_0^2}{\varepsilon_2} + 1 + \frac{1}{\varepsilon_1}\right)^{-1}\frac{1-\kappa}{4},$$

we get that

$$\operatorname{Re}\mathfrak{a}_{\sigma,\psi}(u,u) \geq \frac{1-\kappa}{4}\|u\|_{Q,V}^2 - K_1|\sigma|^{2+2\alpha}\int_\Omega |u|^2 dx, \tag{20}$$

where

$$K_1 = C\left(1 + \frac{2(\kappa^2 + q_0^2)}{1-\kappa}\right)^{\alpha+1}\left(\frac{1-\kappa}{4}\right)^{-\alpha}. \tag{21}$$

Moreover, choosing $\varepsilon = \kappa$, $\varepsilon_1 = \frac{1-\kappa}{2\kappa^2}$, $\varepsilon_2 = \frac{1-\kappa}{2}$, and, if $\sigma \neq 0$,

$$\varepsilon_3 = \sigma^{-2}\left(\frac{q_0^2}{\varepsilon_2} + 1 + \frac{1}{\varepsilon_1}\right)^{-1} \cdot \frac{1-\kappa}{2},$$

we get that

$$\operatorname{Re}\mathfrak{a}_{\sigma,\psi}(u,u) \geq \frac{1-\kappa}{2}\int_\Omega \operatorname{Re}(Q\nabla u, \nabla u)dx - 2^{-\alpha}K_1|\sigma|^{2+2\alpha}\int_\Omega |u|^2 dx. \tag{22}$$

We can thus apply Theorem 2.1 to conclude that the opposite of the operator associated with the form $\mathfrak{a}_{\sigma,\psi}$ generates a C_0-semigroup $(T^{\sigma,\psi}(t))$ on $L^2(\Omega;\mathbb{C})$. Further, (22) and Theorem 2.1 show that

$$\|T^{\sigma,\psi}(t)\|_{2\to 2} \leq e^{K_1 2^{-\alpha}|\sigma|^{2+2\alpha}t}, \qquad t > 0.$$

A straightforward calculation implies that $T^{\sigma,\psi}(t) = e^{\sigma\psi}T(t)(e^{-\sigma\psi}\cdot)$ for every $t \geq 0$, where $(T(t))$ is the strongly continuous, analytic semigroup associated with the form $\mathfrak{a}$ on $L^2(\Omega;\mathbb{C})$.

Our aim is to evaluate the norm $\|T^{\sigma,\psi}(t)\|_{1\to\infty}$ and this will be done through several steps.

Claim 1 For every $p \in [2,+\infty[$ and $t \geq 0$, the operator $T^{\sigma,\psi}(t)$ is bounded on $L^p(\Omega;\mathbb{C})$ and

$$\|T^{\sigma,\psi}(t)\|_{p\to p} \leq e^{-\beta_{p,\sigma}t}, \qquad t \geq 0, \tag{23}$$

where

$$\beta_{p,\sigma} = -C|\sigma|^{2+2\alpha} \times \left[(1+q_0)^{2+2\alpha}\left(\frac{p}{1-2\kappa}\right)^{2\alpha+1} + \left(1+\frac{4\kappa^2+2}{1-2\kappa}\right)^{\alpha+1}\left(\frac{4}{1-2\kappa}\right)^{\alpha}\right]. \tag{24}$$

For every $z \in \mathbb{C}$, with $0 \le \operatorname{Re} z \le 1$, consider the forms $\mathfrak{b}_{\sigma z,\psi}$ and $\mathfrak{c}_{\sigma,\psi}$, defined on $\widetilde{D}_{Q,V} \times \widetilde{D}_{Q,V}$ as follows:

$$\mathfrak{b}_{\sigma z,\psi}(u,w) = \frac{1}{2}\int_\Omega (Q\nabla u, \nabla w)dx + z\int_\Omega (C - \sigma Q\nabla\psi)\cdot\nabla(u\cdot\overline{w})dx + \frac{1}{2}\int_\Omega Vu\overline{w}dx$$

and

$$\begin{aligned}\mathfrak{c}_{\sigma,\psi}(u,w) =& \frac{1}{2}\int_\Omega (Q\nabla u, \nabla w)dx + \int_\Omega \overline{w}(B - C + \sigma(Q^t + Q)\nabla\psi)\cdot\nabla u dx \\ &+ \int_\Omega\left(\frac{1}{2}V - \sigma^2(Q\nabla\psi,\nabla\psi) + \sigma(C-B)\cdot\nabla\psi\right)u\overline{w}dx.\end{aligned}$$

Clearly,

$$\mathfrak{a}_{\sigma,\psi} = \mathfrak{b}_{\sigma\cdot 1,\psi} + \mathfrak{c}_{\sigma,\psi}. \tag{25}$$

Taking advantage of (12)–(14), one can see that $\mathfrak{b}_{\sigma z,\psi}$ is continuous with respect to the norm $\|\cdot\|_{Q,V}$. Moreover, taking (6), Remark 3.1 and Hypothesis (B) into account and recalling that $\operatorname{Re} z \le 1$, we can estimate

$$\begin{aligned}&\operatorname{Re}\mathfrak{b}_{\sigma z,\psi}(u,u)\\ \ge& \frac{1}{2}\int_\Omega \operatorname{Re}(Q\nabla u,\nabla u)dx - \kappa\operatorname{Re} z\int_\Omega (Q\nabla|u|^2,\nabla|u|^2)^{\frac{1}{2}}V^{\frac{1}{2}}dx\\ &-|\sigma|(1+q_0)\operatorname{Re} z\int_\Omega (Q\nabla|u|^2,\nabla|u|^2)^{\frac{1}{2}}(Q\nabla\psi,\nabla\psi)^{\frac{1}{2}}dx + \frac{1}{2}\int_\Omega V|u|^2dx\\ \ge&\left(\frac{1}{2}-\varepsilon_1-\varepsilon_2\right)\int_\Omega \operatorname{Re}(Q\nabla u,\nabla u)dx + \left(\frac{1}{2}-\frac{\kappa^2}{\varepsilon_1}\right)\int_\Omega V|u|^2dx\\ &-\frac{\sigma^2(1+q_0)^2}{\varepsilon_2}\int_\Omega h|u|^2dx\\ \ge&\left(\frac{1}{2}-\varepsilon_1-\varepsilon_2\right)\int_\Omega \operatorname{Re}(Q\nabla u,\nabla u)dx + \left(\frac{1}{2}-\frac{\kappa^2}{\varepsilon_1}-\frac{\sigma^2(1+q_0)^2}{\varepsilon_2}\varepsilon_3\right)\int_\Omega V|u|^2dx\\ &-C\varepsilon_3^{-\alpha}\frac{\sigma^2(1+q_0)^2}{\varepsilon_2}\int_\Omega |u|^2dx\end{aligned} \tag{26}$$

for every $\varepsilon_i > 0$, $i = 1, 2, 3$, and $u \in C_c^\infty(\mathbb{R}^d; \mathbb{C})$. Choosing $\varepsilon_1 = \kappa$, $\varepsilon_2 = \frac{1-2\kappa}{4}$ and, if $\sigma \neq 0$,

$$\varepsilon_3 = \left(\frac{1-2\kappa}{4\sigma(1+q_0)}\right)^2,$$

we deduce that

$$\operatorname{Re} \mathfrak{b}_{\sigma z,\psi}(u,u) \geq \frac{1-2\kappa}{4}\|u\|^2_{Q,V} - C\left(\frac{4}{1-2\kappa}\right)^{2\alpha+1}(1+q_0)^{2+2\alpha}|\sigma|^{2+2\alpha}\int_\Omega |u|^2 dx,$$

where we observe that $1 - 2\kappa > 0$. Consequently, for any $z \in \mathbb{C}$ with $0 \leq \operatorname{Re} z \leq 1$, the opposite of the operator associated with the form $b_{\sigma z,\psi}$ generates a C_0-semigroup $(U^{\sigma z,\psi}(t))$ in $L^2(\Omega; \mathbb{C})$. Further, if we take $\varepsilon_1 = \kappa$, $\varepsilon_2 = \frac{1}{2} - \kappa$ and, if $\sigma \neq 0$,

$$\varepsilon_3 = \left(\frac{1-2\kappa}{2\sigma(1+q_0)}\right)^2$$

in (26), then we get

$$\begin{aligned}\operatorname{Re} \mathfrak{b}_{\sigma z,\psi}(u,u) &\geq -C\left(\frac{2}{1-2\kappa}\right)^{2\alpha+1}(1+q_0)^{2+2\alpha}|\sigma|^{2+2\alpha}\int_\Omega |u|^2 dx \\ &=: -\tilde{K}_1|\sigma|^{2+2\alpha}\|u\|_2^2\end{aligned}$$

and, therefore,

$$\|U^{\sigma z,\psi}(t)\|_{2\to 2} \leq e^{\tilde{K}_1|\sigma|^{2+2\alpha}t}, \qquad t \geq 0.$$

Let us prove that the semigroups $(U^{i\sigma s,\psi}(t))$ are L^∞-contractive for every $s \in \mathbb{R}$. By [27, Proposition 4.11], it follows that

$$\nabla((|u| \wedge 1)\operatorname{sign}(u)) = \frac{i\operatorname{Im}(\operatorname{sign}(\overline{u})\nabla u)}{|u|}\operatorname{sign}(u)\chi_{\{|u|>1\}} + \nabla u\, \chi_{\{|u|\leq 1\}},$$

$$\nabla((|u|-1)^+\operatorname{sign}(u)) = \left(\nabla u - \frac{i\operatorname{Im}(\operatorname{sign}(\overline{u})\nabla u)}{|u|}\operatorname{sign}(u)\right)\chi_{\{|u|>1\}},$$

$$\nabla(|u|-1)^+ = (\nabla|u|)\chi_{|u|>1} = \operatorname{Re}(\nabla\overline{u}\operatorname{sign}(u))\chi_{\{|u|>1\}}$$

for every $u \in C_c^\infty(\mathbb{R}^d; \mathbb{C})$, so that, in particular, $(|u| \wedge 1)\mathrm{sign}(u) \in \widetilde{D}_{Q,V}$. From the formulas above, it follows easily that

$$\begin{aligned}
&\mathrm{Re}\left[\nabla((|u| \wedge 1)\mathrm{sign}(u)) \cdot (|u|-1)^+\mathrm{sign}\,(\overline{u})\right] = 0,\\
&\nabla[(|u| \wedge 1)\mathrm{sign}\,(u)(|u|-1)^+\mathrm{sign}\,(\overline{u})] = \nabla((|u|-1)^+),\\
&\mathrm{Re}\,[(|u| \wedge 1)\mathrm{sign}(u) \cdot \nabla(|u|-1)^+\mathrm{sign}\,(\overline{u}))] = \nabla((|u|-1)^+).
\end{aligned}$$

Therefore, for any $u \in C_c^\infty(\mathbb{R}^d; \mathbb{C})$, it holds that

$$\begin{aligned}
&\mathrm{Re}\, b_{i\sigma s,\psi}\big((|u| \wedge 1)\mathrm{sign}(u), (|u|-1)^+\mathrm{sign}(u)\big)\\
=&\frac{1}{2}\mathrm{Re}\int_{|u|>1}\left(Q\frac{i\mathrm{Im}(\mathrm{sign}(\overline{u})\nabla u)}{|u|}\mathrm{sign}(u)\right)\cdot\left(\frac{i\mathrm{Im}(\mathrm{sign}(\overline{u})\nabla u)}{|u|}\mathrm{sign}(\overline{u})\right)dx\\
&+\frac{1}{2}\mathrm{Re}\int_{|u|>1} Q\frac{i\mathrm{Im}(\mathrm{sign}(\overline{u})\nabla u)}{|u|}\cdot\overline{\mathrm{sign}(\overline{u})\nabla u}\,dx\\
&+\frac{1}{2}\int_{|u|>1} V(|u|-1)dx\\
=&\frac{1}{2}\int_{|u|>1}\mathrm{Re}\left(Q\frac{\mathrm{Im}(\mathrm{sign}(\overline{u})\nabla u)}{|u|}\mathrm{sign}(u)\cdot\left(\frac{\mathrm{Im}\,(\mathrm{sign}(\overline{u})\nabla u)}{|u|}\mathrm{sign}(\overline{u})\right)\right)(|u|-1)dx\\
&+\frac{1}{2}\int_{|u|>1} V(|u|-1)dx \geq 0. \qquad (27)
\end{aligned}$$

This proves the L^∞-contractivity of the semigroup $(U^{i\sigma s,\psi}(t))$.

For every $t \geq 0$, the operators $U^{\sigma z,\psi}(t)$ depend analytically on z, see, e.g., [16, Theorems VII-4.2, IX-2.6]. Hence, Stein's interpolation theorem (see e.g., [29, Chapter 1, Section 18.7, Theorem 1]) applied with $p_0 = q_0 = +\infty$, $p_1 = q_1 = 2$), permits to interpolate the L^2 and L^∞-estimates, obtaining that

$$\|U^{\frac{2}{p}\sigma,\psi}(t)\|_{p\to p} \leq e^{\frac{2\tilde{K}_1}{p}|\sigma|^{2+2\alpha}t}, \qquad t > 0,$$

for every $p \geq 2$. Observing that $\mathfrak{b}_{\sigma z,\psi} = \mathfrak{b}_{\frac{2}{p}\sigma_1 z,\psi}$, where $\sigma_1 = \frac{p}{2}\sigma$, and, therefore, $U^{\sigma,\psi}(t) = U^{\frac{2}{p}\sigma_1,\psi}(t)$ for every $t > 0$, we get

$$\|U^{\sigma,\psi}(t)\|_{p\to p} \leq e^{\tilde{K}_1(\frac{p}{2})^{2\alpha+1}|\sigma|^{2+2\alpha}t}, \qquad t > 0. \qquad (28)$$

We turn now the attention to the form $\mathfrak{c}_{\sigma,\psi} : \widetilde{D}_{Q,V} \times \widetilde{D}_{Q,V} \to \mathbb{C}$. It is continuous with respect to the norm $\|\cdot\|_{Q,V}$, thanks to (12)–(14) and, for every $u \in C_c^\infty(\mathbb{R}^d;\mathbb{C})$ and every ε_i, $i = 1, 2, 3, 4$, it holds that

$$\begin{aligned}
\operatorname{Re}\mathfrak{c}_{\sigma,\psi}(u,u) \geq & \frac{1}{2}\int_\Omega \operatorname{Re}(Q\nabla u, \nabla u)dx - 2\kappa\int_\Omega (\operatorname{Re}(Q\nabla u,\nabla u))^{\frac{1}{2}} V^{\frac{1}{2}}|u|dx \\
& - 2|\sigma|\int_\Omega (Q\nabla\psi,\nabla\psi)^{\frac{1}{2}}(\operatorname{Re}(Q\nabla u,\nabla u))^{\frac{1}{2}}|u|dx \\
& - 2|\sigma|\kappa\int_\Omega V^{\frac{1}{2}}(Q\nabla\psi,\nabla\psi)^{\frac{1}{2}}|u|^2dx \\
& - \sigma^2\int_\Omega (Q\nabla\psi,\nabla\psi)|u|^2dx + \frac{1}{2}\int_\Omega V|u|^2dx \\
\geq & \left(\frac{1}{2} - \varepsilon_1 - \varepsilon_2\right)\int_\Omega \operatorname{Re}(Q\nabla u,\nabla u)dx \\
& + \left(\frac{1}{2} - \frac{\kappa^2}{\varepsilon_1} - \kappa^2\varepsilon_3\right)\int_\Omega V|u|^2dx - \sigma^2\left(1 + \frac{1}{\varepsilon_2} + \frac{1}{\varepsilon_3}\right)\int_\Omega h|u|^2dx \\
\geq & \left(\frac{1}{2} - \varepsilon_1 - \varepsilon_2\right)\int_\Omega \operatorname{Re}(Q\nabla u,\nabla u)dx \\
& + \left[\frac{1}{2} - \frac{\kappa^2}{\varepsilon_1} - \kappa^2\varepsilon_3 - \varepsilon_4\sigma^2\left(1 + \frac{1}{\varepsilon_2} + \frac{1}{\varepsilon_3}\right)\right]\int_\Omega V|u|^2dx \\
& - C\varepsilon_4^{-\alpha}\sigma^2\left(1 + \frac{1}{\varepsilon_2} + \frac{1}{\varepsilon_3}\right)\int_\Omega |u|^2dx,
\end{aligned}$$

where we took advantage of (4).

Choosing $\varepsilon_1 = \kappa$, $\varepsilon_2 = \frac{1-2\kappa}{4}$, $\varepsilon_3 = \frac{1-2\kappa}{4\kappa^2}$ and, if $\sigma \neq 0$,

$$\varepsilon_4 = \sigma^{-2}\left(\frac{1}{\varepsilon_2} + 1 + \frac{1}{\varepsilon_3}\right)^{-1}\frac{1-2\kappa}{8} = \sigma^{-2}\left(1 + \frac{4(\kappa^2+1)}{1-2\kappa}\right)^{-1}\frac{1-2\kappa}{8},$$

we can conclude that

$$\mathfrak{c}_{\sigma,\psi}(u,u) \geq \frac{1-2\kappa}{8}\|u\|_{Q,V}^2 - C|\sigma|^{2+2\alpha}\left(1 + \frac{4(\kappa^2+1)}{1-2\kappa}\right)^{\alpha+1}\left(\frac{8}{1-2\kappa}\right)^\alpha\int_\Omega |u|^2dx.$$

On the other hand, considering $\varepsilon_1 = \kappa$ $\varepsilon_2 = \frac{1-2\kappa}{2}$, $\varepsilon_3 = \frac{1-2\kappa}{4\kappa^2}$, and, only if $\sigma \neq 0$,

$$\varepsilon_4 = \sigma^{-2}\left(\frac{1}{\varepsilon_2} + 1 + \frac{1}{\varepsilon_1}\right)^{-1}\frac{1-2\kappa}{4} = \sigma^{-2}\left(1 + \frac{4\kappa^2+2}{1-2\kappa}\right)^{-1}\frac{1-2\kappa}{4},$$

we get that

$$\mathfrak{c}_{\sigma,\psi}(u,u) \geq -K_2|\sigma|^{2+2\alpha}\|u\|_2^2,$$

where

$$K_2 = C\left(1+\frac{4\kappa^2+2}{1-2\kappa}\right)^{\alpha+1}\left(\frac{4}{1-2\kappa}\right)^{\alpha},$$

so that the opposite of the operator associated to the form $\mathfrak{c}_{\sigma,\psi}$ generates a strongly continuous, analytic semigroup $(S^{\sigma,\psi}(t))$ in $L^2(\Omega;\mathbb{C})$ (by Theorem 2.1), such that

$$\|S^{\sigma,\psi}(t)\|_{2\to 2} \leq e^{|\sigma|^{2+2\alpha}K_2 t}, \qquad t>0. \tag{29}$$

Arguing as in the proof of (27) we can show that, for every $\varepsilon_i>0$, $i=1,2$,

$$\begin{aligned}
&\mathrm{Re}\,\mathfrak{c}_{\sigma,\psi}\big((|u|\wedge 1)\mathrm{sign}(u),(|u|-1)^+\mathrm{sign}(u)\big)\\
=&\frac{1}{2}\int_{|u|>1}Q\frac{\mathrm{Im}\,(\mathrm{sign}(\overline{u})\nabla u)}{|u|}\mathrm{sign}(u)\cdot\left(\frac{\mathrm{Im}\,(\mathrm{sign}(\overline{u})\nabla u)}{|u|}\mathrm{sign}(\overline{u})\right)(|u|-1)dx\\
&+\int_{|u|>1}\left(\frac{1}{2}V-\sigma^2(Q\nabla\psi,\nabla\psi)+\sigma(C-B)\cdot\nabla\psi\right)(|u|-1)dx\\
\geq&\int_{|u|>1}\left(\frac{1}{2}V-\sigma^2(Q\nabla\psi\,\nabla\psi)-2\kappa|\sigma|V^{\frac{1}{2}}(Q\nabla\psi\,\nabla\psi)^{\frac{1}{2}}\right)(|u|-1)dx\\
\geq&\int_{|u|>1}\left(\frac{1}{2}V-\sigma^2h-2\kappa|\sigma|V^{\frac{1}{2}}h^{\frac{1}{2}}\right)(|u|-1)dx\\
\geq&\int_{|u|>1}\left[\left(\frac{1}{2}-\varepsilon_1-\varepsilon_2\sigma^2\left(1+\frac{\kappa^2}{\varepsilon_1}\right)\right)V-C\varepsilon_2^{-\alpha}\sigma^2\left(1+\frac{\kappa^2}{\varepsilon_1}\right)\right](|u|-1)dx.
\end{aligned}$$

Choosing $\varepsilon_1=\frac{1}{4}$ and $\varepsilon_2=\frac{1}{4}\sigma^{-2}(1+4\kappa^2)^{-1}$, only if $\sigma\neq 0$, we conclude that

$$\begin{aligned}
&\mathrm{Re}\,\mathfrak{c}_{\sigma,\psi}\big((|u|\wedge 1)\mathrm{sign}(u),(|u|-1)^+\mathrm{sign}(u)\big)\\
\geq&-4^\alpha C|\sigma|^{2+2\alpha}(1+4\kappa^2)^{\alpha+1}\int_\Omega(|u|-1)dx\\
=&-4^\alpha C|\sigma|^{2+2\alpha}(1+4\kappa^2)^{\alpha+1}\int_\Omega((|u|\wedge 1)\mathrm{sign}(u),(|u|-1)^+\mathrm{sign}(u))dx\\
\geq&-|\sigma|^{2+2\alpha}K_2\int_\Omega((|u|\wedge 1)\mathrm{sign}(u),(|u|-1)^+\mathrm{sign}(u))dx.
\end{aligned}$$

Therefore, the semigroup $(e^{-|\sigma|^{2+2\alpha}K_2 t}S^{\sigma,\psi}(t))$ is L^∞-contractive by [27, Theorem 2.15], i.e.,

$$\|S^{\sigma,\psi}(t)\|_{\infty\to\infty} \le e^{|\sigma|^{2+2\alpha}K_2 t}, \qquad t \ge 0. \tag{30}$$

Hence, by interpolating the estimates (29) and (30), using Riesz-Thorin's theorem, we get that

$$\|S^{\sigma,\psi}(t)\|_{p\to p} \le e^{|\sigma|^{2+2\alpha}\left(\frac{2}{p}K_2+\frac{p-2}{p}K_2\right)t} = e^{|\sigma|^{2+2\alpha}K_2 t}, \qquad t > 0. \tag{31}$$

At this point, taking (25) into account, Trotter product formula yields

$$T^{\sigma,\psi}(t)f = \lim_{n\to+\infty}\left(U^{\sigma,\psi}\left(\frac{t}{n}\right)\circ S^{\sigma,\psi}\left(\frac{t}{n}\right)\right)^n f, \qquad f \in L^p(\Omega;\mathbb{C}),$$

for every $t \ge 0$ and, therefore, due to estimates (28) and (31), we immediately deduce (23).

From now on, we will consider real-valued functions.

Claim 2

1. If $d \ge 3$, then each operator $T^{\sigma,\psi}(t)$ is bounded from $L^2(\Omega)$ into $L^{2^*}(\Omega)$ and

$$\|T^{\sigma,\psi}(t)\|_{2\to 2^*} \le c_d(\nu_0(1-\kappa))^{-\frac{1}{2}}t^{-\frac{1}{2}}e^{-\beta_{2^*,\sigma}t}, \qquad t > 0, \tag{32}$$

   where $2^* = 2d(d-2)^{-1}$ and c_d is a constant such that $\|u\|_{2^*} \le c_d\|\nabla u\|_2$ for every $u \in W^{1,2}(\Omega)$;
2. if $d \le 2$, then $T^{\sigma,\psi}(t)$ is bounded from $L^2(\Omega)$ into $L^q(\Omega)$ for every $q > 2$ and $t > 0$, and there exists a positive constant c_q such that

$$\|T^{\sigma,\psi}(t)\|_{2\to q} \le c_q[\nu_0(1-\kappa)]^{-d\frac{q-2}{4q}}t^{-d\frac{q-2}{4q}}e^{-\beta_{q,\sigma}t}, \qquad t > 0. \tag{33}$$

Let us begin by proving (32). By Hypothesis (C)-(b), there exists a constant $c_d = c_d(\Omega) > 0$ such that for any $u \in \widetilde{D}_{Q,V}$

$$\|u\|_{L^{2^*}(\Omega)} \le c_d\|\nabla u\|_{L^2(\Omega)}. \tag{34}$$

Using (34) and (22), we can estimate

$$\mathfrak{a}_{\sigma,\psi}(u,u) \ge \frac{1-\kappa}{2}\int_\Omega (Q\nabla u,\nabla u)dx - 2^{-\alpha}K_1|\sigma|^{2+2\alpha}\|u\|_2^2 \tag{35}$$

$$\ge \frac{\nu_0(1-\kappa)}{2c_d^2}\|u\|_{2^*}^2 - 2^{-\alpha}K_1|\sigma|^{2+2\alpha}\|u\|_2^2 \tag{36}$$

for every real-valued function $u \in \widetilde{D}_{Q,V}$, where we recall that K_1 is given by (21). Thanks to Claim 1, we deduce that

$$\|T^{\sigma,\psi}(t)\|_{2^*\to 2^*} \le e^{-\beta_{2^*,\sigma}t}, \qquad t \ge 0.$$

For every $f \in L^2(\Omega) \cap L^{2^*}(\Omega)$ and $t \ge 0$, we set $\gamma(t) = \|e^{\beta_{2^*,\sigma}t}T^{\sigma,\psi}(t)f\|_{2^*}^2$.

Since $(e^{\beta_{2^*,\sigma}t}T^{\sigma,\psi}(t))$ is a contraction semigroup on $L^{2^*}(\Omega)$, γ is a decreasing function. Therefore, by applying (36) with $u = e^{\beta_{2^*,\sigma}t}T^{\sigma,\psi}(t)f$, and taking into account that

$$\frac{d}{ds}\|T^{\sigma,\psi}(s)f\|_2^2 = -2\mathfrak{a}_{\sigma,\psi}(T^{\sigma,\psi}(s)f, T^{\sigma,\psi}(s)f)$$

for every $s > 0$, we get that

$$\begin{aligned}
t\gamma(t) \le& \int_0^t \gamma(s)ds\\
\le& \frac{2c_d^2}{\nu_0(1-\kappa)}\bigg[\int_0^t \mathfrak{a}_{\sigma,\psi}(e^{\beta_{2^*,\sigma}s}T^{\sigma,\psi}(s)f, e^{\beta_{2^*,\sigma}s}T^{\sigma,\psi}(s)f)ds\\
&\qquad + 2^{-\alpha}K_1|\sigma|^{2+2\alpha}\int_0^t e^{2\beta_{2^*,\sigma}s}\|T^{\sigma,\psi}(s)f\|_2^2 ds\bigg]\\
=& \frac{2c_d^2}{\nu_0(1-\kappa)}\bigg[-\frac{1}{2}\int_0^t e^{2\beta_{2^*,\sigma}s}\frac{d}{ds}\|T^{\sigma,\psi}(s)f\|_2^2 ds\\
&\qquad + 2^{-\alpha}K_1|\sigma|^{2+2\alpha}\int_0^t e^{2\beta_{2^*,\sigma}s}\|T^{\sigma,\psi}(s)f\|_2^2 ds\bigg]\\
=& \frac{2c_d^2}{\nu_0(1-\kappa)}\bigg[(\beta_{2^*,\sigma} + 2^{-\alpha}K_1|\sigma|^{2+2\alpha})\int_0^t e^{2\beta_{2^*,\sigma}s}\|T^{\sigma,\psi}(s)f\|_2^2 ds\\
&\qquad - \frac{1}{2}e^{2\beta_{2^*,\sigma}t}\|T^{\sigma,\psi}(t)f\|_2^2 + \frac{1}{2}\|f\|_2^2\bigg]\\
\le& \frac{c_d^2}{\nu_0(1-\kappa)}\|f\|_2^2
\end{aligned}$$

for every $t > 0$, where we integrated by parts one integral term and used the fact that for every $p \ge 2$

$$\begin{aligned}
&\beta_{p,\sigma} + 2^{-\alpha}K_1|\sigma|^{2+2\alpha}\\
=& -C|\sigma|^{2+2\alpha}\bigg[(1+q_0)^{2+2\alpha}\left(\frac{p}{1-2\kappa}\right)^{2\alpha+1} + \left(1+\frac{4\kappa^2+2}{1-2\kappa}\right)^{\alpha+1}\left(\frac{4}{1-2\kappa}\right)^{\alpha}
\end{aligned}$$

$$
\begin{aligned}
&\qquad - \left(1+\frac{2(\kappa^2+q_0^2)}{1-\kappa}\right)^{\alpha+1}\left(\frac{2}{1-\kappa}\right)^{\alpha}\Bigg] \\
&\le - C|\sigma|^{2+2\alpha}\left[(1+q_0)^{2+2\alpha}\left(\frac{2}{1-\kappa}\right)^{2\alpha+1} - \left(1+\frac{2(\kappa^2+q_0^2)}{1-\kappa}\right)^{\alpha+1}\left(\frac{2}{1-\kappa}\right)^{\alpha}\right] \\
&\le - C|\sigma|^{2+2\alpha}\frac{2^{\alpha}}{(1-\kappa)^{2\alpha+1}}\left[\left(2(1+q_0)^2\right)^{1+\alpha} - \left(1-\kappa+2(\kappa^2+q_0^2)\right)^{\alpha+1}\right] \le 0
\end{aligned}
\tag{37}
$$

and (32) follows from a straightforward density argument.

If $d \le 2$, Gagliardo-Nirenberg interpolation inequality and the extension property of Ω (Hypothesis (C)-(b)) give that, for every $q > 2$ there exists a positive constant $c_q = c_q(\Omega) > 0$ such that, for every real-valued function $u \in \widetilde{D}_{Q,V} \subseteq W^{1,2}(\Omega)$,

$$
c_q\|u\|_q \le \|u\|_2^{1-d\frac{q-2}{2q}}\|\nabla u\|_2^{d\frac{q-2}{2q}},
$$

or equivalently

$$
\|\nabla u\|_2^2\|u\|_2^{\frac{4q}{d(q-2)}-2} \ge c_q'\|u\|_q^{\frac{4q}{d(q-2)}},
$$

where $c_q' = c_q^{\frac{4q}{d(q-2)}}$.

Observe that $0 < d\frac{q-2}{2q} < 1$. Hence, taking (35) into account, we can write

$$
\begin{aligned}
&\mathfrak{a}_{\sigma,\psi}(u,u)\|u\|_2^{\frac{4q}{d(q-2)}-2} \\
\ge&\|u\|_2^{\frac{4q}{d(q-2)}-2}\left[\frac{1-\kappa}{2}\int_\Omega (Q\nabla u,\nabla u)dx - 2^{-\alpha}K_1|\sigma|^{2+2\alpha}\|u\|_2^2\right] \\
\ge&\frac{\nu_0}{2}(1-\kappa)\|u\|_2^{\frac{4q}{d(q-2)}-2}\|\nabla u\|_2^2 - 2^{-\alpha}K_1|\sigma|^{2+2\alpha}\|u\|_2^{\frac{4q}{d(q-2)}} \\
\ge&\frac{\nu_0}{2}(1-\kappa)c_q'\|u\|_q^{\frac{4q}{d(q-2)}} - 2^{-\alpha}K_1|\sigma|^{2+2\alpha}\|u\|_2^{\frac{4q}{d(q-2)}}.
\end{aligned}
\tag{38}
$$

By Claim 1, the function γ_q, defined by $\gamma_q(t) = \|e^{\beta_{q,\sigma}t}T^{\sigma,\psi}(t)f\|_q$ for $t \ge 0$, is decreasing for every $f \in L^2(\Omega)\cap L^q(\Omega)$, where $\beta_{q,\sigma}$ is given by (24). Also the function $\gamma_{2,q}$, defined by $\gamma_{2,q}(t) = \|e^{\beta_{q,\sigma}t}T^{\sigma,\psi}(t)f\|_2^2$ for every $t > 0$, is decreasing. Indeed,

$$
\gamma_{2,q}(t) = e^{2(\beta_{q,\sigma}-\beta_{2,\sigma})t}(\gamma_2(t))^2, \qquad t \ge 0,
$$

and both the two factors are decreasing functions on $[0,+\infty[$. Note that $\beta_{q,\sigma} < \beta_{2,\sigma}$. Hence, in particular, $\gamma_{2,q}(t) \le \|f\|_2^2$ for every $t \ge 0$. Therefore, taking (38) into account and arguing as in the case $d \ge 3$, we can show that

$$
\begin{aligned}
t\gamma_q(t)^{\frac{4q}{d(q-2)}} \le & \frac{2}{c_q'\nu_0(1-\kappa)}\int_0^t \Big[\mathfrak{a}_{\sigma,\psi}(e^{\beta_{q,\sigma}s}T^{\sigma,\psi}(s)f, e^{\beta_{q,\sigma}s}T^{\sigma,\psi}(s)f) \\
& \qquad + 2^{-\alpha}K_1|\sigma|^{2+2\alpha}\gamma_{2,q}(s)\Big]\gamma_{2,q}(s)^{\frac{2q}{d(q-2)}-1}ds \\
\le & \frac{2}{c_q'\nu_0(1-\kappa)}\|f\|_2^{\frac{4q}{d(q-2)}-2} \\
& \times\Big[\int_0^t \mathfrak{a}_{\sigma,\psi}(e^{\beta_{q,\sigma}s}T^{\sigma,\psi}(s)f, e^{\beta_{q,\sigma}s}T^{\sigma,\psi}(s)f)ds \\
& \qquad + 2^{-\alpha}K_1|\sigma|^{2+2\alpha}\int_0^t \gamma_{2,q}(s)ds\Big] \\
= & \frac{2}{c_q'\nu_0(1-\kappa)}\|f\|_2^{\frac{4q}{d(q-2)}-2}\Big[-\frac{1}{2}\int_0^t e^{2\beta_{q,\sigma}s}\frac{d}{ds}\|T^{\sigma,\psi}(s)f\|_2^2 ds \\
& \qquad + 2^{-\alpha}K_1|\sigma|^{2+2\alpha}\int_0^t \gamma_{2,q}(s)ds\Big] \\
= & \frac{1}{c_q'\nu_0(1-\kappa)}\|f\|_2^{\frac{4q}{d(q-2)}-2} \\
& \times\Big[\frac{1}{2}\|f\|_2^2 - \frac{1}{2}\gamma_{2,q}(t) + \Big(\beta_{q,\sigma} + 2^{-\alpha}K_1|\sigma|^{2+2\alpha}\Big)\int_0^t \gamma_{2,q}(s)ds\Big] \\
\le & \frac{1}{2c_q'\nu_0(1-\kappa)}\|f\|_2^{\frac{4q}{d(q-2)}}
\end{aligned}
$$

for every $t > 0$, where we used (37) with $p = q$. From this chain of inequalities, estimate (33) follows immediately by a density argument.

Claim 3 For every $\sigma \in \mathbb{R}$ and $\psi \in W$ it holds that

$$
\begin{aligned}
\|T^{\sigma,\psi}(t)\|_{2\to\infty} \le & M_{d,\alpha}[(1-\kappa)\nu_0]^{-\frac{d}{4}}t^{-\frac{d}{4}}e^{|\sigma|^{2+2\alpha}K_1t}e^{-\frac{1-\kappa}{4}V_0t} \\
& \times\left(1 + \frac{1-\kappa}{4}V_0t + |\sigma|^{2+2\alpha}c_{d,\alpha,\kappa,q_0}t\right)^{\frac{d}{4}}
\end{aligned}
\tag{39}
$$

for every $t > 0$, where K_1 is given by (21), $M_{d,\alpha}$ is a positive constant depending only on d and α and c_{d,α,κ,q_0} is a positive constant which depends only on the arguments in the subscripts and is explicitly computed in the proof.

If $d \geq 3$, then we start by interpolating the estimates (23) and (32), using Riesz-Thorin theorem, to get

$$\begin{aligned}\|T^{\sigma,\psi}(t)\|_{p\to\frac{pd}{d-1}} &\leq \|T^{\sigma,\psi}(t)\|_{2\to 2^*}^{\frac{1}{p}}\|T^{\sigma,\psi}(t)\|_{2(p-1)\to 2(p-1)}^{1-\frac{1}{p}}\\ &\leq \left(\frac{c_d}{\sqrt{\nu_0(1-\kappa)}}t^{-\frac{1}{2}}e^{-\beta_{2^*,\sigma}t}\right)^{\frac{1}{p}}\left(e^{-\beta_{2(p-1),\sigma}t}\right)^{1-\frac{1}{p}}\\ &= \left(\frac{c_d^2}{\nu_0(1-\kappa)}\right)^{\frac{1}{2p}}t^{-\frac{1}{2p}}e^{C|\sigma|^{2+2\alpha}\left(1+\frac{4\kappa^2+2}{1-2\kappa}\right)^{\alpha+1}\left(\frac{4}{1-2\kappa}\right)^{\alpha}t}\\ &\quad\times e^{\frac{C}{p}|\sigma|^{2+2\alpha}\left[\left(\frac{d}{d-2}\right)^{1+2\alpha}+(p-1)^{2+2\alpha}\right](1+q_0)^{2+2\alpha}\left(\frac{2}{1-2\kappa}\right)^{2\alpha+1}t}.\end{aligned}$$

Now, similarly to the proof of [27, Theorem 6.8], we set $R = (d-1)^{-1}d$, $t_j = \frac{(R^{2\alpha}+1)R-1}{(R^{2\alpha}+1)R}((R^{2\alpha}+1)R)^{-j}$ and $p_j = 2R^j$ for every $j \in \mathbb{N}\cup\{0\}$. Since

$$\sum_{j=0}^{+\infty} t_j = 1, \qquad \sum_{j=0}^{+\infty}\frac{1}{p_j} = \frac{d}{2},$$
$$\sum_{j=0}^{+\infty}\frac{t_j}{p_j} =: A_{d,\alpha} < +\infty, \qquad \prod_{j=0}^{+\infty} t_j^{-\frac{1}{2p_j}} =: B_{d,\alpha} \in \mathbb{R}$$

and

$$\begin{aligned}\sum_{j=0}^{+\infty}\frac{(p_j-1)^{2+2\alpha}}{p_j}t_j \leq& \sum_{j=0}^{+\infty} p_j^{1+2\alpha}t_j\\ =&2^{2\alpha+1}\frac{(R^{2\alpha}+1)R-1}{(R^{2\alpha}+1)R}\sum_{j=0}^{+\infty}\left(\frac{R^{2\alpha}}{R^{2\alpha}+1}\right)^j\\ =&\frac{2^{2\alpha+1}}{R}[(R^{2\alpha}+1)R-1] =: L_{d,\alpha} < +\infty\end{aligned}$$

we can estimate

$$\begin{aligned}\|T^{\sigma,\psi}(t)\|_{2\to\infty} \leq& \prod_{j=0}^{+\infty}\|T(tt_j)\|_{p_j\to p_{j+1}}\\ \leq& \prod_{j=0}^{+\infty}\left(\frac{c_d^2}{\nu_0(1-\kappa)}\right)^{\frac{1}{2p_j}}(t_jt)^{-\frac{1}{2p_j}}e^{C|\sigma|^{2+2\alpha}\left(1+\frac{4\kappa^2+2}{1-2\kappa}\right)^{\alpha+1}\left(\frac{4}{1-2\kappa}\right)^{\alpha}tt_j}\\ &\times e^{\frac{C}{p_j}|\sigma|^{2+2\alpha}\left[\left(\frac{d}{d-2}\right)^{1+2\alpha}+(p_j-1)^{2+2\alpha}\right](1+q_0)^{2+2\alpha}\left(\frac{2}{1-2\kappa}\right)^{2\alpha+1}tt_j}\end{aligned}$$

$$\leq \left(\frac{c_d^2}{\nu_0(1-\kappa)}\right)^{\frac{d}{4}} B_{d,\alpha} t^{-\frac{d}{4}} e^{C|\sigma|^{2+2\alpha}\left(1+\frac{4\kappa^2+2}{1-2\kappa}\right)^{\alpha+1}\left(\frac{4}{1-2\kappa}\right)^{\alpha} t}$$
$$\times e^{C|\sigma|^{2+2\alpha}\left[\left(\frac{d}{d-2}\right)^{1+2\alpha} A_{d,\alpha}+L_{d,\alpha}\right](1+q_0)^{2+2\alpha}\left(\frac{2}{1-2\kappa}\right)^{2\alpha+1} t}$$

for every $t > 0$.

Taking (20) into account, [27, Lemma 6.5] applied to the semigroup $(T^{\sigma,\psi}(t))$ yields that

$$\|T^{\sigma,\psi}(t)\|_{2\to\infty} \leq \left(\frac{c_d^2}{\nu_0(1-\kappa)}\right)^{\frac{d}{4}} B_{d,\alpha} e t^{-\frac{d}{4}} e^{-\frac{1-\kappa}{4}V_0 t} e^{K_1|\sigma|^{2+2\alpha} t}$$
$$\times \left(1+\frac{1-\kappa}{4}V_0 t+|\sigma|^{2+2\alpha}\gamma_{d,q_0,\alpha,\kappa} t\right)^{\frac{d}{4}},$$

for every $t > 0$, where

$$\gamma_{d,q_0,\alpha,\kappa} = 0 \vee C\left\{\left(1+\frac{4\kappa^2+2}{1-2\kappa}\right)^{\alpha+1}\left(\frac{4}{1-2\kappa}\right)^{\alpha} - \left(1+\frac{2(\kappa^2+q_0^2)}{1-\kappa}\right)^{\alpha+1}\left(\frac{4}{1-\kappa}\right)^{\alpha}\right.$$
$$\left. +\left[\left(\frac{d}{d-2}\right)^{1+2\alpha} A_{d,\alpha}+L_{d,\alpha}\right](1+q_0)^{2+2\alpha}\left(\frac{2}{1-2\kappa}\right)^{2\alpha+1}\right\}.$$

If $d \leq 2$, then we apply Claim 2 with $q = \frac{2r}{r-2}$, where $r > 2$ is a fixed number, obtaining the estimate

$$\|T^{\sigma,\psi}(t)\|_{2\to\frac{2r}{r-2}} \leq c_{\frac{2r}{r-2}}[\nu_0(1-\kappa)]^{-\frac{d}{2r}} t^{-\frac{d}{2r}} e^{-\beta_{\frac{2r}{r-2},\sigma} t}, \qquad t > 0.$$

Applying Riesz-Thorin theorem to this estimate and to the estimate (23), with p being replaced by $2(p-1)$, we get

$$\|T^{\sigma,\psi}(t)\|_{p\to\frac{pr}{r-1}} \leq \|T^{\sigma,\psi}(t)\|_{2\to\frac{2r}{r-2}}^{\frac{1}{p}} \|T^{\sigma,\psi}(t)\|_{2(p-1)\to 2(p-1)}^{1-\frac{1}{p}}$$
$$\leq \left(c_{\frac{2r}{r-2}}[\nu_0(1-\kappa)]^{-\frac{d}{2r}} t^{-\frac{d}{2r}} e^{-\beta_{\frac{2r}{r-2},\sigma} t}\right)^{\frac{1}{p}} \left(e^{-\beta_{2(p-1),\sigma} t}\right)^{1-\frac{1}{p}}$$
$$= c_{\frac{2r}{r-2}}^{\frac{1}{p}} [\nu_0(1-\kappa)]^{-\frac{d}{2pr}} t^{-\frac{d}{2pr}} e^{C|\sigma|^{2+2\alpha}\left(1+\frac{4\kappa^2+2}{1-2\kappa}\right)^{\alpha+1}\left(\frac{4}{1-2\kappa}\right)^{\alpha} t}$$
$$\times e^{\frac{C}{p}|\sigma|^{2+2\alpha}\left[\left(\frac{r}{r-2}\right)^{2\alpha+1}+(p-1)^{2\alpha+2}\right](1+q_0)^{2+2\alpha}\left(\frac{2}{1-2\kappa}\right)^{2\alpha+1} t}.$$

Now, we repeat the same procedure as in the case $d \geq 3$, defining the sequences (t_k) and (p_k) in the same way, just replacing d with r. Estimating

$$\begin{aligned}
\|T^{\sigma,\psi}(t)\|_{2\to\infty} \leq & \prod_{j=0}^{+\infty} \|T(tt_j)\|_{p_j\to p_{j+1}} \\
\leq & \prod_{j=0}^{+\infty} c_{\frac{2r}{r-2}}^{\frac{1}{p_j}} [\nu_0(1-\kappa)]^{-\frac{d}{2p_j r}} (t_j t)^{-\frac{d}{2p_j r}} e^{C|\sigma|^{2+2\alpha}\left(1+\frac{4\kappa^2+2}{1-2\kappa}\right)^{\alpha+1}\left(\frac{4}{1-2\kappa}\right)^{\alpha} tt_j} \\
& \times e^{\frac{C}{p_j}|\sigma|^{2+2\alpha}\left[\left(\frac{r}{r-2}\right)^{2\alpha+1}+(p_j-1)^{2\alpha+2}\right](1+q_0)^{2+2\alpha}\left(\frac{2}{1-2\kappa}\right)^{2\alpha+1} tt_j} \\
= & c_{\frac{2r}{r-2}}^{\frac{r}{2}} [\nu_0(1-\kappa)]^{-\frac{d}{4}} B_{r,\alpha}^{\frac{d}{r}} t^{-\frac{d}{4}} e^{C|\sigma|^{2+2\alpha}\left(1+\frac{4\kappa^2+2}{1-2\kappa}\right)^{\alpha+1}\left(\frac{4}{1-2\kappa}\right)^{\alpha} t} \\
& \times e^{C|\sigma|^{2+2\alpha}\left[\left(\frac{r}{r-2}\right)^{1+2\alpha} A_{r,\alpha}+L_{r,\alpha}\right](1+q_0)^{2+2\alpha}\left(\frac{2}{1-2\kappa}\right)^{2\alpha+1} t},
\end{aligned}$$

where $A_{r,\alpha}$, $B_{r,\alpha}$, and $L_{r,\alpha}$ are defined as $A_{d,\alpha}$, $B_{d,\alpha}$ and $L_{d,\alpha}$, with d being replaced by r.

Using again the estimate (20) and the same arguments as above, we conclude that

$$\begin{aligned}
\|T^{\sigma,\psi}(t)\|_{2\to\infty} \leq & c_{\frac{2r}{r-2}}^{\frac{r}{2}} [\nu_0(1-\kappa)]^{-\frac{d}{4}} B_{r,\alpha}^{\frac{d}{r}} e t^{-\frac{d}{4}} e^{-\frac{1-\kappa}{4} V_0 t} e^{K_1|\sigma|^{2+2\alpha} t} \\
& \times \left(1+\frac{1-\kappa}{4} V_0 t + |\sigma|^{2+2\alpha}\gamma_{r,q_0,\alpha,\kappa} t\right)^{\frac{d}{4}},
\end{aligned}$$

where $\gamma_{r,q_0,\alpha,\kappa}$ is defined as $\gamma_{d,q_0,\alpha,\kappa}$, with d being replaced by r. The assertion follows by choosing $M_{d,\alpha} = c_d^{\frac{d}{2}} B_d e$ if $d \geq 3$ and $M_{d,\alpha} = c_6^{\frac{3}{2}} B_3^{\frac{d}{3}} e$ if $d \leq 2$, and $c_{d,\alpha,q_0,\kappa} = \gamma_{d\vee 3,\alpha,q_0,\kappa}$.

Claim 4 There exists a strictly positive constant $H_{d,\alpha}$ depending only on d and α, such that

$$\begin{aligned}
\|T^{\sigma,\psi}(t)\|_{1\to\infty} \leq & H_{d,\alpha}[(1-\kappa)\nu_0]^{-\frac{d}{2}} t^{-\frac{d}{2}} e^{|\sigma|^{2+2\alpha} K_1 t} e^{-\frac{1-\kappa}{4} V_0 t} \\
& \times \left(1+\frac{1-\kappa}{8} V_0 t + |\sigma|^{2+2\alpha}\frac{c_{d,\alpha,q_0,\kappa}}{2} t\right)^{\frac{d}{2}}
\end{aligned} \tag{40}$$

for every $t > 0$, where c_{d,α,κ,q_0} is described in Claim 3.

Since $(\mathfrak{a}_{\sigma,\psi})^* = (\mathfrak{a}^*)_{-\sigma,\psi}$ and the coefficients of $\mathfrak{a}^*$ satisfy the same quantitative and qualitative assumptions of those of $\mathfrak{a}$, we can infer that

$$\begin{aligned}\|(T^{\sigma,\psi})(t)^*\|_{2\to\infty} \leq & M_{d,\alpha}[(1-\kappa)\nu_0]^{-\frac{d}{4}}t^{-\frac{d}{4}}e^{|\sigma|^{2+2\alpha}K_1 t}e^{-\frac{1-\kappa}{4}V_0 t}\\ &\times\left(1+\frac{1-\kappa}{4}V_0 t+|\sigma|^{2+2\alpha}c_{d,\alpha,q_0,\kappa}t\right)^{\frac{d}{4}}\end{aligned}$$

and, therefore,

$$\begin{aligned}\|T^{\sigma,\psi}(t)\|_{1\to 2} \leq & M_{d,\alpha}[(1-\kappa)\nu_0]^{-\frac{d}{4}}t^{-\frac{d}{4}}e^{|\sigma|^{2+2\alpha}K_1 t}e^{-\frac{1-\kappa}{4}V_0 t}\\ &\times\left(1+\frac{1-\kappa}{4}V_0 t+|\sigma|^{2+2\alpha}c_{d,\alpha,q_0,\kappa}t\right)^{\frac{d}{4}}\end{aligned} \tag{41}$$

for every $t > 0$. At this point the estimate (40) follows, with $H_{d,\alpha} = 2^{d/2}M_{d,\alpha}^2$, by writing $\|T^{\sigma,\psi}(t)\|_{1\to\infty} \leq \|T^{\sigma,\psi}(t/2)\|_{2\to\infty}\|T^{\sigma,\psi}(t/2)\|_{1\to 2}$.

Now we can conclude the proof of the theorem. Taking $\sigma = 0$ in (40), it follows that the operator $T(t)$ is bounded from $L^1(\Omega)$ into $L^\infty(\Omega)$ for every $t > 0$. Hence, it is described by a kernel $k(t,\cdot,\cdot)$ Moreover, by (40), the kernel of $T^{\sigma,\psi}(t)$ is $e^{\sigma\psi(x)}k(t,x,y)e^{-\sigma\psi(y)}$ and satisfies the estimate

$$\begin{aligned}0 \leq & e^{\sigma\psi(x)}k(t,x,y)e^{-\sigma\psi(y)}\\ \leq & H_d[(1-\kappa)\nu_0]^{-\frac{d}{2}}t^{-\frac{d}{2}}e^{|\sigma|^{2+2\alpha}K_1 t}e^{-\frac{1-\kappa}{4}V_0 t}\left(1+\frac{1-\kappa}{8}V_0 t+|\sigma|^{2+2\alpha}\frac{c_{d,\alpha,q_0,\kappa}}{2}t\right)^{\frac{d}{2}}\end{aligned}$$

for every $t > 0$, almost every $x, y \in \Omega$ and every $d \geq 1$. Therefore,

$$\begin{aligned}k(t,x,y) \leq & H_d[(1-\kappa)\nu_0]^{-\frac{d}{2}}t^{-\frac{d}{2}}e^{|\sigma|^{2+2\alpha}K_1 t}e^{-\frac{1-\kappa}{4}V_0 t}\\ &\times\left(1+\frac{1-\kappa}{8}V_0 t+|\sigma|^{2+2\alpha}\frac{c_{d,\alpha,q_0,\kappa}}{2}t\right)^{\frac{d}{2}}e^{\sigma(\psi(y)-\psi(x))}\end{aligned}$$

for every $t > 0$. Taking

$$\sigma = (2K_1(\alpha+1)t)^{-\frac{1}{2\alpha+1}}|\psi(x)-\psi(y)|^{\frac{1}{2\alpha+1}-1}(\psi(x)-\psi(y))$$

we get

$$\begin{aligned}&k(t,x,y)\\ &\leq H_d[(1-\kappa)\nu_0]^{-\frac{d}{2}}t^{-\frac{d}{2}}e^{-\frac{1-\kappa}{4}V_0 t}e^{-|\psi(y)-\psi(x)|^{\frac{2\alpha+2}{2\alpha+1}}(K_1 t)^{-\frac{1}{2\alpha+1}}(2\alpha+2)^{-\frac{2+2\alpha}{2\alpha+1}}(2\alpha+1)}\end{aligned}$$

$$\times\left(1+\frac{1-\kappa}{8}V_0t+|\psi(x)-\psi(y)|^{\frac{2+2\alpha}{2\alpha+1}}(2K_1(\alpha+1)t)^{-\frac{2\alpha+2}{2\alpha+1}}\frac{c_{d,\alpha,q_0,\kappa}}{2}t\right)^{\frac{d}{2}}$$

$$\leq H_{d,\alpha}[(1-\kappa)\nu_0]^{-\frac{d}{2}}e^{-\frac{1-\kappa}{4}V_0t}t^{-\frac{d}{2}}e^{-|\psi(y)-\psi(x)|^{\frac{2\alpha+2}{2\alpha+1}}(K_1t)^{-\frac{1}{2\alpha+1}}(2\alpha+2)^{-\frac{2+2\alpha}{2\alpha+1}}(2\alpha+1)}$$

$$\times\left(1+\frac{1-\kappa}{8}V_0t+d_Q(x,y)^{\frac{2+2\alpha}{2\alpha+1}}(2K_1(\alpha+1))^{-\frac{2\alpha+2}{2\alpha+1}}\frac{c_{d,\alpha,q_0,\kappa}}{2t^{\frac{1}{\alpha+1}}}\right)^{\frac{d}{2}}$$

for every $t > 0$. The assertion follows by minimizing over ψ and taking

$$H_1 = H_{d,\alpha}$$

$$H_2 = K_1^{-\frac{1}{2\alpha+1}}(2\alpha+2)^{-\frac{2+2\alpha}{2\alpha+1}}(2\alpha+1)$$

$$H_3 = (2K_1(\alpha+1))^{-\frac{2\alpha+2}{2\alpha+1}}\frac{c_{d,\alpha,q_0,\kappa}}{2}.$$

□

5 The L^p-Analyticity of the Semigroup $(T(t))$

As it is well known, the strongly continuous contractive semigroup $(T(t))$ generated by the opposite of $(A, D(A))$ in $L^2(\Omega, \mathbb{C})$ admits a bounded analytic extension $\{T(z)\}_{z\in\Sigma_{\theta_0}}$ where $\theta_0 := \frac{\pi}{2} - \arctan(M)$, M is the continuity constant of the form $\mathfrak{a}$ defined in (11) and Σ_θ denotes the sector $\{z \in \mathbb{C}, z \neq 0, |\arg z| < \theta\}$ for any $\theta \in (0, \pi/2]$ (see [27, Theorem 1.52]). This means that the function $z \mapsto T(z)f$ is analytic in Σ_{θ_0} for any $f \in L^2(\Omega, \mathbb{C})$ and, in addition, $\{T(z)\}_{z\in\Sigma_\psi}$ is uniformly bounded and strongly continuous in $L^2(\Omega, \mathbb{C})$ for any $\psi \in (0, \theta_0)$.

A classical result concerning semigroups asserts that, if $(T(t))$ is also bounded on $L^q(\Omega, \mathbb{C})$ for some $q \in]2, +\infty[\cup\{+\infty\}$ (as in our case), then the analyticity of the semigroup extends from $L^2(\Omega, \mathbb{C})$ to $L^p(\Omega, \mathbb{C})$ for any $p \in [2, q[$. However, in general, the extended semigroup is analytic on a sector which depends on p and q (see [27, Proposition 3.12]).

In this section, under suitable assumptions on the diffusion matrix Q and on the potential V, we prove that the Gaussian estimates in Theorem 4.1 for real times can be extended to complex times and show how the latter estimates imply that the semigroup $(T(t))$ can be extended to a bounded analytic semigroup on the sector Σ_{θ_0} in $L^p(\Omega, \mathbb{C})$, for any $p \in [1, +\infty[$.

Theorem 5.1 *Under the hypotheses of Theorem 4.1, let us assume further that*

$$\sup_{y\in\Omega}\int_\Omega e^{-\lambda(d_{Q,V}(x,y))^{\frac{2\alpha+2}{2\alpha+1}}}dx < +\infty \tag{42}$$

for every $\lambda > 0$. Then, for any $\theta \in [0, \theta_0[$, the kernel k of the semigroup $(T(t))$ satisfies the estimate

$$|k(z,x,y)| \le C_\theta (\mathrm{Re} z)^{-\frac{d}{2}} \exp\Big(-c_\theta (d_{Q,V}(x,y))^{\frac{2\alpha+2}{2\alpha+1}} |z|^{-\frac{1}{2\alpha+1}}\Big) \tag{43}$$

for any $z \in \Sigma_\theta$ and almost every $x, y \in \Omega$, where C_θ, and c_θ are positive constants, depending on $d, \nu_0, \kappa, \alpha, q_0, V_0$ and θ. Moreover $(T(t))$ extends to a bounded analytic semigroup $(T(z))_{z\in\Sigma_{\theta_0}}$ on $L^p(\Omega, \mathbb{C})$ for any $p \in [1, +\infty[$.

Proof To simplify the notation we denote by C, c two positive constants depending on $d, \nu_0, \kappa, q_0, \alpha, V_0$ and possibly on Ω, that can vary from line to line. First of all, let us observe that estimate (43) holds true for any $z \in]0, +\infty[$. Indeed, for real values of z, estimate (43) follows from (19) observing that

$$\begin{aligned}
&e^{-\frac{1-\kappa}{4}V_0 t} t^{-\frac{d}{2}} e^{-(d_{Q,V}(x,y))^{\frac{2\alpha+2}{2\alpha+1}}(K_1 t)^{-\frac{1}{2\alpha+1}}(2\alpha+2)^{-\frac{2+2\alpha}{2\alpha+1}}}\\
&\qquad\times\left(1+\frac{1-\kappa}{8}V_0 t+\frac{c_{d,q_0,\alpha,\kappa}}{2}\left(d_{Q,V}(x,y)\right)^{\frac{2+2\alpha}{2\alpha+1}}(2K_1(\alpha+1)t)^{-\frac{2\alpha+2}{2\alpha+1}}\right)^{\frac{d}{2}}\\
&\le C t^{-\frac{d}{2}}\exp\Big(-c(d_{Q,V}(x,y))^{\frac{2\alpha+2}{2\alpha+1}} t^{-\frac{1}{2\alpha+1}}\Big)
\end{aligned} \tag{44}$$

for every $t > 0$, $x, y \in \Omega$ and some positive constants c, C.

We now fix $\theta \in (0, \theta_0)$. To prove (43) in the whole sector Σ_θ, for any pair of disjoint sets $E, F \subset \Omega$ and for any $f, g \in L^1(\Omega;\mathbb{C}) \cap L^2(\Omega;\mathbb{C})$ we consider the function $H : \Sigma_{\theta_0} \to \mathbb{C}$, defined by

$$H(z) = \int_\Omega \chi_E(x)(T(z)(\chi_F f))(x)\overline{g(x)}dx, \qquad z \in \Sigma_{\theta_0}.$$

The analyticity of $(T(z))_{z\in\Sigma_{\theta_0}}$ in $L^2(\Omega;\mathbb{C})$ and estimate (44) yield that H is holomorphic on Σ_θ and

$$\begin{aligned}
|H(t)| &\le \int_\Omega\int_\Omega \chi_{E\times F}(x,y)|k(t,x,y)||f(y)||g(x)|dydx\\
&\le Ct^{-\frac{d}{2}}\int_\Omega\int_\Omega \chi_{E\times F}(x,y)\exp\Big(-c\left(d_{Q,V}(x,y)\right)^{\frac{2\alpha+2}{2\alpha+1}}t^{-\frac{1}{2\alpha+1}}\Big)|f(y)||g(x)|dydx\\
&\le Ct^{-\frac{d}{2}}\exp\Big(-cd_{Q,V}(E,F)^{\frac{2\alpha+2}{2\alpha+1}}t^{-\frac{1}{2\alpha+1}}\Big)\|f\|_1\|g\|_1
\end{aligned} \tag{45}$$

for every $t > 0$.

To estimate H on Σ_θ, we begin by observing that, from (39) and (41), both with $\sigma = 0$, we deduce that

$$\|T(t)\|_{1\to 2} + \|T(t)\|_{2\to\infty} \le Ct^{-\frac{d}{4}}, \qquad t > 0. \tag{46}$$

Let us fix $\nu \in (\theta, \theta_0)$. A simple computation shows that we can determine $\varepsilon > 0$ such that, for every $z \in \Sigma_\theta$, $z_0 := z - \varepsilon \mathrm{Re} z$ belongs to Σ_ν. Indeed, if $z = x + iy$, then $z_0 = (1-\varepsilon)x + iy$ and, consequently,

$$\left|\frac{y}{(1-\varepsilon)x}\right| \leq \frac{\tan(\theta)}{1-\varepsilon} < \tan(\nu)$$

provided that $\varepsilon < 1 - (\tan(\nu))^{-1}\tan(\theta)$ and, with this choice of ε, z_0 belongs to Σ_ν.

Thus, we can write $T(z) = T(\varepsilon(\mathrm{Re} z)/2)T(z_0)T(\varepsilon(\mathrm{Re} z)/2)$ and, thanks to (46) and the uniform boundedness of the function $z \mapsto \|T(z)\|_{2\to 2}$ in Σ_ν, we can estimate

$$\|T(z)\|_{1\to\infty} \leq \|T(\varepsilon(\mathrm{Re} z)/2)\|_{2\to\infty}\|T(z_0)\|_{2\to 2}\|T(\varepsilon(\mathrm{Re} z)/2)\|_{1\to 2} \leq C(\mathrm{Re}\, z)^{-\frac{d}{2}}.$$

It follows that

$$|H(z)| \leq \|(T(z)(\chi_F f)\|_\infty \|g\|_1 \leq C(\mathrm{Re} z)^{-\frac{d}{2}}\|f\|_1\|g\|_1 \tag{47}$$

for any $z \in \Sigma_\theta$. Now, using (45), (47) and applying [27, Lemma 6.18] we conclude that

$$|H(z)| \leq C_\theta(\mathrm{Re}\, z)^{-\frac{d}{2}} \exp\Big(- c_\theta \left(d_{Q,V}(E,F)\right)^{\frac{2\alpha+2}{2\alpha+1}} |z|^{-\frac{1}{2\alpha+1}}\Big), \qquad z \in \Sigma_\theta,$$

for some positive constants C_θ, c_θ, depending also on θ. Consequently, $\chi_E T(z)\chi_F$ is a bounded operator from $L^1(\Omega;\mathbb{C})$ to $L^\infty(\Omega;\mathbb{C})$ and

$$\|\chi_E T(z)\chi_F\|_{1\to\infty} \leq C_\theta(\mathrm{Re}\, z)^{-\frac{d}{2}} \exp\Big(- c_\theta \left(d_{Q,V}(E,F)\right)^{\frac{2\alpha+2}{2\alpha+1}} |z|^{-\frac{1}{2\alpha+1}}\Big)$$

for all $z \in \Sigma_\theta$, whence

$$|\chi_E(x)k(z,x,y)\chi_F(y)| \leq C_\theta(\mathrm{Re}\, z)^{-\frac{d}{2}} \exp\Big(- c_\theta \left(d_{Q,V}(E,F)\right)^{\frac{2\alpha+2}{2\alpha+1}} |z|^{-\frac{1}{2\alpha+1}}\Big)$$

for any $z \in \Sigma_\theta$ and almost every $x, y \in \Omega$. Finally, choosing $E = \{w \in \Omega : d_{Q,V}(w,x) \leq r\}$ and $F = \{w \in \Omega : d_Q(w,y) \leq r\}$ with $r = \frac{d_{Q,V}(x,y)}{4}$ we get (43) for any $z \in \Sigma_\theta$. Indeed, since $d_{Q,V}$ is a distance, it follows that

$$\begin{aligned} d_{Q,V}(w_1,w_2) \geq& d_{Q,V}(x,y) - d_{Q,V}(x,w_1) - d_{Q,V}(y,w_2) \\ \geq& d_{Q,V}(x,y) - 2r = \frac{d_{Q,V}(x,y)}{2} \end{aligned}$$

for every $w_1 \in E$ and $w_2 \in F$, so that $d_{Q,V}(E,F) \geq 2^{-1}d_{Q,V}(x,y)$.

Next, we prove that $(T(t))_{t\ge 0}$ extends to a bounded analytic semigroup $(T(z))_{z\in\Sigma_{\theta_0}}$ in $L^1(\Omega;\mathbb{C})$. To this aim, let us fix $f \in L^1(\Omega;\mathbb{C})$ and let (f_n) be a sequence in $L^1(\Omega;\mathbb{C})\cap L^2(\Omega;\mathbb{C})$ converging to f in $L^1(\Omega;\mathbb{C})$. Then, by using estimate (43) and condition (42), we deduce that

$$\begin{aligned}\|\chi_{B_n}T(z)f_n\|_1 =&\int_{B_n}\int_\Omega |k(z,x,y)||f_n(y)|dydx\\ \le& C\int_\Omega |f_n(y)|\int_{B_n} e^{-c(d_{Q,V}(x,y))^{\frac{2\alpha+2}{2\alpha+1}}}dxdy\\ \le& C_1\sup_{n\in\mathbb{N}}\|f_n\|_1\sup_{y\in\Omega}\|e^{-c_1(d_{Q,V}(\cdot,y))^{\frac{2\alpha+2}{2\alpha+1}}}\|_1 =: K,\end{aligned}\tag{48}$$

where c_1 and C_1 depend also on z. Each function $T(\cdot)f_n$ is analytic in Σ_θ with values in $L^2(\Omega)$, so that each function $\chi_{B_n}T(\cdot)f_n$ is analytic in the same sector with values in $L^1(\Omega)$. Since $\chi_{B_n}T(t)f_n$ converges to $T(t)f$ in $L^1(\Omega)$ as n tends to $+\infty$, we can apply Vitali theorem (see [15, Theorem 3.14]) to infer that the function $z\mapsto T(z)f$ is analytic in Σ_θ for any $f\in L^1(\Omega,\mathbb{C})$ and, by estimate (48), $(T(z))_{z\in\Sigma_\theta}$ is uniformly bounded in $L^1(\Omega;\mathbb{C})$.

To conclude the proof, let us show that $(T(z))_{z\in\Sigma_\theta}$ is strongly continuous in $L^1(\Omega;\mathbb{C})$. The uniform boundedness of $\|T(z)\|_{1\to 1}$ for any $z\in\Sigma_\theta$, proved in (48), and the density of $C_c^\infty(\Omega;\mathbb{C})$ in $L^1(\Omega,\mathbb{C})$ allow us considering functions $f\in C_c^\infty(\Omega;\mathbb{C})$. Let B be a bounded subset of Ω containing the support of f and such that $d(\text{supp}f,\Omega\setminus B)>0$, where d stands for the Euclidean distance in $\mathbb{R}^d$. It follows that $d_{Q,V}(\text{supp}f,\Omega\setminus B)=:\delta_{Q,V}>0$. Otherwise there would exists a sequence (x_n) in the support of f and (y_n) in $\Omega\setminus B$ such that $d_{Q,V}(x_n,y_n)$ tends to 0 as n tends to $+\infty$. Up to a subsequence, we can assume that x_n converges to some $x\in\text{supp}f$. Hence, taking (18) into account, we can infer that $d_{Q,V}(x_n,x)$ tends to 0. As a byproduct, $d_{Q,V}(y_n,x)$ converges to 0. This implies that $|y_n-x|$ converges to 0 as n tends to $+\infty$. Indeed, suppose that this is not the case. Then, we can determine $\varepsilon>0$ and a subsequence (y_{n_k}) such that $y_{n_k}\notin B(x,\varepsilon)$ for every $k\in\mathbb{N}$. Let ψ be a smooth function compactly supported in $B(x,\varepsilon)$ and such that $\psi(x)=1$. Finally, let $c>0$ be a positive constant such that $c^2h^{-1}(Q\nabla\psi,\nabla\psi)\le 1$ on Ω and set $\psi_c=c\psi$. Then,

$$d_{Q,V}(x,y_{n_k})\ge|\psi_c(x)-\psi_c(y_{n_k})|=|\psi_c(x)|=c$$

for every $k\in\mathbb{N}$, which is a contradiction. We have so proved that y_{n_k} converges to x as k tends to $+\infty$ and this, of course, cannot be the case since $x\in\text{supp}f\Subset\overline{B}$.

So, let us fix a function $f \in C_c^\infty(\Omega;\mathbb{C})$ and a bounded subset B of Ω containing the support of f and such that $d_{Q,V}(\mathrm{supp} f, \Omega\setminus B) =: \delta_{Q,V} > 0$. Using (43), we can estimate for every $z \in \Sigma_\theta$, with $|z| \le 1$,

$$\begin{aligned}
\|T(z)f - f\|_{L^1(\Omega)} &= \|T(z)f - f\|_{L^1(B)} + \|T(z)f - f\|_{L^1(\Omega\setminus B)}\\
&\le |B|^{\frac12}\|T(z)f - f\|_{L^2(\Omega)} + \|T(z)f\|_{L^1(\Omega\setminus B)}\\
&\le |B|^{\frac12}\|T(z)f - f\|_{L^2(\Omega)} + \int_{\Omega\setminus B}\int_{\mathrm{supp} f}|f(y)||k(z,x,y)|dydx\\
&\le |B|^{\frac12}\|T(z)f - f\|_{L^2(\Omega)}\\
&\quad + C_\theta(\mathrm{Re}\, z)^{-\frac d2}\exp\Big(-\frac{c_\theta}{2}\delta_{Q,V}^{\frac{2\alpha+2}{2\alpha+1}}|z|^{-\frac{1}{2\alpha+1}}\Big)\\
&\quad\times\int_{\mathrm{supp} f}|f(y)|\int_\Omega \exp\Big(-\frac{c_\theta}{2}\left(d_{Q,V}(x,y)\right)^{\frac{2\alpha+2}{2\alpha+1}}\Big)dxdy\\
&\le |B|^{\frac12}\|T(z)f - f\|_{L^2(\Omega)} + C'_\theta(\mathrm{Re} z)^{-\frac d2}\exp\Big(-\frac{c_\theta}{2}\delta_{Q,V}^{\frac{2\alpha+2}{2\alpha+1}}|z|^{-\frac{1}{2\alpha+1}}\Big),
\end{aligned} \tag{49}$$

where $C'_\theta = C_\theta\|f\|_1 \sup_{y\in\Omega}\|e^{-\frac{c_\theta}{2}(d_{Q,V}(\cdot,y))^{\frac{2\alpha+2}{2\alpha+1}}}\|_1$. By the strong continuity of $(T(z))_{z\in\Sigma_\theta}$ in $L^2(\Omega,\mathbb{C})$ and estimate (49), we get the claim.

An interpolation argument extends all the results proved in $L^1(\Omega;\mathbb{C})$ to $L^p(\Omega,\mathbb{C})$ for any $p \in]1,2[$ and, finally, arguing by duality we cover also the case $L^p(\Omega,\mathbb{C})$ for $p\in[2,+\infty[$. □

Corollary 5.1 *Assume that Hypotheses (A), (B), (C) hold true. Let A be the operator associated with the form* $\mathfrak{a}$ *introduced in* (11) *and assume that there exist* $\mu \in]0,+\infty[$ *and* $a \in]0,1[$ *such that*

$$\mathrm{Re}\,(Q(x)\xi,\xi) \le \mu(V(x))^a|\xi|^2 \tag{50}$$

for almost every $x\in\Omega$ *and every* $\xi\in\mathbb{C}^d$. *Then, the semigroup* $(T(t))$ *generated by the operator* $-A$ *is given by a kernel k satisfying the generalized Gaussian estimate*

$$|k(z,x,y)| \le C_\theta(\mathrm{Re} z)^{-\frac d2}e^{-c_\theta(\mu^{-1}|x-y|)^{\frac{2}{1+a}}|z|^{\frac{a-1}{1+a}}}, \tag{51}$$

for almost every $x, y \in \Omega$ *and any* $z\in\Sigma_\theta$ *with* $\theta\in[0,\theta_0[$. *Here,* C_θ*, and* c_θ *are the constants appearing in Theorem 5.1. Moreover,* $(T(t))$ *extends to a bounded analytic semigroup* $(T(z))_{z\in\Sigma_{\theta_0}}$ *on* $L^p(\Omega,\mathbb{C})$ *for any* $p\in[1,+\infty[$. *Finally, if in addition there exist* $\eta\in(0,\mu)$ *such that*

$$\eta(V(x))^a|\xi|^2 \le \mathrm{Re}\,(Q(x)\xi,\xi) \tag{52}$$

for almost every $x \in \Omega$ and every $\xi \in \mathbb{C}^d$, then k satisfies the estimate

$$
\begin{aligned}
k(t,x,y) \leq & H_1[(1-\kappa)\nu_0]^{-\frac{d}{2}} e^{-\frac{1-\kappa}{4}V_0 t} t^{-\frac{d}{2}} e^{-H_2(\mu^{-1}|x-y|)^{\frac{2}{1+a}} t^{-\frac{1-a}{1+a}}} \\
& \times \left(1+\frac{1-\kappa}{8}V_0 t + H_3\big(\eta^{-1}|x-y|\big)^{\frac{2}{1+a}} t^{-\frac{1-a}{1+a}}\right)^{\frac{d}{2}}, \qquad (53)
\end{aligned}
$$

for any $t > 0$ and almost every $x, y \in \Omega$. Here H_i $(i = 1, 2, 3)$ are the constants appearing in Theorem 4.1.

Proof It is enough to observe that we can choose $h = V^a$, and thus estimate (17) is satisfied with $\alpha = \frac{a}{1-a}$. Moreover, since

$$
d_{Q,V}(x,y) = \sup\{|\psi(x)-\psi(y)| \mid \psi \in C_b(\Omega)\cap C^1(\Omega),\ (V^{-a}Q\nabla\psi,\nabla\psi) \leq 1 \text{ a.e. in } \Omega\},
$$

for every $x, y \in \Omega$, by classical computations (see e.g. [10]) it holds that

$$
d_{Q,V}(x,y) \geq \mu^{-1}|x-y|, \qquad x, y \in \Omega
$$

and, consequently, condition (42) is trivially satisfied and Theorem 5.1 can be applied. For the last assertion, it suffices to observe that if (52) holds too, then $d_{Q,V}(x, y)$ is equivalent to the Euclidean distance. More precisely,

$$
\mu^{-1}|x-y| \leq d_{Q,V}(x,y) \leq \eta^{-1}|x-y|, \qquad x, y \in \Omega.
$$

□

Remark 5.1 We recall that, under some additional assumptions on the regularity of the coefficients and their growth, the domain of the generators of the semigroups in $L^p(\mathbb{R}^d)$ appearing in Theorem 5.1 is well described (see e.g., [23, 24])

6 Examples

In this section, we provide some classes of elliptic operators to which the main results of the paper apply.

Example 6.1 Let Ω be a subset of $\mathbb{R}^d$ which enjoys the extension property. To begin with, we first observe that, if $Q(x)$ is a symmetric real-valued matrix for almost every $x \in \Omega$, then condition (7) is satisfied with $q_0 = 0$. Now, let Q be a diagonal perturbation of an antisymmetric matrix-valued function with entries $q_{ij} \in L^\infty_{\rm loc}(\Omega)$. Let us assume that there exist positive constants c_0, C such that

$$
\inf_{x\in\Omega} q_{ii}(x) > c_0, \qquad i = 1, \ldots, d,
$$

and

$$\sum_{j\in\{1,\dots,d\}\setminus\{i\}} |q_{ij}(x)| \le Cq_{ii}(x), \qquad i \in \{1,\dots,d\}, \text{ a.e. } x \in \Omega.$$

In this case, for any $\zeta \in \mathbb{C}^d$ and almost every $x \in \Omega$, it holds that

$$(Q(x)\zeta,\zeta) = \sum_{i,j=1}^{d} q_{ij}\zeta_i\overline{\zeta}_j = \sum_{i=1}^{d} q_{ii}(x)|\zeta_i|^2 + \sum_{i=1}^{d}\sum_{j\in\{1,\dots,d\}\setminus\{i\}} q_{ij}\zeta_i\overline{\zeta}_j$$

whence

$$\operatorname{Re}(Q(x)\zeta,\zeta) = \sum_{i=1}^{d} q_{ii}(x)|\zeta_i|^2, \;\; \operatorname{Im}(Q(x)\zeta,\zeta) = \sum_{i=1}^{d}\sum_{j\in\{1,\dots,d\}\setminus\{i\}} q_{ij}(x)\zeta_i\overline{\zeta}_j.$$

Consequently, Hypothesis (C)-(a) is satisfied with $\nu_0 = c_0$ and, moreover,

$$\begin{aligned}
|\operatorname{Im}(Q(x)\zeta,\zeta)| &= \left|\sum_{i=1}^{d}\sum_{j\in\{1,\dots,d\}\setminus\{i\}} q_{ij}(x)\zeta_i\overline{\zeta}_j\right| \\
&\le \sum_{i=1}^{d}\sum_{j\in\{1,\dots,d\}\setminus\{i\}} |q_{ij}(x)||\zeta_i||\zeta_j| \\
&\le \frac{1}{2}\sum_{i=1}^{d}\sum_{j\in\{1,\dots,d\}\setminus\{i\}} |q_{ij}(x)|(|\zeta_i|^2+|\zeta_j|^2) \\
&= \sum_{i=1}^{d}|\zeta_i|^2\sum_{j\in\{1,\dots,d\}\setminus\{i\}} |q_{ij}(x)| \\
&\le C\sum_{i=1}^{d} q_{ii}(x)|\zeta_i|^2 = C\operatorname{Re}(Q(x)\zeta,\zeta),
\end{aligned}$$

for every $\xi \in \mathbb{C}^d$ and almost every $x \in \Omega$; whence condition (7) is satisfied with $q_0 = C$. Thus, if $V \in L^\infty_{\mathrm{loc}}(\Omega)$ is bounded from below (take Remark 3.3 into account) and B, C satisfy Hypotheses (B), then the semigroup $(T(t))$ generated by the realization of the opposite of the operator $(A, D(A))$ has a kernel k satisfying estimate (19).

If we assume further that $\operatorname{essinf}_\Omega V > 0$ and that there exist constants $C_i > 0$ $(i = 1,\dots,d)$ and $a \in (0,1)$ such that

$$q_{ii}(x) \le C_i(V(x))^a$$

for almost every $x \in \Omega$ and any $i = 1, \dots, d$, then it holds that

$$\mathrm{Re}\,(Q(x)\xi, \xi) \le \mu(V(x))^a |\xi|^2$$

for any $\xi \in \mathbb{C}^d$ and almost every $x \in \Omega$, where $\mu = \max_{i=1,\dots,d} C_i$. Thus, the first assertion in Corollary 5.1 guarantees the validity of the generalized Gaussian estimates (51) and, as consequence, that $(T(t))$ admits a bounded analytic extension in $L^p(\Omega, \mathbb{C})$ for any $p \in [1, +\infty[$ on a sector which does not depend on p. Finally, if there exist positive constants c_i $(i = 1, \dots, d)$ such that $q_{ii}(x) \ge c_i(V(x))^a$ for any $i = 1, \dots, d$ and almost every $x \in \Omega$, then estimate (52) holds too with $\eta = \min_{i=1,\dots,d} c_i$. Thus, again by Corollary 5.1, estimate (53) follows with $\nu_0 = c_0$ and $V_0 = \mathrm{essinf}_\Omega V$.

Example 6.2 Consider $\Omega = \mathbb{R}^2 \setminus [-1, 1] \times [-1, 1]$, $Q = f \begin{pmatrix} 1 & 1 \\ -1 & 1 \end{pmatrix}$ and $V = K \cdot f^\beta$, with $f \in L^\infty_{\mathrm{loc}}(\Omega)$, $K > 0$, $\beta > 1$ and $f(x, y) \ge f_0 > 0$ for almost every $(x, y) \in \Omega$. For example, we could consider $f(x, y) = \max\{|x|, |y|\}^b$ for some $b > 0$. It is immediate to check that

$$\mathrm{Re}\,(Q(x, y)\xi, \xi) = f(x, y)|\xi|^2 \ge |\mathrm{Im}\,(Q(x, y)\xi, \xi)|$$

for almost every $x \in \Omega$ and all $\xi \in \mathbb{C}^2$. Then, choosing $h = V^a$, with $a = \frac{1}{\beta} \in]0, 1[$ and observing that conditions (50) and (52) are both satisfied with $\mu = \eta = K^{-\frac{1}{\beta}}$, by Corollary 5.1 we can find positive constants H_1, H_2 and H_3 such that the semigroup $(T(t))$ generated by the operator $-A$ associated to the form

$$\mathfrak{a}(u, v) = \int_\Omega \left[(Q\nabla u, \nabla v) + V u\overline{v}\right]dx, \qquad u, v \in \widetilde{D}_{Q,V},$$

is given by a kernel k satisfying the estimate

$$\begin{aligned} k(t, x, y) \le & H_1 f_0^{-1} e^{-\frac{1}{4}Kf_0^\beta t} t^{-1} e^{-H_2 K^{\frac{2}{1+\beta}} |x-y|^{\frac{2\beta}{1+\beta}} t^{-\frac{\beta-1}{1+\beta}}} \\ & \times \left(1 + \frac{1}{8}Kf_0^\beta t + H_3 K^{\frac{2}{1+\beta}} |x-y|^{\frac{2\beta}{1+\beta}} t^{-\frac{\beta-1}{1+\beta}}\right) \end{aligned}$$

for every $t > 0$ and almost every $x, y \in \Omega$.

It is worth pointing out that in this case, if $u \in C^2(\mathbb{R}^2) \cap D_{Q,V}$, then by Green's formula it holds that $(\nabla u, Q^t v) = 0$ on $\partial\Omega$ and this means that $u_x = -u_y$ on $\{-1, 1\} \times]-1, 1[$ and $u_x = u_y$ on $]-1, 1[\times \{-1, 1\}$.

Example 6.3 Let $\Omega = \{x \in \mathbb{R}^2 : |x| > 1\}$. Assume that $Q(x) = q(|x|) \begin{pmatrix} 1 & 1 \\ -1 & 1 \end{pmatrix}$, $V(x) = v(|x|)$ for every $x \in \Omega$, where q and v are continuous functions on $[1, +\infty[$ such that $q(r) \ge \nu_0 > 0$ and $v(r) \ge \nu_0 > 0$ for every $r \ge 1$. Moreover, let

$B(x) = b(|x|)x$ for every $x \in \Omega$ and some continuous function $b : [1, +\infty[\to \mathbb{R}$ such that

$$|b(r)| \leq \kappa\sqrt{q(r)v(r)}, \qquad r \geq 1,$$

for some $\kappa < \frac{1}{2}$. Let $a \in]0, 1[$, $h(x) = (V(x))^a$ for every $x \in \Omega$ and

$$\sigma(r) = \int_1^r \sqrt{q(s)^{-1}(v(s))^a ds}, \qquad r > 1.$$

For any $\varphi \in C_b^1(\mathbb{R})$ such that $\|\nabla\varphi\|_\infty \leq 1$, we get that $(Q\nabla(\varphi o\sigma), \nabla(\varphi o\sigma)) \leq V^a$. Therefore,

$$\begin{aligned} d_{Q,V}(x, y) &\geq \sup\{|\varphi(\sigma(|x|)) - \varphi(\sigma(|y|))| \mid \varphi \in C_b^1(\mathbb{R}), \ \|\nabla\varphi\|_\infty \leq 1\} \\ &\geq |\sigma(|x|) - \sigma(|y|)| = \left|\int_{|x|}^{|y|} \sqrt{q(s)^{-1}v(s)^a ds}\right|. \end{aligned}$$

Hence, in this case, there exist positive constants H_1, H_2 and H_3 such that the semigroup $(T(t))$ generated by the operator $-A$ associated to the form

$$\mathfrak{a}(u, v) = \int_\Omega \left[(Q\nabla u, \nabla v) + bx \cdot \nabla u\overline{v} + Vu\overline{v}\right]dx, \qquad u, v \in \widetilde{D}_{Q,V}$$

is given by a kernel k satisfying the estimate

$$\begin{aligned} k(t, x, y) \leq H_1 \left((1 - \kappa)\nu_0\right)^{-1} e^{-\frac{1-\kappa}{4}v_0 t} t^{-1} \exp\left(- H_2\left|\int_{|x|}^{|y|} \sqrt{q(s)^{-1}v(s)^a ds}\right|^{\frac{2}{a+1}} t^{\frac{a-1}{a+1}}\right) \\ \times \left(1 + \frac{1-\kappa}{8} V_0 t + H_3 \left(d_{Q,V}(x, y)\right)^{\frac{2}{a+1}} t^{-\frac{1-a}{1+a}}\right) \end{aligned}$$

for every $t > 0$ and almost every $x, y \in \Omega$. If further $q \geq \eta v^a$ for some $\eta > 0$, then $d_{Q,V}(x, y) \leq \eta^{-1}|x - y|$ for every $(x, y) \in \mathbb{R}^2$, so we deduce that

$$\begin{aligned} k(t, x, y) \leq H_1 \left((1 - \kappa)\nu_0\right)^{-1} e^{-\frac{1-\kappa}{4}v_0 t} t^{-1} e^{-H_2\left|\int_{|x|}^{|y|} \sqrt{q(s)^{-1}v(s)^a ds}\right|^{\frac{2}{a+1}} t^{-\frac{1-a}{a+1}}} \\ \times \left(1 + \frac{1-\kappa}{8} v_0 t + H_3(\eta^{-1}|x - y|)^{\frac{2}{a+1}} t^{-\frac{1-a}{1+a}}\right) \end{aligned}$$

for every $t > 0$ and almost every $x, y \in \Omega$.

Observe that the interesting case holds when $\sqrt{q^{-1}v^a}$ is not summable on $[1, +\infty[$, so that $d_{Q,V}$ is not bounded. For example, if we consider $q(r) = (1+r)^\alpha$, $v(r) = (1 + r)^\beta$, then we can choose $a \in]0, 1[$ such that $\sqrt{q^{-1}v^a}$ is not summable if $\beta > \frac{\alpha-2}{a}$ and $\alpha > 2$.

Concerning the boundary conditions, we observe that writing $u = u(re^{i\theta})$ with $r \geq 1$, we see that a smooth u belongs to $\widetilde{D}_{Q,V}$ if and only if $u_r = u_\theta$ on $\partial\Omega$.

References

1. Angiuli, L., Lorenzi, L.: Compactness and invariance properties of evolution operators associated with Kolmogorov operators with unbounded coefficients. J. Math. Anal. Appl. **379**, 125–149 (2011)
2. Angiuli, L., Lorenzi, L.: Non autonomous parabolic problems with unbounded coefficients in unbounded domains. Adv. Differ. Equ. **20**, 1067–1118 (2015)
3. Angiuli, L., Lorenzi, L., Mangino, E., Rhandi, A.: On vector-valued Schrödinger operators with unbounded diffusion in L^p spaces. J. Evol. Equ. **21**, 3181–3204 (2021)
4. Arendt, W.: Semigroups and evolution equations: functional calculus, regularity and kernel estimates. In: C.M. Dafermos, E. Feireisl (eds.) Evolutionary Equations, vol. I, pp. 1–85. Handbook of Differential Equations. Elsevier, Amsterdam (2004)
5. Arendt, W., Ter Elst, A.F.M.: Gaussian estimates for second order elliptic operators with boundary conditions. J. Oper. Theory **38**, 87–130 (1997)
6. Arendt, W., Metafune, G., Pallara D.: Gaussian estimates for elliptic operators with unbounded drift. J. Math. Anal. Appl. **338**, 505–517 (2008)
7. Boutiah, S.E., Rhandi, A., Tacelli, C.: Kernel estimates for elliptic operators with unbounded diffusion, drift and potential terms. Discr. Contin. Dyn. Syst. **39**, 803–817 (2019)
8. Canale, A., Rhandi, A., Tacelli, C.: Kernel estimates for Schrödinger type operators with unbounded diffusion and potential terms. Z. Anal. Anwend. **36**, 377–392 (2017)
9. Daners, D.: Heat kernel estimates for operators with boundary conditions. Math. Nachr. **217**, 13–41 (2000)
10. Davies, E.B.: Explicit constants for Gaussian upper bounds on heat kernels. Am. Math. J. **109**, 319–333 (1987)
11. Davies, E.B.: Heat Kernels and Spectral Theory, vol. 92. Cambridge Tracts in Mathematics. Cambridge University Press, Cambridge (1989)
12. Davies, E.B., Simon, B.: Ultracontractivity and the heat kernel of Schrödinger operators and Dirichlet Laplacians. J. Funct. Anal. **59**, 335–395 (1984)
13. Durante, T., Manzo, R., Tacelli, C.: Kernel estimates for Schrödinger type operators with unbounded coefficients and critical exponents. Ric. Mat. **65**, 289–305 (2016)
14. Hieber, M., Lorenzi, L., Rhandi, A.: Second-order parabolic equations with unbounded coefficients in exterior domains. Differ. Integr. Equ. **20**, 1253–1284 (2007)
15. Hille, E., Phillips, R.S.: Functional Analysis and Semigroups, vol. 31. College Publications. American Mathematical Society, Providence (1957)
16. Kato, T.: Perturbation Theory of Linear Operators, 2nd edn., vol. 132. Grundlehren der Mathematischen Wissenschaft. Springer-Verlag, Berlin (1976)
17. Kunze, M., Lorenzi, L., Rhandi, A.: Kernel estimates for nonautonomous Kolmogorov equations with potential term. In: New Prospects in Direct, Inverse and Control Problems for Evolution Equations, vol. 10, pp. 229–251. Springer INdAM Series. Springer, Cham (2014)
18. Kunze, M., Lorenzi, L., Rhandi, A.: Kernel estimates for nonautonomous Kolmogorov equations. Adv. Math. **287**, 600–639 (2016)
19. Lorenzi, L.: Analytical Methods for Kolmogorov Equations. Monographs and Research Notes in Mathematics. CRC Press, Boca Raton (2017)
20. Lorenzi, L., Rhandi, A.: On Schrödinger type operators with unbounded coefficients: generation and heat kernel estimates. J. Evol. Equ. **15**, 53–88 (2015)
21. Metafune, G., Spina, C.: Kernel estimates for a class of Schrödinger semigroups. J. Evol. Equ. **7**, 719–742 (2007)

22. Metafune, G., Spina, C.: Heat kernel estimates for some elliptic operators with unbounded diffusion coefficients. Discrete Contin. Dynam. Syst. **32**, 2285–2299 (2012)
23. Metafune, G., Pallara, D., Prüss, J., Schnaubelt, R.: L^p-theory for elliptic operators on $\mathbb{R}^d$ with singular coefficients. Z. Anal. Anwend. **24**, 497–521 (2005)
24. Metafune, G., Prüss, J., Rhandi, A., Schnaubelt, R.: L^p-regularity for elliptic operators with unbounded coefficients. Adv. Differ. Equ. **10**, 1131–1164 (2005)
25. Metafune, G., Pallara, D., Rhandi A.: Kernel estimates for Schrödinger operators. J. Evol. Equ., **6**, 433–457 (2006)
26. Metafune, G., Sobajima, M., Spina, C: Kernel estimates for elliptic operators with second-order discontinuous coefficients. J. Evol. Equ. **17**, 485–522 (2017)
27. Ouhabaz, E.M.: Analysis of Heat Equations on Domains. London Mathematical Society Monographs. Princeton University Press, Princeton (2005)
28. Spina, C.: Kernel estimates for a class of Kolmogorov semigroups. Arch. Math. **91**, 265–279 (2008)
29. Triebel, H.: Interpolation Theory, Function Spaces, Differential Operators. North-Holland Publishing, Amsterdam/New York/Oxford (1978)

An Existence Result for Perturbed (p, q)-Quasilinear Elliptic Problems

Rossella Bartolo, Anna Maria Candela, and Addolorata Salvatore

Dedicated to Francesco Altomare, great mathematician and good friend

Abstract We investigate the existence of solutions of the (p, q)-quasilinear elliptic problem

$$\begin{cases} -\Delta_p u - \Delta_q u \ = \ g(x, u) + \varepsilon\, h(x, u) & \text{in } \Omega, \\ u = 0 & \text{on } \partial\Omega, \end{cases}$$

where Ω is an open bounded domain in $\mathbb{R}^N$, $1 < p < q < +\infty$, the nonlinearity $g(x, u)$ behaves at infinity as $|u|^{q-1}$, $\varepsilon \in \mathbb{R}$ and $h \in C(\overline{\Omega} \times \mathbb{R}, \mathbb{R})$. In spite of the possible lack of a variational structure of this problem, appropriate procedures and estimates allow us to prove the existence of at least one nontrivial solution for small perturbations.

Keywords (p, q)-quasilinear elliptic equation · Asymptotically q-linear problem · q-Laplacian · Variational methods · Essential value · Perturbed problem · Linking

1 Introduction

Classical semilinear and quasilinear equations can be perturbed just by adding continuous functions, with no assumption on their growth or their symmetry, so

R. Bartolo (✉)
Dipartimento di Meccanica, Matematica e Management, Politecnico di Bari, Bari, Italy
e-mail: rossella.bartolo@poliba.it

A. M. Candela · A. Salvatore
Dipartimento di Matematica, Università degli Studi di Bari Aldo Moro, Bari, Italy
e-mail: annamaria.candela@uniba.it; addolorata.salvatore@uniba.it

A. M. Candela et al. (eds.), *Recent Advances in Mathematical Analysis*,
Trends in Mathematics, https://doi.org/10.1007/978-3-031-20021-2_8

that their structure may lose its variational nature. More precisely, here we consider the following class of quasilinear elliptic problems:

$$(P_{\alpha,\varepsilon}) \qquad \begin{cases} -\alpha\Delta_p u - \Delta_q u = g(x,u) + \varepsilon\, h(x,u) & \text{in } \Omega, \\ u = 0 & \text{on } \partial\Omega, \end{cases}$$

with $\alpha \in \{0,1\}$, $1 < p < q < +\infty$, $\Delta_r u = \mathrm{div}(|\nabla u|^{r-2}\nabla u)$ if $r \in \{p,q\}$, $\varepsilon \in \mathbb{R}$, where Ω is an open bounded domain in $\mathbb{R}^N$ with Lipschitz boundary $\partial\Omega$, $N \geq 2$, while $g(x,u)$ and $h(x,u)$ are given functions on $\Omega \times \mathbb{R}$.

If $\alpha = 0$ and $q = 2$, results on multiple solutions of $(P_{0,\varepsilon})$ are stated in [23] for $g(x,\cdot)$ odd, superlinear at infinity, but subcritical (see also [18] for related results). On the other hand, if $g(x,\cdot)$ is asymptotically linear at infinity and both $g(x,\cdot)$ and $h(x,\cdot)$ are odd, a multiplicity theorem is proved in [24, Theorem 1.6] while, by means of the pseudo–index theory stated in [8], in [6, Theorems 1.1, 1.2] existence results are obtained even in presence of resonance as the number of distinct critical values of J is stable under small odd perturbations. Moreover, again in [6], more restrictive multiplicity results are obtained for non-odd functions $h(x,\cdot)$ (see [6, Theorems 1.3, 1.4]).

To our knowledge, for $q \neq 2$ problem $(P_{0,\varepsilon})$ has been studied only in [24, Theorem 1.8], assuming both $g(x,\cdot)$ and $h(x,\cdot)$ odd but $g(x,\cdot)$ "superlinear".

When the variational structure on $W_0^{1,q}(\Omega)$ of the equation in $(P_{0,\varepsilon})$ fails, we use the notion of essential value for perturbations of non-smooth functionals as introduced in [16, 17]; indeed, such values are preserved for small perturbations of a continuous functional. We note that essential values of functionals satisfying the Palais–Smale condition (or its variants) are also critical ones, while the reverse implication does not hold; furthermore, critical values arising from mini–max procedures are essential ones (we refer to Sect. 2.3 for more details).

On the other hand, for $q > p = 2$ problem $(P_{1,0})$ has been studied in [11, 20, 30]; while if $g(x,\cdot)$ is asymptotically "$(q-1)$-linear" at infinity, i.e., there exists

$$\lim_{|t|\to+\infty} \frac{g(x,t)}{|t|^{q-2}t} = \lambda_\infty \in \mathbb{R} \quad \text{uniformly in } \overline{\Omega},$$

and the problem is not resonant, i.e., $\lambda_\infty \notin \sigma(-\Delta_q)$, we refer to [12] for the existence of a nontrivial solution via Morse theory and to [4] for a multiplicity result. A further multiplicity result for $(P_{1,0})$ is contained in the recent paper [14].

At last, we recall that the asymptotically $(q-1)$-linear problem $(P_{0,0})$ has been widely investigated both for $q = 2$ (cf. [1, 3, 6] and references therein) and for $q \neq 2$ (for some existence results see [5, 15, 18, 23, 25, 28] while for some multiplicity ones see [5, 26, 28]). Moreover, for more recent related results we refer to [13].

In this paper, we want to investigate the existence of solutions for problem $(P_{1,\varepsilon})$ when $g(x,\cdot)$ is asymptotically $(q-1)$–linear at infinity and a perturbation term is allowed. More precisely, we consider $\alpha = 1$ and that there exist $\lambda_\infty \in \mathbb{R}$ and

$f : \overline{\Omega} \times \mathbb{R} \to \mathbb{R}$ such that

$$g(x, t) = \lambda_\infty |t|^{q-2}t + f(x, t) \quad \text{for all } (x, t) \in \overline{\Omega} \times \mathbb{R}; \tag{1}$$

hence, problem $(P_{1,\varepsilon})$ reduces to

$$(P_\varepsilon^\infty) \qquad \begin{cases} -\Delta_p u - \Delta_q u = \lambda_\infty |u|^{q-2}u + f(x, u) + \varepsilon\, h(x, u) & \text{in } \Omega, \\ u = 0 & \text{on } \partial\Omega. \end{cases}$$

On function $f : \overline{\Omega} \times \mathbb{R} \to \mathbb{R}$ we assume the following conditions:

(f_1) $f \in C(\overline{\Omega} \times \mathbb{R}, \mathbb{R})$;

(f_2) there exists

$$\lim_{|t| \to +\infty} \frac{f(x, t)}{|t|^{q-1}} = 0 \quad \text{uniformly in } \overline{\Omega};$$

(f_3) there exists

$$\lim_{t \to 0} \frac{f(x, t)}{|t|^{q-2}t} = \lambda_0 \in \mathbb{R} \setminus \{0\} \quad \text{uniformly in } \overline{\Omega}.$$

We note that, if assumption (f_1) is satisfied, then we can define the C^1 real function

$$F(x, t) = \int_0^t f(x, s)\, \mathrm{d}s \quad \text{for all } (x, t) \in \overline{\Omega} \times \mathbb{R} \tag{2}$$

which is so that $F(x, 0) = 0$ for all $x \in \overline{\Omega}$.

The behaviour of the nonlinearity as in (1) calls for a control of the interaction of $g(x, t)$ with the spectrum of $\sigma(-\Delta_q)$ which is mostly unknown for $q \neq 2$. Such a problem was overcome in [5] for $(P_{0,0})$ by taking into account two sequences of quasi-eigenvalues for $-\Delta_q$ in $W_0^{1,q}(\Omega)$ defined as in [10, 26], namely $(\eta_m^0)_m$ and $(\nu_m^0)_m$ (see Sect. 2.1 for their definitions), while here we prefer to use two sequences of quasi-eigenvalues for the (p, q)-Laplacian operator, denoted by $(\eta_m)_m$ and $(\nu_m)_m$, which are introduced in [14] along the lines of [10, 26] (see Sect. 2.1 for more details).

Firstly, we state an existence result which deals with the unperturbed case (P_0^∞).

Theorem 1.1 *Assume that* $(f_1) - (f_3)$ *hold and* $\lambda_\infty \notin \sigma(-\Delta_q)$. *Let* $k \in \mathbb{N}$ *be such that*

(Λ_1) *an integer $\bar{k} \geq k$ exists such that one of following assumptions holds:*

(*i*) *we have that*

$$\lambda_0 + \lambda_\infty < \eta_k, \qquad \frac{q}{p}\nu_k < \lambda_\infty \tag{3}$$

and

$$\nu_{k-1} < \nu_k = \nu_{k+1} = \ldots = \nu_{\bar{k}} < \eta_{\bar{k}+1}; \tag{4}$$

(*ii*) *we have that*

$$\lambda_\infty < \eta_k^0, \qquad \frac{q}{p}\nu_k < \lambda_0 + \lambda_\infty \tag{5}$$

and

$$\nu_{k-1} < \nu_k = \nu_{k+1} = \ldots = \nu_{\bar{k}} < \eta_{\bar{k}+1}^0; \tag{6}$$

(Λ_2) *a constant $\eta > 0$ exists such that*

$$\left[\frac{1}{p}(\nu_{k-1} + \eta) - \frac{\lambda_\infty}{q}\right]|t|^q \leq F(x, t) \quad \textit{for all } (x, t) \in \overline{\Omega} \times \mathbb{R}.$$

Then, problem (P_0^∞) has at least a nontrivial solution

Then, we are able to state the following result concerning the perturbed case.

Theorem 1.2 *Let (f_1) − (f_3) hold and assume that $\lambda_\infty \notin \sigma(-\Delta_q)$ and $k \in \mathbb{N}$ exists such that (Λ_1) − (Λ_2) are satisfied. If $h \in C(\overline{\Omega} \times \mathbb{R}, \mathbb{R})$ then $\bar{\varepsilon} > 0$ exists such that for all $|\varepsilon| \leq \bar{\varepsilon}$ problem (P_ε^∞) has at least one nontrivial solution.*

Remark 1.1 Theorem 1.1 holds even if we replace assumption (f_1) with the weaker hypothesis

(f_1)′ f is a Carathéodory function (i.e., $f(\cdot, t)$ is measurable in Ω for all $t \in \mathbb{R}$ and $f(x, \cdot)$ is continuous in $\mathbb{R}$ for a.e. $x \in \Omega$) and

$$\sup_{|t| \leq a} |f(\cdot, t)| \in L^\infty(\Omega) \qquad \text{for all } a > 0;$$

but such a replacement does not work in Theorem 1.2 if a perturbation term is involved.

It is worth to point out that $q > p$ is not an assumption; indeed, the roles of p and q are interchangeable. Moreover, it is understood that by a solution we mean a weak solution, i.e., a function $u \in W_0^{1,q}(\Omega)$ solving the problems in the sense of distributions. We notice also that, under our assumptions, such weak solutions belong to $C^{1,\beta}(\overline{\Omega})$ for some $\beta \in]0, 1]$ (e.g., see [22, Remark 1.3]).

At last, we point out that the arguments we use for the proof of Theorem 1.2 still apply to the single q-Laplacian perturbed problem $(P_{0,\varepsilon})$; hence, being $\eta_m^0 \leq \nu_m^0$ for all $m \in \mathbb{N}$ (see [5, Proposition 2.9]), we obtain the following new existence result.

Corollary 1.1 *Assume that $(f_1)-(f_3)$ hold, $\lambda_\infty \notin \sigma(-\Delta_q)$ and $h \in C(\overline{\Omega}\times\mathbb{R}, \mathbb{R})$. Moreover, let $k \in \mathbb{N}$ be such that*

$(\Lambda_1)'$ *an integer $\bar{k} \geq k$ exists such that*

$$\min\{\lambda_0 + \lambda_\infty, \lambda_\infty\} < \eta_k^0 \leq \nu_k^0 < \max\{\lambda_0 + \lambda_\infty, \lambda_\infty\}$$

and

$$\nu_{k-1}^0 < \nu_k^0 = \nu_{k+1}^0 = \ldots = \nu_{\bar{k}}^0 < \eta_{\bar{k}+1}^0;$$

$(\Lambda_2)'$ *a constant $\eta > 0$ exists such that*

$$(\nu_{k-1}^0 + \eta - \lambda_\infty)\,\frac{|t|^q}{q} \leq F(x, t) \quad \text{for all } (x, t) \in \overline{\Omega} \times \mathbb{R}.$$

Then, $\bar{\varepsilon} > 0$ exists such that for all $|\varepsilon| \leq \bar{\varepsilon}$ problem $(P_{0,\varepsilon})$ has at least one nontrivial solution.

Remark 1.2

(*a*) We note that, being $p < q$, in (3), respectively (5), it has to be $\eta_k < \frac{q}{p}\nu_k$, respectively $\eta_k^0 < \frac{q}{p}\nu_k$ (see Proposition 2.1). Therefore, the two conditions in (3) can be written as the chain of inequalities:

$$\lambda_0 + \lambda_\infty < \eta_k < \frac{q}{p}\nu_k < \lambda_\infty$$

so that it has to be $\lambda_0 < 0$. Similarly, (5) becomes

$$\lambda_\infty < \eta_k^0 < \frac{q}{p}\nu_k < \lambda_0 + \lambda_\infty,$$

and, then, it has to be $\lambda_0 > 0$. According to [1], we do not know whether an existence result holds for $\lambda_0 = 0$ or not.

(*b*) Instead of (3) and (5), we could state both Theorems 1.1 and 1.2 by requiring

$$\min\{\lambda_0 + \lambda_\infty, \lambda_\infty\} < \eta_k^0 < \frac{q}{p}\nu_k < \max\{\lambda_0 + \lambda_\infty, \lambda_\infty\}$$

but assuming

$$\eta_k^0 \leq \eta_k \quad \text{if } \lambda_0 < 0. \tag{7}$$

Anyway, even if there are not significant changes in the proof, the assumption (7) may be more restrictive since such inequality holds for $k = 1$, but we do not know if it is true also for other $k \geq 2$.

(c) All the existence results in Theorems 1.1, 1.2 and Corollary 1.1 hold also when the limit in (f_3) is infinite (see Propositions 3.2 and 3.3).

This paper is organized as follows: in Sect. 2 we present the tools we are going to use while in Sects. 3 and 4 we prove Theorems 1.1 and 1.2, respectively.

Our strategy is the following: inspired by [23] (see also [6]) we prove the existence of at least one nontrivial solution of (P_0^∞) by using standard critical point theorems. Then, by means of cut-functions, we introduce perturbations of the functional associated to problem (P_0^∞) which have essential values near to the critical level of the solution for ε small enough. Such essential values turn out to be critical ones and suitable procedures—deeply different from those in [6], where the decomposition of $W_0^{1,2}(\Omega)$ by means of eigenvalues is exploited—allow us to prove that the L^∞-norm of the family of critical points of the cut-perturbed functionals is bounded. Hence, finally, solutions of (P_ε^∞) can be found.

2 Preliminary Material

Throughout this paper, we will use the following notations:

- $\bar{\mathbb{R}} = \mathbb{R} \cup \{\pm\infty\}$;
- $(X, \|\cdot\|_X)$ Banach space with $(X', \|\cdot\|_{X'})$ its dual;
- $I : X \to \mathbb{R}$ C^1-functional;
- $I^b = \{u \in X : I(u) \leq b\}$ sublevel of I corresponding to $b \in \mathbb{R}$;
- $I_b = \{u \in X : I(u) \geq b\}$ superlevel of I corresponding to $b \in \mathbb{R}$;
- $|\cdot|_s$ standard norm on the Lebesgue space $L^s(\Omega)$, $1 \leq s \leq +\infty$;
- $\|\cdot\|_q$ standard norm on $W_0^{1,q}(\Omega)$, i.e., $\|u\|_q = |\nabla u|_q$ for all $u \in W_0^{1,q}(\Omega)$;
- $(W^{-1,q'}(\Omega), \|\cdot\|_{W^{-1,q'}})$ dual space of $W_0^{1,q}(\Omega)$;
- $q^* = \frac{qN}{N-q}$ if $q < N$, $q^* = +\infty$ otherwise;
- $B_R = \{u \in W_0^{1,q}(\Omega) : \|u\|_q < R\}$ for any $R > 0$;
- $\overline{B}_R = \{u \in W_0^{1,q}(\Omega) : \|u\|_q \leq R\}$ for any $R > 0$;
- $S_R = \{u \in W_0^{1,q}(\Omega) : \|u\|_q = R\}$ for any $R > 0$.

Moreover, by K_j, $j \in \mathbb{N}$, we denote any positive constant which appears in the proofs and, for simplicity, we denote by $(\beta_m)_m$ any infinitesimal sequence which depends only on a given sequence of functions and by $(\beta_m(\varphi))_m$ any infinitesimal sequence which depends also on a fixed function φ.

2.1 Quasi-Eigenvalues for the Operator $-\Delta_p - \Delta_q$

It is well known that, if $q = 2$, the spectrum $\sigma(-\Delta_2)$ of $-\Delta_2$ in $W_0^{1,2}(\Omega)$ consists of a diverging sequence $(\lambda_m)_m$ of eigenvalues, repeated according to their multiplicity, so that

$$0 < \lambda_1 < \lambda_2 \leq \ldots \leq \lambda_m \leq \ldots,$$

which furnishes a decomposition of the Hilbert space $W_0^{1,2}(\Omega)$. Then, denoting by $(\varphi_m)_m$ the sequence of the corresponding eigenfunctions, for each $m \in \mathbb{N}$ the following inequalities hold:

$$\lambda_m |u|_2^2 \geq |\nabla u|_2^2 \quad \text{for all } u \in V_m$$

and

$$\lambda_m |u|_2^2 \leq |\nabla u|_2^2 \quad \text{for all } u \in W_{m-1}$$

with

$$V_m = \text{span}\{\varphi_1, \ldots, \varphi_m\}, \qquad W_m = V_m^{\perp}.$$

Instead, in the quasilinear case $q \neq 2$, the spectral properties of the q–Laplacian $-\Delta_q$ in the Sobolev space $W_0^{1,q}(\Omega)$ are still mostly unknown; indeed, when $N \geq 2$ it is not known whether the unbounded and increasing sequences of eigenvalues in [2, 21, 27, 28] cover the whole spectrum $\sigma(-\Delta_q)$ of $-\Delta_q$ in $W_0^{1,q}(\Omega)$ or not. Furthermore, unlike the case $q = 2$, the eigenvalues do not furnish a decomposition of the Banach space $W_0^{1,q}(\Omega)$. For these reasons in the $(q-1)$-asymptotically linear case we are studying, it is useful to consider two sequences of quasi-eigenvalues (cf. [5, Section 2]).

The first eigenvalue of $-\Delta_q$, denoted by $\lambda_1^{(q)}$, is characterized by

$$\lambda_1^{(q)} = \inf_{u \in W_0^{1,q}(\Omega) \setminus \{0\}} \frac{|\nabla u|_q^q}{|u|_q^q}$$

and is positive, simple, isolated with a unique positive eigenfunction $\varphi_1^{(q)}$ having unitary L^q-norm (cf., e.g., [27]).

In [10, Section 5], starting from $\eta_1^0 = \lambda_1^{(q)}$ and $\psi_1^0 \equiv \varphi_1^{(q)}$, it is shown the existence of an increasing diverging sequence $(\eta_m^0)_m$ of positive real numbers and a corresponding sequence of functions $(\psi_m^0)_m$, with $\psi_m^0 \neq \psi_n^0$ if $m \neq n$, such that

$$|\psi_m^0|_q = 1 \quad \text{and} \quad \eta_m^0 = |\nabla \psi_m^0|_q^q \qquad \text{for all } m \in \mathbb{N}. \tag{8}$$

Moreover, such a sequence generates the whole space $W_0^{1,q}(\Omega)$ and is such that

$$W_0^{1,q}(\Omega) = Y_m^0 \oplus Z_m^0 \quad \text{for all } m \in \mathbb{N},$$

where $Y_m^0 = \text{span}\{\psi_1^0, \ldots, \psi_m^0\}$ and its complement Z_m^0 can be explicitly described.

We recall that if $Y \subseteq X$ is a closed subspace of a Banach space X, a subspace $Z \subseteq X$ is a *topological complement* of Y, briefly $X = Y \oplus Z$, if Z is closed and every $x \in X$ can be uniquely written as $y + z$, with $y \in Y$ and $z \in Z$; furthermore, the projection operators onto Y and Z are (linear and) continuous and $L = L(Y, Z) > 0$ exists such that

$$\|y\| + \|z\| \le L\|y + z\| \quad \text{for all } y \in Y,\ z \in Z \tag{9}$$

(see, e.g., [9, p. 38]).

Remarkably, for all $m \in \mathbb{N}$ on the infinite dimensional subspace Z_{m-1} the following inequality holds:

$$\eta_m^0 |u|_q^q \le |\nabla u|_q^q \quad \text{for all } u \in Z_{m-1}^0 \tag{10}$$

(cf. [10, Lemma 5.4]).

Unluckily, it is not known whether, by making use of this sequence of quasi-eigenvalues, a reversed inequality holds on finite dimensional subspaces. Then, as in [26], we define another sequence of quasi-eigenvalues. More precisely, for all $m \in \mathbb{N}$ we consider the set of subspaces

$$\mathbb{W}_m^0 = \{Y \subset W_0^{1,q}(\Omega) :\ Y \text{ is a subspace of } W_0^{1,q}(\Omega), \varphi_1^{(q)} \in Y \text{ and } \dim Y \ge m\}$$

and define

$$\nu_m^0 = \inf_{Y \in \mathbb{W}_m^0} \sup_{u \in Y \setminus \{0\}} \frac{|\nabla u|_q^q}{|u|_q^q}.$$

The main properties of such a sequence are the following: $\nu_1^0 = \lambda_1^{(q)}$, $(\nu_m^0)_m$ is an increasing diverging sequence and, if $q = 2$, it agrees with $(\lambda_m)_m$ (cf. [26]). Furthermore, as already pointed out in Sect. 1,

$$\eta_m^0 \le \nu_m^0 \quad \text{for all } m \in \mathbb{N}$$

(see [5, Proposition 2.9])

Moreover, since here we deal with (p, q)-Laplacian problems, it is convenient to use also two sequences of quasi-eigenvalues for the operator $-\Delta_p - \Delta_q$ with zero Dirichlet boundary conditions as introduced in [14, Subsection 2.3] where, overcoming the lack of homogeneity, the previous costructions are extended to the

(p, q)-Laplacian operator. More precisely, starting from

$$\eta_1 := \inf_{\substack{u \in W_0^{1,q}(\Omega) \\ |u|_q = 1}} \left(|\nabla u|_p^p + |\nabla u|_q^q \right) \geq \lambda_1^{(q)},$$

attained by a function $\psi_1 \in W_0^{1,q}(\Omega)$ with $|\psi_1|_q = 1$, it is defined an increasing, diverging sequence $(\eta_m)_m$ of positive real numbers and a corresponding sequence of functions $(\psi_m)_m \subset W_0^{1,q}(\Omega)$ such that $\psi_m \neq \psi_n$ if $m \neq n$ and

$$|\psi_m|_q = 1 \quad \text{and} \quad \eta_m = |\nabla \psi_m|_p^p + |\nabla \psi_m|_q^q \qquad \text{for all } m \in \mathbb{N}. \tag{11}$$

As shown in [14, Lemma 2.6], these functions generate the whole space $W_0^{1,q}(\Omega)$ and for all $m \in \mathbb{N}$ it results

$$W_0^{1,q}(\Omega) = Y_m \oplus Z_m, \tag{12}$$

with $Y_m = \operatorname{span}\{\psi_1, \ldots, \psi_m\}$ and Z_m its topological complement, and the following inequalities hold:

$$\eta_m |u|_q^q \leq |\nabla u|_p^p + |\nabla u|_q^q \qquad \text{for all } u \in Z_{m-1} \cap \{u \in W_0^{1,q}(\Omega) : |u|_q \leq 1\} \tag{13}$$

and

$$\eta_m |u|_q^p \leq |\nabla u|_p^p + |\nabla u|_q^q \qquad \text{for all } u \in Z_{m-1} \setminus \{u \in W_0^{1,q}(\Omega) : |u|_q \leq 1\}.$$

On the other hand, in order to deal with finite dimensional spaces, for all $m \in \mathbb{N}$ we set

$$\mathbb{W}_m = \{Y \subset W_0^{1,q}(\Omega) : Y \text{ subspace of } W_0^{1,q}(\Omega),\ \psi_1 \in Y \text{ and } \dim Y \geq m\} \tag{14}$$

and

$$\nu_m = \inf_{Y \in \mathbb{W}_m} \sup_{u \in Y \setminus \{0\}} \frac{|\nabla u|_p^p + |\nabla u|_q^q}{|u|_q^q}. \tag{15}$$

Again, $(\nu_m)_m$ is increasing and a comparison between such a sequence and the previous ones can be established.

Proposition 2.1 *If $(\eta_m^0)_m$, $(\eta_m)_m$ and $(\nu_m)_m$ are sequences of quasi-eigenvalues defined as above, then it results*

$$\eta_m^0 \leq \nu_m, \quad \eta_m \leq \nu_m \qquad \textit{for all } m \in \mathbb{N}.$$

Proof Fixing $m \in \mathbb{N}$, inequality (10) holds on Z^0_{m-1}, with codim $Z^0_{m-1} = m-1$, while taking any $\sigma > 0$ from (15) a subspace $Y \in \mathbb{W}_m$ exists such that

$$\sup_{u \in Y \setminus \{0\}} \frac{|\nabla u|_p^p + |\nabla u|_q^q}{|u|_q^q} < \nu_m + \sigma.$$

Therefore, since (14) implies $\dim Y \geq m$, an element $\bar{u} \in \left(Y \cap Z^0_{m-1}\right) \setminus \{0\}$ exists such that

$$\eta^0_m \leq \frac{|\nabla \bar{u}|_q^q}{|\bar{u}|_q^q} < \frac{|\nabla \bar{u}|_p^p + |\nabla \bar{u}|_q^q}{|\bar{u}|_q^q} < \nu_m + \sigma.$$

Hence, being σ arbitrary, it has to be $\eta^0_m \leq \nu_m$.

On the other hand, fixing any $\sigma > 0$, reasoning as before but from (13)–(15), an element $\bar{v} \in (Y \cap Z_{m-1}) \setminus \{0\}$ exists, with $|\bar{v}|_q = 1$, which gives $\eta_m < \nu_m + \sigma$ so that, again for the arbitrariness of σ, it follows $\eta_m \leq \nu_m$. □

2.2 *Variational Tools*

In what follows we recall widely known definitions and results which apply to (P_0^∞) under our assumptions.

Firstly, we recall that a functional I satisfies the *Palais–Smale condition at level* c, $c \in \mathbb{R}$, briefly $(PS)_c$, if any sequence $(u_m)_m \subseteq X$ such that

$$\lim_{m \to +\infty} I(u_m) = c \quad \text{and} \quad \lim_{m \to +\infty} \|dI(u_m)\|_{X'} = 0$$

converges in X, up to subsequences.

If $-\infty \leq a < b \leq +\infty$, we say that I satisfies (PS) in $]a, b[$ if so is at each level $c \in]a, b[$.

Then, in order to state a classical existence critical point theorem, we recall the definition of sets which link as follows (e.g., see [31, Section II.8]).

Definition 2.1 Taking a subspace Y of X, let $S \subseteq X$ be a closed subset of X and consider $Q \subseteq Y$ with boundary ∂Q with respect to Y. Then, S and ∂Q *link* if

- $S \cap \partial Q = \emptyset$,
- $\phi(Q) \cap S \neq \emptyset$ for any $\phi \in C(X, X)$ such that $\phi|_{\partial Q} = \text{id}$.

For further use, we recall two examples of linking sets (cf. [31, Examples II.8.2 and II.8.3] and also [3, Propositions 2.1 and 2.2] in the case of an Hilbert space).

Example 2.1 Let V, W be two closed subspaces of X such that $X = V \oplus W$ and $\dim V < +\infty$. Then, setting $Q = \overline{B}_R \cap V$ for $R > 0$ and $S = W$, we have that S and ∂Q link.

Example 2.2 Let V, W be two closed subspaces of X such that $X = V \oplus W$, $\dim V < +\infty$, and fix $e \in W$ with $\|e\|_X = 1$. If $R_1, R_2, \rho > 0$ and

$$S = S_\rho \cap W, \qquad Q = \{te : t \in [0, R_1]\} \oplus \left(\overline{B}_{R_2} \cap V\right), \qquad Y = V \oplus \operatorname{span}\{e\},$$

then S and ∂Q link whenever $R_1 > \rho$.

The following linking theorem holds (cf., e.g., [3, Theorem 2.3] with the weaker Cerami's variant of Palais–Smale condition or [32, Theorem 2.12]).

Theorem 2.1 *Consider $a, b, \alpha, \beta \in \bar{\mathbb{R}}$ such that $a < \alpha < \beta < b$. Assume that:*

(i) the functional I satisfies (PS) in $]a, b[$;
(ii) two subsets S and Q exist such that S is closed in X, $Q \subseteq Y$, with Y subspace of X and ∂Q boundary of Q in Y, and the following assumptions are satisfied:

  (a) $I(u) \le \alpha$ for all $u \in \partial Q$ and $I(u) \ge \beta$ for all $u \in S$;
  (b) S and ∂Q link;
  (c) $\sup_{u \in Q} I(u) < +\infty$.

Then, a critical level c of I exists and is given by

$$c = \inf_{\phi \in \Gamma} \sup_{u \in Q} I(\phi(u)), \qquad \textit{with} \qquad \beta \le c \le \sup_{u \in Q} I(u),$$

where $\Gamma = \left\{\phi \in C(X, X) : \phi|_{\partial Q} = \mathrm{id}\right\}$.

2.3 Essential Values

As already pointed out, we may deal with problems without a variational structure on $W_0^{1,q}(\Omega)$. Hence, following [23], we use the auxiliary notion of essential value as introduced in [17] for the study of perturbations of nonsmooth functionals (see also [16]). We note that: the notion of essential value is topological, an essential value is candidate to be a critical level and is stable under small perturbations, critical levels arising from standard mini–max procedures are essential ones.

Definition 2.2 Let $I : X \to \mathbb{R}$ be continuous and $a, b \in \bar{\mathbb{R}}$, with $a \le b$. The pair (I^b, I^a) is *trivial* if, for each neighbourhood $[\alpha', \alpha'']$ of a and $[\beta', \beta'']$ of b in $\bar{\mathbb{R}}$, a continuous map $\varphi : I^{\beta'} \times [0, 1] \to I^{\beta''}$ exists such that

(i) $\varphi(x, 0) = x$ for each $x \in I^{\beta'}$;
(ii) $\varphi(I^{\beta'} \times \{1\}) \subseteq I^{\alpha''}$;
(iii) $\varphi(I^{\alpha'} \times [0, 1]) \subseteq I^{\alpha''}$.

Since the lack of critical values for a smooth functional may give trivial pairs (see the proof of [17, Theorem 3.1]), the following definition allows one to locate possible critical levels.

Definition 2.3 Let $I : X \to \mathbb{R}$ be a continuous function. A real number c is an *essential value* of I if for each $\varepsilon > 0$ two values $a, b \in]c - \varepsilon, c + \varepsilon[$, $a < b$, exist such that the pair (I^b, I^a) is not trivial.

The following theorem states that small perturbations of a continuous functional preserve the essential values (cf. [17, Theorem 3.1] or also [16, Theorem 2.6]).

Theorem 2.2 *Let $c \in \mathbb{R}$ be an essential value of $I : X \to \mathbb{R}$ continuous function. Then, for every $\eta > 0$ a constant $\delta > 0$ exists such that every functional $G \in C(X, \mathbb{R})$ with*

$$\sup\{|I(u) - G(u)| : u \in X\} < \delta$$

admits an essential value in $]c - \eta, c + \eta[$.

Now, we focus on the setting of smooth functionals and recall some results which link critical and essential values, stating in particular that the critical values arising from mini–max procedures are essential, provided that all the involved deformations are of the "same kind" (see [17, Theorems 3.7 and 3.9]).

Theorem 2.3 *Let $c \in \mathbb{R}$ be an essential value of $I \in C^1(X, \mathbb{R})$. If $(PS)_c$ holds, then c is a critical value of I.*

Remark 2.1 In general, the reverse implication does not hold when the Palais–Smale condition is satisfied since a critical value is not necessarily an essential one (see, e.g., [17, Example 3.12]).

Theorem 2.4 *Taking $I \in C^1(X, \mathbb{R})$, assume that Γ, non empty family of non empty subsets of X, and $d \in \mathbb{R} \cup \{-\infty\}$ are such that*

$$\overline{\varphi(C \times \{1\})} \in \Gamma$$

for every $C \in \Gamma$ and for every continuous deformation $\varphi : X \times [0, 1] \longrightarrow X$ with $\varphi(u, t) = u$ on $I^d \times [0, 1]$. Then, setting

$$c = \inf_{C \in \Gamma} \sup_{u \in C} I(u),$$

if $d < c < +\infty$ we have that c is an essential value of I.

3 The Unperturbed Case

As announced in Sect. 1, at first we deal with the unperturbed problem; it has a variational structure and here we present the variational framework needed in order to study it. Let us consider

$$(P_0^\infty) \qquad \begin{cases} -\Delta_p u - \Delta_q u = \lambda_\infty |u|^{q-2}u + f(x,u) & \text{in } \Omega, \\ u = 0 & \text{on } \partial\Omega. \end{cases}$$

To this aim, we note that from (f_1) and (f_2) for all $\sigma > 0$ a constant $K_\sigma > 0$ exists such that

$$|f(x,t)| \le \sigma |t|^{q-1} + K_\sigma \qquad \text{for all } (x,t) \in \overline{\Omega} \times \mathbb{R}. \tag{16}$$

Hence, taking $F(x,t)$ as in (2), classical variational theorems imply that the weak solutions of problem (P_0^∞) are the critical points of the C^1–functional

$$J(u) = \frac{1}{p}\int_\Omega |\nabla u|^p \, dx + \frac{1}{q}\int_\Omega |\nabla u|^q \, dx - \frac{\lambda_\infty}{q}\int_\Omega |u|^q \, dx \\ - \int_\Omega F(x,u)\, dx \tag{17}$$

on $W_0^{1,q}(\Omega)$, with

$$\langle dJ(u), v\rangle = \int_\Omega |\nabla u|^{p-2}\nabla u \cdot \nabla v \, dx + \int_\Omega |\nabla u|^{q-2}\nabla u \cdot \nabla v \, dx \\ - \lambda_\infty \int_\Omega |u|^{q-2}u\, v\, dx - \int_\Omega f(x,u) v\, dx \tag{18}$$

for all $u, v \in W_0^{1,q}(\Omega)$ (see, e.g., [19, Theorem 9 and p. 355]).

Now, we prove that the functional J satisfies the Palais–Smale condition (cf. also [4, Proposition 3.1]). We point out that here assumption (f_3), i.e. the behaviour of f near 0, is not needed, while it will be crucial in order to obtain the geometric assumptions required in the linking theorem (we refer to [14, Lemma 3.2] for the proof in the resonant case under an additional assumption as in [26]).

Proposition 3.1 *Assume that (f_1)–(f_2) hold and $\lambda_\infty \notin \sigma(-\Delta_q)$. Then, the functional J in* (17) *satisfies (PS) in $\mathbb{R}$.*

Proof Taking $c \in \mathbb{R}$, let $(u_m)_m$ be a sequence in $W_0^{1,q}(\Omega)$ such that

$$\lim_{m\to+\infty} J(u_m) = c \quad \text{and} \quad \lim_{m\to+\infty} \|dJ(u_m)\|_{W^{-1,q'}} = 0. \tag{19}$$

Firstly, we note that from (18) and (19) taking any $\varphi \in W_0^{1,q}(\Omega)$ it has to be

$$\left| \int_\Omega |\nabla u_m|^{p-2} \nabla u_m \cdot \nabla \varphi \, dx + \int_\Omega |\nabla u_m|^{q-2} \nabla u_m \cdot \nabla \varphi \, dx \right. \\ \left. - \lambda_\infty \int_\Omega |u_m|^{q-2} u_m \, \varphi \, dx - \int_\Omega f(x, u_m)\varphi \, dx \right| \leq \beta_m \|\varphi\|_q. \tag{20}$$

Then, since it is enough to show that $(\|u_m\|_q)_m$ is bounded (cf., e.g., [19, Lemma 2]), arguing by contradiction we assume that, up to subsequences, it is

$$\|u_m\|_q \to +\infty \quad \text{as } m \to +\infty. \tag{21}$$

Thus, without loss of generality, for all $m \in \mathbb{N}$ we can consider $\|u_m\|_q > 0$ and set

$$w_m = \frac{u_m}{\|u_m\|_q} \quad \text{with, clearly, } \|w_m\|_q = 1. \tag{22}$$

So, being $(w_m)_m$ bounded in $W_0^{1,q}(\Omega)$, an element $w \in W_0^{1,q}(\Omega)$ exists such that, up to subsequences, we have

$$w_m \rightharpoonup w \quad \text{weakly in } W_0^{1,q}(\Omega), \tag{23}$$

$$w_m \to w \quad \text{strongly in } L^q(\Omega). \tag{24}$$

Now, replacing φ in (20) with $\varphi_m = \frac{w_m - w}{\|u_m\|_q^{q-1}}$, as (21) and (22) imply $\|\varphi_m\|_q \to 0$, we get

$$\int_\Omega \frac{|\nabla w_m|^{p-2}}{\|u_m\|_q^{q-p}} \nabla w_m \cdot \nabla(w_m - w) \, dx + \int_\Omega |\nabla w_m|^{q-2} \nabla w_m \cdot \nabla(w_m - w) \, dx \\ = \lambda_\infty \int_\Omega |w_m|^{q-2} w_m \, (w_m - w) \, dx + \int_\Omega \frac{f(x, u_m)}{\|u_m\|_q^{q-1}} (w_m - w) \, dx + \beta_m,$$

where from Hölder inequality and (24) it follows that

$$\left| \int_\Omega |w_m|^{q-2} w_m (w_m - w) \, dx \right| \leq |w_m|_q^{q-1} |w_m - w|_q = \beta_m,$$

while (16), (21) and, again, (24) imply that

$$\left| \int_\Omega \frac{f(x, u_m)}{\|u_m\|_q^{q-1}} (w_m - w) \, dx \right| \leq \sigma |w_m|_q^{q-1} |w_m - w|_q + \frac{K_\sigma}{\|u_m\|_q^{q-1}} |w_m - w|_1 = \beta_m,$$

and, since $W_0^{1,q}(\Omega) \subset W_0^{1,p}(\Omega)$ being $q > p > 1$, from (21), (22) and direct computations we have that

$$\left| \int_\Omega \frac{|\nabla w_m|^{p-2}}{\|u_m\|_q^{q-p}} \nabla w_m \cdot \nabla(w_m - w)\, dx \right| \leq \frac{1}{\|u_m\|_q^{q-p}} |\nabla w_m|_p^{p-1} |\nabla(w_m - w)|_p$$

$$\leq \frac{K_1}{\|u_m\|_q^{q-p}} \|w_m\|_q^{p-1} \|w_m - w\|_q \leq \frac{K_2}{\|u_m\|_q^{q-p}} = \beta_m.$$

Hence, from all the previous estimates we obtain

$$\int_\Omega |\nabla w_m|^{q-2} \nabla w_m \cdot (\nabla w_m - \nabla w)\, dx = \beta_m,$$

which, together with (23), implies

$$w_m \to w \quad \text{strongly in } W_0^{1,q}(\Omega) \tag{25}$$

(see [19, Theorem 10]) with $w \neq 0$ from definition (22).

Now, taking any $\varphi \in W_0^{1,q}(\Omega)$ and dividing (20) by $\|u_m\|_q^{q-1}$, we have that

$$\begin{aligned} &\int_\Omega \frac{|\nabla w_m|^{p-2}}{\|u_m\|_q^{q-p}} \nabla w_m \cdot \nabla\varphi\, dx + \int_\Omega |\nabla w_m|^{q-2} \nabla w_m \cdot \nabla\varphi\, dx \\ &= \lambda_\infty \int_\Omega |w_m|^{q-2} w_m\, \varphi\, dx + \int_\Omega \frac{f(x, u_m)}{\|u_m\|_q^{q-1}} \varphi\, dx + \beta_m(\varphi), \end{aligned} \tag{26}$$

where, by reasoning as before, (21) and (22) imply

$$\left| \int_\Omega \frac{|\nabla w_m|^{p-2}}{\|u_m\|_q^{q-p}} \nabla w_m \cdot \nabla\varphi\, dx \right| \leq \frac{K_3}{\|u_m\|_q^{q-p}} \|\varphi\|_q = \beta_m \|\varphi\|_q. \tag{27}$$

We claim that

$$\lim_{m \to +\infty} \int_\Omega \frac{f(x, u_m)}{\|u_m\|_q^{q-1}} \varphi\, dx = 0. \tag{28}$$

In fact, taking any $\varepsilon > 0$, since from (24) we have $|w_m|_q^{q-1} \leq K_4$ for all $m \in \mathbb{N}$, for the arbitrariness of the possible choice of $\sigma > 0$ in (16), we can fix $\sigma = \frac{\varepsilon}{2K_4(\|\varphi\|_q + 1)}$ and, for the corresponding K_σ in (16), from (21) an integer $\bar{m} \geq 1$ exists such that

$$\frac{K_\sigma |\varphi|_1}{\|u_m\|_q^{q-1}} < \frac{\varepsilon}{2} \quad \text{for all } m \geq \bar{m}.$$

Thus, from (16), Hölder inequality, all the previous estimates and direct computations it follows that

$$\left|\int_\Omega \frac{f(x,u_m)}{\|u_m\|_q^{q-1}}\varphi\,\mathrm{d}x\right| \le \sigma|w_m|_q^{q-1}|\varphi|_q + \frac{K_\sigma|\varphi|_1}{\|u_m\|_q^{q-1}} < \varepsilon \quad \text{for all } m \ge \bar{m}.$$

Therefore, by means of (27) and (28), passing to the limit in (26), from (24) and (25) it follows that

$$\int_\Omega |\nabla w|^{q-2}\nabla w\cdot\nabla\varphi\,\mathrm{d}x = \lambda_\infty\int_\Omega |w|^{q-2}w\varphi\,\mathrm{d}x.$$

For the arbitrariness of $\varphi \in W_0^{1,q}(\Omega)$ such an equality means that $\lambda_\infty \in \sigma(-\Delta_q)$, in contradiction with our nonresonant assumption. □

Now, in order to prove the existence result for the unperturbed problem (P_0^∞), we need the following technical lemma.

Lemma 3.1 *Assume that assumptions* (f_1)–(f_3) *are satisfied and consider* $F(x,t)$ *as in* (2). *Then, for any* $\sigma > 0$ *and* $s > 0$ *a constant* $k_0^\sigma > 0$, $k_0^\sigma = k_0^\sigma(s)$, *exists such that*

$$-k_0^\sigma|t|^{s+q} + \frac{\lambda_0-\sigma}{q}|t|^q \le F(x,t) \le \frac{\lambda_0+\sigma}{q}|t|^q + k_0^\sigma|t|^{s+q} \tag{29}$$

for all $(x,t) \in \overline{\Omega}\times\mathbb{R}$.

Proof From (f_3) it follows that

$$\lim_{t\to 0}\frac{F(x,t)}{|t|^q} = \frac{\lambda_0}{q} \quad \text{uniformly in } \overline{\Omega}.$$

Therefore, taking any $\sigma > 0$ a constant $\delta_\sigma > 0$ exists such that

$$\left|F(x,t) - \frac{\lambda_0}{q}|t|^q\right| \le \frac{\sigma}{q}|t|^q \qquad \text{for all } x \in \overline{\Omega} \text{ if } |t| < \delta_\sigma. \tag{30}$$

On the other hand, from (f_2) we have that

$$\lim_{|t|\to+\infty}\frac{F(x,t)}{|t|^q} = 0 \quad \text{uniformly in } \overline{\Omega},$$

so, taking any $s > 0$, it results

$$\lim_{|t|\to+\infty}\frac{\left|F(x,t) - \frac{\lambda_0}{q}|t|^q\right|}{|t|^{q+s}} = 0 \qquad \text{uniformly in } \overline{\Omega}.$$

From this last limit, (f_1) and direct computations a constant $k_0^\sigma > 0$ exists such that

$$\frac{\left|F(x,t) - \frac{\lambda_0}{q}|t|^q\right|}{|t|^{s+q}} \leq k_0^\sigma \qquad \text{for all } x \in \overline{\Omega} \text{ if } |t| \geq \delta_\sigma. \tag{31}$$

Hence, from (30) and (31) it follows (29). □

Proof of Theorem 1.1 From Proposition 3.1 functional J in (17) satisfies (PS) in $\mathbb{R}$. Next, we distinguish the proof according to hypothesis (Λ_1) if either case (i) or case (ii) occurs.

Case (i) Taking η as in (Λ_2), from $(\Lambda_1)(i)$ a constant $\sigma \in]0, \eta[$ exists such that

$$\lambda_0 + \lambda_\infty + \sigma < \eta_k, \quad \frac{q}{p}(\nu_k + 2\sigma) < \lambda_\infty, \quad \nu_{\bar{k}} + \sigma < \eta_{\bar{k}+1}. \tag{32}$$

Let us recall that in this setting it has to be $\lambda_0 < 0$ (see Remark 1.2(a)). Moreover, from (14) and (15) with $m = k-1$, a subspace $Y_{k-1}^\sigma \in \mathbb{W}_{k-1}$ exists such that

$$\sup_{u \in Y_{k-1}^\sigma \setminus \{0\}} \frac{|\nabla u|_p^p + |\nabla u|_q^q}{|u|_q^q} < \nu_{k-1} + \sigma. \tag{33}$$

Without loss of generality, it can be chosen so that $\dim Y_{k-1}^\sigma = k-1$.

We claim that

$$J(u) \leq 0 \qquad \text{for all } u \in Y_{k-1}^\sigma. \tag{34}$$

Indeed, from (33) and (Λ_2) we get

$$\begin{aligned} J(u) &\leq \frac{1}{p}\left(|\nabla u|_p^p + |\nabla u|_q^q\right) - \frac{\lambda_\infty}{q}|u|_q^q - \int_\Omega F(x,u)\,\mathrm{d}x \\ &\leq \left[\frac{1}{p}(\nu_{k-1} + \sigma) - \frac{\lambda_\infty}{q}\right]|u|_q^q - \int_\Omega F(x,u)\,\mathrm{d}x \\ &\leq \left[\frac{1}{p}(\nu_{k-1} + \eta) - \frac{\lambda_\infty}{q}\right]|u|_q^q - \int_\Omega F(x,u)\,\mathrm{d}x \leq 0 \qquad \text{for all } u \in Y_{k-1}^\sigma. \end{aligned}$$

On the other hand, from (12) with $m = k-1$, we have that $W_0^{1,q}(\Omega) = Y_{k-1} \oplus Z_{k-1}$, where $Y_{k-1} = \text{span}\{\psi_1, \ldots, \psi_{k-1}\}$ and Z_{k-1} is its complement.

We prove that $\rho > 0$ and $\beta > 0$ exist such that

$$J(u) \geq \beta \quad \text{for all } u \in Z_{k-1} \cap S_\rho. \tag{35}$$

Indeed, taking σ as above and fixing any $s > 0$ such that $s + q < q^*$, from (29) it follows that

$$\int_\Omega F(x,u)\,\mathrm{d}x \le \frac{\lambda_0+\sigma}{q}|u|_q^q + k_0^\sigma |u|_{s+q}^{s+q} \quad \text{for all } u \in W_0^{1,q}(\Omega),$$

which, together with the Sobolev Embedding Theorem, implies the existence of a suitable $k_1^\sigma > 0$ such that

$$J(u) \ge \frac{1}{q}\left(|\nabla u|_p^p + |\nabla u|_q^q\right) - \frac{\lambda_\infty + \lambda_0 + \sigma}{q}|u|_q^q - k_1^\sigma |\nabla u|_q^{s+q} \quad \text{for all } u \in W_0^{1,q}(\Omega).$$

Now, from this last estimate and (13) with $m = k$ we obtain that

$$\begin{aligned} J(u) &\ge \frac{1}{q}\left(|\nabla u|_p^p + |\nabla u|_q^q\right) - \frac{1}{q}\frac{\lambda_\infty + \lambda_0 + \sigma}{\eta_k}\left(|\nabla u|_p^p + |\nabla u|_q^q\right) - k_1^\sigma |\nabla u|_q^{s+q} \\ &\ge \frac{1}{q}\left(1 - \frac{\lambda_\infty + \lambda_0 + \sigma}{\eta_k}\right)|\nabla u|_q^q - k_1^\sigma |\nabla u|_q^{s+q} \end{aligned}$$

for all $u \in Z_{k-1} \cap \{u \in W_0^{1,q}(\Omega) : |u|_q \le 1\}$. Then, this last inequality together with (32), implies that

$$J(u) \ge k_2^\sigma \|u\|_q^q - k_1^\sigma \|u\|_q^{s+q} \quad \text{for all } u \in Z_{k-1} \cap \{u \in W_0^{1,q}(\Omega) : |u|_q \le 1\}$$

for a suitable $k_2^\sigma > 0$. Hence, since $s > 0$, taking $\rho > 0$ small enough such that not only from the Sobolev Embedding Theorem $u \in S_\rho$ gives $|u|_q \le 1$ but also $k_2^\sigma \rho^q - k_1^\sigma \rho^{s+q} > 0$, a constant $\beta > 0$ exists such that (35) holds.

Now, we claim that

$$W_0^{1,q}(\Omega) = Y_{k-1}^\sigma \oplus Z_{k-1}, \tag{36}$$

that is, Y_{k-1} actually is Y_{k-1}^σ. To this aim, firstly we prove that $Y_{k-1}^\sigma \cap Z_{k-1} = \{0\}$. Otherwise, $\bar{u} \in Z_{k-1} \cap Y_{k-1}^\sigma$ exists such that $\bar{u} \ne 0$ and, taking ρ as in (35), it has to be

$$u = \rho \frac{\bar{u}}{\|\bar{u}\|_q} \in Y_{k-1}^\sigma \cap (Z_{k-1} \cap S_\rho),$$

which yields a contradiction as the same u has to satisfy both (34) and (35). Then, $Y_{k-1}^\sigma \subset Y_{k-1}$ and, since the two subspaces have the same dimension, they coincide and (36) follows.

Furthermore, again from (15) but with $m = \bar{k}$, where $\bar{k}$ is as in (4), a subspace $Y^\sigma \in \mathbb{W}_{\bar{k}}$ exists such that $\dim Y^\sigma = \bar{k}$ and

$$\sup_{u\in Y^\sigma\setminus\{0\}} \frac{|\nabla u|_p^p + |\nabla u|_q^q}{|u|_q^q} < \nu_{\bar{k}} + \sigma. \tag{37}$$

Let us show that

$$Y^\sigma = \operatorname{span}\{\psi_1, \dots, \psi_{\bar{k}}\} = Y_{k-1}^\sigma \oplus \operatorname{span}\{\psi_k, \dots, \psi_{\bar{k}}\}, \tag{38}$$

with $(\psi_m)_m$ which generates the whole space $W_0^{1,q}(\Omega)$ and is so that (11) holds. As a matter of fact, if some $j \geq \bar{k}+1$ exists so that $\psi_j \in Y^\sigma$, from (11), (32), (37) and the monotonicity of the sequence $(\eta_m)_m$ we get

$$\eta_j \geq \eta_{\bar{k}+1} > \nu_{\bar{k}} + \sigma > \sup_{u\in Y^\sigma\setminus\{0\}} \frac{|\nabla u|_p^p + |\nabla u|_q^q}{|u|_q^q} \geq |\nabla\psi_j|_p^p + |\nabla\psi_j|_q^q = \eta_j,$$

which is a contradiction. Thus, (38) is proved.

At last, if we consider (16) with the constant σ as in (32), a suitable $k_3^\sigma > 0$ exists such that

$$J(u) \leq \frac{1}{p}\left(|\nabla u|_p^p + |\nabla u|_q^q\right) - \frac{\lambda_\infty}{q}|u|_q^q + \frac{\sigma}{q}|u|_q^q + k_3^\sigma |u|_q \qquad \text{for all } u \in W_0^{1,q}(\Omega).$$

Hence, from (37) it follows that

$$J(u) \leq \left[\frac{1}{p}(\nu_{\bar{k}} + 2\sigma) - \frac{\lambda_\infty}{q}\right]|u|_q^q + k_3^\sigma |u|_q \qquad \text{for all } u \in Y^\sigma.$$

As $\nu_k = \nu_{\bar{k}}$, from (32) we have that

$$J(u) \to -\infty \quad \text{as } |u|_q \to +\infty,\ u \in Y^\sigma,$$

then, since all the norms are equivalent on the finite dimensional space Y^σ, a constant $R_2 > 0$ exists, large enough, such that

$$J(u) \leq 0 \quad \text{if } u \in Y^\sigma,\ \|u\|_q \geq R_2. \tag{39}$$

Finally, setting $V := Y_{k-1}^\sigma$, $W := Z_{k-1}$, $e := \frac{\psi_k}{\|\psi_k\|_q}$, $Y := Y_{k-1}^\sigma \oplus \operatorname{span}\{e\}$ and

$$S = Z_{k-1} \cap S_\rho, \quad Q = \{te : t \in [0, R_1]\} \oplus (\overline{B}_{R_2} \cap Y_{k-1}^\sigma), \tag{40}$$

from Example 2.2, (36) and (38), it results that S and ∂Q, boundary of Q in Y, link just taking $R_1 > \rho$. Then, if we assume also $R_1 \geq R_2$, from (34), (35) and (39) we have that Theorem 2.1 applies and a critical level c exists, with

$$\sup_{u \in Q} J(u) \geq c \geq \beta > 0$$

corresponding to a nontrivial solution of (P_0^∞).
Case (ii) Taking η as in (Λ_2), from $(\Lambda_1)(ii)$ a constant $\sigma \in]0, \eta[$ exists such that

$$\lambda_\infty + \sigma < \eta_k^0, \quad \frac{q}{p}(\nu_k + 2\sigma) < \lambda_0 + \lambda_\infty, \quad \nu_{\bar{k}} + \sigma < \eta_{\bar{k}+1}^0, \tag{41}$$

where it has to be $\lambda_0 > 0$ (see Remark 1.2(a)).

Now, from (16) with such a σ, and the Sobolev inequality, by using (10) with $m = k$, a suitable $k_4^\sigma > 0$ exists such that

$$J(u) \geq \frac{1}{q}\left(1 - \frac{\lambda_\infty + \sigma}{\eta_k^0}\right)|\nabla u|_q^q - k_4^\sigma |\nabla u|_q \quad \text{for all } u \in Z_{k-1}^0$$

and from (41) a constant $\beta < 0$ exists such that

$$J(u) \geq \beta \qquad \text{for all } u \in Z_{k-1}^0. \tag{42}$$

On the other hand, by reasoning as in the previous case, a subspace $Y_{k-1}^\sigma \in \mathbb{W}_{k-1}$ exists such that $\dim Y_{k-1}^\sigma = k-1$ and (33) holds. Then, since for any $u \in W_0^{1,q}(\Omega)$ we can write

$$J(u) = \frac{1}{p}\left(|\nabla u|_p^p + |\nabla u|_q^q\right) - \left(\frac{1}{p} - \frac{1}{q}\right)|\nabla u|_q^q - \frac{\lambda_\infty}{q}|u|_q^q - \int_\Omega F(x, u)\,dx,$$

from (33) (recall that $\sigma < \eta$) it follows that

$$J(u) \leq \int_\Omega \left[\left(\frac{1}{p}(\nu_{k-1} + \eta) - \frac{\lambda_\infty}{q}\right)|u|^q - F(x, u)\right] dx - \left(\frac{1}{p} - \frac{1}{q}\right)|\nabla u|_q^q$$

for all $u \in Y_{k-1}^\sigma$. Hence, setting $\delta_1 = \frac{1}{p} - \frac{1}{q} > 0$, from (Λ_2) this last estimate implies that

$$J(u) \leq -\delta_1 \|u\|_q^q \qquad \text{for all } u \in Y_{k-1}^\sigma. \tag{43}$$

Thus, it results not only that

$$\sup_{u \in Y_{k-1}^\sigma} J(u) = 0, \tag{44}$$

but also that a radius $R > 0$, large enough, and a constant $\alpha < \beta$ exist such that

$$J(u) \le \alpha \qquad \text{for all } u \in Y^{\sigma}_{k-1} \cap S_R. \tag{45}$$

At last, being $\alpha < \beta$, by reasoning as for the proof of (36) but by means of (42) and (45), we have that

$$W_0^{1,q}(\Omega) = Y^{\sigma}_{k-1} \oplus Z^0_{k-1}. \tag{46}$$

Hence, setting

$$S = Z^0_{k-1} \qquad \text{and} \qquad Q = Y^{\sigma}_{k-1} \cap \overline{B}_R, \tag{47}$$

from Example 2.1 and estimates (42), (44), (45), it follows that Theorem 2.1 applies and J has a critical level c such that

$$\beta \le c = \inf_{\phi \in \Gamma} \sup_{u \in Q} J(\phi(u)) \le \sup_{u \in Q} J(u) = 0,$$

where $\Gamma = \left\{\phi \in C(W_0^{1,q}(\Omega), W_0^{1,q}(\Omega)) : \phi|_{\partial Q} = \text{id}\right\}$.

Next, we want to show that $c < 0$; so, (P_0^{∞}) admits a nontrivial solution.

To this aim, it is enough to prove that a function $\bar{\phi} \in C(W_0^{1,q}(\Omega), W_0^{1,q}(\Omega))$ exists, with $\bar{\phi}|_{\partial Q} = \text{id}$, such that

$$\sup_{u \in Q} J(\bar{\phi}(u)) < 0. \tag{48}$$

At first, we observe that from (15) with $m = \bar{k}$ with $\bar{k}$ as in (6), a subspace $Y^{\sigma} \in \mathbb{W}_{\bar{k}}$, with $\dim Y^{\sigma} = \bar{k}$, exists such that (37) holds. Hence, we have that

$$\sup_{u \in Y^{\sigma} \setminus \{0\}} \frac{|\nabla u|_q^q}{|u|_q^q} < \nu_{\bar{k}} + \sigma. \tag{49}$$

We claim that

$$Y^{\sigma} = \text{span}\{\psi_1^0, \dots, \psi_{\bar{k}}^0\} = Y^{\sigma}_{k-1} \oplus \text{span}\{\psi_k^0, \dots, \psi_{\bar{k}}^0\}, \tag{50}$$

where $(\psi_m^0)_m$ generates the whole space $W_0^{1,q}(\Omega)$ and is such that (8) holds. Indeed, if $j \ge \bar{k} + 1$ exists such that $\psi_j^0 \in Y^{\sigma}$, then (8) and estimates (41), (49), together with the monotonicity of the sequence $(\eta_m^0)_m$, imply that

$$\eta_j^0 \ge \eta_{\bar{k}+1}^0 > \nu_{\bar{k}} + \sigma > |\nabla \psi_j^0|_q^q = \eta_j^0,$$

which gives a contradiction.

Now, taking $L > 0$ such that (9) is verified with respect to the decomposition (46), without loss of generality we can suppose $L > 1$. Then, $\rho \in]0, R[$ and $\delta_2 > 0$ exist such that

$$J(u) \le -\delta_2 \qquad \text{for all } u \in Y^\sigma \text{ with } \frac{\rho}{L} \le \|u\|_q \le 2\rho. \tag{51}$$

Indeed, fixing any $s > 0$ and taking $u \in Y^\sigma$, from (29) and (37) it results that

$$\begin{aligned} J(u) &\le \frac{1}{p}\left(|\nabla u|_p^p + |\nabla u|_q^q\right) - \frac{\lambda_\infty}{q}|u|_q^q - \frac{\lambda_0 - \sigma}{q}|u|_q^q + k_0^\sigma |u|_{s+q}^{s+q} \\ &\le \frac{1}{q}\left[\frac{q}{p}(\nu_{\bar{k}} + 2\sigma) - \lambda_\infty - \lambda_0\right]|u|_q^q + k_0^\sigma |u|_{s+q}^{s+q}. \end{aligned}$$

Hence, since all the norms are equivalent on the finite dimension subspace Y^σ, from this last estimates and (41) with $\nu_k = \nu_{\bar{k}}$, we have that two constants $c_1, c_2 > 0$ exist such that

$$J(u) \le -c_1\|u\|_q^q + c_2\|u\|_q^{s+q} \qquad \text{for all } u \in Y^\sigma.$$

Thus, $s > 0$ and direct computations allow us to prove that (51) holds if $\rho > 0$ is small enough, in particular $\rho < R$.

At last, we can define $\bar{\phi} : W_0^{1,q}(\Omega) \to W_0^{1,q}(\Omega)$ as the continuous extension to $W_0^{1,q}(\Omega)$ of function $\bar{\phi} : Y_{k-1}^\sigma \to \mathbb{R}$ such that

$$\bar{\phi}(u) = \begin{cases} u & \text{if } \|u\|_q > \rho \\ \dfrac{\sqrt{\rho^2 - \|u\|_q^2}}{(\eta_k^0)^{\frac{1}{q}}}\,\psi_k^0 + u & \text{if } \|u\|_q \le \rho \end{cases}.$$

We notice that $\bar{\phi}$ satisfies the required assumptions. Indeed, by definition we have that $\bar{\phi} \in C(W_0^{1,q}(\Omega), W_0^{1,q}(\Omega))$ and $\bar{\phi}(u) = u$ for all $u \in \partial Q = Y_{k-1}^\sigma \cap S_R$ as $R > \rho$. Moreover, if $u \in Y_{k-1}^\sigma \cap \bar{B}_R$ two cases may occur: either $\|u\|_q > \rho$ or $\|u\|_q \le \rho$. If $\|u\|_q > \rho$, from the definition of $\bar{\phi}(u)$ and (43) we have that

$$J(\bar{\phi}(u)) = J(u) \le -\delta_1\rho^q < 0.$$

On the other hand, if $\|u\|_q \le \rho$, then (50) implies that $\bar{\phi}(u) \in Y^\sigma$. Furthermore, (8), (9) and direct computations imply that

$$\frac{1}{L}\left(\|u\|_q + \sqrt{\rho^2 - \|u\|_q^2}\right) \le \|\bar{\phi}(u)\|_q \le \|u\|_q + \sqrt{\rho^2 - \|u\|_q^2}$$

with $\rho \leq \|u\|_q + \sqrt{\rho^2 - \|u\|_q^2} \leq 2\rho$; hence, from (51) we have $J(\bar{\phi}(u)) \leq -\delta_2$. Thus, summing up, (48) holds and the proof is complete. □

Finally, we note that an existence result still holds for the unperturbed problem (P_0^∞) if λ_0 as defined in hypothesis (f_3), is infinite. More precisely, the following statements can be proved.

Proposition 3.2 *Suppose that conditions (f_1) and (f_2) hold and $\lambda_\infty \notin \sigma(-\Delta_q)$. If, moreover, we have that*

$(f_3)'$ $\quad \lim\limits_{t\to 0} \dfrac{f(x,t)}{|t|^{q-2}t} = -\infty$ *uniformly in* $\overline{\Omega}$*;*

$(\Lambda_1)''$ $\quad \frac{\nu_1}{p} < \frac{\lambda_\infty}{q}$*;*

then (P_0^∞) has at least a nontrivial solution.

Proof Taking any $\lambda > 0$ and $\sigma > 0$ such that

$$\lambda > \lambda_\infty, \qquad \frac{\nu_1 + 2\sigma}{p} < \frac{\lambda_\infty}{q}, \tag{52}$$

from $(f_3)'$ we have that $\delta_\lambda > 0$ exists so that

$$F(x,t) < -\frac{\lambda}{q}|t|^q \quad \text{for all } x \in \overline{\Omega} \text{ if } |t| \leq \delta_\lambda.$$

On the other hand, from (f_2) a radius $R_\sigma \geq \max\{1, \delta_\lambda\}$ exists such that

$$|F(x,t)| \leq \sigma |t|^q \quad \text{for all } x \in \overline{\Omega} \text{ if } |t| \geq R_\sigma.$$

Then, fixing any $s > 0$ so that $q + s < q^*$, the continuity of $\frac{F(x,t)}{|t|^{q+s}}$ on the compact set $\overline{\Omega} \times [\delta_\lambda, R_\sigma]$ and direct computations allow us to find some constants $k_{\lambda,i} > 0$ large enough so that

$$F(x,t) \leq k_{\lambda,1}|t|^{q+s} \leq -\frac{\lambda}{q}|t|^q + k_{\lambda,2}|t|^{q+s} \quad \text{for all } x \in \overline{\Omega} \text{ if } |t| \geq \delta_\lambda.$$

Hence, summing up it results

$$F(x,t) \leq -\frac{\lambda}{q}|t|^q + k_{\lambda,2}|t|^{q+s} \quad \text{for all } (x,t) \in \overline{\Omega} \times \mathbb{R}$$

which implies

$$J(u) \geq \frac{1}{q}|\nabla u|_q^q + \frac{\lambda - \lambda_\infty}{q}|u|_q^q - k_{\lambda,2}|u|_{q+s}^{q+s} \quad \text{for all } u \in W_0^{1,q}(\Omega),$$

thus, from (52) and the Sobolev Embedding Theorem we obtain that

$$J(u) \geq \beta \quad \text{for all } u \in S_\rho,$$

for suitable constants $\rho > 0$ and $\beta > 0$.

Now, from (15) with $m = 1$, a subspace $Y \in \mathbb{W}_1$ exists such that

$$\sup_{u \in Y \setminus \{0\}} \frac{|\nabla u|_p^p + |\nabla u|_q^q}{|u|_q^q} < \nu_1 + \sigma,$$

where, without loss of generality, we can take $Y = \text{span}\{\psi_1\}$. Thus, from (16) and direct computations it results

$$J(t\psi_1) \leq t^q \left(\frac{\nu_1 + 2\sigma}{p} - \frac{\lambda_\infty}{q} \right) + t K_\sigma \int_\Omega |\psi_1| dx \qquad \text{for all } t > 0,$$

which implies $J(t\psi_1) \to -\infty$ if $t \to +\infty$ as (52) holds.

At last, from Proposition 3.1 and the previous geometrical estimates, the classical Mountain Pass Theorem applies (cf. [29, Theorem 2.2]) and the existence of a nontrivial solution corresponding to a critical level $c \geq \beta > 0$ is proved. □

Proposition 3.3 *Suppose that conditions* (f_1) *and* (f_2) *hold and* $\lambda_\infty \notin \sigma(-\Delta_q)$. *Moreover, assume that*

$(f_3)''$ $\displaystyle\lim_{t \to 0} \frac{f(x,t)}{|t|^{q-2}t} = +\infty$ *uniformly in* $\overline{\Omega}$;

$(\Lambda_1)'''$ *some integers* $1 \leq k \leq \bar{k}$ *exist such that*

$$\lambda_\infty < \eta_k^0, \qquad \nu_{\bar{k}} < \eta_{\bar{k}+1}^0.$$

If (Λ_2) *holds for the same* k *in* $(\Lambda_1)'''$, *then* (P_0^∞) *has at least a nontrivial solution.*

Proof From $(\Lambda_1)'''$ a constant $\sigma > 0$ exists so that

$$\sigma < \eta, \qquad \lambda_\infty + \sigma < \eta_k^0, \qquad \nu_{\bar{k}} + \sigma < \eta_{\bar{k}+1}^0. \tag{53}$$

Firstly, reasoning as in the proof of (42), from (10), (16) and (53) a constant $\beta < 0$ exists such that $J(u) \geq \beta$ for all $u \in Z_{k-1}^0$.

On the other hand, reasoning as in the proof of *Case* (ii) of Theorem 1.1, from (Λ_2) a subspace $Y_{k-1}^\sigma \in \mathbb{W}_{k-1}$ exists such that (43) holds and for a large enough radius $R > 0$ inequality (45) is satisfied with a suitable $\alpha < \beta$.

Hence, (46) is verified and from Example 2.1, Proposition 3.1 and Theorem 2.1 applied to $S = Z_{k-1}^0$ and $Q = Y_{k-1}^\sigma \cap \overline{B}_R$, we get the existence of a critical level $c \leq 0$.

At last, by considering again $\bar{\phi}$ for a suitable $\rho \in]0, R[$ as in the proof of *Case* (ii) of Theorem 1.1, we get that (48) holds, thus $c < 0$ and the corresponding solution is non trivial. Indeed, from (53) both (49) and (50) hold. Furthermore, from $(f_3)''$, taking any $\lambda > \frac{\nu_{\bar{k}}+\sigma}{p} - \frac{\lambda_\infty}{q}$ a constant $\delta_\lambda > 0$ exists such that

$$F(x,t) \geq \lambda|t|^q \quad \text{for all } x \in \overline{\Omega} \text{ if } |t| \leq \delta_\lambda,$$

while from (f_2) and direct computations (as in the proof of Lemma 3.1) taking any $s > 0$ a constant $k_\lambda > 0$ exists such that

$$\frac{|F(x,t) - \lambda|t|^q|}{|t|^{q+s}} \leq k_\lambda \quad \text{for all } x \in \overline{\Omega} \text{ if } |t| \geq \delta_\lambda.$$

Hence,

$$F(x,t) \geq \lambda|t|^q - k_\lambda|t|^{q+s} \quad \text{for all } (x,t) \in \overline{\Omega} \times \mathbb{R}$$

which, together with (53), implies (51) which allows us to prove (48). □

4 The Perturbed Case

Now, we are able to deal with the perturbed problem (P_ε^∞).

Proof of Theorem 1.2 Following [23], for any $j \in \mathbb{N}$ we consider a continuous cut function $\gamma_j : \mathbb{R} \to \mathbb{R}$ such that

$$\gamma_j(t) = \begin{cases} 0 & \text{if } |t| \geq j+1 \\ 1 & \text{if } |t| \leq j \end{cases},$$

and $0 < \gamma_j(t) < 1$ if $j < |t| < j+1$, and set

$$h_j(x,t) = \gamma_j(t)h(x,t), \qquad H_j(x,t) = \int_0^t h_j(x,s)\,\mathrm{d}s.$$

Since for any $j \in \mathbb{N}$ there exists $\varepsilon_1(j) > 0$ such that

$$\varepsilon_1(j)|h_j(x,t)| < 1, \qquad \varepsilon_1(j)|H_j(x,t)| < 1 \quad \text{for all } (x,t) \in \overline{\Omega} \times \mathbb{R}, \tag{54}$$

for any ε, with $|\varepsilon| \leq \varepsilon_1(j)$, we can consider the functionals

$$J_{j,\varepsilon}(u) = J(u) - \varepsilon \int_\Omega H_j(x,u)\,\mathrm{d}x \quad \text{on } W_0^{1,q}(\Omega).$$

Now, taking Q as in the proof of Theorem 1.1 (namely, as in (40) in *Case* (i) or as in (47) in *Case* (ii)), from Theorem 1.1 we have that $c \in [\beta, \sup J(Q)]$ is a critical level of J in $W_0^{1,q}(\Omega)$ with

$$c = \inf_{\phi\in\Gamma} \sup_{u\in Q} J(\phi(u)),$$

where $\Gamma = \left\{\phi \in C(W_0^{1,q}(\Omega), W_0^{1,q}(\Omega)) : \phi|_{\partial Q} = \mathrm{id}\right\}$. Then, from Theorem 2.4 we have that such a level c has to be essential for J; thus, Theorem 2.2 implies the existence of a constant $\varepsilon_2(j) \in]0, \varepsilon_1(j)[$ such that if $|\varepsilon| \le \varepsilon_2(j)$ then $J_{j,\varepsilon}$ has at least one essential value $d^{j,\varepsilon}$ with

$$\frac{\beta}{2} < d^{j,\varepsilon} < \sup_{u\in Q} J(u) + 1.$$

We note that, since for each ε, j the nonlinear term $f(x,t) + \varepsilon h_j(x,t)$ satisfies assumptions (f_1) and (f_2), then from the same arguments in Proposition 3.1 we have that each $J_{j,\varepsilon}$ satisfies the (PS) condition in $\mathbb{R}$. Hence, from Theorem 2.3 it follows that if $|\varepsilon| \le \varepsilon_2(j)$ the level $d^{j,\varepsilon}$ is also critical for $J_{j,\varepsilon}$ and $u^{j,\varepsilon} \in W_0^{1,q}(\Omega)$ exists such that

$$\begin{aligned} &\int_\Omega |\nabla u^{j,\varepsilon}|^{p-2}\nabla u^{j,\varepsilon}\cdot\nabla\varphi\,\mathrm{d}x + \int_\Omega |\nabla u^{j,\varepsilon}|^{q-2}\nabla u^{j,\varepsilon}\cdot\nabla\varphi\,\mathrm{d}x \\ &\quad = \lambda_\infty \int_\Omega |u^{j,\varepsilon}|^{q-2}u^{j,\varepsilon}\,\varphi\,\mathrm{d}x + \int_\Omega f(x,u^{j,\varepsilon})\varphi\,\mathrm{d}x \\ &\qquad + \varepsilon\int_\Omega h_j(x,u^{j,\varepsilon})\varphi\,\mathrm{d}x \end{aligned} \tag{55}$$

for all $\varphi \in W_0^{1,q}(\Omega)$.

We claim that a constant $K_1 > 0$ exists such that

$$\|u^{j,\varepsilon}\|_q \le K_1 \quad \text{for all } j\in\mathbb{N},\ |\varepsilon| \le \varepsilon_2(j). \tag{56}$$

Indeed, arguing by contradiction, let us assume that the set

$$A := \{\|u^{j,\varepsilon}\|_q : j\in\mathbb{N},\ |\varepsilon|\le\varepsilon_2(j)\}$$

is unbounded. Then, a sequence $(u^{j_m,\varepsilon_m})_m \subset W^{1,q}(\Omega)$ exists, with $|\varepsilon_m| \le \varepsilon_2(j_m)$, such that

$$\|u^{j_m,\varepsilon_m}\|_q \to +\infty \quad \text{as } m\to+\infty. \tag{57}$$

Setting $w_{j_m,\varepsilon_m} = \frac{u^{j_m,\varepsilon_m}}{\|u^{j_m,\varepsilon_m}\|_q}$, we have that $(w_{j_m,\varepsilon_m})_m$ is a bounded sequence in $W_0^{1,q}(\Omega)$, so $w \in W_0^{1,q}(\Omega)$ exists such that, up to subsequences, it results

$$w_{j_m,\varepsilon_m} \rightharpoonup w \quad \text{weakly in } W_0^{1,q}(\Omega), \tag{58}$$

$$w_{j_m,\varepsilon_m} \to w \quad \text{strongly in } L^q(\Omega). \tag{59}$$

Now, taking

$$\varphi_{j_m,\varepsilon_m} = \frac{w_{j_m,\varepsilon_m} - w}{\|u^{j_m,\varepsilon_m}\|_q^{q-1}}$$

in (55) with $j = j_m$ and $\varepsilon = \varepsilon_m$, we obtain that

$$\begin{aligned}
&\int_\Omega \frac{|\nabla w_{j_m,\varepsilon_m}|^{p-2}}{\|u^{j_m,\varepsilon_m}\|_q^{q-p}} \nabla w_{j_m,\varepsilon_m} \cdot \nabla(w_{j_m,\varepsilon_m} - w)\, dx \\
&\quad + \int_\Omega |\nabla w_{j_m,\varepsilon_m}|^{q-2} \nabla w_{j_m,\varepsilon_m} \cdot \nabla(w_{j_m,\varepsilon_m} - w)\, dx \\
&= \lambda_\infty \int_\Omega |w_{j_m,\varepsilon_m}|^{q-2} w_{j_m,\varepsilon_m} (w_{j_m,\varepsilon_m} - w)\, dx \\
&\quad + \int_\Omega \frac{f(x, u^{j_m,\varepsilon_m})}{\|u^{j_m,\varepsilon}\|_q^{q-1}} (w_{j_m,\varepsilon_m} - w)\, dx \\
&\quad + \varepsilon_m \int_\Omega \frac{h_j(x, u^{j_m,\varepsilon_m})}{\|u^{j_m,\varepsilon_m}\|_q^{q-1}} (w_{j_m,\varepsilon_m} - w)\, dx.
\end{aligned}$$

Then, from (54), (57) and (59) we have that

$$\varepsilon_m \int_\Omega \frac{h_j(x, u^{j_m,\varepsilon_m})}{\|u^{j_m,\varepsilon_m}\|_q^{q-1}} (w_{j_m,\varepsilon_m} - w)\, dx = \beta_m,$$

and also, by reasoning as in the proof of Proposition 3.1,

$$\int_\Omega \frac{|\nabla w_{j_m,\varepsilon_m}|^{p-2}}{\|u^{j_m,\varepsilon_m}\|_q^{q-p}} \nabla w_{j_m,\varepsilon_m} \cdot \nabla(w_{j_m,\varepsilon_m} - w)\, dx = \beta_m,$$

$$\int_\Omega |w_{j_m,\varepsilon_m}|^{q-2} w_{j_m,\varepsilon_m} (w_{j_m,\varepsilon_m} - w)\, dx = \beta_m,$$

$$\int_\Omega \frac{f(x, u^{j_m,\varepsilon_m})}{\|u^{j_m,\varepsilon}\|_q^{q-1}} (w_{j_m,\varepsilon_m} - w)\, dx = \beta_m,$$

which imply that

$$\int_\Omega |\nabla w_{j_m,\varepsilon_m}|^{q-2}\nabla w_{j_m,\varepsilon_m}\cdot\nabla(w_{j_m,\varepsilon_m}-w)\,dx=\beta_m.$$

Hence, from this last limit and (58) it follows that

$$w_{j_m,\varepsilon_m}\to w\quad\text{strongly in } W_0^{1,q}(\Omega),\tag{60}$$

which gives also $w\neq 0$.

Finally, taking any $\varphi\in W_0^{1,q}(\Omega)$ and applying again (55) with $j=j_m$ and $\varepsilon=\varepsilon_m$ on $\frac{\varphi}{\|u^{j_m,\varepsilon_m}\|_q^{q-1}}$, we obtain

$$\begin{aligned}&\int_\Omega \frac{|\nabla w_{j_m,\varepsilon_m}|^{p-2}}{\|u^{j_m,\varepsilon_m}\|_q^{q-p}}\nabla w_{j_m,\varepsilon_m}\cdot\nabla\varphi\,dx+\int_\Omega|\nabla w_{j_m,\varepsilon_m}|^{q-2}\nabla w_{j_m,\varepsilon_m}\cdot\nabla\varphi\,dx\\&=\lambda_\infty\int_\Omega|w_{j_m,\varepsilon_m}|^{q-2}w_{j_m,\varepsilon_m}\,\varphi\,dx+\int_\Omega\frac{f(x,u^{j_m,\varepsilon_m})}{\|u^{j_m,\varepsilon_m}\|_q^{q-1}}\varphi\,dx\\&\quad+\varepsilon_m\int_\Omega\frac{h_j(x,u^{j_m,\varepsilon_m})}{\|u^{j_m,\varepsilon_m}\|_q^{q-1}}\varphi\,dx.\end{aligned}\tag{61}$$

Thus, since from (54) and (57) we have that

$$\left|\varepsilon_m\int_\Omega\frac{h_j(x,u^{j_m,\varepsilon_m})}{\|u^{j_m,\varepsilon_m}\|_q^{q-1}}\varphi\,dx\right|\le\beta_m\|\varphi\|_q,$$

by reasoning again as in the proof of Proposition 3.1 by means of (57) we are able to prove that

$$\left|\int_\Omega\frac{|\nabla w_{j_m,\varepsilon_m}|^{p-2}}{\|u^{j_m,\varepsilon_m}\|_q^{q-p}}\nabla w_{j_m,\varepsilon_m}\cdot\nabla\varphi\,dx\right|\le\beta_m\|\varphi\|_q,$$

$$\lim_{m\to+\infty}\int_\Omega\frac{f(x,u^{j_m,\varepsilon_m})}{\|u^{j_m,\varepsilon_m}\|_q^{q-1}}\varphi\,dx=0.$$

Hence, from (59), (60) and passing to the limit in (61), for the arbitrariness of φ we get that $\lambda_\infty\in\sigma(-\Delta_q)$, against our assumption. Thus, the claim (56) is proved.
Finally, from [24, Lemmas 4.5 and 4.6] (see [7] for more details) a constant $K_2>0$ exists such that

$$|u^{j,\varepsilon}|_\infty\le K_2\quad\text{for all } j\in\mathbb{N},\ |\varepsilon|\le\varepsilon_2(j);$$

thus, for $j > K_2$ problem (P_ε^∞) has at least a nontrivial solution. □

Acknowledgments The research that led to the present paper was partially supported by MIUR–PRIN Research Project 2017JPCAPN "Qualitative and quantitative aspects of nonlinear PDEs" and by *Fondi di Ricerca di Ateneo* 2017/18 "Problemi differenziali non lineari".

All the authors are members of the Research Group INdAM–GNAMPA.

References

1. Amann, H., Zehnder, E.: Nontrivial solutions for a class of nonresonance problems and applications to nonlinear differential equations. Ann. Scuola Norm. Sup. Pisa Cl. Sci. (4) **7**, 539–603 (1980)
2. Anane, A.: Etude des valeurs propres et de la résonance pour l'opérateur p-laplacien, Thèse de Doctorat, Université Libre de Bruxelles, 1987
3. Bartolo, P., Benci, V., Fortunato, D.: Abstract critical point theorems and applications to some nonlinear problems with "strong" resonance at infinity. Nonlinear Anal. **7**, 981–1012 (1983)
4. Bartolo, R.: Multiplicity results for a class of quasilinear elliptic problems, Mediterr. J. Math. **11**, 1099–1113 (2014)
5. Bartolo, R., Candela, A.M., Salvatore, A.: p-Laplacian problems with nonlinearities interacting with the spectrum. NoDEA **20**, 1701–1721 (2013)
6. Bartolo, R., Candela, A.M., Salvatore, A.: Perturbed asymptotically linear problems, Ann. Mat. Pura Appl. **193**, 89–101 (2014)
7. Bartolo, R., Candela, A.M., Salvatore, A.: Multiple solutions for perturbed quasilinear elliptic problems. Topol. Methods Nonlinear Anal. (to appear)
8. Benci, V.: On the critical point theory for indefinite functionals in the presence of symmetries. Trans. Am. Math. Soc. **274**, 533–572 (1982)
9. Brezis, H.: Functional Analysis, Sobolev Spaces and Partial Differential Equations, Universitext, vol. XIV. Springer, New York (2011)
10. Candela, A.M., Palmieri, G.: Infinitely many solutions of some nonlinear variational equations. Calc. Var. **34**, 495–530 (2009)
11. Chang, K.-C.: Morse theory on Banach space and its applications to partial differential equations. Chinese Ann. Math. Ser. B **4**, 381–399 (1983)
12. Cingolani, S., Degiovanni, M.: Nontrivial solutions for p-Laplace equations with right-hand side having p-linear growth at infinity. Commun. Part. Differ. Equ. **30**, 1191–1203 (2005)
13. Cingolani, S., Degiovanni, M., Vannella G.: Amann–Zehnder type results for p-Laplace problems. Ann. Mat. Pura Appl. **197**, 605–640 (2018)
14. Colasuonno, F.: Multiple solutions for asymptotically q-linear (p, q)-Laplacian problems. Math. Meth. Appl. Sci. 1–19 (2021)
15. Costa, D.G., Magalhães, C.A.: Existence results for perturbations of the p-Laplacian. Nonlinear Anal. **24**, 409–418 (1995)
16. Degiovanni, M., Lancelotti, S.: Perturbations of even nonsmooth functionals. Differ. Integr. Equ. **8**, 981–992 (1995)
17. Degiovanni, M., Lancelotti, S.: Perturbations of critical values in nonsmooth critical point theory. Serdica Math. J. **22**, 427–450 (1996). In: Y. Sonntag (ed.) Well-Posed Problems and Stability in Optimization
18. Degiovanni, M., Rădulescu, V.: Perturbations of nonsmooth symmetric nonlinear eigenvalue problems. C.R. Acad. Sci. Paris Sér. I **329**, 281–286 (1999)
19. Dinca, G., Jebelean, P., Mawhin, J.: Variational and topological methods for Dirichlet problems with p-Laplacian. Portugal. Math. **58**, 339–378 (2001)

20. do Ó, J.M.: Existence of solutions for quasilinear elliptic equations. J. Math. Anal. Appl. **207**, 104–126 (1997)
21. García Azorero, J., Peral Alonso, I.: Existence and nonuniqueness for the p–Laplacian: nonlinear eigenvalues. Commun. Part. Differ. Equ. **12**, 1389–1430 (1987)
22. Guedda, M., Véron, L.: Quasilinear elliptic equations involving critical Sobolev exponents. Nonlinear Anal. **13**, 879–902 (1989)
23. Li, S., Liu, Z.: Perturbations from symmetric elliptic boundary value problems. J. Differ. Equ. **185**, 271–280 (2002)
24. Li, S., Liu Z.: Multiplicity of solutions for some elliptic equations involving critical and supercritical Sobolev exponents. Topol. Methods Nonlinear Anal. **28**, 235–261 (2006)
25. Li, G., Zhou, H.S.: Asymptotically linear Dirichlet problem for the p-Laplacian. Nonlinear Anal. **43**, 1043–1055 (2001)
26. Li, G., Zhou, H.S.: Multiple solutions to p-Laplacian problems with asymptotic nonlinearity as u^{p-1} at infinity. J. Lond. Math. Soc. **65**, 123–138 (2002)
27. Lindqvist, P.: On the equation $\operatorname{div}(|\nabla u|^{p-2}\nabla u) + \lambda|u|^{p-2}u = 0$. Proc. Am. Math. Soc. **109**, 157–164 (1990)
28. Perera, K., Szulkin, A.: p-Laplacian problems where the nonlinearity crosses an eigenvalue. Discrete Contin. Dyn. Syst. **13**, 743–753 (2005)
29. Rabinowitz, P.H.: Minimax Methods in Critical Point Theory with Applications to Differential Equations, CBMS Regional Conference Series in Mathematics, vol. 65. American Mathematical Society, Providence (1986)
30. Sun, M.: Multiplicity of solutions for a class of the quasilinear elliptic equations at resonance. J. Math. Anal. Appl. **386**, 661–668 (2012)
31. Struwe, M.: Variational Methods. Applications to Nonlinear Partial Differential Equations and Hamiltonian Systems, 4rd ed. Ergeb. Math. Grenzgeb. (4), vol. 34. Springer, Berlin (2008)
32. Willem, M.: Minimax Theorems. Birkhäuser, Boston (1996)

Weighted Composition Operators on Weighted Spaces of Banach Valued Analytic Functions

José Bonet and Esther Gómez-Orts

Dedicated to Professor Francesco Altomare on the occasion of his 70th Birthday

Abstract Several properties of weighted composition operators acting between weighted spaces of analytic functions with values on a Banach space are characterized. These results are applied to study weighted composition operators between weighted inductive and projective limits of spaces of vector-valued analytic functions. Operators acting on vector-valued Korenblum type spaces are also considered.

Keywords Weighted composition operator · Vector-valued analytic function · Korenblum type spaces · Fréchet spaces · (LB)-spaces · Weakly compact operator

1 Introduction

The aim of the paper is to present a complete analysis of the continuity, compactness and weak compactness of weighted composition operators $W_{\psi,\varphi}$ acting between weighted spaces of analytic functions with weighted sup-norms defined on the unit disc of the complex plane and with values in a Banach space. The characterizations of continuity and (weak) compactness for weighted Banach spaces of analytic functions are given in Sect. 2. Our theorems complement and extend results for composition operators given in [10]. This extension is one of the main motivations for our research. More results about composition operators on Banach spaces of Banach valued analytic functions can be seen in [10], [21] and the references therein. The characterizations in Theorems 2.1 and 2.2 are utilized in Sects. 3 and 4 to investigate several properties of $W_{\psi,\varphi}$ when it acts on countable inductive and projective limits

J. Bonet (✉) · E. Gómez-Orts
IUMPA, Universitat Politècnica de València, Valencia, Spain
e-mail: jbonet@mat.upv.es; esgoor@upvnet.upv.es

A. M. Candela et al. (eds.), *Recent Advances in Mathematical Analysis*, Trends in Mathematics, https://doi.org/10.1007/978-3-031-20021-2_9

of vector-valued Banach spaces of analytic functions, in particular on Korenblum type spaces of vector-valued analytic functions. Composition operators acting on weighted inductive limits of scalar valued analytic functions were treated in [13]. The case of weighted composition operators on Korenblum type spaces of scalar valued functions has been investigated recently in [16]. Weighted inductive and projective limits of vector-valued functions of the type considered in this paper have been studied by many authors since the paper [8]; see [2], [3], [4], [6] and [28]. Composition operators on weighted Fréchet spaces of analytic functions with values in a Banach space were investigated in [12]. Another motivation for our investigation was to analyze the case of weighted composition operators not only for Fréchet spaces (see Sect. 4), but also for weighted inductive limits in Sect. 3.

A radial, strictly positive continuous function $v : \mathbb{D} \to \mathbb{R}$, which is non-increasing with respect to $|z|$ and is such that $\lim_{|z|\to 1^-} v(z) = 0$, is called a *weight* on $\mathbb{D}$. The associated weight $\tilde{v}$ is defined by

$$\tilde{v}(z) := 1/\sup\{|f(z)| \, : \, |f| \leq 1/v \text{ on } \mathbb{D}\}.$$

The associated weight is also radial, strictly positive, continuous, non-increasing with respect to $|z|$ and satisfies $\lim_{|z|\to 1^-} \tilde{v}(z) = 0$. Moreover $v(z) \leq \tilde{v}(z)$ for all $z \in \mathbb{D}$. See more information in [5]. A weight v is called *essential* if there exists a constant $C > 0$ such that $v(z) \leq \tilde{v}(z) \leq Cv(z)$, for all $z \in \mathbb{D}$.

Let E be a complex Banach space. For each weight v on $\mathbb{D}$, the weighted Banach spaces of vector-valued analytic functions are defined as

$$H_v^\infty(\mathbb{D}, E) := \{F \in H(\mathbb{D}, E) \colon \|F\|_v := \sup_{z\in\mathbb{D}} v(z)\|F(z)\| < \infty\}$$

and

$$H_v^0(\mathbb{D}, E) := \{F \in H(\mathbb{D}, E) \colon \lim_{|z|\to 1^-} v(z)\|F(z)\| = 0\}.$$

They are Banach spaces when endowed with the norm $\|\cdot\|_v$. When the Banach space E is the complex plane $\mathbb{C}$, we denote the spaces by H_v^∞ and H_v^0. The space H_v^0 coincides with the closure of the polynomials in H_v^∞; see e.g. [9]. Moreover, $H_v^\infty = H_{\tilde{v}}^\infty$, $H_v^0 = H_{\tilde{v}}^0$ and $\|f\|_v = \|f\|_{\tilde{v}}$. The space H^∞ of bounded analytic functions on $\mathbb{D}$ is contained in H_v^0 for each weight v.

Let $\varphi : \mathbb{D} \to \mathbb{D}$ be an analytic selfmap on the unit disc $\mathbb{D}$ of the complex plane, and let $\psi : \mathbb{D} \to \mathbb{C}$ be an analytic map. The weighted composition operator is defined by $W_{\psi,\varphi} f := \psi(f \circ \varphi)$ for each analytic function f on the unit disc. Weighted composition operators have been studied by many authors; see the books by Cowen and McCluer [15] and Shapiro [25]. Continuity, compactness and weak compactness of $W_{\psi,\varphi}$ on weighted Banach spaces $H_v^\infty(\mathbb{D})$ of type H^∞ were investigated in [11], [14] and [23].

The space $H(\mathbb{D})$ of all analytic functions on $\mathbb{D}$ is endowed with the Fréchet topology of uniform convergence on compact sets. In the sequel, the word "space" means "Hausdorff locally convex space". We refer the reader to [1], [17], [20], [22] and [27] for results and terminology about functional analysis, and in particular about Fréchet, (LB)-spaces and countable inductive and projective limits. The definition of bounded, compact and weakly compact operator $T : E \to F$ between locally convex spaces E and F is standard and can be seen e.g. in [20, Chapter 42]. An operator $T : E \to F$ is called *Montel* (resp. *reflexive*) if it maps every bounded subset of E into a relatively (resp. weakly) compact subset of F.

2 The Operator $W_{\psi,\varphi}$ on Weighted Banach Spaces of Vector-Valued Functions

Proposition 2.1 *Let v, w be two weights and let E be a Banach space. The weighted composition operator $W_{\psi,\varphi} : H_v^\infty(\mathbb{D}, E) \to H_w^\infty(\mathbb{D}, E)$ is continuous if and only if $W_{\psi,\varphi} : H_v^\infty \to H_w^\infty$ is continuous.*

Proof First, assume $W_{\psi,\varphi} : H_v^\infty \to H_w^\infty$ is continuous. By [14, Proposition 3.1], we have that

$$\sup_{z\in\mathbb{D}} |\psi(z)| \frac{w(z)}{\tilde{v}(\varphi(z))} = M < \infty.$$

That is, $|\psi(z)|w(z) \leq M\tilde{v}(\varphi(z))$ for all $z \in \mathbb{D}$. Then, for each $F \in H_v^\infty(\mathbb{D}, E)$ we obtain that:

$$\|W_{\psi,\varphi}(F)\|_w = \sup_{z\in\mathbb{D}} \|\psi(z)F(\varphi(z))\|w(z) \leq \sup_{z\in\mathbb{D}} M\|F(\varphi(z))\|\tilde{v}(\varphi(z)) \leq M\|F\|_v.$$

Now suppose that $W_{\psi,\varphi} : H_v^\infty(\mathbb{D}, E) \to H_w^\infty(\mathbb{D}, E)$ is continuous. Choose $x_0 \in E$ and $u_0' \in E'$ such that $u_0'(x_0) = 1$. Define $S : H_v^\infty \to H_v^\infty(\mathbb{D}, E)$ by $S(f(z)) := f(z)x_0$ for all $f \in H_v^\infty$, $z \in \mathbb{D}$, and $T : H_w^\infty(\mathbb{D}, E) \to H_w^\infty$ by $T(F) := u_0' \circ F$ for all $F \in H_w^\infty(\mathbb{D}, E)$. The operators S and T are continuous. Moreover, $T \circ W_{\psi,\varphi} \circ S$ is exactly the weighted composition operator $W_{\psi,\varphi}$ in the scalar case. Indeed, for each $f \in H_v^\infty$ and $z \in \mathbb{D}$ we have

$$\begin{aligned}(T \circ W_{\psi,\varphi} \circ S)(f)(z) &= (T \circ W_{\psi,\varphi})(f(z)x_0) = T(\psi(z)f(\varphi(z))x_0) = \\ &= u_0'(\psi(z)f(\varphi(z))x_0) = \psi(z)f(\varphi(z))u_0'(x_0) = W_{\psi,\varphi}(f)(z).\end{aligned}$$

Therefore, $W_{\psi,\varphi} : H_v^\infty \to H_w^\infty$ is continuous since it is the composition of continuous operators. □

Proposition 2.2 *Let v, w be two weights and let E be a Banach space. The weighted composition operator $W_{\psi,\varphi} : H_v^0(\mathbb{D}, E) \to H_w^0(\mathbb{D}, E)$ is continuous if and only if $W_{\psi,\varphi} : H_v^0 \to H_w^0$ is continuous.*

Proof The operator $W_{\psi,\varphi} : H_v^0 \to H_w^0$ is continuous if and only if $\psi \in H_w^0$ and,

$$\sup_{z\in\mathbb{D}} |\psi(z)| \frac{w(z)}{\tilde{v}(\varphi(z))} < \infty,$$

by [14, Proposition 3.2]. The proof of Proposition 2.1 can be easily adapted to get the result. □

Theorem 2.1 *Let v, w be two weights and let E be a Banach space. Assume that $W_{\psi,\varphi} : H_v^0(\mathbb{D}, E) \to H_w^0(\mathbb{D}, E)$ is continuous. The following statements are equivalent:*

(i) $W_{\psi,\varphi} : H_v^\infty(\mathbb{D}, E) \to H_w^\infty(\mathbb{D}, E)$ *is compact,*
(ii) $W_{\psi,\varphi} : H_v^0(\mathbb{D}, E) \to H_w^0(\mathbb{D}, E)$ *is compact,*
(iii) $W_{\psi,\varphi} : H_v^\infty \to H_w^\infty$ *is compact and E has finite dimension,*
(iv) $W_{\psi,\varphi} : H_v^0 \to H_w^0$ *is compact and E has finite dimension.*

Moreover, conditions (i) and (iii) are equivalent under the weaker assumption that $W_{\psi,\varphi} : H_v^\infty(\mathbb{D}, E) \to H_w^\infty(\mathbb{D}, E)$ is continuous.

Proof $(iii) \Rightarrow (i)$. Suppose $W_{\psi,\varphi} : H_v^\infty \to H_w^\infty$ is compact. If $N = \dim E < \infty$, we have the canonical isomorphism $H_v^\infty(\mathbb{D}, E) \cong (H_v^\infty)^N$ and the weighted composition operator acts coordinatewise. Therefore, $W_{\psi,\varphi} : H_v^\infty(\mathbb{D}, E) \to H_w^\infty(\mathbb{D}, E)$ is also compact.

$(i) \Rightarrow (iii)$. Assume that $W_{\psi,\varphi} : H_v^\infty(\mathbb{D}, E) \to H_w^\infty(\mathbb{D}, E)$ is compact. The operators S and T considered in the proof of Proposition 2.1 permits us to conclude that $W_{\psi,\varphi} : H_v^\infty \to H_w^\infty$ is compact. To prove that E must be finite dimensional we proceed as follows. Define $P : E \to H_v^\infty(\mathbb{D}, E)$ by $P(x) = f_x$ such that $f_x : \mathbb{D} \to E$, $f_x(z) := x$ for all $z \in \mathbb{D}$, and $Q : H_w^\infty(\mathbb{D}, E) \to E$ by $Q(F) = F(z_0)/\psi(z_0)$, for some $z_0 \in \mathbb{D}$ with $\psi(z_0) \neq 0$. Both mappings P and Q are well defined, linear and continuous. We have

$$Q \circ W_{\psi,\varphi} \circ P(x) := Q(W_{\psi,\varphi}(f_x)) = W_{\psi,\varphi}(f_x)(z_0)/\psi(z_0) = \psi(z_0) f_x(\varphi(z_0))/\psi(z_0)$$

$$= f_x(\varphi(z_0)) = x.$$

Thus, $Q \circ W_{\psi,\varphi} \circ P = I_E$ and the identity $I_E : E \to E$ is compact, which implies that E is finite dimensional.

Observe that the proof of $(i) \Leftrightarrow (iii)$ only requires the continuity of $W_{\psi,\varphi} : H_v^\infty(\mathbb{D}, E) \to H_w^\infty(\mathbb{D}, E)$. On the other hand, if $W_{\psi,\varphi} : H_v^0(\mathbb{D}, E) \to H_w^0(\mathbb{D}, E)$ is continuous, then the operator $W_{\psi,\varphi}$ is also continuous in H_v^0 (Proposition 2.2), in H_v^∞ ([14, Propositions 3.1 and 3.2]) and in $H_v^\infty(\mathbb{D}, E)$ (Proposition 2.1).

$(i) \Rightarrow (ii)$. Assume $W_{\psi,\varphi} : H_v^\infty(\mathbb{D}, E) \to H_w^\infty(\mathbb{D}, E)$ is compact. Then, $W_{\psi,\varphi} : H_v^0(\mathbb{D}, E) \to H_w^0(\mathbb{D}, E)$ is compact because it is the restriction of a compact operator.

The proof of the equivalence $(ii) \Leftrightarrow (iv)$ is analogous to the one of $(i) \Leftrightarrow (iii)$.

$(iii) \Leftrightarrow (iv)$. This follows from [14, Propositions 3.1 and 3.2 and Corollaries 4.3 and 4.5]. □

The proof of the next Lemma is inspired by [10] and [12].

Lemma 2.1 *Let v, w be two weights and let E be a Banach space. If the operator $W_{\psi,\varphi} : H_v^\infty \to H_w^\infty$ is compact and E is reflexive, then $W_{\psi,\varphi} : H_v^\infty(\mathbb{D}, E) \to H_w^\infty(\mathbb{D}, E)$ is weakly compact.*

Proof Consider the transpose operator $W_{\psi,\varphi}^t : (H_w^\infty)' \to (H_v^\infty)'$. The predual of H_v^∞, defined as in [10, 5(c)], is denoted by ${}'H_v^\infty$. Applying [10, 5(d)] to our spaces, we have that

$$ {}'H_v^\infty = \overline{\operatorname{span}\{\delta_z : z \in \mathbb{D}\}}^{(H_v^\infty)'}, \quad {}'H_w^\infty = \overline{\operatorname{span}\{\delta_z : z \in \mathbb{D}\}}^{(H_w^\infty)'}. \tag{1} $$

Since $W_{\psi,\varphi}^t(\delta_z) = \psi(z)\delta_{\varphi(z)}$, we have

$$ W_{\psi,\varphi}^t(\operatorname{span}\{\delta_z : z \in \mathbb{D}\}) \subseteq \operatorname{span}\{\delta_z : z \in \mathbb{D}\}. \tag{2} $$

Now, by applying (1) and (2) and the continuity of $W_{\psi,\varphi}^t$, we get

$$ W_{\psi,\varphi}^t({}'H_w^\infty) = W_{\psi,\varphi}^t(\overline{\operatorname{span}\{\delta_z : z \in \mathbb{D}\}}) \subseteq \overline{W_{\psi,\varphi}^t(\operatorname{span}\{\delta_z : z \in \mathbb{D}\})} $$
$$ \subseteq \overline{\operatorname{span}\{\delta_z : z \in \mathbb{D}\}} = {}'H_v^\infty. $$

Thus, the restricted operator $W_{\psi,\varphi}^t|_{{}'H_w^\infty} : {}'H_w^\infty \to {}'H_v^\infty$ is well defined and continuous. Moreover, it is also compact, by Schauder's Theorem [22, Theorem 15.3], since its transpose map coincides with $W_{\psi,\varphi} : H_v^\infty \to H_w^\infty$, which is compact by assumption.

We now consider the operators χ and ϕ given in [10, Lemma 10], which are defined as follows: The operator $\chi : \mathcal{L}({}'H_w^\infty, E) \to H_w^\infty(\mathbb{D}, E)$ is defined by $\chi(T) := T \circ \Delta$, where $\Delta : \mathbb{D} \to {}'H_w^\infty$ is given by $\Delta(z) = \delta_z$. The operator $\phi : H_v^\infty(\mathbb{D}, E) \to \mathcal{L}({}'H_v^\infty, E)$ is defined, for $F \in H_v^\infty(\mathbb{D}, E)$, by $(\phi(F)(g))(u') := g(u' \circ F)$ for all $g \in ({}'H_v^\infty)$ and $u' \in E'$. Both operators are well defined, linear, continuous and their norms are less or equal to 1.

The wedge operator

$$ W_{\psi,\varphi}^t|_{{}'H_w^\infty} \wedge I_E : \mathcal{L}({}'H_v^\infty, E) \to \mathcal{L}({}'H_w^\infty, E) $$

is defined, for each $X \in \mathcal{L}('H_v^\infty, E)$, by

$$(W^t_{\psi,\varphi}|_{'H_w^\infty} \wedge I_E)(X) := I_E \circ X \circ W^t_{\psi,\varphi}|_{'H_w^\infty}.$$

Since $W^t_{\psi,\varphi}|_{'H_w^\infty}$ is compact and I_E is weakly compact, we can apply [12, Corollary 2.11] (or [24, Theorem 2.9] applied to four different spaces, as mentioned in the comment below that theorem) to conclude that $W^t_{\psi,\varphi}|_{'H_w^\infty} \wedge I_E$ is weakly compact.

Now, for every $F \in H_v^\infty(\mathbb{D}, E)$ and $z \in \mathbb{D}$, we have

$$\begin{aligned}(\chi \circ (W^t_{\psi,\varphi}|_{'H_w^\infty} \wedge I_E) \circ \phi)(F)(z) &= \chi \circ (I_E \circ \phi(F) \circ W^t_{\psi,\varphi}|_{'H_w^\infty})(z) = \\ &= (I_E \circ \phi(F) \circ W^t_{\psi,\varphi}|_{'H_w^\infty}) \circ \Delta(z) = \\ &= I_E \circ \phi(F) \circ (W^t_{\psi,\varphi}|_{'H_w^\infty}(\delta_z)) = \phi(F)(\psi(z)\delta_{\varphi(z)}) = \\ &= \psi(z)\phi(F)(\delta_{\varphi(z)}) = \psi(z)F(\varphi(z)) = W_{\psi,\varphi}(F)(z).\end{aligned}$$

Therefore $W_{\psi,\varphi} = \chi \circ (W^t_{\psi,\varphi}|_{'H_w^\infty} \wedge I_E) \circ \phi$ and $W_{\psi,\varphi} : H_v^\infty(\mathbb{D}, E) \to H_w^\infty(\mathbb{D}, E)$ is weakly compact. □

Theorem 2.2 *Let v, w be two weights and let E be a Banach space. Suppose $W_{\psi,\varphi} : H_v^0(\mathbb{D}, E) \to H_w^0(\mathbb{D}, E)$ is continuous. The following statements are equivalent:*

(i) $W_{\psi,\varphi} : H_v^\infty(\mathbb{D}, E) \to H_w^\infty(\mathbb{D}, E)$ *is weakly compact,*
(ii) $W_{\psi,\varphi} : H_v^0(\mathbb{D}, E) \to H_w^0(\mathbb{D}, E)$ *is weakly compact,*
(iii) $W_{\psi,\varphi} : H_v^\infty \to H_w^\infty$ *is compact and E is reflexive,*
(iv) $W_{\psi,\varphi} : H_v^0 \to H_w^0$ *is compact and E is reflexive.*

Moreover, conditions (i) and (iii) are equivalent under the weaker assumption that $W_{\psi,\varphi} : H_v^\infty(\mathbb{D}, E) \to H_w^\infty(\mathbb{D}, E)$ is continuous.

Proof The implication $(iii) \Rightarrow (i)$ was the content of Lemma 2.1.

$(i) \Rightarrow (iii)$. The operators S and T defined in Proposition 2.1 permit us to conclude that $W_{\psi,\varphi} : H_v^\infty \to H_w^\infty$ is the composition of the weakly compact weighted composition operator in the vector-valued case and continuous operators, thus it is also weakly compact. But, it is also compact. Indeed, if it is not compact, we apply [14, Theorem 5.2] to find a subspace $H \subset H_v^\infty$ isomorphic to ℓ_∞ such that $W_{\psi,\varphi}|_H : F \to H$ is an isomorphism. Since $W_{\psi,\varphi}|_F$ is weakly compact, we conclude that ℓ_∞ is reflexive, a contradiction.

To prove that E is reflexive, we consider the operators P and Q as in the proof of Proposition 2.1. Since $Q \circ W_{\psi,\varphi} \circ P = I_E$, the identity I_E of E is weakly compact, as the composition of weakly compact and continuous operators. This implies that E is reflexive.

$(i) \Rightarrow (ii)$. If $W_{\psi,\varphi} : H_v^\infty(\mathbb{D}, E) \to H_w^\infty(\mathbb{D}, E)$ is weakly compact then $W_{\psi,\varphi} : H_v^0(\mathbb{D}, E) \to H_w^0(\mathbb{D}, E)$ is also weakly compact.

$(ii) \Rightarrow (iv)$. The proof is similar to $(i) \Rightarrow (iii)$ using [14, Theorem 5.1].

$(iii) \Leftrightarrow (iv)$. This follows from [14, Proposition 3.1 and 3.2 and Corollaries 4.3 and 4.5]. □

3 Inductive Limits of Weighted Banach Spaces of Vector-Valued Functions

Let $V = (v_n)_n$ be a sequence of weights on $\mathbb{D}$ such that $v_n(z) \geq v_{n+1}(z)$ for each $n \in \mathbb{N}$ and $z \in \mathbb{D}$. Let E be a Banach space. The weighted inductive limits of spaces of vector-valued analytic functions are defined as the countable inductive limit of the Banach spaces $H^\infty_{v_n}(\mathbb{D}, E)$.

$$VH(\mathbb{D}, E) := \operatorname*{ind}_n H^\infty_{v_n}(\mathbb{D}, E).$$

Analogously, we define

$$V_0H(\mathbb{D}, E) := \operatorname*{ind}_n H^0_{v_n}(\mathbb{D}, E).$$

Both spaces are (LB)-spaces. Spaces of this type were studied in [8]. See also [1] and [3] and the references therein. The following condition (S) was introduced in [8, Section 0.4]: For every $n \in \mathbb{N}$ there exists $m > n$ such that

$$\lim_{r \to 1^-} \frac{v_m(r)}{v_n(r)} = 0. \qquad (S)$$

If condition (S) holds, then $VH(\mathbb{D}, E) = V_0H(\mathbb{D}, E)$. Moreover, if $m > n$ is selected for n according to (S), the spaces $H^\infty_{v_m}(\mathbb{D}, E)$, $VH(\mathbb{D}, E)$ and the compact open topology all induce the same topology on the bounded subsets of $H^\infty_{v_n}(\mathbb{D}, E)$.

The definitions of regular, compactly regular, (strongly) boundedly retractive inductive limits, as well as conditions (M) and (M_0), can be seen in [1], [26] and [27].

Proposition 3.1 *Let $E = ind_n E_n$ be a strongly boundedly retractive (LB)-space. Then every weakly compact subset of E is contained and weakly compact in some E_n.*

Proof If E is strongly boundedly retractive then E has condition (M) by Neus' Theorem (see [27, Theorem 6.4] or [26, Section 9.5, Theorem (9)]), hence condition (M_0) (see [1, p. 105]). By condition (M_0), there is an increasing sequence $(U_n)_n$ such that U_n is an absolutely convex 0-neighbourhood in E_n such that for every $n \in \mathbb{N}$ there exists $m > n$ with $\sigma(E_m, E'_m)|_{U_n} = \sigma(E, E')|_{U_n}$, see Chapter 1, Section 9.4 in [26]. Now take $B \subset E$ which is $\sigma(E, E')$-compact. Since E is regular, there is $n \in \mathbb{N}$ such that $B \subset E_n$ and bounded. Find $\lambda > 0$ with $B \subset \lambda U_n$. Since

$\sigma(E, E')|_{\lambda U_n} = \sigma(E_m, E'_m)|_{\lambda U_n}$ and B is $\sigma(E, E')$-compact, it follows that B is $\sigma(E_m, E'_m)$-compact. □

Proposition 3.2 *Let E be a Banach space and let $V = (v_n)_n$ be a decreasing sequence of weights.*

(i) $VH(\mathbb{D}, E)$ is a regular (LB)-space.
(ii) If $V = (v_n)_n$ satisfies condition (S), then $VH(\mathbb{D}, E) = V_0H(\mathbb{D}, E)$ is strongly boundedly retractive. In particular, every (weakly) relatively compact subset of $VH(\mathbb{D}, E)$ is contained and (weakly) relatively compact in some step $H_{v_n}(\mathbb{D}, E)$.

Proof

(i). Consider the following inductive limit of Banach spaces of continuous functions

$$VC(\mathbb{D}, E) := \operatorname{ind}_n C_{v_n}(\mathbb{D}, E)$$

where $C_{v_n}(\mathbb{D}, E) := \{F \in C(\mathbb{D}, E) : \sup_{z\in\mathbb{D}} v_n(z)\|F(z)\| < \infty\}$ for all $n \in \mathbb{N}$. By [2, Corollary 2.5], $VC(\mathbb{D}, E)$ is a regular inductive limit. Since $H^\infty_{v_n}(\mathbb{D}, E) \subseteq C_{v_n}(\mathbb{D}, E)$ with continuous inclusion for all $n \in \mathbb{N}$, then $VH(\mathbb{D}, E)$ is also regular.

(ii). $VH(\mathbb{D}, E)$ is strongly boundedly retractive by [8, p. 114]. The statement about relatively (weakly) compact sets follows from Proposition 3.1.

□

Proposition 3.3 *Let $T : F \to G$ be a continuous linear operator between two (LB)-spaces $F = ind_n$ and $G = ind_m G_m$.*

(a) Assume that G is regular. Then T is bounded if, and only if, there is m such that $T : F_n \to G_m$ is continuous for all $n \in \mathbb{N}$.
(b) Assume that F is regular and that G is strongly boundedly retractive.

(i) T is Montel if, and only if, for all $n \in \mathbb{N}$ there exists $m \in \mathbb{N}$ such that $T(F_n) \subseteq G_m$ and $T : F_n \to G_m$ is compact.
(ii) T is reflexive if, and only if, for all $n \in \mathbb{N}$ there exists $m \in \mathbb{N}$ such that $T(F_n) \subseteq G_m$ and $T : F_n \to G_m$ is weakly compact.
(iii) T is compact if, and only if, there exists $m \in \mathbb{N}$ such that $T(F_n) \subseteq G_m$ and $T : F_n \to G_m$ is compact for all $n \in \mathbb{N}$.
(iv) T is weakly compact if, and only if, there exists $m \in \mathbb{N}$ such that $T(F_n) \subseteq G_m$ and $T : F_n \to G_m$ is weakly compact for all $n \in \mathbb{N}$.

Proof The closed unit ball of a Banach space X is denoted by $\overline{B}_X$.

(a) The proof is easy using Theorem 1 (1) in Chapter 4, Part 2 in [17] about sequences of bounded sets in a metrizable space.

$(b.i)$ and $(b.ii)$ The first result is stated in [13, Lemma 13]. The proof follows from the fact that relatively (weakly) compact subsets of G are localized in a step. Apply Proposition 3.1 for the weakly compact case.

$(b.iii)$ Firstly, suppose that T is compact. That is, there exists a 0-neighbourhood $U \subseteq F$ such that $T(U)$ is relatively compact in G. Since G is strongly boundedly retractive, there is $m \in \mathbb{N}$ such that $T(U)$ is relatively compact in G_m. Now, $T(F) \subseteq G_m$ and $T(U \cap F_n)$ is relatively compact in G_m, which implies that $T : F_n \to G_m$ is compact. For the converse, we have, by assumption, that there is $m \in \mathbb{N}$ such that for each $n \in \mathbb{N}$ $T(\overline{B}_{F_n})$ is relatively compact in G_m. There are $\varepsilon_n > 0, n \in \mathbb{N}$, such that $\bigcup_n \varepsilon_n T(\overline{B}_{F_n})$ is relatively compact in G_m. If A denotes the absolutely convex hull of $\bigcup_n \varepsilon_n T(\overline{B}_{F_n})$, A is also relatively compact in G_m, by Krein's Theorem ([20, 24.5(4) p. 325]). Now, the absolutely convex hull U of $\bigcup_n \varepsilon_n \overline{B}_{F_n}$ is a 0-neighbourhood in F and $T(U) \subseteq A$ is relatively compact.

$(b.iv)$ The proof is similar to the one of (b.iii) using Proposition 3.1, and [20, 24.5(4') p. 325] to ensure that the absolutely convex hull of a relatively weakly compact subset of a Banach space is also relatively weakly compact.

□

Proposition 3.4 *Let E be a Banach space. Let $V = (v_n)_n$ be a decreasing sequence of essential weights.*

(a) The following are equivalent:

(i) $W_{\psi,\varphi} : VH(\mathbb{D}, E) \to VH(\mathbb{D}, E)$ is continuous.
(ii) For all $n \in \mathbb{N}$ there is $m > n$ such that $\sup_{z \in \mathbb{D}} |\psi(z)| v_m(z)/v_n(\varphi(z)) < +\infty$.

(b) The following are equivalent:

(i) $W_{\psi,\varphi} : VH(\mathbb{D}, E) \to VH(\mathbb{D}, E)$ is bounded.
(ii) There is $m \in \mathbb{N}$ such that $\sup_{z \in \mathbb{D}} |\psi(z)| v_m(z)/v_n(\varphi(z)) < +\infty$ for all $n \in \mathbb{N}$.

Proof The inductive limit $VH(\mathbb{D}, E)$ is regular by Proposition 3.2. The proof now follows from Grothendieck's Factorization Theorem ([22, Theorem 24.33]), Propositions 2.1 and 3.3 (a) and [14, Proposition 3.1]. □

Proposition 3.5 *Let E be a Banach space. Suppose that $V = (v_n)_n$ is a decreasing sequence of essential weights with condition (S) such that $W_{\psi,\varphi} : VH(\mathbb{D}, E) \to VH(\mathbb{D}, E)$ is continuous.*

(a) The following are equivalent:

(i) $W_{\psi,\varphi} : VH(\mathbb{D}, E) \to VH(\mathbb{D}, E)$ is Montel (resp. reflexive),

(ii) *E has finite dimension (resp. reflexive) and for all $n \in \mathbb{N}$ there is $m > n$ such that*

$$\lim_{r \longrightarrow 1^-} \sup_{|\varphi(z)|>r} |\psi(z)| \frac{v_m(z)}{v_n(\varphi(z))} = 0.$$

(b) *The following are equivalent:*

(i) $W_{\psi,\varphi} : VH(\mathbb{D}, E) \to VH(\mathbb{D}, E)$ *is compact (resp. weakly compact),*
(ii) *E has finite dimension (resp. reflexive) and there is $m \in \mathbb{N}$ such that*

$$\lim_{r \longrightarrow 1^-} \sup_{|\varphi(z)|>r} |\psi(z)| \frac{v_m(z)}{v_n(\varphi(z))} = 0 \quad \textit{for all } n \in \mathbb{N}.$$

Proof Since $VH(\mathbb{D}, E)$ is strongly boundedly retractive by Proposition 3.2), the proof follows from Proposition 3.3(b), Theorems 2.1 and 2.2 and [14, Corollaries 4.3 and 4.5]. □

We denote by $A^{-\infty}(E)$ the vector-valued Korenblum space

$$A^{-\infty}(E) := \bigcup_{n \in \mathbb{N}} H_n^\infty(\mathbb{D}, E)\ .$$

Here, $H_n^\infty(\mathbb{D}, E)$ denotes the weighted Banach space of vector-valued functions where the weights are $v_n(z) = (1 - |z|)^n$. It is endowed with the inductive limit topology: $A^{-\infty}(E) = \operatorname{ind}_n H_n^\infty(\mathbb{D}, E)$. Observe that $A^{-\infty}(E) = VH(\mathbb{D}, E)$ when $V = (v_n)_n = ((1 - |z|)^n)_n$ is the decreasing sequence of weights, which are essential. The inductive limit $A^{-\infty}(E)$ is regular. Moreover, $(v_n)_n$ has condition (S). Thus, $A^{-\infty}(E)$ is also strongly boundedly retractive. The space $A^{-\infty}$ was introduced in the scalar case by Korenblum in [19]. More information can be seen in [18, Section 4.3]. We mention the following consequences of Propositions 3.4 and 3.5.

Corollary 3.1 *Let E be a Banach space.*

(a) *The following are equivalent:*

(i) $W_{\psi,\varphi} : A^{-\infty}(E) \to A^{-\infty}(E)$ *is continuous.*
(ii) *For all $n \in \mathbb{N}$ there is $m > n$ such that*

$$\sup_{z \in \mathbb{D}} |\psi(z)| \frac{(1 - |z|)^m}{(1 - |\varphi(z)|)^n} < +\infty.$$

(b) *The following are equivalent:*

(i) $W_{\psi,\varphi} : A^{-\infty}(E) \to A^{-\infty}(E)$ *is bounded.*

(ii) There is $m \in \mathbb{N}$ such that for all $n \in \mathbb{N}$

$$\sup_{z\in\mathbb{D}} |\psi(z)| \frac{(1-|z|)^m}{(1-|\varphi(z)|)^n} < +\infty.$$

(c) The following are equivalent:

(i) $W_{\psi,\varphi} : A^{-\infty}(E) \to A^{-\infty}(E)$ is Montel (resp. reflexive).
(ii) E has finite dimension (resp. reflexive) and for all $n \in \mathbb{N}$ there is $m > n$ such that

$$\lim_{r\longrightarrow 1^-} \sup_{|\varphi(z)|>r} |\psi(z)| \frac{(1-|z|)^m}{(1-|\varphi(z)|)^n} = 0.$$

(d) The following are equivalent:

(i) $W_{\psi,\varphi} : A^{-\infty}(E) \to A^{-\infty}(E)$ is compact (resp. weakly compact).
(ii) E has finite dimension (resp. reflexive) and there is $m \in \mathbb{N}$ such that for all $n \in \mathbb{N}$

$$\lim_{r\longrightarrow 1^-} \sup_{|\varphi(z)|>r} |\psi(z)| \frac{(1-|z|)^m}{(1-|\varphi(z)|)^n} = 0.$$

Similar results can be stated for weighted composition operators on the inductive limits

$$A_-^{-\alpha}(E) = \mathrm{ind}_n H_{\alpha_n}^\infty(\mathbb{D}, E) ,$$

where $H_{\alpha_n}^\infty(\mathbb{D}, E)$ denotes the weighted Banach space of vector-valued functions with weights $v_{\alpha_n}(z) = (1-|z|)^{\alpha-\frac{1}{n}}$, where $n \geq n_0$ such that $\alpha - \frac{1}{n_0} > 0$.

4 Projective Limits of Weighted Banach Spaces of Vector-Valued Functions

Let $W = (w_n)_n$ be a sequence of weights on $\mathbb{D}$ such that $w_n(z) \leq w_{n+1}(z)$ for each $n \in \mathbb{N}$ and $z \in \mathbb{D}$. Let E be a Banach space. The weighted Fréchet space $HW(\mathbb{D}, E)$ is defined as the projective limit of the Banach spaces $H_{w_n}^\infty(\mathbb{D}, E)$. That is,

$$HW(\mathbb{D}, E) = \mathrm{proj}_n\, H_{w_n}^\infty(\mathbb{D}, E),$$

and it is endowed with the norms

$$\|F\|_{w_n} := \sup_{z\in\mathbb{D}} w_n(z)\|F(z)\|, \quad n \in \mathbb{N}.$$

We associate with $W = (w_n)_n$ the following family of weights on $\mathbb{D}$:

$$V(W) := \{v : \mathbb{D} \to]0, +\infty[\ ;\ v \text{ is continuous, radial and}$$
$$w_n v \text{ is bounded in } \mathbb{D} \text{ for each } n\}.$$

Lemma 4.1 *A subset $B \subseteq HW(\mathbb{D}, E)$ is bounded if and only if there is $v \in V(W)$ such that*

$$B \subseteq B_v := \{F \in HW(\mathbb{D}, E)\ :\ \|F(z)\| \leq v(z) \text{ for all } z \in \mathbb{D}\}.$$

Proof Since for each $F \in B_v$

$$\sup_{z\in\mathbb{D}} w_n(z)\|F(z)\| \leq \sup_{z\in\mathbb{D}} w_n(z)v(z) < +\infty,$$

we have that B_v is bounded in $HW(\mathbb{D}, E)$.

Given a bounded set $B \subseteq HW(\mathbb{D}, E)$, for each $n \in \mathbb{N}$ there is M_n such that

$$\sup_{F\in B}\sup_{z\in\mathbb{D}} w_n(z)\|F(z)\| \leq M_n.$$

Then, the set $\tilde{B} := \{\|F(\cdot)\| : \mathbb{D} \to \mathbb{C}\ ;\ F \in B\}$ is bounded in the Köthe echelon space $\lambda_\infty(\mathbb{D}, W)$. By the characterization of bounded subsets in the Köthe echelon space of infinite order, [7, Proposition 2.5], there exists $\overline{w} : \mathbb{D} \to]0, +\infty[$ such that $w_n\overline{w}$ is bounded for each $n \in \mathbb{N}$ and

$$\sup_{z\in\mathbb{D}} \frac{\|F(z)\|}{\overline{w}(z)} \leq 1 \quad \text{for each } F \in B.$$

We can apply [8, Proposition in p. 112] to show that $\overline{w}$ is dominated by $v \in V(W)$. This implies that $\|F(z)\| \leq v(z)$ for all $z \in \mathbb{D}$, and $B \subseteq B_v$. □

Given a holomorphic function $F \in H(\mathbb{D}, E)$ there is $(x_k)_k \subseteq E$ such that

$$F(z) = \sum_{k=0}^{\infty} x_k z^k, \quad z \in \mathbb{D},$$

and the series converges uniformly on compact subsets of $\mathbb{D}$. The k-th Taylor polynomial of F is denoted by P_k:

$$P_k(z) := \sum_{j=0}^{k} x_j z^j.$$

If w is a weight on $\mathbb{D}$, then every (vector-valued) polynomial $P(z) = \sum_{j=0}^{k} x_j z^j$ belongs to $H_w^\infty(\mathbb{D}, E)$. Given $G \in H(\mathbb{D}, E)$, $G(z) = \sum_{k=0}^{\infty} x_k z^k$, $z \in \mathbb{D}$, we denote by $C_n(G)$ the *Cesàro sums* of the Taylor polynomials of G. That is,

$$C_n(G)(z) := \frac{1}{n+1} \sum_{j=0}^{n} \left(\sum_{k=0}^{j} x_k z^k \right).$$

The sequence $(C_n(G))_n$ tends to G uniformly on compact sets of $\mathbb{D}$ whenever $n \to +\infty$.

Corollary 4.1 *For each $F \in H_{w_n}^\infty(\mathbb{D}, E)$ the sequence $(C_k(F))_k \subset HW(\mathbb{D}, E)$ satisfies*

$$\sup_{z \in \mathbb{D}} w_n(z) \|C_k(F)(z)\| \leq \sup_{z \in \mathbb{D}} w_n(z) \|F(z)\|$$

and $C_k(F) \to F$ uniformly on the compact subsets of $\mathbb{D}$.

Proof This is a consequence of [4, Proposition 1.2] and Hahn-Banach Theorem. □

Proposition 4.1 *Let H be a Hausdorff locally convex space and $T : H \to H$ continuous. Let $E, F \subseteq H$ be two Fréchet spaces $E = proj_n E_n$, $F = proj_n F_n$, where $E_{n+1} \subseteq E_n \subseteq E_1 \subseteq H$ and $F_{n+1} \subseteq F_n \subseteq F_1 \subseteq H$ with continuous inclusions for all $n \in \mathbb{N}$. Assume also that the following conditions (C1) and (C2) are satisfied:*

(C1) For all $n \in \mathbb{N}$, $x \in E_n$ there is $(y_k)_k \subseteq E$ such that $y_k \to x$ in H and $\|y_k\|_{E_n} \leq \|x\|_{E_n}$ for each k.
(C2) Each C_{F_n} is closed in H.

(a) The following conditions are equivalent:

(i) $T(E) \subseteq F$.
(ii) $T : E \to F$ is continuous.
(iii) For all $n \in \mathbb{N}$ there exists $m \in \mathbb{N}$ with $T(E_m) \subseteq F_n$.
(iv) For all $n \in \mathbb{N}$ there exists $m \in \mathbb{N}$ such that $T : E_m \to F_n$ is well-defined and continuous.

(b) *The following conditions are equivalent:*

(i) $T : E \to F$ *is bounded.*
(ii) *There exists* $m \in \mathbb{N}$ *such that* $T(E_m) \subseteq F_n$ *for all* $n \in \mathbb{N}$.
(iii) *There exists* $m \in \mathbb{N}$ *such that* $T : E_m \to F_n$ *is well-defined and continuous for all* $n \in \mathbb{N}$.

(c) *The following conditions are equivalent:*

(i) $T : E \to F$ *is Montel (respectively reflexive).*
(ii) *For all absolutely convex closed bounded subset* $B \subseteq E$, $T : E_B \to F_n$ *is compact (resp. weakly compact) for all* $n \in \mathbb{N}$.

(d) *The following conditions are equivalent:*

(i) $T : E \to F$ *is compact (respectively weakly compact).*
(ii) *There exists* $m \in \mathbb{N}$ *such that* $T : E_m \to F_n$ *is compact (resp. weakly compact) for all* $n \in \mathbb{N}$.

Proof The proof is the same as in [12, Proposition 4.2], keeping in mind that it is enough to assume that each C_{F_n} is closed in H, condition (C2), instead of the compactness of these balls for the topology of H. □

In the applications we have in mind, the space H in Proposition 4.1 is $H(\mathbb{D}, E)$ with the compact-open topology, $E_n = F_n = H^\infty_{w_n}(\mathbb{D}, E)$ for each $n \in \mathbb{N}$, and

$$C_{E_n} = C_{F_n} = \{f \in H(\mathbb{D}, E) : \sup_{z\in\mathbb{D}} w_n(z)\|F(z)\| \le 1\}.$$

Observe that condition (C1) follows from Corollary 4.1. It is easy to see that C_{F_n} is closed in $H(\mathbb{D}, E)$ with the τ_{co} topology. So, condition (C2) is also satisfied.

Our next result should be compared with [12, Theorem 4.3]. The proofs of the result and the corollary follow from Propositions 4.1 and 2.1, Theorems 2.1 and 2.2 and [14, Proposition 3.1 and Corollaries 4.3 and 4.5].

Proposition 4.2 *Let* $W = (w_n)_n$ *be an increasing sequence of weights on* $\mathbb{D}$. *Let* E *be a Banach space.*

(a) *The following conditions are equivalent:*

(i) $W_{\psi,\varphi} : HW(\mathbb{D}, E) \to HW(\mathbb{D}, E)$ *is continuous,*
(ii) *for all* $n \in \mathbb{N}$ *there exists* $m \in \mathbb{N}$ *such that*

$$\sup_{z\in\mathbb{D}} \frac{w_n(z)}{\tilde{w}_m(\varphi(z))}|\psi(z)| < \infty.$$

(b) *The following conditions are equivalent:*

(i) $W_{\psi,\varphi} : HW(\mathbb{D}, E) \to HW(\mathbb{D}, E)$ *is bounded,*

(ii) there exists $m \in \mathbb{N}$ such that

$$\sup_{z\in\mathbb{D}} \frac{w_n(z)}{\tilde{w}_m(\varphi(z))}|\psi(z)| < \infty \quad \textit{for all } n \in \mathbb{N}.$$

(c) The following conditions are equivalent:

(i) $W_{\psi,\varphi} : HW(\mathbb{D}, E) \to HW(\mathbb{D}, E)$ is Montel (reflexive),
(ii) E has finite dimension (resp. E is reflexive), $\psi \in HW(\mathbb{D}$ and for all $v \in V(W)$ there is $n \in \mathbb{N}$ such that

$$\lim_{r\longrightarrow 1^-} \sup_{|\varphi(z)|>r} \frac{w_n(z)}{\widetilde{(1/v)}(\varphi(z))}|\psi(z)| = 0.$$

(d) The following conditions are equivalent:

(i) $W_{\psi,\varphi} : HW(\mathbb{D}, E) \to HW(\mathbb{D}, E)$ is (weakly) compact,
(ii) E has finite dimension (resp. E is reflexive) and there exists $m \in \mathbb{N}$ such that $\psi \in H^\infty_{w_m}(\mathbb{D})$ and

$$\lim_{r\longrightarrow 1^-} \sup_{|\varphi(z)|>r} \frac{w_n(z)}{\tilde{w}_m(\varphi(z))}|\psi(z)| = 0 \quad \textit{for all } n \in \mathbb{N}.$$

For a fixed $\alpha \geq 0$, we denote by $A_+^{-\alpha}(E)$ the Fréchet space $A_+^{-\alpha}(E) := \operatorname{proj}_n H^\infty_{\alpha_n}(\mathbb{D}, E)$, where, $H^\infty_{\alpha_n}(\mathbb{D}, E)$ is the weighted Banach space of vector-valued functions for the weight $v_{\alpha_n}(z) = (1-|z|)^{\alpha+\frac{1}{n}}$. Observe that $A_+^{-\alpha}(E) = HW(\mathbb{D}, E)$ for the increasing sequence of essential weights $W = (v_{\alpha_n})_n$.

Corollary 4.2 *Let E be a Banach space.*

(a) The following conditions are equivalent:

(i) $W_{\psi,\varphi} : A_+^{-\alpha}(E) \to A_+^{-\alpha}(E)$ is continuous.
(ii) For all $n \in \mathbb{N}$ there exists $m \in \mathbb{N}$ such that

$$\sup_{z\in\mathbb{D}} \frac{(1-|z|)^{\alpha+\frac{1}{n}}}{(1-|\varphi(z)|)^{\alpha+\frac{1}{m}}}|\psi(z)| < \infty.$$

(b) The following conditions are equivalent:

(i) $W_{\psi,\varphi} : A_+^{-\alpha}(E) \to A_+^{-\alpha}(E)$ is bounded.
(ii) There exists $m \in \mathbb{N}$ such that

$$\sup_{z\in\mathbb{D}} \frac{(1-|z|)^{\alpha+\frac{1}{n}}}{(1-|\varphi(z)|)^{\alpha+\frac{1}{m}}}|\psi(z)| < \infty \quad \textit{for all } n \in \mathbb{N}.$$

(c) The following conditions are equivalent:

(i) $W_{\psi,\varphi} : A_{+}^{-\alpha}(E) \to A_{+}^{-\alpha}(E)$ *is Montel (reflexive).*
(ii) E has finite dimension (resp. E is reflexive) and for all $v \in V((v_{\alpha_n})_n)$ *there is* $n \in \mathbb{N}$ *such that*

$$\lim_{|z|\longrightarrow 1^-} \frac{(1-|z|)^{\alpha+\frac{1}{n}}}{\widetilde{(1/v)}(\varphi(z))}|\psi(z)| = 0.$$

(d) The following conditions are equivalent:

(i) $W_{\psi,\varphi} : A_{+}^{-\alpha}(E) \to A_{+}^{-\alpha}(E)$ *is (weakly) compact.*
(ii) E has finite dimension (resp. E is reflexive) and there exists $m \in \mathbb{N}$ *such that*

$$\lim_{|z|\longrightarrow 1^-} \frac{(1-|z|)^{\alpha+\frac{1}{n}}}{(1-|\varphi(z)|)^{\alpha+\frac{1}{m}}}|\psi(z)| = 0 \quad \textit{for all } n \in \mathbb{N}.$$

Acknowledgments This research was partially supported by the research project MCIN PID2020-119457GB-I00/AEI/10.13039/501100011033. The research of Gómez-Orts was also supported by the grant BES-2017-081200.

The authors are thankful to the referee for the remarks which improved the presentation of the paper.

References

1. Bierstedt, K.D.: An introduction to locally convex inductive limits. In: Functional Analysis and Its Applications (Nice, 1986). ICPAM Lecture Notes, pp. 35–133. World Sci. Publishing, Singapore (1988)
2. Bierstedt, K.D., Bonet, J.: Projective descriptions of weighted inductive limits: the vector-valued cases. In: Advances in the Theory of Fréchet Spaces, Istanbul, 1988, pp. 195–221 (1989)
3. Bierstedt, K.D., Bonet, J.: Weighted (LB)-spaces of holomorphic functions: $\mathcal{V}H(G) = \mathcal{V}_0 H(G)$ and completeness of $\mathcal{V}_0 H(G)$. J. Math. Anal. Appl. **323**(2), 747–767 (2006)
4. Bierstedt, K.D., Bonet, J., Galbis, A.: Weighted spaces of holomorphic functions on balanced domains. Michigan Math. J. **40**(2), 271–297 (1993)
5. Bierstedt, K.D., Bonet, J., Taskinen, J.: Associated weights and spaces of holomorphic functions. Studia Math. **127**(2), 137–168 (1998)
6. Bierstedt, K.D., Holtmanns, S.: An operator representation for weighted inductive limits of spaces of vector valued holomorphic functions. Bull. Belg. Math. Soc. Simon Stevin **8**(4), 577–589 (2001)
7. Bierstedt, K.D., Meise, R., Summers, W.H.: Köthe sets and Köthe sequence spaces. In: Functional Analysis, Holomorphy and Approximation Theory (Rio de Janeiro, 1980), vol. 71, pp. 27–91. North-Holland Math. Stud., North-Holland, Amsterdam, New York (1982)
8. Bierstedt, K.D., Meise, R., Summers, W. H.: A projective description of weighted inductive limits. Trans. Am. Math. Soc. **272**(1), 107–160 (1982)

9. Bierstedt, K.D., Summers, W.H.: Biduals of weighted Banach spaces of analytic functions. J. Austral. Math. Soc. **54**, Series A, 70–79 (1993)
10. Bonet, J., Domański, P., Lindström, M.: Weakly compact composition operators on analytic vector-valued function spaces. Ann. Acad. Sci. Fenn. Math. **26**(1), 233–248 (2001)
11. Bonet, J., Domański, P., Lindström, M., Taskinen, J.: Composition operators between weighted Banach spaces of analytic functions. J. Austral. Math. Soc. Ser. A **64**(1), 101–118 (1998)
12. Bonet, J., Friz, M.:Weakly compact composition operators on locally convex spaces. Math. Nachr **245**, 26–44 (2002)
13. Bonet, J., Friz, M., Jordá, E.: Composition operators between weighted inductive limits of spaces of holomorphic functions. Publ. Math. Debrecen **67**(3–4), 333–348 (2005)
14. Contreras, M.D., Hernández-Díaz, A.G.: Weighted composition operators in weighted Banach spaces of analytic functions. J. Austral. Math. Soc. Ser. A **69**(1), 41–60 (2000)
15. Cowen, C.C., Maccluer, B.M.: Composition operators on spaces of analytic functions. In: Studies in Advanced Mathematics. CRC Press, Boca Raton, FL (1995)
16. Gómez-Orts, E.: Weighted composition operators on Korenblum type spaces of analytic functions. Rev. R. Acad. Cienc. Exactas Fís. Nat. Ser. A Mat. RACSAM **114**(4), 15 pp. (2020). Paper No. 199
17. Grothendieck, A.: Topological Vector Spaces. Gordon and Breach, New York (1973)
18. Hedenmalm, H., Korenblum, B., Zhu, K.: Theory of Bergman Spaces. Grad. Texts in Math., vol. 199. Springer, New York (2000)
19. Korenblum, B.: An extension of the Nevanlinna theory. Acta Math. **135**, 187–219 (1975)
20. Köthe, G.: Topological vector spaces I and II. Springer, Berlin, Heidelberg, New York (1969 and 1979)
21. Laitila, J., Tylli, H.-O., Wang, M.: Composition operators from weak to strong spaces of vector-valued analytic functions. J. Oper. Theory **62**(2), 281–295 (2009)
22. Meise, R., Vogt, D.: Introduction to Functional Analysis. Clarendon Press, Oxford (1997)
23. Montes-Rodríguez, A.: Weighted composition operators on weighted Banach spaces of analytic functions. J. London Math. Soc. (2) **61**(3), 872–884 (2000)
24. Saksman, E., Tylli, H.-O.: Weak compactness of multiplication operators on spaces of bounded linear operators. Math. Scand. **70**, 91–111 (1992)
25. Shapiro, J.H.: Composition Operators and Classical Function Theory. Springer, New York (1993)
26. Valdivia, M.: Topics in Locally Convex Spaces. North-Holland, Amsterdam (1982)
27. Wengenroth, J.: Derived Functors in Functional Analysis. Springer, Berlin, Heidelberg (2003)
28. Wolf, E.: Weighted Fréchet spaces of holomorphic functions. Studia Math. **174**(3), 255–275 (2006)

Existence of Positive Solutions of Nonlinear Second Order Dirichlet Problems Perturbed by Integral Boundary Conditions

Alberto Cabada, Lucía López-Somoza, and Mouhcine Yousfi

Abstract In this paper we study a second-order nonlinear perturbed Dirichlet problem with integral boundary conditions. We obtain the exact expression of the Green's function related to the perturbed problem in terms of the Green's function of the homogeneous Dirichlet problem. Moreover, we characterize the set of parameters where the Green's function has constant sign (which, contrary to the homogeneous case, can be either positive or negative) on its square of definition. Finally, as an application, the existence of positive solutions is derived from fixed point theory applied to related operators defined on suitable cones in Banach spaces.

Keywords Integral boundary conditions · Green's functions · Comparison results · Nonlinear boundary value problems

1 Introduction

In the present paper, we consider the following nonlinear equations

$$u''(t) + \gamma\, u(t) + f(t, u(t)) = 0, \quad t \in I \equiv [0, 1], \tag{1}$$

and

$$u''(t) + \gamma\, u(t) - f(t, u(t)) = 0, \quad t \in I, \tag{2}$$

Partially supported by Xunta de Galicia (Spain), project EM2014/032 and AIE, Spain and FEDER, grant PID2020-113275GB-I00.

A. Cabada (✉) · L. López-Somoza · M. Yousfi
CITMAga, Santiago de Compostela, Spain

Departamento de Estatística, Análise Matemática e Optimización, Facultade de Matemáticas, Universidade de Santiago de Compostela, Santiago de Compostela, Spain
e-mail: alberto.cabada@usc.es; lucia.lopez.somoza@usc.es; yousfi.mouhcine@usc.es

A. M. Candela et al. (eds.), *Recent Advances in Mathematical Analysis*, Trends in Mathematics, https://doi.org/10.1007/978-3-031-20021-2_10

subject to the integral boundary conditions

$$u(0) = \delta_1 \int_0^1 u(s)\, ds, \quad u(1) = \delta_2 \int_0^1 u(s) ds. \tag{3}$$

We will study these nonlinear problems for $\gamma \leq \pi^2$, $\delta_1 \geq 0$ and $\delta_2 \geq 0$.

The existence of positive solutions will depend on the regularity of the function f and the different values of the real parameters γ, δ_1 and δ_2. More concisely, if the Green's function associated to the related linear part of the considered problem is positive, then we will deduce existence of positive solutions of problem (1), (3). We will ensure the existence of solutions of problem (2), (3) when such function is negative. So, first of all, we will do an exhaustive study of the sign of the Green's function depending on the values of the parameters γ, δ_1 and δ_2. We mention that problem (1), (3) with either $\delta_1 = 0$ or $\delta_2 = 0$, has been studied in [4]. So our results generalize the ones obtained in that paper. The main difference with the properties of the Green's function in this general case, is that in [4] it is proved that, depending on the values of the real parameters, either the Green's function is positive on $(0, 1) \times (0, 1)$ or it changes its sign on $I \times I$. This property is, in some sense, natural and expected, because it is the situation for the homogeneous case $\delta_1 = \delta_2 = 0$.

However, as we will see on this paper, we will prove that, in addition to being positive for small enough values of $\delta_1 + \delta_2$, the Green's function related to our problem is negative on $(0, 1) \times (0, 1)$ for some large enough values of $\delta_1 + \delta_2$ with $\delta_1 > 0$ and $\delta_2 > 0$. This is a, a priori, non expected situation, but it may be interpreted by the fact that in such a case we are considering a strong perturbation of the homogeneous case.

We mention that the interest in considering this boundary conditions relies on the fact that they appear in different real phenomena such as, among others, problems of blood flow, chemical engineering, thermoelastic or population dynamics, see for instance, [2, 6, 7, 9, 10].

The paper is divided in three sections. After the introduction we will study in Sect. 2 the linear problem and prove the main properties of the related Green's function. Section 3 is devoted to ensure the existence of solutions of nonlinear problems (1) and (2) coupled to (3). Moreover, some examples of the applicability of the obtained results are shown at the end of the paper.

2 Linear Part: Green's Function

In this section, we will study the sign of the Green's function related to the linear problem

$$u''(t) + \gamma\, u(t) + \sigma(t) = 0, \quad t \in I, \tag{4}$$

coupled to (3). To this end, in what follows we do the study distinguishing three cases depending on the sign of the value γ.

First, it is very simple to verify that problem $u''(t) + \gamma\, u(t) = 0$, coupled to boundary conditions (3) has a trivial solution if and only if $\gamma = (2k\pi)^2$, $k = 1, 2, \ldots$ or

$$\delta_1 + \delta_2 = \Delta(\gamma) := \begin{cases} \dfrac{\sqrt{-\gamma}}{\tanh(\frac{\sqrt{-\gamma}}{2})}, & \gamma < 0, \\ 2, & \gamma = 0, \\ \dfrac{\sqrt{\gamma}}{\tan(\frac{\sqrt{\gamma}}{2})}, & \gamma > 0. \end{cases} \tag{5}$$

We point out that in the homogeneous case ($\delta_1 = \delta_2 = 0$) the second condition ($\delta_1 + \delta_2 = \Delta(\gamma)$) is rewritten as $\Delta(\gamma) = 0$, which is equivalent to $\gamma = (k\pi)^2$ with k odd. However, the eigenvalues of the homogeneous case are $\gamma = (k\pi)^2$, $k = 1, 2, \ldots$. Therefore, we have two cases regarding the eigenvalues of the homogeneous problem: those with k even remain eigenvalues of the perturbed problem for any value of δ_1 and δ_2, meanwhile those with k odd are eigenvalues only when $\delta_1 + \delta_2 = 0$.

As we will see, we must take this in account when obtaining the expression of the Green's function for $\gamma > 0$.

In the sequel we derive the following property of symmetry with respect to the parameters δ_1 and δ_2.

Theorem 2.1 *Assume that the linear problem* (4) *coupled to the homogeneous boundary conditions* $u(0) = u(1) = 0$, *has a unique solution for any* $\sigma \in C(I)$, *that is,* $\gamma \neq (k\pi)^2$, $k = 1, 2, \ldots$ *and that* $\delta_1 + \delta_2 \neq \Delta(\gamma)$. *Let* $G_{\gamma,\delta_1,\delta_2}$ *be the Green's function related to problem* (4), (3). *Then the following symmetry property holds*

$$G_{\gamma,\delta_1,\delta_2}(t,s) = G_{\gamma,\delta_2,\delta_1}(1-t, 1-s), \quad \forall\, (t,s) \in I \times I. \tag{6}$$

Proof From [5, Theorem 2.6], taking $C(u) = \displaystyle\int_0^1 u(s)\, ds$, it follows that the Green's function of problem (4), (3) is given by the expression

$$G_{\gamma,\delta_1,\delta_2}(t,s) = G_{\gamma,0,0}(t,s) + \frac{\delta_1\, w_1(t) + \delta_2\, w_2(t)}{1 - \delta_1\, C(w_1) - \delta_2\, C(w_2)} \int_0^1 G_{\gamma,0,0}(r,s)\, dr, \tag{7}$$

where $G_{\gamma,0,0}$ is the Green's function of the homogeneous Dirichlet problem

$$u''(t) + \gamma\, u(t) + \sigma(t) = 0, \quad t \in I, \quad u(0) = u(1) = 0,$$

w_1 is the unique solution of

$$u''(t) + \gamma\, u(t) = 0, \quad t \in I, \quad u(0) = 1,\ u(1) = 0,$$

and w_2 is the unique solution of

$$u''(t) + \gamma\, u(t) = 0, \quad t \in I, \quad u(0) = 0,\ u(1) = 1.$$

It is immediate to verify that $w_2(t) = w_1(1-t)$ for all $t \in I$, therefore $C(w_1) = C(w_2)$.

First, note that if u is a solution of the homogeneous problem

$$u''(t) + \gamma\, u(t) + \sigma(t) = 0,\ t \in I, \quad u(0) = u(1) = 0,$$

then the function $v(t) = u(1-t)$ satisfies

$$v''(t) + \gamma\, v(t) + \sigma(1-t) = 0,\ t \in I, \quad v(0) = v(1) = 0.$$

Thus,

$$u(1-t) = \int_0^1 G_{\gamma,0,0}(t,s)\,\sigma(1-s)\,ds = \int_0^1 G_{\gamma,0,0}(t,1-s)\,\sigma(s)\,ds.$$

As a consequence

$$u(t) = \int_0^1 G_{\gamma,0,0}(1-t,1-s)\,\sigma(s)\,ds$$

or, which is the same,

$$G_{\gamma,0,0}(t,s) = G_{\gamma,0,0}(1-t,1-s),\ \forall (t,s) \in I \times I.$$

Then, using formula (7), we infer that

$$\begin{aligned}
&G_{\gamma,\delta_2,\delta_1}(1-t,1-s) = G_{\gamma,0,0}(1-t,1-s) \\
&+ \frac{\delta_2\, w_1(1-t) + \delta_1\, w_2(1-t)}{1 - \delta_2\, C(w_1) - \delta_1\, C(w_2)} \int_0^1 G_{\gamma,0,0}(r,1-s)\,dr \\
&= G_{\gamma,0,0}(t,s) + \frac{\delta_1\, w_1(t) + \delta_2\, w_2(t)}{1 - \delta_1\, C(w_1) - \delta_2\, C(w_2)} \int_0^1 G_{\gamma,0,0}(1-r,1-s)\,dr \\
&= G_{\gamma,0,0}(t,s) + \frac{\delta_1\, w_1(t) + \delta_2\, w_2(t)}{1 - \delta_1\, C(w_1) - \delta_2\, C(w_2)} \int_0^1 G_{\gamma,0,0}(r,s)\,dr \\
&= G_{\gamma,\delta_1,\delta_2}(t,s), \quad \text{for all } (t,s) \in I \times I.
\end{aligned}$$

□

Remark 2.1 In the case of $\gamma = (k\pi)^2$, for some $k \in \mathbb{N}$ odd, and $\delta_1 + \delta_2 \neq 0$, we have that

$$G_{(k\pi)^2,\delta_1,\delta_2}(t,s) = \lim_{\gamma \to (k\pi)^2} G_{\gamma,\delta_1,\delta_2}(t,s) \quad \forall\, (t,s) \in I \times I.$$

So, as a direct consequence, we have that equality (6) is also true in this situation.

2.1 Case $\gamma = 0$

In this subsection, we obtain the expression of the Green's function related to the linear problem

$$u''(t) + \sigma(t) = 0, \quad t \in I, \tag{8}$$

subject to the integral boundary conditions (3). Moreover, we deduce some suitable properties of such function as, among others, the range of values of δ_1 and δ_2 for which it has constant sign on $I \times I$.

Theorem 2.2 *Let $\delta_1 + \delta_2 \neq 2\,(= \Delta(0))$ and $\sigma \in C(I)$, then problem (8), (3) has a unique solution $u \in C^2(I)$, which is given by the following expression*

$$u(t) = \int_0^1 G_{0,\delta_1,\delta_2}(t,s)\,\sigma(s)ds,$$

where

$$G_{0,\delta_1,\delta_2}(t,s) = \begin{cases} s\,\frac{(1-t)(2-\delta_2-\delta_1 s)+\delta_2(1-s)t}{2-\delta_1-\delta_2}, & 0 \le s \le t \le 1, \\ (1-s)\,\frac{t(2-\delta_1-\delta_2+\delta_2 s)+\delta_1 s(1-t)}{2-\delta_1-\delta_2}, & 0 \le t < s \le 1. \end{cases} \tag{9}$$

Proof According to Eq. (7), the Green's function of problem (8), (3) is given by

$$G_{0,\delta_1,\delta_2}(t,s) = G_{0,0,0}(t,s) + \frac{\delta_1\, w_1(t) + \delta_2\, w_2(t)}{1 - \delta_1\, C(w_1) - \delta_2\, C(w_2)} \int_0^1 G_{0,0,0}(t,s)\,dt, \tag{10}$$

where

$$G_{0,0,0}(t,s) = \begin{cases} s\,(1-t), & 0 \le s \le t \le 1, \\ t\,(1-s), & 0 \le t < s \le 1, \end{cases}$$

is the Green's function related to the homogeneous Dirichlet problem

$$u''(t) + \sigma(t) = 0, \quad t \in I, \quad u(0) = u(1) = 0,$$

$w_1(t) = 1 - t$ and $w_2(t) = t, t \in I$.

The result is obtained by taking into account that $C(w_1) = C(w_2) = \displaystyle\int_0^1 t\,dt = \frac{1}{2}$, $C(G_{0,0,0}(\cdot, s)) = \frac{1}{2}s(1-s)$ and expression (10). □

Next, we will state two lemmas related to the properties of the Green's function that will be useful to prove the existence of a positive solution of the nonlinear problems (1), (3) and (2), (3) with $\gamma = 0$.

Lemma 2.1 *Let G_{0,δ_1,δ_2} be the Green's function related to problem (8), (3), given by expression (9). Then, for all $\delta_1 + \delta_2 \neq 2\,(= \Delta(0))$ the following properties are fulfilled:*

1. $G_{0,\delta_1,\delta_2}(t, 0) = G_{0,\delta_1,\delta_2}(t, 1) = 0$, *for all* $t \in I$.
2. $G_{0,\delta_1,\delta_2}(t, s)$ *is continuous on* $I \times I$.
3. $G_{0,\delta_1,\delta_2}(0, s) = 0$ *for all* $s \in (0, 1)$ *if and only if* $\delta_1 = 0$.
4. $(2 - \delta_1 - \delta_2)\, G_{0,\delta_1,\delta_2}(0, s) > 0$ *for all* $s \in (0, 1)$ *if and only if* $\delta_1 > 0$.
5. $G_{0,\delta_1,\delta_2}(1, s) = 0$ *for all* $s \in (0, 1)$ *if and only if* $\delta_2 = 0$.
6. $(2 - \delta_1 - \delta_2)\, G_{0,\delta_1,\delta_2}(1, s) > 0$ *for all* $s \in (0, 1)$ *if and only if* $\delta_2 > 0$.
7. $(2 - \delta_1 - \delta_2)\, G_{0,\delta_1,\delta_2}(s, s) > 0$ *for all* $s \in (0, 1)$ *if and only if* $\delta_1 \leq 2$ *and* $\delta_2 \leq 2$.
8. $G_{0,\delta_1,\delta_2}(t, s) > 0$ *for all* $t, s \in (0, 1)$ *if and only if* $0 \leq \delta_1 + \delta_2 < 2$, $0 \leq \delta_1 \leq 2$ *and* $0 \leq \delta_2 \leq 2$.
9. $G_{0,\delta_1,\delta_2}(t, s) < 0$ *for all* $(t, s) \in (0, 1) \times (0, 1)$ *if and only if* $\delta_1 + \delta_2 > 2$, $0 < \delta_1 \leq 2$ *and* $0 < \delta_2 \leq 2$.
10. G_{0,δ_1,δ_2} *changes its sign on* $(0, 1) \times (0, 1)$ *if and only if one of the following situations holds:*

    (a) $\delta_1 < 0$.
    (b) $\delta_2 < 0$.
    (c) $\delta_1 > 2$.
    (d) $\delta_2 > 2$.

11. *If* $0 \leq \delta_1 + \delta_2 < 2$, $0 \leq \delta_1 \leq 2$ *and* $0 \leq \delta_2 \leq 2$, *then* $G_{0,\delta_1,\delta_2}(t, s) \leq \frac{1}{2(2-\delta_1-\delta_2)}$, $\forall t, s \in I$.
12. *If* $\delta_1 + \delta_2 > 2$, $0 < \delta_1 \leq 2$ *and* $0 < \delta_2 \leq 2$, *then* $G_{0,\delta_1,\delta_2}(t, s) \geq \frac{1}{2(2-\delta_1-\delta_2)}$, $\forall t, s \in I$.

Proof Properties 1. and 2. are immediate from the expression of G_{0,δ_1,δ_2}. Let's now prove the remaining properties:

3. It follows from the equality $G_{0,\delta_1,\delta_2}(0, s) = \delta_1 \frac{s\,(1-s)}{2-\delta_1-\delta_2}$.

4. This statement follows from the following fact

$$(2-\delta_1-\delta_2)\, G_{0,\delta_1,\delta_2}(0,s) = \delta_1\, s\,(1-s), \quad \text{for all } s \in (0,1). \tag{11}$$

5. It is immediate from the fact that $G_{0,\delta_1,\delta_2}(1,s) = \delta_2 \frac{s\,(1-s)}{2-\delta_1-\delta_2}$.
6. This statement follows from previous equality.
7. The following equality is fulfilled for all $s \in (0,1)$:

$$(2-\delta_1-\delta_2)\, G_{0,\delta_1,\delta_2}(s,s) = s(1-s)\,(2-\delta_2+(\delta_2-\delta_1)s). \tag{12}$$

   Therefore, $(2-\delta_1-\delta_2)\, G_{0,\delta_1,\delta_2}(s,s) > 0$ if and only if $h(s) = 2-\delta_2+(\delta_2-\delta_1)s > 0$ for all $s \in (0,1)$, which is true if and only if $\delta_1 \leq 2$ and $\delta_2 \leq 2$.
8. Since $G_{0,\delta_1,\delta_2}(t,s)$ is linear on t, for all $s \in I$ fixed, $G_{0,\delta_1,\delta_2}(\cdot,s)$ attains its maximum and minimum either at $t=0$, $t=s$ or $t=1$.
   From Property 7., we have that $G_{0,\delta_1,\delta_2}(s,s) > 0$ for all $s \in (0,1)$, if and only if $\delta_1 \leq 2$, $\delta_2 \leq 2$ and $\delta_1+\delta_2 < 2$.
   From Property 4., we have that $G_{0,\delta_1,\delta_2}(0,s) > 0$ for all $s \in (0,1)$, if and only if $\delta_1 > 0$ and $\delta_1+\delta_2 < 2$.
   From Property 6., we have that $G_{0,\delta_1,\delta_2}(1,s) > 0$ for all $s \in (0,1)$, if and only if $\delta_2 > 0$ and $\delta_1+\delta_2 < 2$.
   As a consequence of the three previous assertions this property holds.
9. The proof is analogous to the proof of Property 8.
10. The proof is a direct consequence of Properties 8. and 9.
11. From Property 8., we know that $G_{0,\delta_1,\delta_2}(t,s) > 0$ for all $(t,s) \in (0,1) \times (0,1)$. As in Property 8., we know that the maximum values will be attained at $G_{0,\delta_1,\delta_2}(s,s)$, $G_{0,\delta_1,\delta_2}(0,s)$ and/or $G_{0,\delta_1,\delta_2}(1,s)$. Now, since

$$\begin{aligned}(2-\delta_1-\delta_2)G_{0,\delta_1,\delta_2}(s,s) &= s(1-s)\,(2-\delta_2-\delta_1 s+\delta_2 s)\\ &= s(1-s)(2-\delta_1-\delta_2)+\delta_1\, s(1-s)^2+\delta_2\, s^2(1-s)\\ &\leq s(1-s)(2-\delta_1-\delta_2)+\delta_1\, s(1-s)+\delta_2\, s(1-s)\\ &= 2s(1-s) \leq \frac{1}{2},\end{aligned}$$

$$(2-\delta_1-\delta_2)\, G_{0,\delta_1,\delta_2}(0,s) = \delta_1 s(1-s) \leq \frac{\delta_1}{4} \leq \frac{1}{2},$$

   and

$$(2-\delta_1-\delta_2)\, G_{0,\delta_1,\delta_2}(1,s) = \delta_2 s(1-s) \leq \frac{\delta_2}{4} \leq \frac{1}{2},$$

   the proof is concluded.

12. Making the same argument for the negative case ($\delta_1 + \delta_2 > 2$, $0 < \delta_1 \le 2$ and $0 < \delta_2 \le 2$) we obtain that $G_{0,\delta_1,\delta_2}(t,s) \ge \frac{1}{2(2-\delta_1-\delta_2)}$.

□

Remark 2.2 Properties 1, 2, 3, 4, 5, 6, 7, 8, 10 and 11 in previous lemma generalize the ones given in [4, Lemma 4] for the case $\delta_1 = 0$.

We point out that properties 9 and 12 have no sense for either $\delta_1 = 0$ or $\delta_2 = 0$ and cover the new situation in which the considered Green's function is negative on $(0, 1) \times (0, 1)$.

Now two inequalities for the positiveness of the Green's function are derived.

Lemma 2.2 *Assume that* $0 < \delta_1 + \delta_2 < 2 (= \Delta(0))$, $0 < \delta_1 < 2$ *and* $0 < \delta_2 < 2$. *Let* G_{0,δ_1,δ_2} *be the Green's function related to problem (8), (3), given by expression (9). Then, there are two real constants* $c_1 \le 1 \le c_2$ *such that:*

$$c_1\, G_{0,\delta_1,\delta_2}(1,s) \le G_{0,\delta_1,\delta_2}(t,s) \le c_2\, G_{0,\delta_1,\delta_2}(1,s), \quad \textit{for all } t, s \in I. \tag{13}$$

Proof Since there exist

$$\lim_{s \to 0^+} \frac{G_{0,\delta_1,\delta_2}(t,s)}{G_{0,\delta_1,\delta_2}(1,s)} = t + (1-t)\frac{2-\delta_1-\delta_2}{\delta_2} > 0, \quad \text{for all } t \in I$$

and

$$\lim_{s \to 1^-} \frac{G_{0,\delta_1,\delta_2}(t,s)}{G_{0,\delta_1,\delta_2}(1,s)} = \frac{t(2-\delta_1) + \delta_1(1-t)}{\delta_2} > 0, \quad \text{for all } t \in I,$$

we can continuously extend the strictly positive function $\frac{G_{0,\delta_1,\delta_2}(t,s)}{G_{0,\delta_1,\delta_2}(1,s)}$ defined on $I \times (0, 1)$ to the compact set $I \times I$. Thus, we have that the continuous positive expansion $\tilde{H}_{0,\delta_1,\delta_2}(t,s)$ attains a positive maximum c_2 and a positive minimum c_1 on $I \times I$, that is

$$c_1 \le \tilde{H}_{0,\delta_1,\delta_2}(t,s) \le c_2, \quad \text{for all } t,\ s \in I.$$

Now, since $G_{0,\delta_1,\delta_2}(1,s) > 0$ for all $s \in (0, 1)$, we deduce inequalities (13). □

Remark 2.3 In the above proof we can take $c_1 = \min\left\{1, \frac{\delta_1}{\delta_2}\right\}$ and $c_2 = \frac{2}{\delta_2}$. Indeed, on the one hand we have that

$$\lim_{t \to s^+} \frac{G_{0,\delta_1,\delta_2}(t,s)}{G_{0,\delta_1,\delta_2}(1,s)} = \frac{2 + (s-1)\delta_2 - \delta_1 s}{\delta_2} \le \frac{2}{\delta_2},$$

and, on the other hand,

$$\lim_{t\to 0^+} \frac{G_{0,\delta_1,\delta_2}(t,s)}{G_{0,\delta_1,\delta_2}(1,s)} = \frac{\delta_1}{\delta_2}, \quad \lim_{t\to 1^-} \frac{G_{0,\delta_1,\delta_2}(t,s)}{G_{0,\delta_1,\delta_2}(1,s)} = 1.$$

Moreover, since $\frac{G_{0,\delta_1,\delta_2}(t,s)}{G_{0,\delta_1,\delta_2}(1,s)}$ is a non-negative, continuous and linear function with respect to t, we deduce

$$\frac{G_{0,\delta_1,\delta_2}(t,s)}{G_{0,\delta_1,\delta_2}(1,s)} \le \max\left\{1, \frac{\delta_1}{\delta_2}, \frac{2}{\delta_2}\right\} = \frac{2}{\delta_2},$$

and

$$\frac{G_{0,\delta_1,\delta_2}(t,s)}{G_{0,\delta_1,\delta_2}(1,s)} \ge \min\left\{1, \frac{\delta_1}{\delta_2}, \frac{2}{\delta_2}\right\} = \min\left\{1, \frac{\delta_1}{\delta_2}\right\}.$$

As a corollary of the previous result we arrive at the following one:

Corollary 2.1 *Assume that* $0 < \delta_1 + \delta_2 < 2 (= \Delta(0))$, $0 < \delta_1 < 2$ *and* $0 < \delta_2 < 2$. *Let* G_{0,δ_1,δ_2} *be the Green's function related to problem (8), (3), given by expression (9). Then, it holds that:*

$$\min_{t\in I} G_{0,\delta_1,\delta_2}(t,s) \ge \frac{c_1}{c_2} \max_{t\in I} G_{0,\delta_1,\delta_2}(t,s), \qquad \textit{for all } s \in I.$$

Proof From Lemma 2.2 we have that

$$c_1\, G_{0,\delta_1,\delta_2}(1,s) \le G_{0,\delta_1,\delta_2}(t,s), \qquad \text{for all } t, s \in I,$$

and

$$\frac{1}{c_2}\, G_{0,\delta_1,\delta_2}(t,s) \le G_{0,\delta_1,\delta_2}(1,s), \qquad \text{for all } t, s \in I.$$

Then,

$$\min_{t\in I} G_{0,\delta_1,\delta_2}(t,s) \ge c_1\, G_{0,\delta_1,\delta_2}(1,s), \qquad \text{for all } s \in I,$$

and

$$\frac{1}{c_2} \max_{t\in I} G_{0,\delta_1,\delta_2}(t,s) \le G_{0,\delta_1,\delta_2}(1,s), \qquad \text{for all } s \in I.$$

Therefore,

$$\min_{t\in I} G_{0,\delta_1,\delta_2}(t,s) \geq c_1\, G_{0,\delta_1,\delta_2}(1,s) \geq \frac{c_1}{c_2} \max_{t\in I} G_{0,\delta_1,\delta_2}(t,s), \quad \text{for all}\, s\in I,$$

and the result holds. □

2.2 Case $\gamma > 0$

The aim of this subsection is to study the constant sign of the Green's function related to the following linear problem

$$u''(t) + m^2\, u(t) + \sigma(t) = 0, \quad t\in I, \tag{14}$$

subject to the integral boundary conditions (3).

On the following result the expression of the corresponding Green's function is obtained.

Theorem 2.3 *Let* $\delta_1+\delta_2 \neq \frac{m\cos(\frac{m}{2})}{\sin(\frac{m}{2})}$ $(=\Delta(m^2))$, $m>0$, $m\neq 2k\pi$, $k=1,2,\ldots$, *and* $\sigma\in C(I)$. *Then problem (14), (3) has a unique solution* $u\in C^2(I)$, *which is given by the expression*

$$u(t) = \int_0^1 G_{m,\delta_1,\delta_2}(t,s)\,\sigma(s)\,ds,$$

where

$$G_{m,\delta_1,\delta_2}(t,s) = \begin{cases} G^1_{m,\delta_1,\delta_2}(t,s), & 0\leq s\leq t\leq 1,\\ G^2_{m,\delta_1,\delta_2}(t,s), & 0\leq t< s\leq 1. \end{cases} \tag{15}$$

Here, if $m\neq k\pi$, $k\in\mathbb{N}$ *odd,*

$$\begin{aligned} G^1_{m,\delta_1,\delta_2}(t,s) =& \frac{\csc(m)\,\sin(ms)\,\sin(m-mt)}{m}\\ &+\frac{(-1+w_1(s)+w_2(s))(\delta_1\, w_1(t)+\delta_2\, w_2(t))}{m(m-(\delta_1+\delta_2)\tan(\frac{m}{2}))}, \end{aligned}$$

and

$$\begin{aligned} G^2_{m,\delta_1,\delta_2}(t,s) =& \frac{\csc(m)\,\sin(mt)\,\sin(m-ms)}{m}\\ &+\frac{(-1+w_1(s)+w_2(s))(\delta_1\, w_1(t)+\delta_2\, w_2(t))}{m(m-(\delta_1+\delta_2)\tan(\frac{m}{2}))}, \end{aligned}$$

being $w_1(t) = \csc(m)\,\sin(m(1-t))$, $t \in I$ *and* $w_2(t) = \csc(m)\,\sin(mt)$, $t \in I$.

In the case of $m = k\pi$, *for some* $k \in \mathbb{N}$ *odd, we have that*

$$G^1_{k\pi,\delta_1,\delta_2}(t,s) = \lim_{m \to k\pi} G^1_{m,\delta_1,\delta_2}(t,s)$$

and

$$G^2_{k\pi,\delta_1,\delta_2}(t,s) = \lim_{m \to k\pi} G^2_{m,\delta_1,\delta_2}(t,s).$$

Proof Consider the case $m \neq k\pi$, $k \in \mathbb{N}$, and let w_1 be the unique solution of

$$w_1''(t) + m^2\, w_1(t) = 0, \quad t \in I, \qquad w_1(0) = 1, \quad w_1(1) = 0,$$

and w_2 be defined as the unique solution of

$$w_2''(t) + m^2\, w_2(t) = 0, \quad t \in I, \qquad w_2(0) = 0, \quad w_2(1) = 1.$$

It is immediate to verify that

$$w_2(t) = \csc(m)\,\sin(mt),$$

and

$$w_1(t) = w_2(1-t) = \csc(m)\,\sin(m - mt).$$

It is very well-known, see [3], that the Green's function related to the homogeneous case ($\delta_1 = \delta_2 = 0$) is given by the expression

$$G_{m,0,0}(t,s) = \begin{cases} \dfrac{\csc(m)\sin(ms)\sin(m-mt)}{m}, & 0 \le s \le t \le 1, \\ \dfrac{\csc(m)\sin(m-ms)\sin(mt)}{m}, & 0 \le t < s \le 1. \end{cases}$$

On the other hand, it is not difficult to verify that $C(w_1) = C(w_2) = \frac{\tan(\frac{m}{2})}{m}$ and

$$C(G_{m,0,0}(\cdot,s)) = \frac{-1 + w_1(s) + w_2(s)}{m^2}.$$

Substituting those values in (7), we obtain the result.

Finally, for $m = k\pi$ with k odd, it is easy to see using the axiomatic definition of Green's function (see [3, Definition 1.4.1]) that the expression obtained by taking limits on G^1_m and G^2_m when m tends to $k\pi$ is a Green's function of problem (14), (3). □

Remark 2.4 We point out that in previous theorem the expression of the Green's function when $\gamma = (k\pi)^2$, with k odd and $\delta_1 + \delta_2 \neq 0 (= \Delta(\gamma))$, can not been obtained from expression (10) (as for the case $\gamma = (k\pi)^2$, k even) because in this situation, since $(k\pi)^2$ is an eigenvalue of problem $u''(t) = 0$, $t \in I$, $u(0) = u(1) = 0$, we have that $G_{\gamma,0,0}$ does not exist.

In the following result we describe the points in $I \times I$ where a constant sign Green's function of problem (14), (3) can vanish.

Lemma 2.3 *Let $\gamma < \pi^2$ and assume that problem (14), (3) has a unique solution for any $\sigma \in C(I)$. If G_{m,δ_1,δ_2} has constant sign on $I \times I$ and vanishes at some point (t_0, s_0), then either $s_0 = 0$, $s_0 = 1$, $t_0 = 0$, $t_0 = 1$ or $t_0 = s_0$.*

Proof Let us suppose that $(t_0, s_0) \in (0,1) \times (0,1)$, with $t_0 > s_0$. In such a case, from the definition of Green's function, we have that $u(t) := G_{\gamma,\delta_1,\delta_2}(t, s_0)$, $t \in I$, is the unique solution of the problem

$$u''(t) + \gamma\, u(t) = 0, \quad t \in (s_0, 1], \qquad u(t_0) = u'(t_0) = 0,$$

and so $G_{\gamma,\delta_1,\delta_2}(t, s_0) = 0$ for all $t \in (s_0, 1]$.

Now, from the properties of the Green's function, we have that

$$u(1) = \delta_2 \int_0^1 u(s)\, ds = 0$$

which implies that either $u(t) = 0$ for all $t \in I$, which contradicts that $u'(s_0^-) = 1 < 0 = u'(s_0^+)$, or u changes its sign on I, which contradicts the constant sign of the Green's function.

In the case $(t_0, s_0) \in (0,1) \times (0,1)$, with $t_0 < s_0$, from Theorem 2.1 and Remark 2.1 we deduce that if G_{m,δ_1,δ_2} has constant sign on $I \times I$ and vanishes at (t_0, s_0), then G_{m,δ_2,δ_1} will also have constant sign on $I \times I$ and vanish at $(1 - t_0, 1 - s_0)$ with $1 - t_0 > 1 - s_0$. Therefore, we are in the previous case and the proof is concluded. □

In the sequel we deduce some properties of the Green's function G_{m,δ_1,δ_2}.

Lemma 2.4 *Let G_{m,δ_1,δ_2} be the Green's function related to problem (14), (3), given by the expression (15). Then for all $\delta_1 + \delta_2 \neq \frac{m}{\tan(\frac{m}{2})} (= \Delta(m^2))$, $m > 0$, $m \neq 2k\pi$, $k = 1, 2, \ldots$ (we understand that $\Delta(\pi^2)(\equiv \frac{\pi}{\tan(\pi/2)} = 0)$, the following properties hold:*

1. *$G_{m,\delta_1,\delta_2}(t, 1) = G_{m,\delta_1,\delta_2}(t, 0) = 0$, for all $t \in I$.*
2. *$G_{m,\delta_1,\delta_2}(t, s)$ is continuous at $(t, s) \in I \times I$.*
3. *$G_{m,\delta_1,\delta_2}(0, s) = 0$ for all $s \in [0, 1]$ if and only if $\delta_1 = 0$.*
4. *$\left(m - (\delta_1 + \delta_2) \tan\left(\frac{m}{2}\right)\right) G_{m,\delta_1,\delta_2}(0, s) > 0$ for all $s \in (0, 1)$ if and only if $m \in (0, \pi)$ and $\delta_1 > 0$.*
5. *$G_{m,\delta_1,\delta_2}(1, s) = 0$ for all $s \in (0, 1)$ if and only if $\delta_2 = 0$.*

6. $\left(m - (\delta_1 + \delta_2)\tan\left(\frac{m}{2}\right)\right) G_{m,\delta_1,\delta_2}(1, s) > 0$ *for all* $s \in (0, 1)$ *if and only if* $m \in (0, \pi)$ *and* $\delta_2 > 0$.
7. *If* $m \in (0, \pi)$, *then* $\left(m - (\delta_1 + \delta_2)\tan\left(\frac{m}{2}\right)\right) G_{m,\delta_1,\delta_2}(s, s) > 0$ *for all* $s \in (0, 1)$, $\delta_1 \le \frac{m}{\tan(\frac{m}{2})}$ *and* $\delta_2 \le \frac{m}{\tan(\frac{m}{2})}$.
8. $G_{m,\delta_1,\delta_2}(t, s) > 0$ *for all* $t, s \in (0, 1)$ *if and only if* $0 \le \delta_1 + \delta_2 < \frac{m}{\tan(\frac{m}{2})}$, $m \in (0, \pi]$, $\delta_1 \ge 0$ *and* $\delta_2 \ge 0$.
9. $G_{m,\delta_1,\delta_2}(t, s) < 0$ *for all* $t, s \in (0, 1)$ *if and only if* $\delta_1 + \delta_2 > \frac{m}{\tan(\frac{m}{2})}$, $0 < \delta_1 \le \frac{m}{\tan(\frac{m}{2})}$, $0 < \delta_2 \le \frac{m}{\tan(\frac{m}{2})}$ *and* $m \in (0, \pi)$.
10. G_{m,δ_1,δ_2} *changes its sign on* $(0, 1) \times (0, 1)$ *if and only if one the following properties is fulfilled for* $m > 0$, $m \ne 2k\pi$ *and* $k = 1, 2, \ldots$:

    a. $m > \pi$,
    b. $\delta_1 > \frac{m}{\tan(\frac{m}{2})}$,
    c. $\delta_2 > \frac{m}{\tan(\frac{m}{2})}$,
    d. $\delta_1 < 0$,
    e. $\delta_2 < 0$.

Proof Properties 1. and 2. are immediate. Let's now see the others:

3. Let $s \in (0, 1)$, then $G_{m,\delta_1,\delta_2}(0, s) = 0$ if and only if

$$\delta_1(-1 + w_1(s) + w_2(s)) = 0.$$

It is easy to see that the function

$$r_m(s) := -1 + w_1(s) + w_2(s),$$

is positive on $(0, 1)$ if and only if $m \in (0, \pi]$. Moreover, it is not identically zero on $(0, 1)$ for any $m \ne 2k\pi$, $k = 1, 2, \ldots$.
Therefore $G_{m,\delta_1,\delta_2}(0, s) = 0$ if and only if $\delta_1 = 0$.

4. Using expression (15), we have that

$$\left(m - (\delta_1 + \delta_2)\tan\left(\frac{m}{2}\right)\right) G_{m,\delta_1,\delta_2}(0, s) = \frac{\delta_1}{m} r_m(s) > 0,$$

for all $s \in (0, 1)$, if and only if $m \in (0, \pi)$ and $\delta_1 > 0$.

5. 6. These properties are immediately deduced from Properties 3. and 4. and the following equality deduced from (6) and Remark 2.1:

$$G_{m,\delta_1,\delta_2}(1, s) = G_{m,\delta_2,\delta_1}(0, 1 - s), \quad \forall s \in I.$$

7. Let us define the function

$$h(s, m, \delta_1, \delta_2) := m\,\sin(m)\,\left(-m + (\delta_1 + \delta_2)\tan\left(\frac{m}{2}\right)\right) G_{m,\delta_1,\delta_2}(s, s).$$

So,

$$\frac{\partial}{\partial \delta_1} h(s,m,\delta_1,\delta_2) = (1-\cos(m))\sin(m-ms) > 0,$$

and

$$\frac{\partial}{\partial \delta_2} h(s,m,\delta_1,\delta_2) = 2\sin(ms)\sin^2\left(\frac{1}{2}(m-ms)\right) > 0,$$

for all $s \in (0,1)$ and $m \in (0,\pi]$.
Moreover,

$$h\left(s,m,\frac{m}{\tan\left(\frac{m}{2}\right)},\frac{m}{\tan\left(\frac{m}{2}\right)}\right) = -4\sin^2\left(\frac{ms}{2}\right)\sin^2\left(\frac{m(1-s)}{2}\right) < 0,$$

for all $s \in (0,1)$ and $m \in (0,\pi]$.
Therefore, $h(s,m,\delta_1,\delta_2) < 0$ for all $s \in (0,1)$, $m \in (0,\pi]$, $\delta_1 \le \frac{m}{\tan(\frac{m}{2})}$ and $\delta_2 \le \frac{m}{\tan(\frac{m}{2})}$, and the property is fulfilled.

8. From Lemma 2.3, for any $s \in (0,1)$ fixed, if function $G_{m,\delta_1,\delta_2}(\cdot,s)$ has constant sign, then it attains its maximum at minimum values at $t=0$, $t=s$ or $t=1$.
   From Property 4., we have that $G_{m,\delta_1,\delta_2}(0,s) > 0$ for all $s \in (0,1)$, if and only if $\delta_1+\delta_2 < \frac{m}{\tan(\frac{m}{2})}$ and $\delta_1 > 0$.
   From Property 6., we have that $G_{0,\delta_1,\delta_2}(1,s) > 0$ for all $s \in (0,1)$, if and only if $\delta_1+\delta_2 < \frac{m}{\tan(\frac{m}{2})}$ and $\delta_2 > 0$.
   From Property 7. we have that if $m \in (0,\pi]$ and $\delta_1+\delta_2 < \frac{m}{\tan(\frac{m}{2})}$ then $G_{m,\delta_1,\delta_2}(s,s) > 0$ for all $s \in (0,1)$, provided that $\delta_1 \le \frac{m}{\tan(\frac{m}{2})}$ and $\delta_2 \le \frac{m}{\tan(\frac{m}{2})}$.
   Notice that if $\delta_1 > 0$ and $\delta_2 > 0$, the fact that $\delta_1+\delta_2 < \frac{m}{\tan(\frac{m}{2})}$ implies that $\delta_1 \le \frac{m}{\tan(\frac{m}{2})}$ and $\delta_2 \le \frac{m}{\tan(\frac{m}{2})}$. So, we deduce that $G_{m,\delta_1,\delta_2}(t,s) > 0$ for all $(t,s) \in (0,1)\times(0,1)$ for such values of the parameters.
   If $G_{m,\delta_1,\delta_2}(s,s)$ is positive for other range of values we have that either $G_{m,\delta_1,\delta_2}(0,s)$ or $G_{m,\delta_1,\delta_2}(1,s)$ will be negative, and Property 8. holds.
9. The proof in this case is analogous to the previous one.
10. Arguing as in proof of [3, Theorems 1.8.5 and 1.8.6], one can verify that for any fixed values of t, s, δ_1 and δ_2, we have that $G_{m,\delta_1,\delta_2}(t,s)$ is monotone increasing with respect to m on the intervals of m where such function has constant sign on $I \times I$.
    Thus, let $\delta_1 > 0$ and $\delta_2 > 0$ be such that $\delta_1+\delta_2 < \frac{m_1}{\tan\left(\frac{m_1}{2}\right)}$ for some $m_1 \in (0,\pi)$, as a consequence, since function $\Delta(m^2) = \frac{m}{\tan(\frac{m}{2})}$ is strictly positive

and strictly monotone decreasing on $m \in (0, \pi)$, we have that there is a unique value $m_0 \in (m_1, \pi)$ such that

$$\delta_1 + \delta_2 = \frac{m_0}{\tan\left(\frac{m_0}{2}\right)} = \Delta(m_0^2)$$

and

$$0 < \delta_1 + \delta_2 < \frac{m}{\tan\left(\frac{m}{2}\right)} \quad \text{for all } m \in (0, m_0).$$

Thus $G_{m,\delta_1,\delta_2}(t,s) > 0$ on $(0,1) \times (0,1)$ for all $m \in (0, m_0)$.

Therefore, if $\delta_1 > 0$ and $\delta_2 > 0$, we have that there is $m_2 \in (m_0, \pi)$ such that $\delta_1 \leq \frac{m}{\tan(\frac{m}{2})}$ and $\delta_2 \leq \frac{m}{\tan(\frac{m}{2})}$ for all $m \in (m_0, m_2]$. In consequence, $G_{m,\delta_1,\delta_2}(t,s) < 0$ on $(0,1) \times (0,1)$ if and only if $m \in (m_0, m_2]$ (see Fig. 1). For $m > m_2$ we have that G_{m,δ_1,δ_2} must change its sign on $I \times I$, on the contrary it contradicts the monotone increasing property of this function with respect to m.

In the case of $\delta_1 = 0$, we will be have $\delta_2 = \frac{m_0}{\tan\left(\frac{m_0}{2}\right)}$ and so, the Green's function G_{m,δ_1,δ_2} can never be negative on $(0,1) \times (0,1)$ and must change its sign for all $m > m_0$. The same holds for $\delta_2 = 0$ (see Fig. 2). Both cases have been considered in [4].

□

Remark 2.5 We mention that Properties 1, 2, 5, 6, 8 and 10 are a generalization of the ones given in [4, Lemma 8]. Despite this, it is important to mention that the proof of Property 10 here is completely different to the simpler one given in that reference.

The rest of the cases have no sense for that situation.

Now we deduce some inequalities on the Green's function analogous to the ones given in Lemma 2.2.

Lemma 2.5 *Let* $0 < m < \pi$, $0 < \delta_1 + \delta_2 < \frac{m}{\tan(\frac{m}{2})}$ $(= \Delta(m^2))$ *and* G_{m,δ_1,δ_2} *be the Green's function of problem (14), (3) given by expression (15). Then, there are real constants* $c_3 > 0$ *and* $c_4 > 0$ *such that:*

$$c_3\, G_{m,\delta_1,\delta_2}(1,s) \leq G_{m,\delta_1,\delta_2}(t,s) \leq c_4\, G_{m,\delta_1,\delta_2}(1,s), \qquad \textit{for all } t, s \in I. \tag{16}$$

Proof If $s = 0$ or $s = 1$ the result follows from Lemma 2.4. Let then $t \in I$ be arbitrarily set.

On the one hand, we have that

$$\lim_{s \to 0^+} \frac{G_{m,\delta_1,\delta_2}(t,s)}{G_{m,\delta_1,\delta_2}(1,s)} = \frac{w_1(t)\left(m - (\delta_1+\delta_2)\tan\left(\frac{m}{2}\right)\right)}{\delta_2 \csc(m)\,(1-\cos(m))} + \frac{\delta_1\, w_1(t) + \delta_2\, w_2(t)}{\delta_2} > 0,$$

and on the other hand, since

$$w_1'(1) + w_2'(1) < 0$$

we have that

$$\lim_{s \to 1^-} \frac{G_{m,\delta_1,\delta_2}(t,s)}{G_{m,\delta_1,\delta_2}(1,s)} = \frac{-w_1(t)\left(m - (\delta_1+\delta_2)\tan\left(\frac{m}{2}\right)\right)}{\delta_2(w_1'(1)+w_2'(1))} + \frac{\delta_1\, w_1(t) + \delta_2\, w_2(t)}{\delta_2} > 0.$$

Taking into account Properties 6. and 7. of Lemma 2.4, we have that the strictly positive function $\frac{G_{m,\delta_1,\delta_2}(\cdot,s)}{G_{m,\delta_1,\delta_2}(1,s)}$ defined on $I \times (0,1)$ extends continuously to the compact set $I \times I$. Therefore, there exists $c_3 > 0$ and $c_4 > 0$, the minimum and the maximum, respectively, of such extension on $I \times I$. Therefore the inequalities (16) are fulfilled and we conclude the proof. □

Corollary 2.2 *Let* $0 < m < \pi$, $0 < \delta_1 + \delta_2 < \frac{m}{\tan(\frac{m}{2})}$ $(= \Delta(m^2))$ *and* G_{m,δ_1,δ_2} *be the Green's function of problem (14), (3) given by expression (15). Then, it holds that:*

$$\min_{t \in I} G_{m,\delta_1,\delta_2}(t,s) \geq \frac{c_3}{c_4} \max_{t \in I} G_{m,\delta_1,\delta_2}(t,s), \qquad \textit{for all } t, s \in I.$$

Proof The proof is analogous to the one made in Corollary 2.1. □

2.3 Case $\gamma < 0$

Finally, we deal in this subsection with the study of the linear problem

$$u''(t) - m^2\, u(t) + \sigma(t) = 0, \quad t \in I, \tag{17}$$

subject to the integral boundary conditions (3).

For the construction of the Green's function we obtain the following result:

Theorem 2.4 *Let* $\delta_1 + \delta_2 \neq \frac{m}{\tanh(\frac{m}{2})}$ $(= \Delta(-m^2))$ *and* $\sigma \in C(I)$, *then problem (17), (3) has a unique solution* $u \in C^2(I)$, *which is given by the expression*

$$u(t) = \int_0^1 G_{m,\delta_1,\delta_2}(t,s)\,\sigma(s)ds,$$

where

$$G_{m,\delta_1,\delta}(t,s) = \begin{cases} G^1_{m,\delta_1,\delta_2}(t,s), & 0 \leq s \leq t \leq 1, \\ G^2_{m,\delta_1,\delta_2}(t,s), & 0 \leq t < s \leq 1, \end{cases} \tag{18}$$

with

$$G^1_{m,\delta_1,\delta_2}(t,s) = \frac{\sinh(ms)\,\sinh(m-mt)}{m\sinh(m)} - \frac{(-1+w_1(s)+w_2(s))(\delta_1\,w_1(t)+\delta_2\,w_2(t))}{m(m-(\delta_1+\delta_2)\tanh(m/2))},$$

and

$$G^2_{m,\delta_1,\delta_2}(t,s) = \frac{\sinh(mt)\,\sinh(m-ms)}{m\sinh(m)} - \frac{(-1+w_1(s)+w_2(s))(\delta_1\,w_1(t)+\delta_2\,w_2(t))}{m(m-(\delta_1+\delta_2)\tanh(\frac{m}{2}))},$$

being $w_1(t) = \frac{\sinh(m-mt)}{\sinh(m)}$, $t \in I$ *and* $w_2(t) = \frac{\sinh(mt)}{\sinh(m)}$, $t \in I$.

Proof It is proved analogously to Theorem 2.3 taking into account that in this case $w_1(t) = \frac{\sinh(m-mt)}{\sinh(m)}$, $w_2(t) = \frac{\sinh(mt)}{\sinh(m)}$, $C(w_1) = C(w_2) = \frac{\tanh(\frac{m}{2})}{m}$ and

$$G_{m,0,0}(t,s) = \begin{cases} \frac{\sinh(ms)\,\sinh(m-mt)}{m\sinh(m)}, & 0 \le s \le t \le 1, \\ \frac{\sinh(mt)\,\sinh(m-ms)}{m\sinh(m)}, & 0 \le t < s \le 1, \end{cases}$$

with

$$C(G_{m,0,0}(\cdot,s)) = -\frac{-1+w_1(s)+w_2(s)}{m^2}.$$

□

We will now state the following results that characterize the constant sign of the Green's function on $I \times I$. We omit the proofs because they are analogous to the previous case.

Lemma 2.6 *Let G_{m,δ_1,δ_2} be the Green's function associated with problem (17), (3), given by expression (18). Then for all $\delta_1+\delta_2 \neq \frac{m}{\tanh(\frac{m}{2})}$ $(= \Delta(-m^2))$, $m > 0$, the following properties hold:*

1. $G_{m,\delta_1,\delta_2}(t,0) = G_{m,\delta_1,\delta_2}(t,1) = 0$, *for all* $t \in I$.
2. $G_{m,\delta_1,\delta_2}(t,s)$ *is continuous on* $I \times I$.
3. $G_{m,\delta_1,\delta_2}(0,s) = 0$, *for all* $s \in (0,1)$ *if and only if* $\delta_1 = 0$.
4. $\left(m-(\delta_1+\delta_2)\tanh\left(\frac{m}{2}\right)\right) G_{m,\delta_1,\delta_2}(0,s) > 0$ *for all* $s \in (0,1)$ *and* $m > 0$ *if and only if* $\delta_1 > 0$.
5. $G_{m,\delta_1,\delta_2}(1,s) = 0$, *for all* $s \in (0,1)$ *if and only if* $\delta_2 = 0$.
6. $\left(m-(\delta_1+\delta_2)\tanh\left(\frac{m}{2}\right)\right) G_{m,\delta_1,\delta_2}(1,s) > 0$ *for all* $s \in (0,1)$ *and* $m > 0$ *if and only if* $\delta_2 > 0$.
7. $\left(m-(\delta_1+\delta_2)\tanh\left(\frac{m}{2}\right)\right) G_{m,\delta_1,\delta_2}(s,s) > 0$ *for all* $s \in (0,1)$, $m > 0$, $\delta_1 \le \frac{m}{\tanh(\frac{m}{2})}$ *and* $\delta_2 \le \frac{m}{\tanh(\frac{m}{2})}$.
8. $G_{m,\delta_1,\delta_2}(t,s) > 0$ *for all* $t, s \in (0,1)$ *if and only if* $0 \le \delta_1+\delta_2 < \frac{m}{\tanh(\frac{m}{2})}$, $0 \le \delta_1$, $0 \le \delta_2$ *and* $m > 0$.

9. $G_{m,\delta_1,\delta_2}(t,s) < 0$ *for all* $t, s \in (0,1)$ *if and only if* $\delta_1 + \delta_2 > \frac{m}{\tanh(\frac{m}{2})}$, $0 < \delta_1 \leq \frac{m}{\tanh(\frac{m}{2})}$, $0 < \delta_2 \leq \frac{m}{\tanh(\frac{m}{2})}$ *and* $m > 0$.
10. G_{m,δ_1,δ_2} *changes its sign on* $(0,1) \times (0,1)$ *if and only if one the following properties is fulfilled:*
    a. $\delta_1 > \frac{m}{\tanh(\frac{m}{2})}$,
    b. $\delta_2 > \frac{m}{\tanh(\frac{m}{2})}$,
    c. $\delta_1 < 0$,
    d. $\delta_2 < 0$.

Remark 2.6 We point out that in order to prove Property 10 in previous Lemma, we must argue in a similar manner to Property 10 in Lemma 2.4. In this case, we must take into account that function Δ is strictly positive and strictly decreasing in $(-\infty, 0)$ and (see Figs. 1 and 2) that

$$\lim_{\gamma \to -\infty} \Delta(\gamma) = +\infty.$$

Lemma 2.7 *Let* $m > 0$, $\delta_1 > 0$, $\delta_2 > 0$ *and* $0 < \delta_1 + \delta_2 < \frac{m}{\tanh(\frac{m}{2})}$ $(= \Delta(-m^2))$ *and* G_{m,δ_1,δ_2} *be the Green's function of problem (17), (3) given by expression (18). Then there are two positive real constants* c_5 *and* c_6 *such that:*

$$c_5\, G_{m,\delta_1,\delta_2}(1,s) \leq G_{m,\delta_1,\delta_2}(t,s) \leq c_6\, G_{m,\delta_1,\delta_2}(1,s), \qquad \textit{for all } t, s \in I. \tag{19}$$

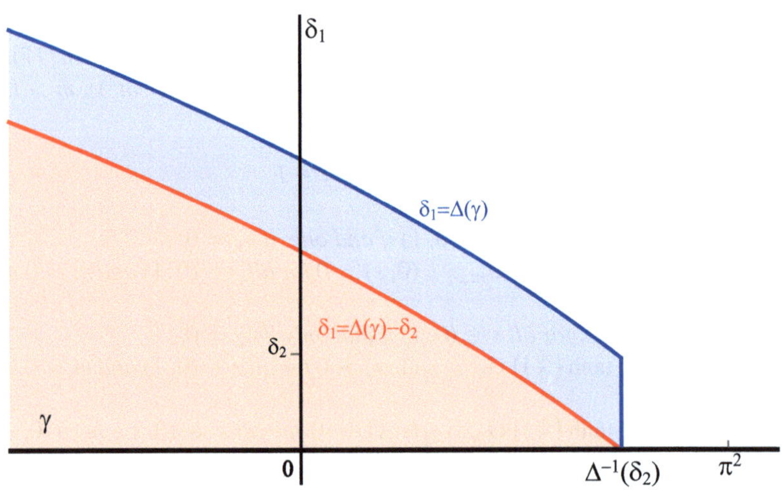

Fig. 1 Given $\delta_2 > 0$, the Green's function is positive on the lower region, negative on the upper one, and changes its sign otherwise

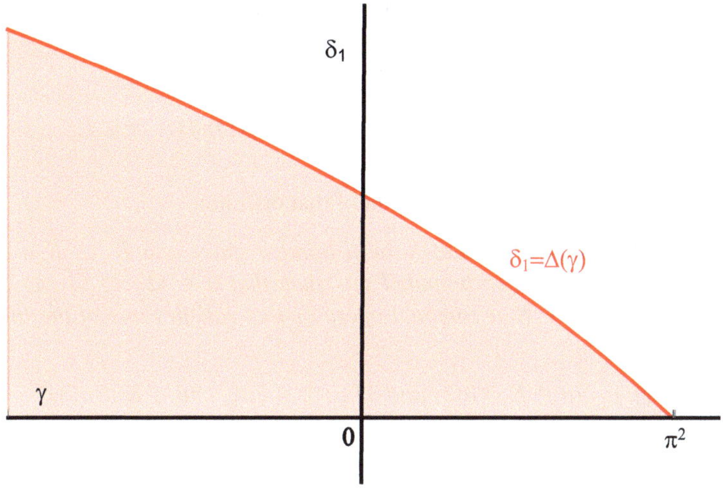

Fig. 2 If $\delta_2 = 0$, the Green's function is positive on the colored region and changes its sign otherwise

Corollary 2.3 *Let* $m > 0$, $\delta_1 > 0$, $\delta_2 > 0$ *and* $0 < \delta_1 + \delta_2 < \frac{m}{\tanh(\frac{m}{2})}$ $(= \Delta(-m^2))$ *and* G_{m,δ_1,δ_2} *be the Green's function of problem (17), (3) given by expression (18). Then, it holds that:*

$$\min_{t\in[0,1]} G_{m,\delta_1,\delta_2}(t,s) \geq \frac{c_5}{c_6} \max_{t\in[0,1]} G_{m,\delta_1,\delta_2}(t,s), \qquad \textit{for all } t, s \in I.$$

3 Nonlinear Problem

This section studies the existence of positive solutions of the nonlinear problems (1), (3) and (2), (3). We will ensure the existence of positive solutions of problem (1), (3) when the related Green's function is positive and of problem (2), (3) if it is negative. The existence results will be deduced from fixed point theory of integral operators defined in suitable cones. More concisely, we will use the classical Krasnoselskii's fixed point Theorem. The arguments are similar to the ones developed in [1].

We assume that the nonlinear part of equation satisfies the following regularity and constant sign condition:

$(f) \quad f : I \times [0,\infty) \to [0,\infty)$ is a continuous function.

Let $X \equiv (C(I), \|\cdot\|_\infty)$ be the real Banach space equipped with the supremum norm

$$\|u\|_\infty = \sup_{t\in I} |u(t)|, \qquad \text{for all } u \in X.$$

The following Krasnoselskii's fixed point Theorem [8] will be applied to the operator $T_{\gamma,\delta_1,\delta_2} : X \to X$, defined as

$$T_{\gamma,\delta_1,\delta_2} u(t) := \int_0^1 G_{\gamma,\delta_1,\delta_2}(t,s)\, f(s,u(s))\, ds, \quad t \in I, \tag{20}$$

to guarantee the existence of a fixed point of that operator.

Theorem 3.1 (Krasnoselskii) *Let X be a Banach space and $K \subset X$ a cone in X. Let $\Omega_1, \Omega_2 \subset X$ be open bounded sets such that $0 \in \Omega_1 \subset \overline{\Omega_1} \subset \Omega_2$ and $T : K \cap (\overline{\Omega_2} \setminus \Omega_1) \to K$ a compact operator that satisfies one of the following properties:*

1. $\|T(u)\| \geq \|u\|, \ \forall u \in K \cap \partial\Omega_1$ *and* $\|T(u)\| \leq \|u\|, \ \forall u \in K \cap \partial\Omega_2$.
2. $\|T(u)\| \leq \|u\|, \ \forall u \in K \cap \partial\Omega_1$ *and* $\|T(u)\| \geq \|u\|, \ \forall u \in K \cap \partial\Omega_2$.

Then T has a fixed point at $K \cap (\overline{\Omega_2} \setminus \Omega_1)$.

Note that the fixed points of operator $T_{\gamma,\delta_1,\delta_2}$ coincide with the solutions of problem (1), (3).

First, we consider the situation in which the related Green's function $G_{\gamma,\delta_1,\delta_2}$ is positive, that is:

$$\gamma < \pi^2 \text{ and } \delta_1,\ \delta_2 > 0 \text{ are such that } 0 < \delta_1 + \delta_2 < \Delta(\gamma), \tag{21}$$

where $\Delta : (-\infty, \pi^2) \to \mathbb{R}$ is the function defined by expression (5).

Next we define the cone K where we will apply Krasnoselskii's Theorem:

$$K = \{u \in X \,/\, \min\{u(t) : t \in I\} \geq \bar{c}\, \|u\|_\infty\},$$

where

$$\bar{c} = \begin{cases} \frac{c_1}{c_2}, & \gamma = 0, \\ \frac{c_3}{c_4}, & \gamma > 0, \\ \frac{c_5}{c_6}, & \gamma < 0, \end{cases}$$

with c_1, c_2, c_3, c_4, c_5 and c_6 given in inequalities (13), (16) and (19).

In this case, the following inequality is satisfied:

$$\min_{t\in I} G_{\gamma,\delta_1,\delta_2}(t,s) \geq \bar{c} \max_{t \in I} G_{\gamma,\delta_1,\delta_2}(t,s), \quad \text{for all } s,t \in I. \tag{22}$$

Using inequalities (13), (16) and (19) again, we can guarantee that there are two real positive constants

$$h = \begin{cases} c_1, & \gamma = 0, \\ c_3, & \gamma > 0, \\ c_5, & \gamma < 0, \end{cases} \quad \text{and} \quad k = \begin{cases} c_2, & \gamma = 0, \\ c_4, & \gamma > 0, \\ c_6, & \gamma < 0, \end{cases}$$

such that

$$h\, G_{\gamma,\delta_1,\delta_2}(1,s) \leq G_{\gamma,\delta_1,\delta_2}(t,s) \leq k\, G_{\gamma,\delta_1,\delta_2}(1,s), \qquad \text{for all } t, s \in I. \tag{23}$$

Moreover, we will denote by

$$R = \int_0^1 G_{\gamma,\delta_1,\delta_2}(1,s)\, ds = \begin{cases} \frac{\delta_2}{6(2-\delta_1-\delta_2)}, & \gamma = 0, \\ \frac{\delta_2\left(2\tan\left(\frac{\sqrt{\gamma}}{2}\right)-\sqrt{\gamma}\right)}{\gamma\left(\sqrt{\gamma}-(\delta_1+\delta_2)\tan\left(\frac{\sqrt{\gamma}}{2}\right)\right)}, & \gamma > 0, \\ \frac{\delta_2\left(2\tanh\left(\frac{\sqrt{-\gamma}}{2}\right)-\sqrt{-\gamma}\right)}{\gamma\left(\sqrt{-\gamma}-(\delta_1+\delta_2)\tanh\left(\frac{\sqrt{-\gamma}}{2}\right)\right)}, & \gamma < 0. \end{cases} \tag{24}$$

Remark 3.1 It is obvious that the value of $\bar{c}$, h, k and R will depend on those of γ, δ_1 and δ_2, but we will omit such dependence in the notation for the sake of simplicity. Moreover, it is clear that, when γ, δ_1 and δ_2 are fixed then $\bar{c}$, h, k and R are constants too.

Next, we will prove the following theorem to ensure the existence of positive solutions.

Theorem 3.2 *Assume that the positiveness condition of the Green's function* (21) *is fulfilled. Moreover, suppose that the following conditions hold:*

1. *There exists $p > 0$ such that*

$$f(t,u) \leq \frac{p}{k\,R}, \qquad \text{for all } t \in I \text{ and } u \in [0,p].$$

2. *There exists $q > 0$, with $q \neq p$ such that*

$$f(t,u) \geq \frac{q}{h\,R}, \qquad \text{for all } t \in I \text{ and } u \in [\bar{c}\,q, q].$$

Then problem (1), (3) has at least one positive solution $u \in K$, such that $\|u\|_\infty$ lies between p and q.

Proof First, since $G_{\gamma,\delta_1,\delta_2}(t,s) > 0$ for all $t, s \in (0,1)$, $f \geq 0$ and the fixed points of the operator $T_{\gamma,\delta_1,\delta_2}$ coincide with the solutions of problem (1), (3), we deduce that these solutions are nonnegative.

Now, we show that the operator $T_{\gamma,\delta_1,\delta_2}$ defined in (20) is compact and maps K into K.

Let us see that $T_{\gamma,\delta_1,\delta_2}$ maps K into K. For all $u \in K$, using inequality (22) we infer that, for all $t \in I$,

$$T_{\gamma,\delta_1,\delta_2}\, u(t) = \int_0^1 G_{\gamma,\delta_1,\delta_2}(t,s)\, f(s,u(s))\, ds \geq \bar{c} \int_0^1 \max_{t\in I} G_{\gamma,\delta_1,\delta_2}(t,s)\, f(s,u(s))\, ds$$
$$\geq \bar{c} \max_{t\in I} \int_0^1 G_{\gamma,\delta_1,\delta_2}(t,s)\, f(s,u(s))\, ds.$$

So, $T_{\gamma,\delta_1,\delta_2}\, u(t) \geq \bar{c}\, \|T_{\gamma,\delta_1,\delta_2}\, u\|_\infty$ for all $t \in I$, that is, $T_{\gamma,\delta_1,\delta_2}\, u \in K$.

On the other hand, since $G_{\gamma,\delta_1,\delta_2}$ and f are continuous, we have that operator $T_{\gamma,\delta_1,\delta_2}$ is continuous too.

Finally, we will prove that $T_{\gamma,\delta_1,\delta_2}$ maps bounded sets into relatively compact sets. Let $H \subset K$ be a bounded set. Then, using (23), it is easy to see that $T_{\gamma,\delta_1,\delta_2}(H)$ is bounded.

Let us show then the equicontinuity of $T_{\gamma,\delta_1,\delta_2}(H)$. Since H is bounded, there exists $r \in \mathbb{R}$, $r > 0$ such that $\|u\|_\infty \leq r$ for all $u \in H$. Let us take

$$M = \max_{t\in I, 0\leq u\leq r} |f(t,u)|.$$

So, for all $t \in I$ and $u \in H$, we have that

$$|(T_{\gamma,\delta_1,\delta_2}\, u)'(t)| = \left| \int_0^1 \frac{\partial G_{\gamma,\delta_1,\delta_2}}{\partial t}(t,s)\, f(s,u(s))\, ds \right|$$
$$\leq \int_0^1 \left| \frac{\partial G_{\gamma,\delta_1,\delta_2}}{\partial t}(t,s) \right| |f(s,u(s))|\, ds \leq M \int_0^1 \left| \frac{\partial G_{\gamma,\delta_1,\delta_2}}{\partial t}(t,s) \right| ds.$$

Using the regularity of the Green's function $G_{\gamma,\delta_1,\delta_2}$ we deduce that there exists $N \in \mathbb{R}$, $N > 0$ such that

$$M \int_0^1 \left| \frac{\partial G_{\gamma,\delta_1,\delta_2}}{\partial t}(t,s) \right| ds \leq N.$$

So, for all $t_1, t_2 \in I$, $t_1 < t_2$, the following inequality holds

$$|(T_{\gamma,\delta_1,\delta_2}u)(t_2) - (T_{\gamma,\delta_1,\delta_2}u)(t_1)| = \left| \int_{t_1}^{t_2} (T_{\gamma,\delta_1,\delta_2}u)'(s)ds \right|$$
$$\leq \int_{t_1}^{t_2} |(T_{\gamma,\delta_1,\delta_2}u)'(s)|ds \leq N(t_2 - t_1).$$

Therefore, $T_{\gamma,\delta_1,\delta_2}(H)$ is an equicontinuous set in X. By Arzelà-Ascoli's Theorem, we deduce that $T_{\gamma,\delta_1,\delta_2}(H)$ is relatively compact, that is, $T_{\gamma,\delta_1,\delta_2} : K \to K$ is a compact operator.

Next, let us define the following sets of K

$$K_p = \{u \in X \,/\, \|u\|_\infty < p\} \quad \text{and} \quad K_q = \{u \in X \,/\, \|u\|_\infty < q\}.$$

From inequality (23) we infer that

$$\begin{aligned}\|T_{\gamma,\delta_1,\delta_2}\, u(t)\|_\infty &= \max_{t\in I} \int_0^1 G_{\gamma,\delta_1,\delta_2}(t,s)\, f(s,u(s))\, ds \\ &\leq k \int_0^1 G_{\gamma,\delta_1,\delta_2}(1,s)\, f(s,u(s))\, ds.\end{aligned}$$

Using Condition 1. we have, for all $u \in K \cap \partial K_p$, that

$$\|T_{\gamma,\delta_1,\delta_2}\, u(t)\|_\infty \leq k \int_0^1 G_{\gamma,\delta_1,\delta_2}(1,s)\, f(s,u(s))\, ds \leq p = \|u\|_\infty.$$

On the other hand, for $u \in K \cap \partial K_q$, using Condition 2. we have that $q \geq u(t) \geq \bar{c}\|u\|_\infty = \bar{c}\, q$ for all $t \in I$ and from (23) we deduce that

$$\|T_{\gamma,\delta_1,\delta_2}\, u(t)\|_\infty \geq h \int_0^1 G_{\gamma,\delta_1,\delta_2}(1,s)\, f(s,u(s))\, ds \geq q = \|u\|_\infty.$$

Thus, applying Krasnoselskii's Theorem 3.1, we obtain that $T_{\gamma,\delta_1,\delta_2}$ has a fixed point $u \in K$ such that $\|u\|_\infty$ lies between p and q. □

In the sequel, we present an example to illustrate the previous result.

Example 3.1 Consider the following problem

$$\begin{cases} u''(t) + u(t)\, e^{u^2(t)} = 0, \; t \in I, \\ u(0) = \dfrac{1}{3} \displaystyle\int_0^1 u(s)\, ds, \\ u(1) = \dfrac{1}{6} \displaystyle\int_0^1 u(s)\, ds. \end{cases} \tag{25}$$

In this case, $\gamma = 0$, $\delta_1 = \frac{1}{3}$, $\delta_2 = \frac{1}{6}$ and $\delta_1 + \delta_2 = \frac{1}{2} < 2$. As we have said in Remark 2.3 in this case we may consider $h = c_1 = 1$ and $k = c_2 = 12$ and, consequently, $\bar{c} = \frac{1}{12}$. Moreover, from (24) we know that

$$\int_0^1 G_{0,\frac{1}{3},\frac{1}{6}}(1,s)\, ds = \frac{1}{54}.$$

Moreover, $f(t,u) = u\,e^{u^2}$ is continuous on $I \times [0,\infty)$. Now, for $p \leq \sqrt{\log\left(\frac{9}{2}\right)} \approx 1.226$ it holds that

$$f(t,u) \leq p\,e^{p^2} \leq \frac{9}{2}\,p \left(= \frac{p}{k\int_0^1 G_{0,\frac{1}{3},\frac{1}{6}}(1,s)\,ds} \right) \quad \forall t \in [0,1],\ u \in [0,p].$$

Similarly, for $q \geq 12\sqrt{\log 648} \approx 30.533$ it holds that

$$f(t,u) \geq \frac{1}{12}\,q\,e^{\frac{q^2}{144}} \geq 54\,q \left(= \frac{q}{h\int_0^1 G_{0,\frac{1}{3},\frac{1}{6}}(1,s)\,ds} \right) \quad \forall t \in [0,1],\ u \in \left[\frac{q}{12}, q\right].$$

Therefore, Theorem 3.2 guarantees the existence of a positive solution of problem (25).

3.1 Problem (2), (3)

To study the nonlinear problem (2), (3), we must use the Green's function related to

$$u''(t) + \gamma\,u(t) - \sigma(t) = 0, \quad t \in I,$$

coupled to the integral boundary conditions (3).

It is immediate to verify that such Green's function is indeed $-G_{\gamma,\delta_1,\delta_2}$.

So, by assuming the following condition:

$$\gamma < \pi^2,\ \delta_1 + \delta_2 > \Delta(\gamma),\ \delta_1 \leq \Delta(\gamma) \text{and } \delta_2 \leq \Delta(\gamma), \tag{26}$$

we can ensure the positiveness of function $-G_{\gamma,\delta_1,\delta_2}$.

Therefore, the inequalities that we have used in the previous case, in Lemmas 2.2, 2.5 and 2.7, remain valid since it is clear that

$$c_1 \leq \frac{-G_{0,\delta_1,\delta_2}(t,s)}{-G_{0,\delta_1,\delta_2}(1,s)} \leq c_2, \qquad \text{for all } t \in I \text{ and } s \in (0,1),$$

$$c_3 \leq \frac{-G_{\gamma,\delta_1,\delta_2}(t,s)}{-G_{\gamma,\delta_1,\delta_2}(1,s)} \leq c_4, \quad \text{for } \gamma > 0 \qquad \text{and all } t \in I \text{ and } s \in (0,1),$$

and

$$c_5 \leq \frac{-G_{\gamma,\delta_1,\delta_2}(t,s)}{-G_{\gamma,\delta_1,\delta_2}(1,s)} \leq c_6, \quad \text{for } \gamma < 0 \qquad \text{and all } t \in I \text{ and } s \in (0,1).$$

Thus, we may define the constants $\bar{c}$, h, k and R (< 0) in the same way than before and inequalities (22) and (23) hold by simply changing $G_{\gamma,\delta_1,\delta_2}$ by $-G_{\gamma,\delta_1,\delta_2}$.

In this case, an existence result can also be proved by considering now values of γ, δ_1 and δ_2 for which $G_{\gamma,\delta_1,\delta_2}$ is negative (that is, $-G_{\gamma,\delta_1,\delta_2}$ is positive). The result is the following one.

Theorem 3.3 *Assume that condition* (26) *is fulfilled. Moreover, suppose that the following conditions hold:*

1. *There exists* $p' > 0$ *such that*

$$f(t,u) \leq \frac{p'}{-k\,R}, \quad \text{for all } t \in I \text{ and } u \in [0, p'].$$

2. *There exists* $q' > 0$*, with* $q' \neq p'$ *such that*

$$f(t,u) \geq \frac{q'}{-h\,R}, \quad \text{for all } t \in I \text{ and } u \in [\bar{c}\,q', q'].$$

Then problem (2), (3) has at least one positive solution $u \in K$*, such that* $||u||_\infty$ *lies between* p' *and* q'.

References

1. Anderson, D.R., Hoffacker, J.: Existence of solutions for a cantilever beam problem. J. Math. Anal. Appl. **323**, 958–973 (2006)
2. Ahmad, B., Hamdan, S., Alsaedi, A., Ntouyas, S.K.: On a nonlinear mixed-order coupled fractional differential system with new integral boundary conditions. AIMS Math. **6**(6), 5801–5816 (2021)
3. Cabada, A.: Green's Functions in the Theory of Ordinary Differential Equations. Springer Briefs Math. Springer, New York (2014)
4. Cabada, A., Iglesias, J.: Nonlinear differential equations with perturbed dirichlet integral boundary conditions. Bound. Value Problems **2021**, 66 (2021)
5. Cabada, A., López-Somoza, L., Yousfi, M.: Green's function related to a n order linear differential equation coupled to arbitrary linear non local boundary conditions. Mathematics **1948**, 9 (2021)
6. Chandran, K., Gopalan, K., Tasneem, Z. S., Abdeljawad, T.: A fixed point approach to the solution of singular fractional differential equations with integral boundary conditions. Adv. Differ. Equ. **2021**, 16 pp. Paper No. 56
7. Duraisamy, P., Nandha, G.T., Subramanian, M.: Analysis of fractional integro-differential equations with nonlocal Erdélyi-Kober type integral boundary conditions. Fract. Calc. Appl. Anal. **23**(5), 1401–1415 (2020)
8. Guo, D., Lakshmikantham, L.: Nonlinear Problems in Abstract Cones. Academic Press, San Diego (1988)

9. Hu, Q-Q., Yan, B.: Existence of multiple solutions for second-order problem with stieltjes integral boundary condition. J. Funct. Spaces 7 pp. (2021). Art. ID 6632236
10. Zhang, Y., Abdella, K., Feng, W.: Positive solutions for second-order differential equations with singularities and separated integral boundary conditions. Electron. J. Qual. Theory Differ. Equ. 12 pp. (2020). Paper No. 75

A Degenerate Operator in Non Divergence Form

Alessandro Camasta and Genni Fragnelli

In honour of Francesco Altomare, with deep affection on his 70th birthday

Abstract In this paper we consider a fourth order operator in non divergence form $Au := au''''$, where $a : [0, 1] \to \mathbb{R}_+$ is a function that degenerates somewhere in the interval. We prove that the operator generates an analytic semigroup, under suitable assumptions on the function a. We extend these results to a general operator $A_n u := au^{(2n)}$.

Keywords Degenerate operators in non divergence form · Linear differential operators of order $2n$ · Interior and boundary degeneracy · Analytic semigroups

1 Introduction

In this paper we analyze the properties of a degenerate fourth order differential operator in non divergence form under Dirichlet boundary conditions in the real setting. More precisely, we consider the operator $Au := au''''$ with a suitable domain, where we denote with $'$ the derivative of a function depending only on one variable x, which we assume to vary in $[0, 1]$. The coefficient a is a function for which the degeneracy may occur in the interior of the interval or on the boundary of it.

A. Camasta (✉)
Mathematics Department, University of Bari Aldo Moro, Bari, Italy
e-mail: alessandro.camasta@uniba.it

G. Fragnelli
Department of Ecology and Biology, Tuscia University, Largo dell'Università, Viterbo, Italy
e-mail: genni.fragnelli@unitus.it

A. M. Candela et al. (eds.), *Recent Advances in Mathematical Analysis*,
Trends in Mathematics, https://doi.org/10.1007/978-3-031-20021-2_11

In fact, we shall admit two types of degeneracy for a, namely weak and strong degeneracy. In particular, following [1], [15] or [17], we say that a function $g : [0, 1] \to \mathbb{R}$ is

- *weakly degenerate at* $x_0 \in [0, 1]$ if $g \in C[0, 1]$, $g(x_0) = 0$, $g > 0$ on $[0, 1] \setminus \{x_0\}$ and $\frac{1}{g} \in L^1(0, 1)$;
- *strongly degenerate at* $x_0 \in [0, 1]$ if $g \in C^1([0, 1]$, $g(x_0) = 0$, $g > 0$ on $[0, 1] \setminus \{x_0\}$ and $\frac{1}{g} \notin L^1(0, 1)$.

We are interested in this type of operators since many problems that are relevant for applications are described by fourth order operators. Among these applications we can find dealloying (corrosion processes, see, e.g., [14]), population dynamics (see, e.g., [10]), bacterial films (see, e.g., [23]), thin film (see, e.g., [27]), Chemistry (see, e.g., [29]), tumor growth (see, e.g., [2], [21]), image processing (denoising, inpainting, see, e.g., [5], [9], [12]), Astronomy (rings of Saturn, see, e.g., [28]), Ecology (surprisingly, the clustering of mussels can be perfectly well described by the Cahn–Hilliard equation, see, e.g, [26]) and so on.

Let us present very briefly some interesting results about the existence and uniqueness of solutions for problems associated to the operators under consideration. In [22] the authors study the epitaxial growth of nanoscale thin films which can be described by a parabolic equation of the form

$$\frac{\partial u}{\partial t} + \Delta^2 u - \nabla \cdot (f(\nabla u)) = g$$

in $(0, T) \times (0, L)$, where f and g belong to $C^1(\mathbb{R}^N, \mathbb{R}^N)$, $N \geq 2$, and $L^2((0, T) \times (0, L))$, respectively. The authors show existence, uniqueness and regularity of solutions in suitable functional spaces. In [11] the authors consider a degenerate fourth order operator of the form $\nabla \cdot (m(u)\nabla\Delta u)$ where m is a specific function, proving existence and non uniqueness results for the parabolic equation associated to this operator (see also [3, 4, 6, 13, 19] or [24]). In [20] the existence of a weak solution for the following equation is proved

$$\frac{\partial u}{\partial t} + \nabla \cdot (|\nabla\Delta u|^{p(x)-2}\nabla\Delta u) = f(x, u),$$

in $(0, T) \times \Omega$, under the conditions $u = \Delta u = 0$ on $\partial\Omega$ and $u(0, x) = u_0(x)$, $x \in \Omega \subset \mathbb{R}^N$, $N \geq 2$. Here p and f are specific functions and u_0 is an initial datum. Observe that for $p \equiv 2$ the parabolic problem associated to the previous operator becomes the classical Cahn–Hilliard problem, which has been extensively studied (see, e.g., [25]). The previous model can describe some properties of medical magnetic resonance images in space and time. In particular, if $f(x, u) := u(t, x) - a(x)$, then u represents a digital image and a its observation.

Recently, in [16] the general operator $\tilde{A}_n u := (au^{(n)})^{(n)}$ is considered, where $a \in C[0, 1]$ degenerates in an interior point x_0. The authors give sufficient conditions on the function a so that the operator $(\tilde{A}_n, D(\tilde{A}_n))$ generates a contractive analytic semigroup on $L^2(0, 1)$.

As for the second order operator (see, e.g., [7], [8] or [18] and the references therein), the generation property for the operator in non divergence form cannot be deduced by the one of the operator in divergence form without assuming other assumptions on the function a. Moreover, another difference between the operator in divergence form and the one in non divergence form is the fact that the natural space to study the equation associated to $\tilde{A}u := (au'')''$, or $\tilde{A}_n u := (au^{(n)})^{(n)}$, is $L^2(0,1)$ whereas the problem associated to $Au := au''''$, or more in general to $A_n u := au^{(2n)}$, is more conveniently set in the weighted space

$$L^2_{\frac{1}{a}}(0,1) := \left\{u \in L^2(0,1) : \int_0^1 \frac{u^2(x)}{a(x)}\,dx \in \mathbb{R}\right\}.$$

In such a space we will prove that the fourth order operator A and, in general, the operator A_n generate an analytic semigroup.

The paper is organized in the following way. In Sect. 2 we assume that the degeneracy point belongs to the boundary of the space domain and we consider the fourth order operator, proving some preliminary results that will be crucial to prove the generation property of it in Theorem 2.1. In Sect. 3 we characterize the domain of the operator in the weakly and in the strongly degenerate case under additional assumptions on the degenerate function a. Thanks to the characterization of the domain we prove again the generation property, if the degeneracy point is in the interior of the domain. In Sect. 4 we extend the previous results to the general operator $A_n u = au^{(2n)}$, $n \geq 3$.

A final comment on the notation: by C we shall denote universal positive constants, which are allowed to vary from line to line.

This paper is a tribute to Professor Francesco Altomare for celebrating his 70th birthday and for thanking him for the wonderful teaching and research activities realized with great efficiency, accuracy and passion.

2 The Fourth Order Operator if the Degeneracy Point belongs to the Boundary

In this section we introduce the operator $Au := au''''$, where $a : [0,1] \to \mathbb{R}_+$ is a given function that degenerates somewhere in the space domain, and we consider the following (weighted) Hilbert spaces:

$$L^2_{\frac{1}{a}}(0,1) := \left\{u \in L^2(0,1) : \int_0^1 \frac{u^2}{a}\,dx < +\infty\right\}$$

and

$$H^i_{\frac{1}{a}}(0,1) := L^2_{\frac{1}{a}}(0,1) \cap H^i_0(0,1), \quad i = 1, 2,$$

with the norms

$$\|u\|^2_{L^2_{\frac{1}{a}}(0,1)} := \int_0^1 \frac{u^2}{a}\,dx \qquad \forall\, u \in L^2_{\frac{1}{a}}(0,1)$$

and

$$\|u\|^2_{H^i_{\frac{1}{a}}(0,1)} := \|u\|^2_{L^2_{\frac{1}{a}}(0,1)} + \sum_{k=1}^{i} \|u^{(k)}\|^2_{L^2(0,1)} \qquad \forall\, u \in H^i_{\frac{1}{a}}(0,1),$$

$i = 1, 2$, respectively. We recall that $H^i_0(0,1) := \{u \in H^i(0,1) : u^{(k)}(j) = 0,\ j = 0, 1,\ k = 0, \ldots, i-1\}$, with $u^{(0)} = u$ and $i = 1, 2$.

Observe that for all $u \in H^i_{\frac{1}{a}}(0,1)$, using the fact that $u^{(k)}(j) = 0$ for all $k = 0, \ldots, i-1$ and $j = 0, 1$, it is easy to prove that $\|u\|^2_{H^i_{\frac{1}{a}}(0,1)}$ is equivalent to the following one

$$\|u\|_i^2 := \|u\|^2_{L^2_{\frac{1}{a}}(0,1)} + \|u^{(i)}\|^2_{L^2(0,1)}.$$

Thus, for simplicity, in the rest of the paper we will use $\|\cdot\|_i$ in place of $\|\cdot\|_{H^i_{\frac{1}{a}}(0,1)}$.

Using the previous spaces, it is possible to define the operator A by

$$Au := au'''' \quad \text{for all } u \in D(A) := \left\{u \in H^2_{\frac{1}{a}}(0,1) : au'''' \in L^2_{\frac{1}{a}}(0,1)\right\}, \tag{1}$$

if $x_0 \in \{0, 1\}$. The case $x_0 \in (0, 1)$ will be considered in the next section. In order to prove that $(A, D(A))$ generates a semigroup, we assume that a satisfies the following hypothesis:

Hypothesis 2.1 The function a belongs to the space of continuous functions $C[0, 1]$ and there exists a point $x_0 \in \{0, 1\}$ such that $a(x_0) = 0$ and $a > 0$ on $[0, 1] \setminus \{x_0\}$.

Proposition 2.1 (Green's Formula)
Assume Hypothesis 2.1. For all $(u, v) \in D(A) \times H^2_{\frac{1}{a}}(0,1)$ *one has*

$$\int_0^1 u''''v\,dx = \int_0^1 u''v''dx. \tag{2}$$

The proof of the previous proposition is based on the next result which is standard, but here we give it for the reader's convenience.

Lemma 2.1 *Let $I := (\alpha, \beta)$, with $\alpha, \beta \in \mathbb{R}$, $\alpha < \beta$, $p \geq 1$ and $\mathscr{D}'(I)$ the space of distributions defined on I. If $f \in \mathscr{D}'(I)$ has n-th derivative which is a function belonging to $L^p(I)$, then $f \in W^{n,p}(I)$, where $n \in \mathbb{N}$, $n \geq 1$.*

Proof Let $x \in I$ and define

$$\begin{cases} v_1(x) := \int_\alpha^x f^{(n)}(s)ds, \\ v_i(x) := \int_\alpha^x v_{i-1}(s)ds, \end{cases}$$

$i = 2, \dots n$. Then, for all $i = 1, \dots n$,

$$v_i \in W^{i,p}(I) \quad \text{and} \quad v_i^{(i)} = f^{(n)}. \tag{3}$$

Indeed, for $i = 1$ the thesis is obvious. Now, we prove it for $i = 2$:

$$v_2'' = (v_2')' = (v_1)' = f^{(n)}.$$

Hence, iterating the procedure, one has that (3) holds.

Now, define $\psi := v_n - f$. Hence the distributional n-th derivative of ψ is given by $\psi^{(n)} = v_n^{(n)} - f^{(n)} = 0$. Thus, there exists a constant $c \in \mathbb{R}$ such that

$$f^{(n-1)} = v_n^{(n-1)} + c$$

a.e. in I; this implies that $f^{(n-1)} \in W^{1,p}(I)$. In particular, one has $f^{(n-1)} \in L^p(I)$. Proceeding as in the first part of the proof and iterating the procedure, one has that $f^{(i)} \in L^p(I)$ for all $i = 1, 2, \dots, n-2$.

Now, define $z(x) := \int_\alpha^x f'(s)ds$ and $w(x) := z(x) - f(x)$. Clearly, $z \in W^{2,p}(I)$ and $w'(x) = 0$ a.e. in I. This implies that there exists a constant $C \in \mathbb{R}$ such that

$$f = z + C.$$

In particular, $f \in L^p(I)$ and the thesis follows. □

Proof of Proposition 2.1 Following the idea of [8, Lemma 2.1], one can prove that the space $H_c^2(0,1) := \{v \in H^2(0,1) : \operatorname{supp} v \subset (0,1)\}$ is dense in $H^2_{\frac{1}{a}}(0,1)$.

Indeed, we can consider the sequence $(v_n)_{n \geq 4}$, where $v_n := \xi_n v$ for a fixed function $v \in H^2_{\frac{1}{a}}(0,1)$ and

$$\xi_n(x) := \begin{cases} 0, & x \in [0, 1/n] \cup [1 - 1/n, 1], \\ 1, & x \in [2/n, 1 - 2/n], \\ -2n^3x^3 + 9n^2x^2 - 12nx + 5, & x \in (1/n, 2/n), \\ f(n,x), & x \in (1 - 2/n, 1 - 1/n). \end{cases}$$

Here $f(n,x) := a_n x^3 + b_n x^2 + c_n x + d_n$, being

$$a_n := \frac{2}{-\left(1-\frac{2}{n}\right)^3 + 3\left(1-\frac{1}{n}\right)\left(1-\frac{2}{n}\right)^2 - 3\left(1-\frac{1}{n}\right)^2\left(1-\frac{2}{n}\right) + \left(1-\frac{1}{n}\right)^3},$$

$$b_n := \frac{-3\left(1-\frac{2}{n}\right) - 3\left(1-\frac{1}{n}\right)}{-\left(1-\frac{2}{n}\right)^3 + 3\left(1-\frac{1}{n}\right)\left(1-\frac{2}{n}\right)^2 - 3\left(1-\frac{1}{n}\right)^2\left(1-\frac{2}{n}\right) + \left(1-\frac{1}{n}\right)^3},$$

$$c_n := \frac{6\left(1-\frac{1}{n}\right)\left(1-\frac{2}{n}\right)}{-\left(1-\frac{2}{n}\right)^3 + 3\left(1-\frac{1}{n}\right)\left(1-\frac{2}{n}\right)^2 - 3\left(1-\frac{1}{n}\right)^2\left(1-\frac{2}{n}\right) + \left(1-\frac{1}{n}\right)^3}$$

and

$$d_n := \frac{\left(1-\frac{1}{n}\right)^3 - 3\left(1-\frac{1}{n}\right)^2\left(1-\frac{2}{n}\right)}{-\left(1-\frac{2}{n}\right)^3 + 3\left(1-\frac{1}{n}\right)\left(1-\frac{2}{n}\right)^2 - 3\left(1-\frac{1}{n}\right)^2\left(1-\frac{2}{n}\right) + \left(1-\frac{1}{n}\right)^3}.$$

It is easy to see that $v_n \to v$ in $L^2_{\frac{1}{a}}(0,1)$. Indeed, setting $g_n := v_n - v$, one has that $\lim_{n\to+\infty} g_n = 0$ a.e. and $|g_n| \le 2|v| \in L^2_{\frac{1}{a}}(0,1)$ for all $n \in \mathbb{N}$, $n \ge 4$. Hence, by the Lebesgue Theorem, one can conclude that $v_n \to v$ in $L^2_{\frac{1}{a}}(0,1)$. Moreover, one has that

$$\begin{aligned}\int_0^1 ((v_n - v)'')^2 dx &\le 2\int_0^1 (1-\xi_n)^2 (v'')^2 dx + 2\int_{\frac{1}{n}}^{\frac{2}{n}} (\xi_n'' v)^2 dx \\ &+ 2\int_{1-\frac{2}{n}}^{1-\frac{1}{n}} (\xi_n'' v)^2 dx + 8\left(\int_{\frac{1}{n}}^{\frac{2}{n}} (\xi_n' v')^2 dx + \int_{1-\frac{2}{n}}^{1-\frac{1}{n}} (\xi_n' v')^2 dx\right).\end{aligned} \tag{4}$$

Obviously, proceeding as before, the first term in the last member of (4) converges to zero. Furthermore, since $v, v' \in H^1_0(0,1)$, by Hölder's inequality one has that

$$v^2(x) \le x\int_0^x (v')^2(y)dy \qquad \forall\, x \in [0,1]$$

and

$$(v')^2(y) \le y\int_0^y (v'')^2(z)dz \qquad \forall\, y \in [0,1].$$

Hence,

$$v^2(x) \le x \int_0^x \left(y \int_0^y (v'')^2(z)dz \right) dy \le x^2 \int_0^x \int_0^x (v'')^2(z)dz\, dy$$
$$\le x^3 \int_0^x (v'')^2(z)dz.$$

Using this inequality, one can prove that there exists a positive constant C such that

$$\int_{\frac{1}{n}}^{\frac{2}{n}} (\xi_n'' v)^2 dx \le C \int_{\frac{1}{n}}^{\frac{2}{n}} (n^6 x^2 + n^4) v^2(x) dx$$
$$\le C \int_{\frac{1}{n}}^{\frac{2}{n}} (n^6 x^2 + n^4) x^3 \left(\int_0^x (v'')^2(z)dz \right) dx$$
$$= C \int_{\frac{1}{n}}^{\frac{2}{n}} (n^6 x^5 + n^4 x^3) \int_0^x (v'')^2(z)dz\, dx$$
$$= C \int_0^{\frac{2}{n}} (v'')^2(z) \left(\int_{\frac{1}{n}}^{\frac{2}{n}} (n^6 x^5 + n^4 x^3) dx \right) dz \to 0 \quad \text{as } n \to +\infty.$$

Analogously the term $\int_{1-\frac{2}{n}}^{1-\frac{1}{n}} (\xi_n'' v)^2 dx$ tends to 0 as $n \to +\infty$. Since the remaining terms in (4) can be similarly estimated, one has that

$$\lim_{n \to +\infty} \int_0^1 ((v_n - v)'')^2 dx = 0$$

and our preliminary claim is proved.

Now, fixed $u \in D(A)$, set $\Phi(v) := \int_0^1 u'''' v\, dx - \int_0^1 u'' v'' dx$, with $v \in H^2_{\frac{1}{a}}(0, 1)$ $\left(\text{observe that } u'''' v \in L^1(0, 1) \text{ since } u'''' v = \sqrt{a} u'''' \frac{v}{\sqrt{a}} \right)$. By definition of Φ, it follows that

$$\Phi(v) = 0$$

for all $v \in H^2_c(0, 1)$.

In order to prove this fact, we assume $x_0 = 0$, the case $x_0 = 1$ being treated in analogous way. Now, let $v \in H^2_c(0, 1)$ and let $\delta > 0$ be such that $\operatorname{supp} v \subset \mathcal{K}$, where $\mathcal{K} := [\delta, 1]$ (or $\mathcal{K} := [0, 1 - \delta]$ if $x_0 = 1$). By definition of $D(A)$, $u'''' \in L^2(\mathcal{K})$, thus $u'' \in H^2(\mathcal{K})$ (by Lemma 2.1 with $n = p = 2$ and $f = u''$) and, in particular, $u \in H^4(\mathcal{K})$. Hence, we can integrate by parts, obtaining

$$\begin{aligned} \int_0^1 u'''' v \, dx &= \int_0^\delta u'''' v \, dx + \int_\delta^1 u'''' v \, dx \\ &= \int_0^\delta u'''' v \, dx + \int_\delta^1 u'' v'' dx, \end{aligned} \tag{5}$$

since $v(\delta) = v'(\delta) = 0$. Now we prove that

$$\lim_{\delta \to 0} \int_\delta^1 u'' v'' dx = \int_0^1 u'' v'' dx \tag{6}$$

and

$$\lim_{\delta \to 0} \int_0^\delta u'''' v \, dx = 0. \tag{7}$$

To this aim, observe that

$$\int_\delta^1 u'' v'' dx = \int_0^1 u'' v'' dx - \int_0^\delta u'' v'' dx.$$

Moreover, as before for $u''''v$, also $u''v''$ belongs to $L^1(0, 1)$ using Hölder's inequality. Thus, for any $\varepsilon > 0$, by the absolute continuity of the Lebesgue integral, there exists $\delta := \delta(\varepsilon) > 0$ such that

$$\left| \int_0^\delta u'' v'' dx \right| \le \left| \int_0^\delta |u'' v''| \, dx \right| < \varepsilon,$$

$$\left| \int_0^\delta u'''' v \, dx \right| \le \left| \int_0^\delta |u'''' v| \, dx \right| < \varepsilon.$$

Taking such a δ in (5), from the arbitrariness of ε, we can deduce

$$\lim_{\delta \to 0} \int_0^\delta u'''' v \, dx = \lim_{\delta \to 0} \int_0^\delta u'' v'' dx = 0.$$

Thus, by the previous equalities and by (5), (6) and (7), it follows that

$$\int_0^1 u'''' v \, dx = \int_0^1 u'' v'' dx \iff \Phi(v) = \int_0^1 (u'''' v - u'' v'') dx = 0,$$

for all $v \in H^2_c(0,1)$. Then, Φ is a bounded linear functional on $H^2_{\frac{1}{a}}(0,1)$ such that $\Phi = 0$ on $H^2_c(0,1)$, hence $\Phi = 0$ on $H^2_{\frac{1}{a}}(0,1)$, i.e., (2) holds. □

Actually, in the weakly degenerate case, one can proceed as done for the case $x_0 \in (0,1)$ (see below) making the proof simpler.

As a consequence of Proposition 2.1 one has the next theorem.

Theorem 2.1 *Assume Hypothesis 2.1. The operator* $(A, D(A))$ *defined in* (1) *is self-adjoint and non negative on* $L^2_{\frac{1}{a}}(0,1)$*. Hence* $-A$ *generates a contractive analytic semigroup of angle* $\frac{\pi}{2}$ *on* $L^2_{\frac{1}{a}}(0,1)$*.*

Proof Observe that $D(A)$ is dense in $L^2_{\frac{1}{a}}(0,1)$. In order to show that A is self-adjoint it is sufficient to prove that A is symmetric, non negative and $(I + A)(D(A)) = L^2_{\frac{1}{a}}(0,1)$. Indeed, if A is non negative and $I + A$ is surjective on $D(A)$, then A is maximal monotone and in this case A is symmetric if and only if A is self-adjoint.

A is symmetric: thanks to Proposition 2.1, for any $u, v \in D(A)$, one has

$$\langle v, Au\rangle_{L^2_{\frac{1}{a}}(0,1)} = \int_0^1 \frac{vau''''}{a}\,dx = \int_0^1 v''''u\,dx = \langle Av, u\rangle_{L^2_{\frac{1}{a}}(0,1)}\,.$$

A is non negative: using again Proposition 2.1, for any $u \in D(A)$

$$\langle Au, u\rangle_{L^2_{\frac{1}{a}}(0,1)} = \int_0^1 \frac{au''''u}{a}\,dx = \int_0^1 (u'')^2dx \geq 0.$$

$I + A$ is surjective: observe that $H^2_{\frac{1}{a}}(0,1)$, equipped with the inner product

$$\langle u, v\rangle_{H^2_{\frac{1}{a}}(0,1)} := \int_0^1 \left(\frac{uv}{a} + u''v''\right)dx \qquad \forall\, u, v \in H^2_{\frac{1}{a}}(0,1),$$

is a Hilbert space. Moreover

$$H^2_{\frac{1}{a}}(0,1) \hookrightarrow L^2_{\frac{1}{a}}(0,1) \hookrightarrow \left(H^2_{\frac{1}{a}}(0,1)\right)^*,$$

where $\left(H^2_{\frac{1}{a}}(0,1)\right)^*$ is the dual space of $H^2_{\frac{1}{a}}(0,1)$ with respect to $L^2_{\frac{1}{a}}(0,1)$. Indeed, the continuous embedding of $H^2_{\frac{1}{a}}(0,1)$ in $L^2_{\frac{1}{a}}(0,1)$ is readily seen. In addition, for any $f \in H^2_{\frac{1}{a}}(0,1)$ and $\varphi \in L^2_{\frac{1}{a}}(0,1)$

$$\left|\langle f,\varphi\rangle_{L^2_{\frac{1}{a}}(0,1)}\right| = \left|\int_0^1 \frac{f\varphi}{a}\,dx\right| = \left|\int_0^1 \frac{f}{\sqrt{a}}\frac{\varphi}{\sqrt{a}}\,dx\right| \le \|f\|_{L^2_{\frac{1}{a}}(0,1)}\|\varphi\|_{L^2_{\frac{1}{a}}(0,1)}$$
$$\le \|f\|_2\|\varphi\|_{L^2_{\frac{1}{a}}(0,1)}.$$

Hence, $L^2_{\frac{1}{a}}(0,1) \hookrightarrow \left(H^2_{\frac{1}{a}}(0,1)\right)^*$. Then, $\left(H^2_{\frac{1}{a}}(0,1)\right)^*$ is the completion of $L^2_{\frac{1}{a}}(0,1)$ with respect to the norm of $\left(H^2_{\frac{1}{a}}(0,1)\right)^*$. Now, for $f \in L^2_{\frac{1}{a}}(0,1)$, define the functional $F \in \left(H^2_{\frac{1}{a}}(0,1)\right)^*$ given by

$$F(v) := \int_0^1 \frac{fv}{a}\,dx \qquad \forall\, v \in H^2_{\frac{1}{a}}(0,1).$$

Consequently, by the Lax-Milgram Theorem, there exists a unique $u \in H^2_{\frac{1}{a}}(0,1)$ such that for all $v \in H^2_{\frac{1}{a}}(0,1)$:

$$\langle u,v\rangle_{H^2_{\frac{1}{a}}(0,1)} = \int_0^1 \frac{fv}{a}\,dx. \tag{8}$$

In particular, since $C_c^\infty(0,1) \subset H^2_{\frac{1}{a}}(0,1)$, the integral representation (8) holds for all $v \in C_c^\infty(0,1)$, i.e.,

$$\int_0^1 u''v''dx = \int_0^1 \frac{f-u}{a}\,v\,dx \qquad \forall\, v \in C_c^\infty(0,1).$$

Thus, the distributional second derivative of u'' is equal to $\dfrac{f-u}{a}$ a.e. in $(0,1)$. Since $f-u \in L^2_{\frac{1}{a}}(0,1)$ and

$$au'''' = f-u \quad \text{a.e. in } (0,1),$$

$u \in D(A)$. Hence $(I+A)(u) = f$. As an immediate consequence of the Stone-von Neumann Spectral Theorem and functional calculus associated with the Spectral

Theorem, one has that the operator $(A, D(A))$ generates a cosine family and an analytic semigroup of angle $\frac{\pi}{2}$ on $L^2_{\frac{1}{a}}(0,1)$. □

The previous result can be used to prove that the one-dimensional fourth order parabolic systems associated to the operator $Au := au''''$ are well posed. More precisely, fixed $T > 0$, $u_0 \in L^2_{\frac{1}{a}}(0,1)$ and $f \in L^2(0,T; L^2_{\frac{1}{a}}(0,1))$, the problem

$$\begin{cases} \dfrac{\partial u}{\partial t}(t,x) + a(x)\dfrac{\partial^4 u}{\partial x^4}(t,x) = f(t,x), & (t,x) \in (0,T) \times (0,1), \\ u(t,0) = u(t,1) = 0, & t \in (0,T), \\ \dfrac{\partial u}{\partial x}(t,0) = \dfrac{\partial u}{\partial x}(t,1) = 0, & t \in (0,T), \\ u(0,x) = u_0(x), & x \in (0,1), \end{cases} \tag{9}$$

admits a unique solution $u \in C([0,T]; L^2_{\frac{1}{a}}(0,1)) \cap L^2(0,T; H^2_{\frac{1}{a}}(0,1))$ (see also [18]). As we will see, the same result holds in the case $x_0 \in (0,1)$. These preliminary considerations will be the starting point to study the observability and the null controllability for this kind of problems.

Observe that in the non degenerate case, i.e., if $a(x) > 0$ for all $x \in [0,1]$, the previous problem is also known as Cahn–Hilliard type problem.

3 The Domain of the Operator A and the Case $x_0 \in (0, 1)$

Assuming further hypotheses on the function a, we can prove some characterizations for $D(A)$ in both the weak and the strong case. More precisely, we assume the following:

Hypothesis 3.1 (Weakly Degenerate Function) The function $a \in C[0,1]$ is weakly degenerate, i.e., there exists a point $x_0 \in [0,1]$ such that $a(x_0) = 0$, $a > 0$ on $[0,1] \setminus \{x_0\}$ and $\frac{1}{a} \in L^1(0,1)$.

For example, as a, we can consider $a(x) = |x - x_0|^\alpha$, $0 < \alpha < 1$.

Hypothesis 3.2 (Strongly Degenerate Function) The function $a \in C^1([0,1]$ is strongly degenerate, i.e., there exists a point $x_0 \in [0,1]$ such that $a(x_0) = 0$, $a > 0$ on $[0,1] \setminus \{x_0\}$ and $\frac{1}{a} \notin L^1(0,1)$.

For example, as a, we can consider $a(x) = |x - x_0|^\alpha$, $\alpha \geq 1$.

Thanks to the previous assumptions, we can characterize the spaces introduced in Sect. 2. In particular, we have the following results.

Proposition 3.1 *Assume Hypothesis 3.1. Then, the spaces* $H^i_{\frac{1}{a}}(0,1)$ *and* $H^i_0(0,1)$, $i=1,2$, *coincide algebraically and the two norms are equivalent. Moreover, if* $u \in H^2_0(0,1)$, *i.e.,* $u \in H^2_{\frac{1}{a}}(0,1)$, *then* $(au)(x_0) = (au')(x_0) = 0$.

Proof By definition $H^i_{\frac{1}{a}}(0,1) \subseteq H^i_0(0,1)$, $i=1,2$. Moreover, if $u \in H^i_0(0,1)$ then $u \in C[0,1]$ and, using the fact that $\frac{1}{a} \in L^1(0,1)$, one has

$$\int_0^1 \frac{u^2}{a}\,dx \le \max_{[0,1]} u^2 \int_0^1 \frac{1}{a}\,dx \in \mathbb{R},$$

i.e., $u \in L^2_{\frac{1}{a}}(0,1)$. This implies $u \in H^i_{\frac{1}{a}}(0,1)$.

Furthermore, since $H^i_0(0,1)$ is continuously embedded in $C[0,1]$, one has that for all $u \in H^i_0(0,1)$

$$\|u\|^2_{L^2_{\frac{1}{a}}(0,1)} = \int_0^1 \frac{u^2}{a}\,dx \le \|u\|^2_{C[0,1]} \left\|\frac{1}{a}\right\|_{L^1(0,1)} \le C\|u\|^2_{H^i_0(0,1)},$$

where C is a positive constant. In this way, for all $u \in H^i_0(0,1)$,

$$\|u\|_i \le (C+1)\|u\|_{H^i_0(0,1)} \le (C+1)\|u\|_i,$$

for a positive constant C.

Now, take $u \in H^2_0(0,1)$. Hence $u, u' \in C[0,1]$ and, using the fact that $a(x_0)=0$, one has $(au)(x_0) = (au')(x_0) = 0$. □

Thus, the space $C^\infty_c(0,1)$ is dense in $H^i_{\frac{1}{a}}(0,1)$, $i=1,2$. Observe that, if $x_0 \in \{0,1\}$, the conditions $(au)(x_0) = (au')(x_0) = 0$ are clearly satisfied and the domain $D(A)$ defined in (1) can be rewritten as

$$D(A) = \left\{u \in H^2_0(0,1) : au'''' \in L^2_{\frac{1}{a}}(0,1)\right\}.$$

If $x_0 \in (0,1)$ and Hypothesis 3.1 is satisfied, we consider as $D(A)$ the same domain given in (1), i.e.

$$D(A) := \left\{u \in H^2_{\frac{1}{a}}(0,1) : au'''' \in L^2_{\frac{1}{a}}(0,1)\right\}.$$

Clearly thanks to Proposition 3.1, it can be rewritten as

$$D(A) = \left\{u \in H^2_0(0,1) : (au)(x_0) = (au')(x_0) = 0,\ au'''' \in L^2_{\frac{1}{a}}(0,1)\right\}.$$

In this case (2) follows immediately for all $(u, v) \in D(A) \times H^2_{\frac{1}{a}}(0,1)$. Indeed, if $au'''' \in L^2_{\frac{1}{a}}(0,1)$, then $u'''' \in L^1(0,1)$ and $u \in W^{4,1}(0,1) \hookrightarrow C^3[0,1]$. Hence, under Hypothesis 3.1, Theorem 2.1 still holds and problem (9) admits a unique solution u.

In the strongly degenerate case, if $x_0 \in (0,1)$, the domain of A is given again by (1), but in order to characterize it, we introduce the following space

$$X := \left\{ u \in H^2_{\frac{1}{a}}(0,1) : u(x_0) = (au')(x_0) = 0 \right\}.$$

Notice that, when $x_0 \in \{0, 1\}$, then

$$X = H^2_{\frac{1}{a}}(0,1) \tag{10}$$

in a trivial way.

Actually, the equality (10) can also be proved if $x_0 \in (0,1)$ adding a further assumption on the function a, as we will see in Proposition 3.2.

Hypothesis 3.3 Assume that there exist $K \in [1,2]$ and $C > 0$ such that

$$\frac{1}{a(x)} \leq \frac{C}{|x - x_0|^K}$$

for all $x \in [0,1] \setminus \{x_0\}$.

Observe that the previous hypothesis is obviously satisfied by the prototype $|x - x_0|^K$, where $K \in [1,2]$.

Proceeding as in [15, Lemma 3.7], if $x_0 \in (0,1)$, or as in [8, Proposition 2.6], if $x_0 = 0$ or $x_0 = 1$, one can prove the following result.

Lemma 3.1 *Assume Hypotheses 3.2 and 3.3. Then there exists a positive constant C such that*

$$\int_0^1 \frac{v^2}{a} dx \leq C \int_0^1 (v')^2 dx \qquad \forall\, v \in X.$$

Actually Lemma 3.1 is proved in [8, Proposition 2.6] if $x_0 = 0$ or $x_0 = 1$ under a stronger assumption, but it is evident by the proof that it holds under Hypothesis 3.3.

Proposition 3.2 *If Hypotheses 3.2 and 3.3 are satisfied and $x_0 \in (0,1)$, then* (10) *holds and the norms $\|u\|_2^2$ and $\int_0^1 (u'')^2(x)dx$ are equivalent.*

Proof Obviously, $X \subseteq H^2_{\frac{1}{a}}(0,1)$. Now, take $u \in H^2_{\frac{1}{a}}(0,1)$. In order to have the thesis, it is sufficient to prove that $u(x_0) = (au')(x_0) = 0$. Indeed, since $u \in H^2_0(0,1)$, then u, u', and hence au', belong to $C[0,1]$. This implies that there exists

$$\lim_{x\to x_0} u(x) = u(x_0) = L \in \mathbb{R}.$$

Clearly, $L = 0$. Indeed, if $L \neq 0$, then there exists $C > 0$ such that

$$|u(x)| \geq C,$$

for all x in a neighbourhood of x_0, $x \neq x_0$. Thus,

$$\frac{u^2(x)}{a(x)} \geq \frac{C^2}{a(x)},$$

for all x in a neighbourhood of x_0, $x \neq x_0$. Since, by hypothesis, $\frac{1}{a} \notin L^1(0,1)$, we obtain $u \notin L^2_{\frac{1}{a}}(0,1)$. Hence $L = 0$. Moreover, arguing as before, there exists

$$\lim_{x\to x_0} u'(x) = u'(x_0) = M \in \mathbb{R}$$

and hence, using the fact that $a(x_0) = 0$, one has

$$\lim_{x\to x_0} (au')(x) = 0.$$

Now, we prove that the two norms are equivalent. To this aim, take $u \in X$; thus, by Lemma 3.1 and using the fact that $u'(0) = u'(1) = 0$, one can prove that there exists a positive constant C such that

$$\|u''\|^2_{L^2(0,1)} \leq \|u\|^2_2 \leq C(\|u'\|^2_{L^2(0,1)} + \|u''\|^2_{L^2(0,1)}) \leq C\|u''\|^2_{L^2(0,1)}.$$

Hence the two norms are equivalent. □

As a consequence of the previous proposition one has that if Hypotheses 3.2 and 3.3 are satisfied and $x_0 \in (0,1)$, then the domain of the operator can be written as

$$D(A) := \left\{u \in H^2_{\frac{1}{a}}(0,1) : u(x_0) = (au')(x_0) = 0 \text{ and } au'''' \in L^2_{\frac{1}{a}}(0,1)\right\}.$$

Hence, proceeding as in the proof of Proposition 2.1, one can prove that (2) is satisfied. Indeed, in this case, in order to prove that $H^2_c(0,1)$ is dense in $H^2_{\frac{1}{a}}(0,1)$ we have to modify the definition of ξ_n in a suitable way (see also [15]). Then, one can consider $v \in H^2_c(0,1)$ and take $\delta > 0$ such that $\operatorname{supp} v \subset \mathcal{K}$, where $\mathcal{K} := [0, x_0 - \delta) \cup [x_0 + \delta, 1]$. By definition of $D(A)$, $u'''' \in L^2(\mathcal{K})$, thus $u'' \in H^2(\mathcal{K})$ (by Lemma 2.1 with $n = p = 2$ and $f = u''$) and, in particular, $u \in H^4(\mathcal{K})$. Hence, we can integrate by parts, obtaining

$$\int_0^1 u''''v\,dx = \int_0^{x_0-\delta} u''''v\,dx + \int_{x_0-\delta}^{x_0+\delta} u''''v\,dx + \int_{x_0+\delta}^1 u''''v\,dx$$
$$= \int_0^{x_0-\delta} u''v''dx + \int_{x_0-\delta}^{x_0+\delta} u''''v\,dx + \int_{x_0+\delta}^1 u''v''dx,$$

since $v \in H^2_c(0,1)$. The rest of the proof is similar to the case $x_0 \in \{0,1\}$, so we omit it. Hence, under Hypotheses 3.2 and 3.3, Theorem 2.1 still holds and problem (9) admits a unique solution u.

Now, we come back to Proposition 3.2, observing that if we know a priori that $u' \in L^2_{\frac{1}{a}}(0,1)$, then we could prove that $u'(x_0) = 0$. Indeed the next result holds.

Proposition 3.3 *Assume Hypotheses 3.2 and 3.3 and $x_0 \in (0,1)$. If $u \in H^2_{\frac{1}{a}}(0,1)$ is such that $u' \in L^2_{\frac{1}{a}}(0,1)$, then $u'(x_0) = 0$.*

Proof Let $u \in H^2_{\frac{1}{a}}(0,1)$ so that $u' \in L^2_{\frac{1}{a}}(0,1)$. In order to obtain the thesis, one can proceed as in Proposition 3.2. Indeed, as before, $u' \in C[0,1]$; this implies that there exists

$$\lim_{x \to x_0} u'(x) = u'(x_0) = M \in \mathbb{R}.$$

Clearly, $M = 0$. Indeed, if $M \neq 0$, then there exists $C > 0$ such that

$$|u'(x)| \geq C,$$

for all x in a neighbourhood of x_0, $x \neq x_0$. Thus,

$$\frac{(u')^2(x)}{a(x)} \geq \frac{C^2}{a(x)},$$

for all x in a neighbourhood of x_0, $x \neq x_0$. As in Proposition 3.2, one can conclude that $M = 0$. □

Observe that if $x_0 \in \{0,1\}$ and $u \in H^2_{\frac{1}{a}}(0,1)$, then $u'(x_0) = 0$ without additional assumptions on a or on u.

Thanks to the previous proposition, one can prove the next result. To this purpose, define

$$\widetilde{H}^2_{\frac{1}{a}}(0,1) := \left\{ u \in H^2_{\frac{1}{a}}(0,1) : u' \in L^2_{\frac{1}{a}}(0,1) \right\}$$

and

$$\widetilde{X} := \left\{ u \in H^2_{\frac{1}{a}}(0,1) : u(x_0) = u'(x_0) = 0 \right\}.$$

Proposition 3.4 *If Hypotheses 3.2 and 3.3 are satisfied, then there exists a positive constant C such that*

$$\int_0^1 \frac{(u')^2}{a} dx \le C \int_0^1 (u'')^2 dx, \tag{11}$$

for all $u \in \widetilde{X}$. *Hence*

$$\widetilde{H}^2_{\frac{1}{a}}(0,1) = \widetilde{X}. \tag{12}$$

Proof Take $u \in \widetilde{X}$ and assume, first of all, that $x_0 \in \{0, 1\}$. Then (11) follows by [8, Lemma 3.7] applied to $w := u' \in H^1_0(0,1)$.

Now, assume $x_0 \in (0,1)$ and define again $w := u' \in H^1_0(0,1)$. Clearly, $w(x_0) = 0$. By Hypothesis 3.3, there exists $C > 0$ such that $\frac{1}{a(x)} \le \frac{C}{|x - x_0|^2}$ for all $x \in [0,1] \setminus \{x_0\}$. Then, for a suitable $\varepsilon > 0$ and using the assumptions on a and Hardy's inequality, one has

$$\begin{aligned}
\int_0^1 w^2 \frac{1}{a} dx &= \int_0^{x_0-\epsilon} w^2 \frac{1}{a} dx + \int_{x_0-\epsilon}^{x_0+\epsilon} w^2 \frac{1}{a} dx + \int_{x_0+\epsilon}^1 w^2 \frac{1}{a} dx \\
&\le \frac{1}{\min_{[0,x_0-\epsilon]} a} \int_0^{x_0-\epsilon} w^2 dx + \int_{x_0-\epsilon}^{x_0+\epsilon} w^2 \frac{1}{a} dx \\
&+ \frac{1}{\min_{[x_0+\epsilon,1]} a} \int_{x_0+\epsilon}^1 w^2 dx \\
&\le \frac{1}{\min_{[0,x_0-\epsilon]\cup[x_0+\epsilon,1]} a} \int_0^1 w^2 dx + \int_{x_0-\epsilon}^{x_0} w^2 \frac{1}{a} dx + \int_{x_0}^{x_0+\epsilon} w^2 \frac{1}{a} dx \\
&\le \frac{1}{\min_{[0,x_0-\epsilon]\cup[x_0+\epsilon,1]} a} \int_0^1 w^2 dx + C \int_{x_0-\epsilon}^{x_0} w^2 \frac{1}{|x - x_0|^2} dx \\
&+ C \int_{x_0}^{x_0+\epsilon} w^2 \frac{1}{|x - x_0|^2} dx \\
&\le \frac{1}{\min_{[0,x_0-\epsilon]\cup[x_0+\epsilon,1]} a} \int_0^1 w^2 dx + C_H \int_{x_0-\epsilon}^{x_0} (w')^2 dx \\
&+ C_H \int_{x_0}^{x_0+\epsilon} (w')^2 dx \\
&\le C \left(\int_0^1 w^2 dx + \int_0^1 (w')^2 dx \right),
\end{aligned}$$

for a positive constant C; here C_H is the Hardy constant. By Poincaré's inequality, it follows that

$$\int_0^1 w^2 \frac{1}{a} dx \leq C \int_0^1 (w')^2 dx$$

for a suitable constant C. Hence (11) holds. This implies that $\widetilde{X} \subseteq \widetilde{H}^2_{\frac{1}{a}}(0,1)$. By Proposition 3.3 the other inclusion follows immediately, hence (12) holds. □

Remark 3.1 Clearly, if Hypotheses 3.2 and 3.3 are satisfied and $u \in H^2_{\frac{1}{a}}(0,1)$ satisfies (11), then $u' \in L^2_{\frac{1}{a}}(0,1)$ and again $u'(x_0) = 0$ by Proposition 3.4.

Adding an additional assumption on the function a, one can prove a characterization on $D(A)$.

Hypothesis 3.4 Assume that the function a belongs to $W^{1,\infty}(0,1)$ and is

$$\begin{cases} \text{non increasing on the left and non decreasing on the right of } x_0, & \text{if } x_0 \in (0,1), \\ \text{non decreasing on the right of } x_0, & \text{if } x_0 = 0, \\ \text{non increasing on the left of } x_0, & \text{if } x_0 = 1. \end{cases}$$

Proposition 3.5 *Let*

$$D := \left\{ u \in H^2_{\frac{1}{a}}(0,1) : au'''' \in L^2_{\frac{1}{a}}(0,1),\ u(x_0) = (au')(x_0) = (au'')(x_0) = 0 \right\}.$$

If Hypotheses 3.2, 3.3 and 3.4 are satisfied, then $D(A) = D$.

Proof Evidently $D \subseteq D(A)$. Now, we take $u \in D(A)$ and we prove that $u \in D$. By Proposition 3.2, $u(x_0) = (au')(x_0) = 0$. Thus, it is sufficient to prove that $(au'')(x_0) = 0$ (observe that, since $u \in H^2(0,1)$ and $a \in W^{1,\infty}(0,1)$, then $u'' \in L^2(0,1)$, $\sqrt{a}u'' \in L^2(0,1)$ and $au'' \in L^2(0,1)$).

Let $\delta > 0$ sufficiently small and $x = x_0 - \delta$. From the proof of Theorem 2.1 it follows that $u \in H^4(\mathcal{K})$, where $\mathcal{K} := [0,x]$. In particular, since au'' is a continuous function in $\mathcal{K}$, the following formula holds:

$$(au'')(x) - (au'')(0) = \int_0^x (au'')'(t)dt = \int_0^x (a'u'')(t)dt + \int_0^x (au''')(t)dt. \tag{13}$$

Now we will estimate the last two terms in (13). To this aim, observe that

$$\int_0^x (a'u'')(t)dt = \int_0^{x_0} (a'u'')(t)dt - \int_x^{x_0} (a'u'')(t)dt$$

and

$$\begin{aligned}\int_0^x (au''')(t)dt &= \int_0^x a(t)\left(\int_0^t u''''(s)ds + u'''(0)\right)dt \\ &= \int_0^x a(t)\int_0^t u''''(s)dsdt + \int_0^x a(t)u'''(0)dt.\end{aligned} \tag{14}$$

In particular, we have

$$\lim_{x\to x_0}\int_0^x (a'u'')(t)dt = \int_0^{x_0}(a'u'')(t)dt.$$

Indeed, Hypothesis 3.2 and $u'' \in L^2(\mathcal{K}) \subset L^1(\mathcal{K})$ imply that

$$\left|\int_x^{x_0}(a'u'')(t)dt\right| \le \int_x^{x_0}|(a'u'')(t)|dt \le \|a'\|_{L^\infty(0,1)}\int_x^{x_0}|u''(t)|dt.$$

Hence

$$\lim_{x\to x_0}\int_x^{x_0}(a'u'')(t)dt = 0$$

by the absolute continuity of the Lebesgue integral. As far as the two terms in (14) are concerned, we have

$$\lim_{x\to x_0}\int_0^x a(t)u'''(0)dt = u'''(0)\lim_{x\to x_0}\int_0^x a(t)dt = u'''(0)\int_0^{x_0}a(t)dt,$$

where we recall that $\int_0^x a(t)\,dt$ can be written as $\int_0^{x_0} a(t)\,dt - \int_x^{x_0} a(t)\,dt$ and, thanks to the hypotheses on a, $\lim_{x\to x_0}\int_x^{x_0} a(t)\,dt = 0$. Moreover, using the monotonicity condition of a on the left of x_0, one has

$$\begin{aligned}\left|a(t)\int_0^t u''''(s)ds\right| &= \left|\sqrt{a(t)}\int_0^t \sqrt{a(t)}u''''(s)ds\right| \\ &\le \sqrt{a(t)}\int_0^t \sqrt{a(s)}|u''''(s)|ds\end{aligned}$$

for all $t \in (0,x)$. Using again the assumptions on a and the fact that $au'''' \in L^2_{\frac{1}{a}}(0,1)$, one can conclude that

$$f(t) := a(t)\int_0^t u''''(s)ds \in L^1(0,x_0).$$

Arguing as before,

$$\lim_{x\to x_0}\int_0^x f(t)dt = \int_0^{x_0} f(t)dt$$

and hence

$$\exists \lim_{x\to x_0^-}(au'')(x) = (au'')(x_0^-) = L \in \mathbb{R}.$$

In a similar way, one can prove that

$$\exists \lim_{x\to x_0^+}(au'')(x) = (au'')(x_0^+) = M \in \mathbb{R}.$$

In order to complete the proof it remains to prove that $L = M = 0$. Indeed, if $L \neq 0$, then there exist $C > 0$ and a left neighbourhood I_- of x_0 such that

$$a(x)|u''(x)| \geq C \qquad \forall x \in I_-.$$

Thus,

$$|u''(x)| \geq \frac{C}{a(x)} \qquad \forall x \in I_-.$$

But $\frac{1}{a} \notin L^1(0,1)$, thus $u'' \notin L^1(0,1)$ and this contradicts the fact that $u'' \in L^2(0,1) \subset L^1(0,1)$. Hence $L = 0$; analogously one can prove $M = 0$. Thus, we obtain $L = M = 0$. In this way $\lim_{x\to x_0}(au'')(x)$ exists and it is equal to 0, i.e., $(au'')(x_0) = 0$. □

4 The General Operator of Order 2*n*

In this section we will extend the previous results to a general operator $A_n u := au^{(2n)}$. To this aim, taking $n \in \mathbb{N}$ with $n \geq 3$ (the case $n = 2$ is considered in the previous sections), we introduce the following spaces

$$H^i_{\frac{1}{a}}(0,1) := L^2_{\frac{1}{a}}(0,1) \cap H^i_0(0,1),$$

with the norm

$$\|u\|^2_{H^i_{\frac{1}{a}}(0,1)} := \|u\|^2_{L^2_{\frac{1}{a}}(0,1)} + \sum_{k=1}^{i} \|u^{(k)}\|^2_{L^2(0,1)} \qquad \forall\, u \in H^i_{\frac{1}{a}}(0,1),$$

$i = 3, \ldots, n$.

As before, $H^i_0(0,1) := \{u \in H^i(0,1) : u^{(k)}(j) = 0,\ j = 0,1,\ k = 0,1,\ldots,i-1\}$, $i = 3,\ldots,n$, and for all $u \in H^i_{\frac{1}{a}}(0,1)$ the norms $\|u\|^2_{H^i_{\frac{1}{a}}(0,1)}$ and

$$\|u\|_i^2 := \|u\|^2_{L^2_{\frac{1}{a}}(0,1)} + \|u^{(i)}\|^2_{L^2(0,1)}$$

are equivalent; hence in the following we will use $\|\cdot\|_i$.

Define the operator A_n by

$$A_n u := a u^{(2n)}, \quad \forall\, u \in D(A_n) := \left\{u \in H^n_{\frac{1}{a}}(0,1) : a u^{(2n)} \in L^2_{\frac{1}{a}}(0,1)\right\},$$

if $x_0 \in \{0,1\}$, or in the strongly degenerate case when $x_0 \in (0,1)$. The next general Green's formula holds.

Proposition 4.1 (Green's Formula) *Assume Hypothesis 2.1 if $x_0 \in \{0,1\}$ or Hypotheses 3.2 and 3.3 if $x_0 \in (0,1)$. Then, for all $n \in \mathbb{N}$ with $n \geq 3$ and for all $(u,v) \in D(A_n) \times H^n_{\frac{1}{a}}(0,1)$ one has*

$$\int_0^1 u^{(2n)} v\, dx = (-1)^n \int_0^1 u^{(n)} v^{(n)} dx.$$

Proof The proof is similar to the one of Proposition 2.1, so we sketch it. Actually, it is sufficient to prove that, fixed $u \in D(A_n)$ and defined $\Phi(v) := \int_0^1 u^{(2n)} v\, dx + (-1)^{n+1} \int_0^1 u^{(n)} v^{(n)} dx$, with $v \in H^n_{\frac{1}{a}}(0,1)$, it follows that

$$\Phi(v) = 0 \tag{15}$$

for all $v \in H^n_{\frac{1}{a}}(0,1)$. As in Proposition 2.1, it is sufficient to prove (15) for all $v \in H^n_c(0,1) := \{v \in H^n(0,1) : \operatorname{supp} v \subset (0,1) \setminus \{x_0\}\}$. The thesis will follow using the density of $H^n_c(0,1)$ in $H^n_{\frac{1}{a}}(0,1)$, which can be proved as in Proposition 2.1.

Indeed, assume $x_0 \in (0,1)$, let $\delta > 0$ and set $\mathcal{K} := [0, x_0 - \delta] \cup [x_0 + \delta, 1]$. By definition of $D(A_n)$, $u^{(2n)} \in L^2(\mathcal{K})$, thus $u^{(n)} \in H^n(\mathcal{K})$ (by Lemma 2.1 with $p = 2$) and, in particular, $u \in H^{2n}(\mathcal{K})$. Observe that, fixed $v \in H^n_c(0,1)$, one can easily prove that

$$\int_{\mathcal{K}} u^{(2n)} v\, dx = (-1)^n \int_{\mathcal{K}} u^{(n)} v^{(n)} dx.$$

Hence

$$\int_0^1 u^{(2n)} v\,dx = \int_0^{x_0-\delta} u^{(2n)} v\,dx + \int_{x_0-\delta}^{x_0+\delta} u^{(2n)} v\,dx + \int_{x_0+\delta}^1 u^{(2n)} v\,dx$$
$$= (-1)^n \int_0^{x_0-\delta} u^{(n)} v^{(n)} dx + \int_{x_0-\delta}^{x_0+\delta} u^{(2n)} v\,dx + (-1)^n \int_{x_0+\delta}^1 u^{(n)} v^{(n)} dx. \tag{16}$$

As in (6) and in (7), one can prove

$$\lim_{\delta\to 0} \int_0^{x_0-\delta} u^{(n)} v^{(n)} dx = \int_0^{x_0} u^{(n)} v^{(n)} dx,$$

$$\lim_{\delta\to 0} \int_{x_0+\delta}^1 u^{(n)} v^{(n)} dx = \int_{x_0}^1 u^{(n)} v^{(n)} dx$$

and

$$\lim_{\delta\to 0} \int_{x_0-\delta}^{x_0+\delta} u^{(2n)} v\,dx = 0.$$

Thus, by the previous limits and (16), it follows

$$\int_0^1 u^{(2n)} v\,dx = (-1)^n \int_0^1 u^{(n)} v^{(n)} dx \iff$$
$$\Phi(v) = \int_0^1 (u^{(2n)} v + (-1)^{n+1} u^{(n)} v^{(n)}) dx = 0,$$

for all $v \in H^n_c(0,1)$. For the rest of the proof, one can proceed as in Proposition 2.1. The case $x_0 \in \{0,1\}$ is similar, so we omit it. □

In the weakly degenerate case, if $x_0 \in (0,1)$, we consider the same domain as before

$$A_n u := a u^{(2n)}, \ \forall\, u \in D(A_n) := \left\{ u \in H^n_{\frac{1}{a}}(0,1) : a u^{(2n)} \in L^2_{\frac{1}{a}}(0,1) \right\},$$

and we underline that, as for the case $n = 2$, if $a u^{(2n)} \in L^2_{\frac{1}{a}}(0,1)$, then $u \in W^{2n,1}(0,1)$. Thus, the Proposition 4.1 follows immediately.

A natural consequence of the previous proposition is collected in the next theorem, which establishes the generation property in the general case.

Theorem 4.1 *If $x_0 \in \{0,1\}$, assume Hypothesis 2.1; if $x_0 \in (0,1)$ assume Hypothesis 3.1 or Hypotheses 3.2 and 3.3. Then, the operator $\tilde{A}_n : D(A_n) \to$*

$L^2_{\frac{1}{a}}(0,1)$, defined by $\tilde{A}_n u := (-1)^n a u^{(2n)}$ for all $u \in D(A_n)$, is self-adjoint and non negative on $L^2_{\frac{1}{a}}(0,1)$. Hence $-\tilde{A}_n$ generates a contractive analytic semigroup of angle $\frac{\pi}{2}$ on $L^2_{\frac{1}{a}}(0,1)$ for any $n \in \mathbb{N}$, $n \geq 3$.

Proof Observe that $D(A_n)$ is dense in $L^2_{\frac{1}{a}}(0,1)$. It is also clear that $\tilde{A}_n$ is symmetric and non negative. Indeed,

$\tilde{A}_n$ is symmetric: for any $u, v \in D(A_n)$

$$\langle \tilde{A}_n u, v\rangle_{L^2_{\frac{1}{a}}(0,1)} = \int_0^1 \frac{(-1)^n a u^{(2n)} v}{a} dx = \int_0^1 u^{(n)} v^{(n)} dx = \int_0^1 \frac{(-1)^n a v^{(2n)} u}{a} dx$$
$$= \langle u, \tilde{A}_n v\rangle_{L^2_{\frac{1}{a}}(0,1)}.$$

$\tilde{A}_n$ is non negative: for any $u \in D(A_n)$

$$\langle \tilde{A}_n u, u\rangle_{L^2_{\frac{1}{a}}(0,1)} = \int_0^1 \frac{(-1)^n a u^{(2n)} u}{a} dx = \int_0^1 u^{(n)} u^{(n)} dx = \int_0^1 (u^{(n)})^2 dx \geq 0.$$

$I + \tilde{A}_n$ is surjective: as in Theorem 2.1, one can prove that for $f \in L^2_{\frac{1}{a}}(0,1)$ there exists a unique $u \in H^n_{\frac{1}{a}}(0,1)$ such that for all $v \in H^n_{\frac{1}{a}}(0,1)$:

$$\langle u, v\rangle_{H^n_{\frac{1}{a}}(0,1)} = \int_0^1 \frac{fv}{a} dx, \tag{17}$$

where

$$\langle u, v\rangle_{H^n_{\frac{1}{a}}(0,1)} := \int_0^1 \left(\frac{uv}{a} + u^{(n)} v^{(n)}\right) dx \qquad \forall\, u, v \in H^n_{\frac{1}{a}}(0,1).$$

More precisely, since $C_c^\infty(0,1) \subset H^n_{\frac{1}{a}}(0,1)$, the relation (17) holds for all $v \in C_c^\infty(0,1)$, i.e.,

$$\int_0^1 u^{(n)} v^{(n)} dx = \int_0^1 \frac{f-u}{a} v\, dx \qquad \forall\, v \in C_c^\infty(0,1).$$

However, the previous equality is equivalent to

$$(-1)^n \int_0^1 u^{(2n)} v\, dx = \int_0^1 \frac{f-u}{a} v\, dx \Longleftrightarrow \int_0^1 \frac{(-1)^n a u^{(2n)} - f + u}{a} v\, dx = 0,$$

for all $v \in C_c^\infty(0,1)$. Evidently,

$$u \in D(A_n) \quad \text{and} \quad u + (-1)^n a u^{(2n)} = f,$$

i.e., $(I + \tilde{A}_n)(u) = f$. As a consequence, the operator $(\tilde{A}_n, D(A_n))$ generates a cosine family and an analytic semigroup of angle $\frac{\pi}{2}$ on $L^2_{\frac{1}{a}}(0,1)$. □

As in Sect. 2, fixed a natural number $n \geq 3$, $T > 0$, $u_0 \in L^2_{\frac{1}{a}}(0,1)$ and $f \in L^2(0,T; L^2_{\frac{1}{a}}(0,1))$, we have that the following problem

$$\begin{cases} \dfrac{\partial u}{\partial t}(t,x) + a(x)\dfrac{\partial^{2n} u}{\partial x^{2n}}(t,x) = f(t,x), & (t,x) \in (0,T) \times (0,1), \\ u(t,0) = u(t,1) = 0, & t \in (0,T), \\ \dfrac{\partial^i u}{\partial x^i}(t,0) = \dfrac{\partial^i u}{\partial x^i}(t,1) = 0, & t \in (0,T), i = 1, \ldots, n-1, \\ u(0,x) = u_0(x), & x \in (0,1), \end{cases}$$

has a unique solution as a consequence of the previous theorem.

Finally, fixed $n \in \mathbb{N}$, $n \geq 3$, it is possible to extend the results of Sect. 3 in a natural way. More precisely, considering the same assumptions on the function a, we have the next general proposition which is useful to characterize $D(A_n)$ in the weakly degenerate setting.

Proposition 4.2 *Fix $n \in \mathbb{N}$, $n \geq 3$, and assume Hypothesis 3.1. Then, we have the following properties*

1. *the spaces $H^n_{\frac{1}{a}}(0,1)$ and $H^n_0(0,1)$ coincide algebraically;*
2. *the norms $\|\cdot\|_i$ and $\|\cdot\|_{H^i_0(0,1)}$, $i = 1, \ldots, n$, are equivalent;*
3. *if $u \in H^n_0(0,1)$, i.e., $u \in H^n_{\frac{1}{a}}(0,1)$, then*

$$(au^{(i)})(x_0) = 0, \qquad i = 0, 1, \ldots, n-1.$$

Proof One can prove the first two points as in Proposition 3.1; thus we omit it. We will prove only the last point. To this purpose, we take $u \in H^n_0(0,1)$. Hence $u, u', u'', \ldots u^{(n-1)} \in C[0,1]$ and, using the fact that $a(x_0) = 0$, one has $(au^{(i)})(x_0) = 0$ for all $i = 0, 1, \ldots, n-1$. Hence, the thesis follows. □

We underline that the previous proposition holds if $x_0 \in [0,1]$.

Consequently, the space $C_c^\infty(0,1)$ is dense in $H^i_{\frac{1}{a}}(0,1)$, $i \geq 3$, and, in particular, $D(A_n)$ can be rewritten as

$$D(A_n) = \left\{u \in H^n_0(0,1) : (au^{(i)})(x_0) = 0,\ i = 0, 1, \ldots, n-1,\ au^{(2n)} \in L^2_{\frac{1}{a}}(0,1)\right\},$$

$n \geq 3$.

The discussion in the strongly degenerate case is based on the introduction of the space

$$X_n := \left\{u \in H^n_{\frac{1}{a}}(0,1) : u(x_0) = 0,\ (au^{(i)})(x_0) = 0,\ i = 1, \ldots, n-1\right\}, \quad n \geq 3.$$

Observe that X_2 coincides with the space X introduced in the previous section.

Naturally, if $x_0 \in \{0,1\}$, then

$$X_n = H^n_{\frac{1}{a}}(0,1). \tag{18}$$

The following result is crucial to characterize the domain of the operator in the strongly degenerate case.

Proposition 4.3 *Fix $n \in \mathbb{N}$, $n \geq 3$. Assume Hypotheses 3.2 and 3.3 and $x_0 \in (0,1)$. Then* (18) *holds and the norms $\|u\|_n^2$ and $\int_0^1 (u^{(n)})^2 dx$ are equivalent for all $u \in H^n_{\frac{1}{a}}(0,1)$.*

Proof Obviously, $X_n \subseteq H^n_{\frac{1}{a}}(0,1)$. Now, take $u \in H^n_{\frac{1}{a}}(0,1)$. By Proposition 3.2, it holds $u(x_0) = (au')(x_0) = 0$. Thus, in order to have the thesis, it is sufficient to prove that $(au'')(x_0) = \ldots = (au^{(n-1)})(x_0) = 0$. To this purpose, since $u \in H^n_0(0,1)$, then u'',, $u^{(n-1)}$ and hence au'',..., $au^{(n-1)}$ belong to $C[0,1]$. Hence, using the fact that $a(x_0) = 0$, one has

$$\lim_{x \to x_0} (au^{(i)})(x) = 0, \quad i = 2, \ldots, n-1.$$

Now, we prove that the two norms are equivalent. To this purpose, take $u \in X_n$. By Lemma 3.1, there exists a positive constant C such that

$$\|u^{(n)}\|^2_{L^2(0,1)} \leq \|u\|_n^2 \leq C\left(\|u'\|^2_{L^2(0,1)} + \|u^{(n)}\|^2_{L^2(0,1)}\right). \tag{19}$$

Using Jensen's inequality and the fact that $u^{(i)}(j) = 0$, for all $i = 1, \dots, n-1$ and $j = 0, 1$, one can prove that

$$\int_0^1 (u^{(i)})^2(x)dx \le \int_0^1 (u^{(i+1)})^2(x)dx, \quad i = 1, \dots, n-1.$$

Thus, by the previous inequality in (19), one can conclude that the two norms are equivalent. □

Also in this context we underline that, if we know a priori that $u'', \dots, u^{(n-1)} \in L^2_{\frac{1}{a}}(0,1)$, using the same strategy employed in the previous section, we could prove that $u''(x_0) = \dots = u^{(n-1)}(x_0) = 0$. Hence, the next characterizations hold.

Define

$$\widetilde{H}^n_{\frac{1}{a}}(0,1) := \left\{u \in H^n_{\frac{1}{a}}(0,1) : u^{(i)} \in L^2_{\frac{1}{a}}(0,1) \quad \forall\, i = 1, \dots, n-1\right\}$$

and

$$\widetilde{X}_n := \left\{u \in H^n_{\frac{1}{a}}(0,1) : u^{(i)}(x_0) = 0 \quad \forall\, i = 0, \dots, n-1\right\}.$$

Proposition 4.4 *Fix $n \in \mathbb{N}$, $n \ge 3$, and assume Hypotheses 3.2 and 3.3. Then there exists a positive constant C such that, for all $u \in \widetilde{X}_n$,*

$$\int_0^1 \frac{(u^{(i)})^2}{a} dx \le C \int_0^1 (u^{(i+1)})^2 dx \tag{20}$$

for all $i = 2, \dots, n-1$. Hence

$$\widetilde{H}^n_{\frac{1}{a}}(0,1) = \widetilde{X}_n.$$

Remark 4.1 If Hypotheses 3.2 and 3.3 are satisfied and $u \in H^n_{\frac{1}{a}}(0,1)$ satisfies the relations (20), then $u'', \dots, u^{(n-1)} \in L^2_{\frac{1}{a}}(0,1)$ and again $u''(x_0) = \dots = u^{(n-1)}(x_0) = 0$ by the previous proposition. Clearly, if $u' \in L^2_{\frac{1}{a}}(0,1)$, then $u'(x_0) = 0$ by Remark 3.1.

Proposition 4.5 *Fix $n \in \mathbb{N}$, $n \ge 3$, and let*

$$D_n := \left\{u \in H^n_{\frac{1}{a}}(0,1) : au^{(2n)} \in L^2_{\frac{1}{a}}(0,1), u(x_0) = 0, \right.$$
$$\left. (au^{(i)})(x_0) = 0 \quad \forall\, i = 1, \dots, n\right\}.$$

If Hypotheses 3.2, 3.3 and 3.4 are satisfied, then $D(A_n) = D_n$.

Acknowledgments Alessandro Camasta is a member of the *Gruppo Nazionale per l'Analisi Matematica, la Probabilità e le loro Applicazioni (GNAMPA)* of the Istituto Nazionale di Alta Matematica (INdAM) and a member of UMI "Modellistica Socio-Epidemiologica (MSE)". He is partially supported by PRIN 2017–2019 *Qualitative and quantitative aspects of nonlinear PDEs.*

Genni Fragnelli is a member of the *Gruppo Nazionale per l'Analisi Matematica, la Probabilità e le loro Applicazioni (GNAMPA)* of the Istituto Nazionale di Alta Matematica (INdAM) and a member of *UMI "Modellistica Socio-Epidemiologica (MSE)"*. She is partially supported by the FFABR *Fondo per il finanziamento delle attività base di ricerca* 2017 and by PRIN 2017–2019 *Qualitative and quantitative aspects of nonlinear PDEs.*

References

1. Alabau-Boussouira, F., Cannarsa, P., Fragnelli, G.: Carleman estimates for degenerate parabolic operators with application to null controllability. J. Evol. Equ. **6**, 161–204 (2006)
2. Aristotelous, A.C., Karakashian, O.A., Wise, S.M.: Adaptive, second-order in time, primitive-variable discontinuous Galerkin schemes for a Cahn-Hilliard equation with a mass source. IMA J. Numer. Anal. **35**, 1167–1198 (2015)
3. Beretta, E., Bertsch, M., Dal Passo, R.: Nonnegative solutions of a fourth-order nonlinear degenerate parabolic equation. Arch. Ration. Mech. Anal. **129**, 175–200 (1995)
4. Bernis, F., Friedman, A.: Higher order nonlinear degenerate parabolic equations. J. Differ. Equ. **83**, 179–206 (1990)
5. Bertozzi, A., Esedoglu, S., Gillette, A.: Inpainting of binary images using the Cahn-Hilliard equation. IEEE Trans. Image Process. **16**, 285–291 (2007)
6. Bertozzi, A.L., Pugh, M.: The lubrication approximation for thin viscous films: regularity and long-time behavior of weak solutions. Commun. Pure Appl. Math. **49**, 85–123 (1996)
7. Cannarsa, P., Fragnelli, G., Rocchetti, D.: Null controllability of degenerate parabolic operators with drift. Netw. Heterog. Media **2**, 693–713 (2007)
8. Cannarsa, P., Fragnelli, G., Rocchetti, D.: Controllability results for a class of one-dimensional degenerate parabolic problems in nondivergence form. J. Evol. Equ. **8**, 583–616 (2008)
9. Chalupecki, V.: Numerical studies of Cahn-Hilliard equations and applications in image processing. In: Proceedings of Czech-Japanese Seminar in Applied Mathematics, August 4–7, 2004, Czech Technical University in Prague, 2004
10. Cohen, D., Murray, J.M.: A generalized diffusion model for growth and dispersion in a population. J. Math. Biol. **12**, 237–248 (1981)
11. Dal Passo, R., Garcke, H., Grün, G.: On a fourth-order degenerate parabolic equation: global entropy estimates, existence, and qualitative behavior of solutions. SIAM J. Math. Anal. **29**, 321–342 (1998)
12. Dolcetta, I.C., Vita, S.F.: Area-preserving curve-shortening flows: from phase separation to image processing. Inter-faces Free Bound. **4**, 325–343 (2002)
13. Elliott, C.M., Garcke, H.: On the Cahn–Hilliard equation with degenerate mobility. SIAM J. Math. Anal. **27**, 404–423 (1996)
14. Erlebacher, J., Aziz, M.J., Karma, A., Dimitrov, N., Sieradzki, K.: Evolution of nanoporosity in dealloying. Nature **410**, 450–453 (2001)
15. Fragnelli, G., Goldstein, G.R., Goldstein, J.A., Romanelli, S.: Generators with interior degeneracy on spaces of L^2 type. Electron. J. Differ. Equ. **2012**, 1–30 (2012)
16. Fragnelli, G., Goldstein, J.A., Mininni, R.M., Romanelli, S.: Operators of order $2n$ with interior degeneracy. Discrete Contin. Dyn. Syst.-S **13**, 3417–3426 (2020)
17. Fragnelli, G., Mugnai, D.: Carleman estimates and observability inequalities for parabolic equations with interior degeneracy. Adv. Nonlinear Anal. **2**, 339–378 (2013)
18. Fragnelli, G., Mugnai, D.: Control of Degenerate and Singular Parabolic Equations. In: Carleman Estimates and Observability. SpringerBriefs in Mathematics. Springer International Publishing, Cham (2021)

19. Grün, G.: Degenerate parabolic differential equations of fourth order and a plasticity model with non-local hardening. Z. Anal. Anwend. **14**, 541–574 (1995)
20. Guo, B., Gao, W.: Study of weak solutions for a fourth-order parabolic equation with variable exponent of nonlinearity. Z. Angew. Math. Phys. **62**, 909–926 (2011)
21. Khain, E., Sander, L.M.: A generalized Cahn-Hilliard equation for biological applications. Phys. Rev. E **77**, 051129 (2008)
22. King, B.B., Stein, O., Winkler, M.: A fourth-order parabolic equation modeling epitaxial thin film growth. J. Math. Anal. Appl. **286**, 459–490 (2003)
23. Klapper, I., Dockery, J.: Role of cohesion in the material description of biofilms. Phys. Rev. E **74**, 0319021 (2006)
24. Li, B., Liu, J.G.: Thin film epitaxy with or without slope selection. Eur. J. Appl. Math. **14**, 713–743 (2003)
25. Liang, B., Zheng, S.: Existence and asymptotic behavior of solutions to a nonlinear parabolic equation of fourth order. J. Math. Anal. Appl. **348**, 234–243 (2008)
26. Liu, Q.-X., Doelman, A., Rottschäfer, V., de Jager, M., Herman, P.M.J., Rietkerk, M., van de Koppel, J.: Phase separation explains a new class of self-organized spatial patterns in ecological systems. Proc. Natl. Acad. Sci. **110**, 11905–11910 (2013)
27. Oron, A., Davis, S.H., Bankoff, S.G.: Long-scale evolution of thin liquid films. Rev. Mod. Phys. **69**, 931–980 (1997)
28. Tremaine, S.: On the origin of irregular structure in Saturn's rings. Astron. J. **125**, 894–901 (2003)
29. Verdasca, J., Borckmans, P., Dewel, G.: Chemically frozen phase separation in an adsorbed layer. Phys. Rev. E **52**, 4616–4619 (1995)

Korovkin Approximation of Set-Valued Integrable Functions

Michele Campiti

This paper is dedicated to Professor Francesco Altomare on the occasion of the meeting in his honor held in Bari in September 23–24, 2021

Abstract In this paper we consider the convergence of monotone linear operators defined on a cone of integrable set-valued functions with compact and convex values and we establish some Korovkin-type result for these operators. Some applications are given concerning in particular the sequences Kantorovich and Bernstein-Durrmeyer type operators in a set-valued setting.

Keywords Korovkin-type theorems · Kantorovich operators · Bernstein-Durrmeyer operators · Set-valued operators · Integral of set-valued functions

1 Introduction and Notation

This paper is devoted to the study of Korovkin-type approximation in cones of integrable set-valued functions. It is also considered the extension of some classical approximation processes in L^1 to the setting of set-valued integrable function and it is obtained the convergence of these new sequences.

Although Korovkin approximation of set-valued functions has been deeply investigated in the last decades, until now the setting of cones of integrable set-valued functions has not been yet considered.

Indeed, the first result on the Korovkin approximation dates back to the paper [13] in 1988, where some classical Korovkin-type theorems (see e.g. [1, 2]) have been extended to the setting of cones of Hausdorff continuous functions.

M. Campiti (✉)
Dipartimento di Matematica e Fisica "E. De Giorgi", Università del Salento, Lecce, Italy
e-mail: michele.campiti@unisalento.it

A. M. Candela et al. (eds.), *Recent Advances in Mathematical Analysis*, Trends in Mathematics, https://doi.org/10.1007/978-3-031-20021-2_12

After the paper by Keimel and Roth, many Korovkin-type results have been obtained in cones of set-valued Hausdorff continuous functions. An extension of some characterizations for single-valued continuous functions has been obtained in [4], while in [5, 6] it has been introduced a particular class of operators, namely the so-called convexity monotone operators, in order to obtain more deep results (see also [7–9]). These operators have been considered more recently [10] even in the case where the limit operator has not been assigned. Finally, the analysis of Korovkin approximation in cones of set-valued functions has been continued through the papers [11, 12] but again in the Hausdorff continuous set-valued setting.

On the contrary, in this paper we consider cones of integrable set-valued functions and obtain the existence of Korovkin systems which may include integrable set-valued functions which are not Hausdorff continuous.

In the last section, we give some applications to classical sequences of Kantorovich and Bernstein-Durrmeyer type operators in the setting of integrable set-valued functions.

We begin by introducing some notation.

We shall denote by $\mathcal{K}(\mathbb{R}^d)$ the cone of all non-empty compact convex subsets of $\mathbb{R}^d$ endowed with the natural addition and multiplication by positive scalars; in $\mathcal{K}(\mathbb{R}^d)$ it is defined the Hausdorff distance

$$d_H(A, B) := \max\left\{\sup_{x\in A}\inf_{y\in B}\|x-y\|, \sup_{y\in B}\inf_{x\in A}\|x-y\|\right\}.$$

If $a, b \in \mathbb{R}$ satisfy $a < b$, we denote by $\mathcal{F}([a, b], \mathbb{R}^d)$ the space of all functions defined on $[a, b]$ with values in $\mathbb{R}^d$ and by $\mathcal{F}([a, b], \mathcal{K}(\mathbb{R}^d))$ the cone of all set-valued functions defined on the interval $[a, b]$ with values in $\mathcal{K}(\mathbb{R}^d)$.

A function $F \in \mathcal{F}([a, b], \mathcal{K}(\mathbb{R}^d))$ is Hausdorff continuous if it is continuous with respect to the Hausdorff distance on $\mathcal{K}(\mathbb{R}^d)$ at each $x_0 \in [a, b]$, i.e., for every $\varepsilon > 0$ there exists $\delta > 0$ such that $d_H(F(x), F(x_0)) < \varepsilon$ (or equivalently $F(x) \subset F(x_0)+\varepsilon\cdot\mathbf{B}$ and $F(x_0) \subset F(x)+\varepsilon\cdot\mathbf{B}$) for every $x \in [a, b]$ satisfying $|x-x_0| < \delta$.

Here, we have denoted by **B** the closed unit ball with center 0 in $\mathbb{R}^d$.

The cone of all Hausdorff continuous set-valued functions on $[a, b]$ will be denoted by $C([a, b], \mathcal{K}(\mathbb{R}^d))$.

We recall that if $F \in \mathcal{F}([a, b], \mathcal{K}(\mathbb{R}^d))$, a selection of F is a Lebesgue measurable function $\varphi : [a, b] \to \mathbb{R}^q$ such that $\varphi(x) \in F(x)$ almost everywhere in $[a, b]$.

We shall denote by $\mathrm{Sel}(F)$ the convex subset of $\mathcal{F}([a, b], \mathbb{R}^d)$ consisting of all selections of F.

Then the classical Aumann integral of F is defined as follows

$$(A)\int_a^b F(x)\,dx := \left\{\int_a^b \varphi \mid \varphi \in \mathrm{Sel}(F)\right\}.$$

Observe that the Aumann integral is defined for every set-valued function and may be empty.

Many properties and applications of Aumann integrals have been deeply studied together with different extensions in more abstract settings.

However, for our purposes we need only to use some classical properties of Aumann integrals also in connection with the Riemann integrability of set-valued functions.

In order to point out these properties, we recall the definition of Riemann integral of a set-valued function.

For a subdivision $\Delta := \{x_0, \ldots, x_m\}$ of $[a, b]$ ($x_0 = a$, $x_m = b$ and $x_0 < x_1 < \cdots < x_m$), we shall denote by $|\Delta| := \max_{i=0,\ldots,m-1}(x_{i+1} - x_i)$ its diameter.

Then, a bounded set-valued function $F : [a, b] \to \mathcal{K}(\mathbb{R}^d)$ is Riemann integrable if there exists a non-empty subset I of $\mathbb{R}^d$ satisfying the following property:

(R) For every $\varepsilon > 0$ there exists $\delta > 0$ such that

$$d_H\left(I, \sum_{i=1}^{m}(x_i - x_{i-1})F(t_i)\right) < \varepsilon$$

whenever $\Delta := \{x_0, \ldots, x_m\}$ is a subdivision of $[a, b]$ with $|\Delta| < \delta$ and $t_i \in [x_{i-1}, x_i]$ for every $i = 1, \ldots, m$.

In this case the subset I is called the Riemann integral of F and is denoted by $(R)\int_a^b F(x)\,dx$.

We shall denote by $\mathcal{R}([a, b], \mathcal{K}(\mathbb{R}^d))$ the cone of all bounded Riemann integrable set-valued functions.

It is well-known that a bounded set-valued function $F : [a, b] \to \mathcal{K}(\mathbb{R}^d)$ is Riemann integrable if and only if it is a.e. continuous [14, Theorem 1].

In particular, we have $C([a, b], \mathcal{K}(\mathbb{R}^d)) \subset \mathcal{R}([a, b], \mathcal{K}(\mathbb{R}^d))$.

Since we are considering set-valued functions with values in $\mathcal{K}(\mathbb{R}^d)$, for such set-valued functions there is no distinction between Riemann and Aumann integrals; hence we shall call them simply the integral of F and we shall denote both by $\int_a^b F(x)\,dx$ whenever $F \in \mathcal{R}([a, b], \mathcal{K}(\mathbb{R}^d))$.

From the classical properties of the Aumann integral (see [3, Theorems 2 and 4] and [15]) we have that $\int_a^b F(x)\,dx \in \mathcal{K}(\mathbb{R}^d)$ for every $F \in \mathcal{R}([a, b], \mathcal{K}(\mathbb{R}^d))$.

The convergence in the space $\mathcal{R}([a, b], \mathcal{K}(\mathbb{R}^d))$ is intended with respect to the Hausdorff distance of the integrals, i.e., a net $(F_i)_{i\in I}^{\leq}$ in $\mathcal{R}([a, b], \mathcal{K}(\mathbb{R}^d))$ converges to $F \in \mathcal{R}([a, b], \mathcal{K}(\mathbb{R}^d))$ if

$$\lim_{i\in I^{\leq}} d_H\left(\int_a^b F_i(x)\,dx, \int_a^b F(x)\,dx\right) = 0\,.$$

Now, we consider a subcone C of $\mathcal{R}([a, b], \mathcal{K}(\mathbb{R}^d))$ and an operator $T : C \to \mathcal{R}([a, b], \mathcal{K}(\mathbb{R}^d))$.

Then T is said to be *linear* if it preserves addition and multiplication with positive scalars and *monotone* if $T(F) \le T(G)$ whenever $F, G \in C$ and $F \le G$. Here $F \le G$ means that $F(x) \subset G(x)$ almost everywhere.

The continuity of T is obviously intended with respect to the convergence in $\mathcal{R}([a,b], \mathcal{K}(\mathbb{R}^d))$, i.e. the condition

$$\lim_{i \in I^{\le}} d_H \left(\int_a^b F_i(x)\, dx, \int_a^b F(x)\, dx \right) = 0$$

implies

$$\lim_{i \in I^{\le}} d_H \left(\int_a^b T(F_i)(x)\, dx, \int_a^b T(F)(x)\, dx \right) = 0,$$

whenever $(F_i)_{i \in I}^{\le}$ is a net in C and $F \in C$.

2 Korovkin-Type Results in the Cone of Set-Valued Integrable Functions

We begin with the definition of Korovkin system.

Let C be a subcone of $\mathcal{R}([a,b], \mathcal{K}(\mathbb{R}^d))$ and let $\mathcal{M}$ be a subset of C. Then $\mathcal{M}$ is said to be a *Korovkin system* in C with respect to equicontinuous nets of monotone linear continuous operators if whenever $(T_i)_{i \in I}^{\le}$ is an equicontinuous net of monotone linear operators from C into $\mathcal{R}([a,b], \mathcal{K}(\mathbb{R}^d))$ satisfying

$$\lim_{i \in I^{\le}} T_i(H) = H \qquad \text{for every } H \in \mathcal{M}$$

(i.e., $\lim_{i \in I^{\le}} d_H \left(\int_a^b T_i(H)(x)\, dx, \int_a^b H(x)\, dx \right) = 0$), we also have

$$\lim_{i \in I^{\le}} T_i(F) = F \qquad \text{for every } F \in C\,.$$

If no confusion arises, we shall simply call $\mathcal{M}$ a Korovkin system in C or a Korovkin system if $C = \mathcal{R}([a,b], \mathcal{K}(\mathbb{R}^d))$.

Now, we can state the following general result.

In cones of Hausdorff continuous set-valued functions a similar result has been established only under stronger assumptions in [5, 4, Theorem 2.4 and Corollary 2.5] (see also [12, Theorem 2.1]).

Theorem 2.1 *Let C be a subcone of $\mathcal{R}([a,b], \mathcal{K}(\mathbb{R}^d))$ and let $\mathcal{M}$ be a subset of C containing the set-valued functions $\{\varphi\} : x \mapsto \{\varphi(x)\}$ for every $\varphi \in L^1(a,b)$.*

If $\mathcal{M}$ satisfies the following condition:

(H) *For every* $F \in C$, $\varepsilon > 0$ *and* $x_0 \in [a,b]$ *there exist* $H \in \mathcal{M}$ *and* $\delta > 0$ *such that*

$$F \le H \qquad (\text{i.e., } F(x) \subset H(x) \text{almost everywhere}),$$

$$\int_{\max\{a,x_0-\delta\}}^{\min\{b,x_0+\delta\}} H(x)\,dx \subset \int_{\max\{a,x_0-\delta\}}^{\min\{b,x_0+\delta\}} F(x)\,dx + \varepsilon \cdot \mathbf{B},$$

then $\mathcal{M}$ *is a Korovkin system in* C.

Proof Let $(T_i)_{i\in I}^{\le}$ be an equicontinuous net of monotone linear operators from C into $\mathcal{R}([a,b],\mathcal{K}(\mathbb{R}^d))$ such that

$$\lim_{i\in I^{\le}} T_i(H) = H \qquad \text{for every } H \in \mathcal{M}.$$

Let $F \in C$ and $\varepsilon > 0$. First, we shall show that there exist $i_0 \in I$ such that $\int_a^b T_i(F)(x)\,dx \subset \int_a^b F(x)\,dx + \varepsilon \cdot \mathbf{B}$ for every $i \ge i_0$ and after, in a similar way, we shall show the existence of $i_0 \in I$ such that $\int_a^b F(x)\,dx \subset \int_a^b T_i(F)(x)\,dx + \varepsilon \cdot \mathbf{B}$ for every $i \ge i_0$.

For every $x_0 \in [a,b]$, from condition (H) we can find $H \in \mathcal{M}$ and $\delta > 0$ such that

$$F \le H, \qquad \int_{\max\{a,x_0-\delta\}}^{\min\{b,x_0+\delta\}} H(x)\,dx \subset \int_{\max\{a,x_0-\delta\}}^{\min\{b,x_0+\delta\}} F(x)\,dx + \frac{\varepsilon}{2} \cdot \mathbf{B}.$$

Using a straightforward argument on the compactness of $[a,b]$, we can find $y_1,\dots,y_s \in [a,b]$, $K_1,\dots,K_s \in \mathcal{M}$ and $\delta_1,\dots,\delta_s > 0$ such that $y_1 < \dots < y_s$,

$$[a,b] = \bigcup_{j=1}^{s} [\max\{a, y_j - \delta_j\}, \min\{b, y_j + \delta_j\}]$$

and

$$F \le K_j, \qquad \int_{\max\{a,y_j-\delta_j\}}^{\min\{b,y_j+\delta_j\}} K_j(x)\,dx \subset \int_{\max\{a,y_j-\delta_j\}}^{\min\{b,y_j+\delta_j\}} F(x)\,dx + \frac{\varepsilon}{2} \cdot \mathbf{B}.$$

for every $j = 1,\dots,s$. Rearranging the above intervals, we obtain a subdivision $\Delta := \{x_0,\dots,x_m\}$ of $[a,b]$ (with diameter less or equal to 2δ) and $H_1,\dots,H_m \in \mathcal{M}$ such that

$$F \le H_j \text{ for every } j = 1,\dots,m, \qquad \sum_{j=1}^{m} \int_{x_{j-1}}^{x_j} H_j(x)\,dx \subset \int_a^b F(x)\,dx + \frac{\varepsilon}{2} \cdot \mathbf{B}.$$

For every $j = 1\dots, m$, the net $(T_i(H_j))_{i\in I}^{\leq}$ converges to H_j and therefore there exists $i_0 \in I$ such that, for every $i \geq i_0$ and $j = 1, \dots, m$,

$$\int_a^b T_i(H_j)(x)\,dx \subset \int_a^b H_j(x)\,dx + \frac{\varepsilon}{2}\cdot \mathbf{B}\,, \quad \int_a^b H_j(x)\,dx \subset \int_a^b T_i(H_j)(x)\,dx + \frac{\varepsilon}{2}\cdot \mathbf{B}\,.$$

For every $j = 1, \dots, m$, from the monotonicity of T_i we have $T_i(F) \leq T_i(H_j)$ and therefore it follows

$$\int_{x_{j-1}}^{x_j} T_i(F)(x)\,dx \subset \int_{x_{j-1}}^{x_j} T_i(H_j)(x)\,dx\,.$$

Hence we obtain

$$\begin{aligned}\int_a^b T_i(F)(x)\,dx &\subset \sum_{j=1}^m \int_{x_{j-1}}^{x_j} T_i(H_j)(x)\,dx \\ &\subset \sum_{j=1}^m \int_{x_{j-1}}^{x_j} H_j(x)\,dx + \frac{\varepsilon}{2}\cdot \mathbf{B} \\ &\subset \int_a^b F(x)\,dx + \varepsilon\cdot \mathbf{B}\end{aligned}$$

and this completes the first part of the proof.

For the second part, we observe that $\int_a^b F(x)\,dx = \bigcup_{\varphi\in \mathrm{Sel}(f)} \left\{\int_a^b \varphi(x)\,dx\right\}$ and since $\int_a^b F(x)\,dx$ is compact, we can find $\varphi_1, \dots, \varphi_m \in \mathrm{Sel}(F)$ such that

$$\int_a^b F(x)\,dx \subset \bigcup_{j=1}^m \left\{\int_a^b \varphi_j(x)\,dx\right\} + \frac{\varepsilon}{2}\cdot \mathbf{B}\,.$$

For every $j = 1, \dots, m$ we have $\varphi_j \in \mathcal{M}$ and hence there exists $i_0 \in I$ such that, for every $i \geq i_0$,

$$\bigcup_{j=1}^m \left\{\int_a^b \varphi_j(x)\,dx\right\} \subset \bigcup_{j=1}^m \left\{\int_a^b T_i(\{\varphi_j\})(x)\,dx\right\} + \frac{\varepsilon}{2}\cdot \mathbf{B}\,.$$

Since every selection of $T_i(\{\varphi\})$ is also a selection of $T_i(F)$, putting together the above inequalities, we conclude that

$$\int_a^b F(x)\,dx \subset \bigcup_{j=1}^m \left\{\int_a^b T_i(\{\varphi_j\})(x)\,dx\right\} + \varepsilon\cdot \mathbf{B} \subset \int_a^b T_i(F)(x)\,dx + \varepsilon\cdot \mathbf{B}$$

and this completes the second part of the proof. □

The above theorem is quite general and includes the results stated in [5, 4, Theorem 2.4 and Corollary 2.5] and [12, Theorem 2.1] in the case where the limit operator is the identity operator.

Indeed, condition *(H)* allows to consider set-valued functions in $\mathcal{M}$ which are not necessarily Hausdorff continuous.

Further, as shown in the preceding proof, condition *(H)* may be reformulated as follows

$(H)_1$ *For every $F \in C$ and $\varepsilon > 0$, there exist a subdivision* $\{x_0, x_1, \ldots, x_m\}$ of $[a, b]$ *and* $H_1, \ldots, H_m \in \mathcal{M}$ *such that*

$$F \le H_j \text{ for every } j = 1, \ldots, m\,,$$

$$\sum_{j=1}^{m} \int_{x_{j-1}}^{x_j} H_j(x)\, dx \subset \int_a^b F(x)\, dx + \varepsilon \cdot \mathbf{B}\,.$$

Moreover, we observe that the requirement that $\mathcal{M}$ contains the set-valued functions $\{\varphi\}$ for every $\varphi \in L^1(a, b)$ is quite natural in applications since approximation processes for set-valued integrable functions are in general obtained from approximation processes in $L^1(a, b)$ for which the convergence to the identity operator is already known, as we shall see in the examples in the next section.

However, this assumption may also be weakened by assuming that $\mathcal{M}$ contains the set-valued functions $\{\psi\} : x \mapsto \{\psi(x)\}$ for every ψ in a Korovkin system in $L^1(a, b)$ for single-valued L^1-functions (see [1, 2] for more details on Korovkin systems in $L^1(a, b)$).

Now, we can state some consequences of Theorem 2.1. The first consequence is the analogous of the particular cases considered in [5, 4, Theorem 2.4 and Corollary 2.5].

Corollary 2.1 *Let C be a subcone of $\mathcal{R}([a, b], \mathcal{K}(\mathbb{R}^d))$ and let $\mathcal{M}$ be a subset of C containing the set-valued functions $\{\varphi\} : x \mapsto \{\varphi(x)\}$ for every $\varphi \in L^1(a, b)$.*

If $\mathcal{M}$ satisfies the following condition:

$(H)_2$ *For every $F \in C$, $\varepsilon > 0$ and $x_0 \in [a, b]$ there exist $H \in \mathcal{M}$ and $\delta > 0$ such that*

$$F \le H\,, \qquad H(x) \subset F(x) + \varepsilon \cdot \mathbf{B} \quad \text{a.e. in the interval } [a, b] \cap [x_0 - \delta, x_0 + \delta]\,,$$

then $\mathcal{M}$ is a Korovkin system in C.

The proof is immediate since condition $(H)_2$ obviously implies condition *(H)*.

Of course, we have also in this case the following alternative reformulation of $(H)_2$:

(H)$_3$ *For every* $F \in C$ *and* $\varepsilon > 0$, *there exist a subdivision* $\{x_0, x_1, \ldots, x_m\}$ of $[a, b]$ *and* $H_1, \ldots, H_m \in \mathcal{M}$ *such that, for every* $j = 1, \ldots, m$

$$F \leq H_j\,, \qquad H_j(x) \subset F(x) + \varepsilon \cdot \mathbf{B} \quad \textit{a.e. in } [x_{j-1}, x_j]\,.$$

The preceding results allow us to state different consequences which are more suitable in applications.

Theorem 2.2 *Let* C *be a subcone of* $\mathcal{R}([a, b], \mathcal{K}(\mathbb{R}^d))$ *and let* $\mathcal{M}$ *be a subset of* C *containing the following set-valued functions*

(i) the single-valued functions $\{\varphi\} : x \mapsto \{\varphi(x)\}$ *for every* $\varphi \in L^1(a, b)$;
(ii) the constant functions $K : x \mapsto K$ *for every* $K \in \mathcal{K}(\mathbb{R}^d)$;
(iii) the functions $x \mapsto |x - x_0| \cdot \mathbf{B}$ *for every* $x_0 \in [a, b]$.

Then $\mathcal{M}$ *is a Korovkin system in* C.

Proof Let $F \in C$, $\varepsilon > 0$ and $x_0 \in [a, b]$. Since F is a.e. continuous, we can find $\delta > 0$ such that

$$F(x) \subset \overline{\mathrm{co}}\left(\operatorname*{ess\,inf}_{t \in J_\delta(x_0)} F(t)\right) + \frac{\varepsilon}{2} \cdot \mathbf{B}$$

a.e. in $J_\delta(x_0) := [a, b] \cap [x_0 - \delta, x_0 + \delta]$, where

$$\operatorname*{ess\,inf}_{t \in J_\delta(x_0)} F(t) := \bigcup \left\{K \in \mathcal{K}(\mathbb{R}^d) \mid \mu(\{x \in J_\delta(x_0) \mid F(x) \not\subset K\}) = 0\right\},$$

μ denotes the Lebesgue measure in $[a, b]$ and $\overline{\mathrm{co}}(K)$ denotes the closed convex hull of K.

Moreover F is also a.e. bounded and therefore there exists $M > 0$ such that

$$F(x) \subset \overline{\mathrm{co}}\left(\operatorname*{ess\,inf}_{t \in J_\delta(x_0)} F(t)\right) + M \cdot \mathbf{B}$$

almost everywhere in $[a, b]$.

Now, consider the set-valued function $H : [a, b] \to \mathcal{K}(\mathbb{R}^d)$ defined by setting, for every $x \in [a, b]$,

$$H(x) := \overline{\mathrm{co}}\left(\operatorname*{ess\,inf}_{t \in J_\delta(x_0)} F(t)\right) + \frac{\varepsilon}{2} \cdot \mathbf{B} + \frac{M}{\delta}|x - x_0| \cdot \mathbf{B}\,.$$

From our assumption we obviously have $H \in \mathcal{M}$.

Now, we show that $F \leq H$. Indeed, if $|x - x_0| \leq \delta$ we have almost everywhere

$$F(x) \subset \overline{\mathrm{co}}\left(\operatorname*{ess\,inf}_{t \in J_\delta(x_0)} F(t)\right) + \frac{\varepsilon}{2} \cdot \mathbf{B} \subset H(x)$$

and similarly, if $|x - x_0| > \delta$, we have almost everywhere

$$F(x) \subset \overline{\text{co}}\left(\operatorname*{ess\,inf}_{t \in J_\delta(x_0)} F(t)\right) + M \cdot \mathbf{B}$$

$$\subset \overline{\text{co}}\left(\operatorname*{ess\,inf}_{t \in J_\delta(x_0)} F(t)\right) + \frac{M}{\delta}|x - x_0| \cdot \mathbf{B} \subset H(x)\,.$$

Hence condition *(H)*$_2$ of Corollary 2.1 is satisfied and we can conclude that $\mathcal{M}$ is a Korovkin system in C. □

The proof of the following result is similar by considering the function $K : [a, b] \to \mathcal{K}(\mathbb{R}^d)$ defined by setting, for every $x \in [a, b]$,

$$K(x) := \overline{\text{co}}\left(\operatorname*{ess\,inf}_{t \in J_\delta(x_0)} F(t)\right) + \frac{\varepsilon}{2} \cdot \mathbf{B} + \frac{M}{\delta^2}(x - x_0)^2 \cdot \mathbf{B}\,.$$

in place of H. For the sake of brevity we omit the details.

Theorem 2.3 *Let C be a subcone of $\mathcal{R}([a, b], \mathcal{K}(\mathbb{R}^d))$ and let $\mathcal{M}$ be a subset of C containing the following set-valued functions*

(i) the single-valued functions $\{\varphi\} : x \mapsto \{\varphi(x)\}$ for every $\varphi \in L^1(a, b)$;
(ii) the constant functions $K : x \mapsto K$ for every $K \in \mathcal{K}(\mathbb{R}^d)$;
(iii) the functions $x \mapsto (x, x_0)^2 \cdot \mathbf{B}$ for every $x_0 \in [a, b]$.

Then $\mathcal{M}$ is a Korovkin system in C.

In particular, if $d = 1$, the subcone $\mathcal{M}$ of $\mathcal{R}([a, b], \mathcal{K}(\mathbb{R}))$ containing the single-valued functions, the constant set-valued functions and the set-valued functions

$$x \mapsto |x - x_0| \cdot [-1, 1],$$

for every $x_0 \in [a, b]$, is a Korovkin system in $\mathcal{R}([a, b], \mathcal{K}(\mathbb{R}^d))$.

In the alternative formulation, we have to require that the single-valued functions, the constant set-valued functions and the set-valued functions

$$x \mapsto (x - x_0)^2 \cdot [-1, 1], \qquad x_0 \in [a, b],$$

belong to $\mathcal{M}$.

Hence we find the classical Korovkin system in $\mathcal{R}([a, b], \mathcal{K}(\mathbb{R}))$ consisting of the functions

$$x \mapsto \mathbf{B}, \qquad x \mapsto x \cdot \mathbf{B}, \qquad x \mapsto x^2 \cdot \mathbf{B}\,.$$

(Indeed, the preceding system also ensure the convergence on the single-valued functions from the classical case.)

In the next section we shall apply this last result to particular sequences of set-valued operators.

3 Sequences of Set-Valued Kantorovich and Bernstein-Durrmeyer Operators

In this section we define the Kantorovich and the Bernstein-Durrmeyer operators in spaces of set-valued functions. For the sake of simplicity, we consider only the classical definition of these operators, but the convergence result can be obtained in a similar way also for many generalizations and modifications of these operators.

For every $n \geq 1$, consider the operators $K_n : \mathcal{R}([0,1], \mathcal{K}(\mathbb{R}^d)) \to \mathcal{R}([0,1], \mathcal{K}(\mathbb{R}^d))$ defined by setting, for every $F \in \mathcal{R}([0,1], \mathcal{K}(\mathbb{R}^d))$ and $x \in [0,1]$,

$$K_n(F)(x) := \sum_{k=0}^{n} \binom{n}{k} x^k (1-x)^{n-k} (n+1) \int_{k/(n+1)}^{(k+1)/(n+1)} F(t)\, dt\,, \tag{1}$$

and

$$D_n(F)(x) := \sum_{k=0}^{n} \binom{n}{k} x^k (1-x)^{n-k} (n+1) \int_0^1 \binom{n}{k} t^k (1-t)^{n-k} F(t)\, dt\,, \tag{2}$$

The operators K_n and D_n are obviously linear monotone and continuous and from the classical properties of these operators in the single-valued setting, we easily obtain

$$K_n(\mathbf{1} \cdot \mathbf{B})(x) = 1 \cdot \mathbf{B}\,,$$

$$K_n(\mathrm{id} \cdot \mathbf{B})(x) = \frac{n}{n+1} x \cdot \mathbf{B} + \frac{1}{2(n+1)} \cdot \mathbf{B}\,,$$

$$K_n(\mathrm{id}^2 \cdot \mathbf{B})(x) = \frac{n(n-1)}{(n+1)^2} x^2 \cdot \mathbf{B} + \frac{2n}{(n+1)^2} x \cdot \mathbf{B} + \frac{1}{3(n+1)^2} \cdot \mathbf{B}\,,$$

and similarly

$$D_n(\mathbf{1} \cdot \mathbf{B})(x) = 1 \cdot \mathbf{B}\,,$$

$$D_n(\mathrm{id} \cdot \mathbf{B})(x) = \frac{n}{n+2} x \cdot \mathbf{B} + \frac{1}{n+2} \cdot \mathbf{B}\,,$$

$$D_n(\mathrm{id}^2 \cdot \mathbf{B})(x) = \frac{n(n-1)}{(n+2)(n+3)} x^2 \cdot \mathbf{B} + \frac{4n}{(n+2)(n+3)} x \cdot \mathbf{B} + \frac{2}{(n+2)(n+3)} \cdot \mathbf{B}\,.$$

Hence, using the last result in the preceding section we can state the following result.

Theorem 3.1 *For every* $F \in \mathcal{R}([0,1], \mathcal{K}(\mathbb{R}^d))$, *we have*

$$\lim_{n\to+\infty} K_n(F) = F\,, \quad \lim_{n\to+\infty} D_n(F) = F\,.$$

Acknowledgments Work performed under the auspices of G.N.A.M.P.A. (I.N.d.A.M.) and the UMI Group TAA "Approximation Theory and Applications".

References

1. Altomare, F., Campiti, M.: Korovkin-type Approximation Theory and Its Applications. De Gruyter Studies in Mathematics, vol. 17. De Gruyter, Berlin, New York (1994)
2. Altomare, F., Cappelletti Montano, M., Leonessa, V., Raşa, I.: Markov Operators, Positive Semigroups and Approximation Processes. De Gruyter Studies in Mathematics, vol. 61. De Gruyter, Berlin, Munich, Boston (2014)
3. Aumann, R.J.: Integrals of set-valued functions. J. Math. Anal. Appl. **12**, 1–12 (1965)
4. Campiti, M.: A Korovkin-type theorem for set-valued Hausdorff continuous functions. Le Mathematiche **42**, 29–35 (1987)
5. Campiti, M.: Approximation of continuous set-valued functions in Fréchet spaces I. Rev. Anal. Numér. Théor. Approx. **20**, 15–23 (1991)
6. Campiti, M.: Approximation of continuous set-valued functions in Fréchet spaces II. Rev. Anal. Numér. Théor. Approx. **20**, 24–38 (1991)
7. Campiti, M.: Korovkin theorems for vector-valued continuous functions. In: Approximation Theory, Spline Functions and Applications, Internat. Conf., Maratea, May 1991, pp. 293–302. Kluwer Acad. Publ., Dordrecht (1992). Nato Adv. Sci. Inst. Ser. C: Math. Phys. Sci. **356**
8. Campiti, M.: Convergence of nets of monotone operators between cones of set-valued functions. Atti dell'Accademia delle Scienze di Torino **126**, 39–54 (1992)
9. Campiti, M.: Convexity-monotone operators in Korovkin theory, Rend. Circ. Mat. Palermo **33**, 229–238 (1993)
10. Campiti, M., Korovkin-type approximation in spaces of vector-valued and set-valued functions. Appl. Anal. **98**, 2486–2496 (2019)
11. Campiti, M.: On the Korovkin-type approximation of set-valued continuous functions. Construct. Math. Anal. **4**, 119–134 (2021)
12. Campiti, M.: Korovkin-type approximation of set-valued and vector-valued functions. Math. Found. Comput. (2021). http://doi.org/10.3934/mfc.2021032. http://aimsciences.org//article/id/a687e0f8-e559-4a09-ad59-f86fc1ece577
13. Keimel, K., Roth, W.: A Korovkin type approximation theorem for set-valued functions. Proc. Am. Math. Soc. **104**, 819–824 (1988)
14. Polovinkin, E.: Riemannian integral of set-valued function. In: Optimization Techniques, IFIP Technical Conference, Novosibirsk, USSR, July 1–7, pp. 405–410 (1974)
15. Richter, H.: Verallgemeinerung eines in der Statistik benötigten Satzes der Masstheorie. Math. Annalen **150**, 85–90 (1963). Errata corrige: Math. Annalen **150**, 440–441 (1963)

Convergence of a Class of Generalized Sampling Kantorovich Operators Perturbed by Multiplicative Noise

Marco Cantarini, Danilo Costarelli, and Gianluca Vinti

Dedicated to Professor Francesco Altomare on the occasion of his 70th birthday with deep esteem and warm friendship

Abstract In this paper a new family of sampling type series is introduced. From the mathematical point of view, the present definition generalizes the notion of the well-known sampling Kantorovich operators, in fact providing a weighted version of the original family of operators by functions $g_{k,w}$, $k \in \mathbb{Z}$, $w > 0$, called noise functions. From the application point of view, this situation represents the reconstruction problem of signals perturbed by linear or nonlinear multiplicative noise sources. In this respect, approximation results have been established in various contexts. First, pointwise and uniform approximation theorems have been proved. Then, convergence theorems have been derived in the general setting of Orlicz spaces. The latter context allows us to deduce, in particular, an L^p-convergence theorem. Finally, the concept of delta convergent sequence is introduced and also used in order to prove that the above family of sampling type operators extend the well-known generalized sampling series of P.L. Butzer.

Keywords Generalized sampling Kantorovich series · Noise functions · Orlicz spaces · Modular convergence · Uniform convergence · Delta convergence

1 Introduction

In recent years, the theory of approximating functions by means of families of positive linear operators has been deeply studied. Concerning this topic one of the main scientists is, of course, Prof. Francesco Altomare; among his contributions to the above theory we can see, e.g., [2–6].

M. Cantarini, D. Costarelli · G. Vinti (✉)
Department of Mathematics and Computer Science, University of Perugia, Perugia, Italy
e-mail: marco.cantarini@unipg.it; danilo.costarelli@unipg.it; gianluca.vinti@unipg.it

A. M. Candela et al. (eds.), *Recent Advances in Mathematical Analysis*,
Trends in Mathematics, https://doi.org/10.1007/978-3-031-20021-2_13

Among the positive linear operators, we can find those of the sampling type when positive kernels are chosen [7, 13–15, 24–26, 31]. These mathematical operators have been introduced by P.L. Butzer and his school with the main purpose to consider an approximate version of the well-known Whittaker-Kotel'nikov-Shannon sampling theorem for band-limited functions. The operators introduced by P.L. Butzer have the following general form:

$$(S_w^\chi f)(x) := \sum_{k\in\mathbb{Z}} f\left(\frac{k}{w}\right) \chi\,(wx-k) \tag{1}$$

where $w > 0$, $\chi : \mathbb{R} \to \mathbb{R}$ is a suitable kernel function, and $f : \mathbb{R} \to \mathbb{R}$ is any bounded function.

Recently, the above operators have been further generalized with the introduction of the corresponding Kantorovich and Durrmeyer versions of the above (sampling) operators.

The first version, i.e., the Kantorovich type one K_w^χ, consists in the replacement of the sample values $f(k/w)$ with the following integrals $w\int_{k/w}^{(k+1)/w} f(u)\,du$. The main advantage of this replacement is the possibility to obtain a family of operators that are not depending from the pointwise values of the signal f, and that are, in fact, suitable to approximate not necessarily continuous signals [10, 16, 29]. Moreover, from the point of view of the signal theory, the operators K_w^χ allow to reduce the so-called "time-jitter error" occurring when a signal is not sampled exactly at the theoretical sample node k/w. In fact, in practice, the image acquisition process takes place by averaging the signal around the node rather than taking it from the single node.

The second version, i.e., the Durrmeyer type one $D_w^{\chi,\psi}$, consists in the replacement of the sample values $f(k/w)$ with the convolution integrals:

$$w\int_{\mathbb{R}} \psi(wu-k)\, f(u)\,du,$$

where the function $\psi : \mathbb{R} \to \mathbb{R}$ is a suitable approximate identity. Here, the function ψ in the convolution has the task to filter the signal f before the reconstruction. One of the main advantages provided by the Durrmeyer sampling type series is, from the mathematical point of view, the possibility to have a family of sampling type operators which generalize both the classical and the sampling Kantorovich series [8, 11, 22].

From the point of view of signal analysis, in all the above families of operators the possible presence of perturbations of the signal f that we need to reconstruct have not been taken into account; but very often, signals can be affected by several kind of noises.

A typical example of noice sources $g_{k,w}$, $k \in \mathbb{Z}$, $w > 0$, is provided by a "Speckle" type noise: this is a disturbance that typically affects SAR (Synthetic Aperture Radar) remote sensing systems. It is a noise produced by interference phenomena that are generated whenever images of complex objects are acquired through the use of highly coherent waves. This is a "multiplicative" noise, resulting therefore dependent on the signal itself to which it is superimposed; as example, images disturbed by this type of interference have a typical "graininess", due to the irregularity of the surfaces that are illuminated by the pulses of electromagnetic energy. Another example of multiplicative noise is represented by the fading, i.e. the time-varying attenuation to which are subject radio signals in the presence of irregular propagation (for example, variations in the electron density of the reflective layers which ensure certain forms of propagation) as is sometimes the case when listening to a very distant radio station. This situation, or similar ones, are not covered (as previously mentioned) by the operators S_w^χ, K_w^χ and $D_w^{\chi,\psi}$, which not allow to reconstruct f eliminating the effects provided by certain noises represented, e.g., by $g_{k,w}$, $k \in \mathbb{Z}$, $w > 0$.

For the latter reasons, here we introduce a new family of sampling type operators (below denoted by $K_w^{\chi,\mathcal{G}}$), in which the sample values $f(k/w)$ are replaced by the following integrals:

$$\frac{\displaystyle\int_{k/w}^{(k+1)/w} g_{k,w}(u)\, f(u)\, du}{\displaystyle\int_{k/w}^{(k+1)/w} g_{k,w}(u)\, du}$$

where we can find the presence of noise functions $g_{k,w}$, which represent, in fact, multiplicative noise sources.

In the next sections, we establish approximation/reconstruction results by means of the operators $K_w^{\chi,\mathcal{G}}$ in various spaces of functions.

First, we prove a pointwise and uniform convergence theorem in case of bounded functions which are continuous and uniformly continuous, respectively. Further, in order to deal with not necessarily continuous signals, we consider the above theory in the general setting of Orlicz spaces. It is well-known that Orlicz spaces arise since the first half of the 1900 by the Polish mathematician W. Orlicz and provide an abstract generalization of many known spaces of functions, such as the L^p-spaces. Other important instances of Orlicz spaces are provided by the Zygmund spaces, the exponential spaces, and several others.

In this sense, in order to obtain a modular convergence theorem in Orlicz spaces, we first establish a modular continuity property for the above operators, a Luxembug norm convergence theorem in case of continuous functions with compact support, and finally (using density arguments) a modular convergence theorem for functions in Orlicz spaces.

At the end of the paper, the concept of delta convergent sequences is recalled. Such notion is used in order to prove that the operators $K_w^{\chi,\mathcal{G}}$ extend, other than the classical sampling Kantorovich operators, also the generalized sampling series (1) of P.L. Butzer.

2 Notations and Preliminary Results

We denote by $C(\mathbb{R})$ the space of bounded and uniformly continuous functions $f : \mathbb{R} \to \mathbb{R}$ and with $C_c(\mathbb{R}) \subset C(\mathbb{R})$ the subspace of $C(\mathbb{R})$ of functions having compact support. For any $f \in C(\mathbb{R})$, we will denote by $\|f\|_\infty$ the usual sup-norm. Furthermore, we denote by $M(\mathbb{R})$ the linear space of all Lebesgue measurable and bounded real functions defined on $\mathbb{R}$.

Definition 2.1 We call *kernel* a function $\chi : \mathbb{R} \to \mathbb{R}$ belonging to $L^1(\mathbb{R})$, that is bounded in a neighbourhood of the origin and such that the following conditions hold:

$(\chi 1)$ for every $u \in \mathbb{R}$, we have

$$\sum_{k\in\mathbb{Z}} \chi(u-k) = 1;$$

$(\chi 2)$ there exists $\beta > 0$ such that

$$m_\beta(\chi) := \sup_{x\in\mathbb{R}} \sum_{k\in\mathbb{Z}} |\chi(x-k)|\,|x-k|^\beta < +\infty.$$

Note that, several examples of kernel functions χ, both with bounded or unbounded support, can be found, e.g., in [1, 12].

We recall that from Definition 2.1, it is possible to prove the following properties (see [10]):

(i) $m_0(\chi) := \sup_{x\in\mathbb{R}} \sum_{k\in\mathbb{Z}} |\chi(x-k)| < +\infty$;
(ii) for every $\gamma > 0$ we have

$$\lim_{w\to+\infty} \sum_{|u-k|>\gamma w} |\chi(u-k)| = 0 \tag{2}$$

uniformly with respect to $u \in \mathbb{R}$;
(iii) for every $\gamma > 0$ and $\varepsilon > 0$, there exists a constant $M > 0$ such that

$$\int_{|x|>M} w\,|\chi(wx-k)|\,dx < \varepsilon$$

for sufficiently large $w > 0$ and $k \in \mathbb{Z}$ such that $k/w \in [-\gamma, \gamma]$.

Now we are able to define the following class of operators.

Definition 2.2 We define by $\left(K_w^{\chi,\mathcal{G}}\right)_{w>0}$ the family of generalized sampling Kantorovich operators perturbed by multiplicative noise, by

$$\left(K_w^{\chi,\mathcal{G}} f\right)(x) := \sum_{k\in\mathbb{Z}} \chi\,(wx-k)\,\frac{\displaystyle\int_{k/w}^{(k+1)/w} g_{k,w}\,(u)\, f(u)\, du}{\displaystyle\int_{k/w}^{(k+1)/w} g_{k,w}\,(u)\; du}$$

where $\mathcal{G} := (\mathcal{G}_w)_{w>0}$, is a family of noise sequences, with $\mathcal{G}_w = (g_{k,w})_{k\in\mathbb{Z}}$, $g_{k,w} : \mathbb{R} \to \mathbb{R}^+$ are locally integrable noise functions, $f : \mathbb{R} \to \mathbb{R}$ is such that $g_{k,w} f$ are locally integrable, and the above series is convergent for every $x \in \mathbb{R}$.

From now on, we simply call the operators $K_w^{\chi,\mathcal{G}}$ as the *perturbed sampling Kantorovich operators*.

Note that these operators are well-defined, e.g., for every bounded f since

$$\left|\left(K_w^{\chi,\mathcal{G}} f\right)(x)\right| \leq m_0\,(\chi)\, \|f\|_\infty\,, \qquad x\in\mathbb{R}. \tag{3}$$

Concerning the expression of the operators $K_w^{\chi,\mathcal{G}}$, one can simply construct examples of noise sequences $g_{k,w}$, for every $k \in \mathbb{Z}$ and $w > 0$, of the form:

$$g^1_{k,w}(x) \;:=\; \sin\left(\frac{k\,x}{w}\right) + \frac{3}{2}, \quad x\in\mathbb{R},$$

$$g^2_{k,w}(x) \;:=\; e^{\cos(k\,w\,x)}, \quad x\in\mathbb{R},$$

and several others. Of course, we can also consider not necessarily continuous functions $g_{k,w}$.

Now, we recall one of the general contexts in which we will study the above operators $K_w^{\chi,\mathcal{G}}$.

From now on, we denote by $\phi : \mathbb{R}_0^+ \to \mathbb{R}_0^+$ a convex ϕ-function, that is, a function that satisfies the following assumptions:

(ϕ1) ϕ is convex on $\mathbb{R}_0^+$;
(ϕ2) $\phi(0) = 0$, $\phi(u) > 0$ for every $u > 0$.

Consider now the functional

$$I^\phi[f] := \int_{\mathbb{R}} \phi\,(|f\,(x)|)\, dx$$

with $f \in M(\mathbb{R})$. It is well known that (see [9, 27, 30]) I^ϕ is a convex modular functional on $M(\mathbb{R})$. We now define the Orlicz space generated by ϕ by

$$L^\phi(\mathbb{R}) := \left\{ f \in M(\mathbb{R}) : I^\phi[\lambda f] < +\infty \text{ for some } \lambda > 0 \right\}.$$

We recall that the space

$$E^\phi(\mathbb{R}) := \left\{ f \in M(\mathbb{R}) : I^\phi[\lambda f] < +\infty \text{ for every } \lambda > 0 \right\},$$

that is known with the name of *space of the finite elements of* $L^\phi(\mathbb{R})$, is strictly contained in $L^\phi(\mathbb{R})$; these two spaces coincides if and only if ϕ satisfies the so called Δ_2- condition, that is, if there exists a constant $M > 0$ such that

$$\phi(2u) \le M\phi(u), \ u \in \mathbb{R}_0^+. \tag{4}$$

In Orlicz spaces we usually deal with the notion of modular convergence. We can say that a family of functions $(f_w)_{w>0}$ is *modularly convergent* to a function $f \in L^\phi(\mathbb{R})$ if

$$\lim_{w \to +\infty} I^\phi[\lambda(f_w - f)] = 0, \tag{5}$$

for some $\lambda > 0$. We also recall that it is possible to introduce an F-norm, e.g., the well-known Luxemburg norm, on $L^\phi(\mathbb{R})$, as follows:

$$\|f\|_\phi := \inf\left\{\lambda > 0 : I^\phi[f/\lambda] \le 1\right\}.$$

It is well known that $\|f_w - f\|_\phi \to 0$ if and only if $\lim_{w\to+\infty} I^\phi[\lambda(f_w - f)] = 0$ for every $\lambda > 0$. Then, we immediately see that, in general, the convergence that can be deduced from the Luxemburg norm is stronger than the modular convergence, and both the modular and the Luxemburg norm convergences are equivalent if and only if (4) holds.

3 Convergence Theorems

We first prove a pointwise and uniform convergence theorem.

Theorem 3.1 *Let $f \in M(\mathbb{R})$ be a function which is continuous at $x \in \mathbb{R}$. Then*

$$\lim_{w\to+\infty} \left(K_w^{\chi,\mathcal{G}} f\right)(x) = f(x).$$

Furthermore, if $f \in C(\mathbb{R})$, then

$$\left\| \left(K_w^{\chi,\mathcal{G}} f\right)(\cdot) - f(\cdot) \right\|_\infty \to 0$$

as $w \to +\infty$.

Proof We prove only the second part of the thesis, since the first one follows by similar techniques. Let $f \in C(\mathbb{R})$ and $x \in \mathbb{R}$ be fixed. For every $\varepsilon > 0$ there exists some $\gamma > 0$ such that if $|x-y| \leq \gamma$ then $|f(x) - f(y)| \leq \varepsilon$. Now, using condition $(\chi 1)$ of Sect. 2, we can observe that:

$$\left(K_w^{\chi,\mathcal{G}} \mathbf{1}\right)(x) = 1, \text{ for every } x \in \mathbb{R},\ w > 0,$$

where $\mathbf{1}(x) = 1$, $x \in \mathbb{R}$, then we can easily write what follows

$$\left|\left(K_w^{\chi,\mathcal{G}} f\right)(x) - f(x)\right| \leq \sum_{k\in\mathbb{Z}} |\chi(wx-k)| \frac{\displaystyle\int_{k/w}^{(k+1)/w} g_{k,w}(u)\,|f(u) - f(x)|\,du}{\displaystyle\int_{k/w}^{(k+1)/w} g_{k,w}(u)\,du}.$$

Consider now the sets:

$$S_1 := \{k \in \mathbb{Z} : |x - k/w| \leq \gamma/2\},$$

$$S_2 := \{k \in \mathbb{Z} : |x - k/w| > \gamma/2\}$$

then

$$\left|\left(K_w^{\chi,\mathcal{G}} f\right)(x) - f(x)\right| \leq \sum_{k\in S_1} |\chi(wx-k)| \frac{\displaystyle\int_{k/w}^{(k+1)/w} g_{k,w}(u)\,|f(u) - f(x)|\,du}{\displaystyle\int_{k/w}^{(k+1)/w} g_{k,w}(u)\,du}$$

$$+ \sum_{k\in S_2} |\chi(wx-k)| \frac{\displaystyle\int_{k/w}^{(k+1)/w} g_{k,w}(u)\,|f(u) - f(x)|\,du}{\displaystyle\int_{k/w}^{(k+1)/w} g_{k,w}(u)\,du} =: I_1 + I_2.$$

Let us analyze I_1. If $u \in (k/w, (k+1)/w)$, we have that

$$|u - x| \leq |u - k/w| + |k/w - x| \leq \frac{1}{w} + \frac{\gamma}{2} < \gamma,$$

for a sufficiently large w, so we have

$$I_1 \leq \varepsilon \sum_{k \in \mathbb{Z}} |\chi(wx-k)| \frac{\displaystyle\int_{k/w}^{(k+1)/w} g_{k,w}(u)\,du}{\displaystyle\int_{k/w}^{(k+1)/w} g_{k,w}(u)\,du} \leq \varepsilon\, m_0(\chi).$$

Now we consider I_2. Since there exists $M > 0$ for which $|f(x)| \leq M$, for every $x \in \mathbb{R}$, we obtain

$$I_2 \leq 2M \sum_{k \in S_2} |\chi(wx-k)|,$$

and from property (2), it turns out that $I_2 \leq \varepsilon$, for $w > 0$ sufficiently large and so the thesis immediately follows. □

Now we study the above operators in the general setting of Orlicz spaces.
First, the following Luxemburg norm convergence theorem can be established.

Theorem 3.2 *For every $f \in C_c(\mathbb{R})$ and $\lambda > 0$, we have*

$$\lim_{w \to +\infty} I^{\phi}[\lambda\left(K_w^{\chi,\mathcal{G}} f - f\right)] = 0.$$

Proof We have to prove that

$$\lim_{w \to +\infty} I^{\phi}\left[\lambda\left(K_w^{\chi,\mathcal{G}} f - f\right)\right] = 0$$

for every fixed $\lambda > 0$. In order to establish the above claim, we will use the well-known Vitali convergence theorem. From Theorem 3.1, we already know that

$$\lim_{w \to +\infty} \phi\left(\lambda \left\| K_w^{\chi,\mathcal{G}} f - f \right\|_{\infty}\right) = 0.$$

Let now $\varepsilon > 0$ be fixed, and let $[-\overline{\gamma}, \overline{\gamma}]$ be a interval containing the support of f. Now, we also fix $\gamma > \overline{\gamma} + 1$; hence, for $k \notin [-w\gamma, w\gamma]$, we have that $\left[\frac{k}{w}, \frac{k+1}{w}\right] \cap [-\gamma, \gamma] = \emptyset$ and therefore we get

$$\int_{k/w}^{(k+1)/w} g_{k,w}(u)\, |f(u)|\, du = 0,$$

for every (sufficiently large) $w > 0$. Now, since χ is a kernel, in view of condition (iii) of Sect. 2 there exists a constant $M > 0$ such that

$$\int_{|x|>M} w\, |\chi\,(wx - k)|\, dx < \varepsilon$$

for sufficiently large $w > 0$ and $k \in [-w\gamma, w\gamma]$. Then, by Jensen inequality (see, e.g., [18]) and Fubini-Tonelli theorem, we have

$$\int_{|x|>M} \phi\left(\lambda \left|\left(K_w^{\chi,\mathcal{G}} f\right)(x)\right|\right) dx$$

$$= \int_{|x|>M} \phi\left(\lambda \left|\sum_{k\in\mathbb{Z}} \chi\,(wx - k)\, \frac{\int_{k/w}^{(k+1)/w} g_{k,w}\,(u)\, f(u) du}{\int_{k/w}^{(k+1)/w} g_{k,w}\,(u)\, du}\right|\right) dx$$

$$\leq \int_{|x|>M} \phi\left(\lambda \sum_{k\in[-w\gamma, w\gamma]} |\chi\,(wx - k)|\, \frac{\int_{k/w}^{(k+1)/w} g_{k,w}\,(u)\, |f(u)|\, du}{\int_{k/w}^{(k+1)/w} g_{k,w}\,(u)\, du}\right) dx$$

$$\leq \frac{1}{m_0(\chi)} \sum_{k\in[-w\gamma, w\gamma]} \int_{|x|>M} \phi\left(\lambda\, m_0\,(\chi)\, \frac{\int_{k/w}^{(k+1)/w} g_{k,w}\,(u)\, |f(u)|\, du}{\int_{k/w}^{(k+1)/w} g_{k,w}\,(u)\, du}\right) |\chi\,(wx - k)|\, dx$$

$$\leq \frac{\phi\left(\lambda\, m_0\,(\chi)\, \|f\|_\infty\right)}{w\, m_0\,(\chi)} \sum_{k\in[-w\gamma, w\gamma]} \int_{|x|>M} w\, |\chi\,(wx - k)|\, dx$$

$$\leq \frac{\phi\left(\lambda m_0\,(\chi)\, \|f\|_\infty\right)}{w\, m_0\,(\chi)} \sum_{k\in[-w\gamma, w\gamma]} \varepsilon \leq \frac{2\,\varepsilon\, \phi\left(\lambda m_0\,(\chi)\, \|f\|_\infty\right)}{m_0\,(\chi)}\, (\gamma + 1)\,.$$

Finally, let $B \subset \mathbb{R}$ be a measurable set with Lebesgue measure

$$|B| \leq \frac{\varepsilon}{\phi\left(2\,\lambda\, m_0\,(\chi)\, \|f\|_\infty\right)}.$$

Then, by the convexity of ϕ, the inequality in (3), and noting that $m_0(\chi) \geq 1$, we can obtain what follows:

$$\int_B \phi\left(\lambda \left|\left(K_w^{\chi,\mathcal{G}} f\right)(x) - f(x)\right|\right) dx$$

$$\leq \frac{1}{2}\int_B \phi\left(2\lambda \left|\left(K_w^{\chi,\mathcal{G}} f\right)(x)\right|\right) dx + \frac{1}{2}\int_B \phi\left(2\lambda\, |f(x)|\right) dx$$

$$\leq \int_B \phi\left(2\lambda\, m_0(\chi)\, \|f\|_\infty\right) dx < \varepsilon.$$

Then the integrals:

$$\int_{(\cdot)} \phi\left(\lambda \left|\left(K_w^{\chi,\mathcal{G}} f\right)(x) - f(x)\right|\right) dx$$

are equi-absolute continuous, and thus the thesis follows as a consequence of the Vitali convergence theorem. This completes the proof. □

In order to prove the modular convergence in $L^\phi(\mathbb{R})$ for the above operators, we need of a modular continuity property that also shows that $K_w^{\chi,\mathcal{G}}$ are well-defined in the general setting of Orlicz spaces.

Hereafter, we need of an additional assumption on the noise functions $g_{k,w}$ belonging to the family $\mathcal{G}$ of noise sequences; we suppose that there exists two positive numbers δ, σ such that $0 < \delta \leq g_{k,w}(u) \leq \sigma$, for every $u \in \mathbb{R}$, $k \in \mathbb{Z}$, $w > 0$.

Note that, all the noise functions $g_{k,w}$ mentioned in Sect. 2 satisfy the above inequality for suitable values of δ and σ.

Theorem 3.3 *For every $f \in L^\phi(\mathbb{R})$ there holds*

$$I^\phi\left[\lambda\, K_w^{\chi,\mathcal{G}} f\right] \leq C_{\mathcal{G}} \frac{\|\chi\|_1}{m_0(\chi)} I^\phi\left[\lambda\, m_0(\chi)\, f\right], \tag{6}$$

where the above constant $C_{\mathcal{G}} := \sigma/\delta$ depends only on $\mathcal{G}$.

Proof Let $\lambda > 0$ be fixed. Proceeding with similar computations to those given in the proof of Theorem 3.2, we have

$$\int_{\mathbb{R}} \phi\left(\lambda \left|\left(K_w^{\chi,\mathcal{G}} f\right)(x)\right|\right) dx$$

$$\leq \frac{1}{m_0(\chi)} \int_{\mathbb{R}} \sum_{k\in\mathbb{Z}} |\chi(wx-k)|\, \phi\left(\lambda\, m_0(\chi) \frac{\displaystyle\int_{k/w}^{(k+1)/w} g_{k,w}(u)\, |f(u)|\, du}{\displaystyle\int_{k/w}^{(k+1)/w} g_{k,w}(u)\, du}\right) dx.$$

Now, using Fubini-Tonelli theorem and Jensen inequality again we get:

$$\int_{\mathbb{R}} \phi\left(\lambda \left|\left(K_w^{\chi,\mathcal{G}} f\right)(x)\right|\right) dx$$

$$\leq \frac{1}{m_0(\chi)} \sum_{k\in\mathbb{Z}} \frac{\displaystyle\int_{k/w}^{(k+1)/w} g_{k,w}(u)\,\phi\left(\lambda\, m_0(\chi)\,|f(u)|\right) du}{\displaystyle\int_{k/w}^{(k+1)/w} g_{k,w}(u)\, du} \int_{\mathbb{R}} |\chi(wx-k)|\, dx.$$

Further, recalling the properties of the noise functions $g_{k,w}$ and tacking the change of variables $s = wx - k$ in all the above integrals, we get

$$\int_{\mathbb{R}} \phi\left(\lambda \left|\left(K_w^{\chi,\mathcal{G}} f\right)(x)\right|\right) dx$$

$$\leq \frac{w}{\delta\, m_0(\chi)} \sum_{k\in\mathbb{Z}} \int_{k/w}^{(k+1)/w} g_{k,w}(u)\,\phi\left(\lambda\, m_0(\chi)\,|f(u)|\right) du \int_{\mathbb{R}} |\chi(wx-k)|\, dx$$

$$\leq \frac{\|\chi\|_1}{\delta\, m_0(\chi)} \sum_{k\in\mathbb{Z}} \int_{k/w}^{(k+1)/w} g_{k,w}(u)\,\phi\left(\lambda\, m_0(\chi)\,|f(u)|\right) du$$

$$\leq \frac{\sigma\,\|\chi\|_1}{\delta\, m_0(\chi)} \sum_{k\in\mathbb{Z}} \int_{k/w}^{(k+1)/w} \phi\left(\lambda\, m_0(\chi)\,|f(u)|\right) du$$

$$= \frac{\sigma\,\|\chi\|_1}{\delta\, m_0(\chi)} \int_{\mathbb{R}} \phi\left(\lambda\, m_0(\chi)\,|f(u)|\right) du.$$

This completes the proof. □

Note that, assuming $\lambda > 0$ sufficiently small, from Theorem 3.3 it turns out that the operator $K_w^{\chi,\mathcal{G}}$ maps $L^\phi(\mathbb{R})$ into itself, i.e., $K_w^{\chi,\mathcal{G}} f \in L^\phi(\mathbb{R})$ whenever f belongs to $L^\phi(\mathbb{R})$.

Now we can finally prove the modular convergence theorem for the operators $K_w^{\chi,\mathcal{G}}$, that is, in fact, one of the main theorems of this section.

Theorem 3.4 *For every $f \in L^\phi(\mathbb{R})$, there exists a $\lambda > 0$ such that*

$$\lim_{w\to+\infty} I^\phi\left[\lambda\left(K_w^{\chi,\mathcal{G}} f - f\right)\right] = 0.$$

Proof Let $f \in L^\phi(\mathbb{R})$. Since it is well-known that the space $C_c(\mathbb{R})$ is modularly dense in $L^\phi(\mathbb{R})$, for every fixed $\varepsilon > 0$ there exists a function $h \in C_c(\mathbb{R})$ and a parameter $\overline{\lambda} > 0$ such that $I^{\phi}\left[\overline{\lambda}(f-h)\right] < \varepsilon$. So let $\lambda > 0$ be such that $3\lambda(1+m_0(\chi)) < \overline{\lambda}$. By the properties of ϕ, the linearity of the operators $K_w^{\chi,\mathcal{G}}$ and by Theorem 3.3, we have

$$\begin{aligned} I^{\phi}\left[\lambda\left(K_w^{\chi,\mathcal{G}}f - f\right)\right] &\leq I^{\phi}\left[3\lambda\left(K_w^{\chi,\mathcal{G}}f - K_w^{\chi,\mathcal{G}}h\right)\right] + I^{\phi}\left[3\lambda\left(K_w^{\chi,\mathcal{G}}h - h\right)\right] \\ &\quad + I^{\phi}\left[3\lambda(h-f)\right] \\ &\leq \left(C_{\mathcal{G}}\frac{\|\chi\|_1}{m_0(\chi)} + 1\right)\varepsilon + I^{\phi}\left[3\lambda\left(K_w^{\chi,\mathcal{G}}h - h\right)\right], \end{aligned}$$

where again $C_{\mathcal{G}} = \sigma/\delta$. The proof can be completed using Theorem 3.2, i.e.,

$$I^{\phi}\left[\lambda\left(K_w^{\chi,\mathcal{G}}f - f\right)\right] \leq \left(C_{\mathcal{G}}\frac{\|\chi\|_1}{m_0(\chi)} + 2\right)\varepsilon,$$

for $w > 0$ sufficiently large. □

We now apply these results to particular and important cases. It is well known that if we take $\phi(u) = u^p$, $u \in \mathbb{R}_0^+$, $1 \leq p < +\infty$, then $L^{\phi}(\mathbb{R}) = E^{\phi}(\mathbb{R}) = L^p(\mathbb{R})$ and the modular convergence is equivalent to the usual convergence with respect to the L^p-norm. Hence, in the above case we immediately have the following corollaries as a straightforward application of Theorem 3.3 and Theorem 3.4.

Corollary 3.1 *For every $f \in L^p(\mathbb{R})$, $1 \leq p < +\infty$, we have*

$$\left\|K_w^{\chi,\mathcal{G}}f\right\|_p \leq C_{\mathcal{G}}^{1/p}\frac{\|\chi\|_1^{1/p}}{m_0(\chi)^{(1/p)-1}}\|f\|_p,$$

i.e., $K_w^{\chi,\mathcal{G}} : L^p(\mathbb{R}) \to L^p(\mathbb{R})$, $w > 0$.

Corollary 3.2 *For every $f \in L^p(\mathbb{R})$, $1 \leq p < +\infty$, we have*

$$\lim_{w\to+\infty}\left\|K_w^{\chi,\mathcal{G}}f - f\right\|_p = 0.$$

Another interesting case is when $\phi(u) = \phi_{\alpha,\beta}(u) = u^{\alpha}\log^{\beta}(u+e)$, $u \geq 0$, $\alpha \geq 1$ and $\beta > 0$. The corresponding Orlicz space is the set of functions $f \in M(\mathbb{R})$ such that

$$I^{\phi_{\alpha,\beta}}[f] = \int_{\mathbb{R}}(\lambda|f(x)|)^{\alpha}\log^{\beta}(e+\lambda|f(x)|)\,dx < +\infty$$

for some $\lambda > 0$, and it is denoted by the symbol $L^{\alpha}\log^{\beta}L(\mathbb{R})$ (such spaces are called Zygmund spaces). It is well-known that $\phi_{\alpha,\beta}(u)$ satisfies the Δ_2 condition and so $L^\alpha \log^\beta L(\mathbb{R})$ coincides with $E^{\phi_{\alpha,\beta}}$. We now show the following corollary for the most interesting case $\alpha = \beta = 1$.

Corollary 3.3 *For every $f \in L\log L(\mathbb{R})$ and $\lambda > 0$ we have*

$$\int_{\mathbb{R}} \left(\lambda \left|K_w^{\chi,\mathcal{G}} f(x)\right|\right) \log\left(e + \lambda \left|K_w^{\chi,\mathcal{G}} f(x)\right|\right) dx$$

$$\leq C_{\mathcal{G}}\, \|\chi\|_1 \int_{\mathbb{R}} (\lambda\, |f(x)|) \log(e + \lambda\, m_0(\chi)\, |f(x)|)\, dx.$$

In particular, $K_w^{\chi,\mathcal{G}} f$ is well-defined in $L\log L(\mathbb{R})$ and $K_w^{\chi,\mathcal{G}} f \in L\log L(\mathbb{R})$ whenever $f \in L\log L(\mathbb{R})$, $w > 0$.

Recalling that when in the Orlicz spaces the Δ_2-condition holds, then the modular convergence and the norm convergence are equivalent, we get the following.

Corollary 3.4 *For every $f \in L\log L(\mathbb{R})$ and $\lambda > 0$, we have*

$$I^{\phi_{1,1}}\left[\lambda \left|K_w^{\chi,\mathcal{G}} f - f\right|\right] = 0.$$

As a last interesting example, we can consider $\phi(u) = \phi_\alpha(u) = \exp(u^\alpha) - 1$, $u \geq 0$, $\alpha > 0$. In this case the Orlicz space is the set of functions $f \in M(\mathbb{R})$ such that

$$I^{\phi_\alpha}[f] = \int_{\mathbb{R}} \left[\exp\left(\lambda\, |f(x)|^\alpha\right) - 1\right] dx < +\infty$$

for some $\lambda > 0$ i.e., the so-called exponential spaces. Note that since $\phi_\alpha(u)$ does not satisfy the Δ_2-condition, then $L^{\phi_\alpha}(\mathbb{R})$ does not coincide with $E^{\phi_\alpha}(\mathbb{R})$, and the modular convergence does not imply norm convergence; hence we can prove the following corollaries:

Corollary 3.5 *For every $f \in L^{\phi_\alpha}(\mathbb{R})$, we have*

$$\int_{\mathbb{R}} \left[\exp\left(\lambda \left|K_w^{\chi,\mathcal{G}} f(x)\right|^\alpha\right) - 1\right] dx \leq C_{\mathcal{G}} \frac{\|\chi\|_1}{m_0(\chi)} \int_{\mathbb{R}} \left[\exp\left(\lambda m_0(\chi)\, |f(x)|^\alpha\right) - 1\right] dx,$$

for $\lambda > 0$. In particular, $K_w^{\chi,\mathcal{G}} f$ is well defined in $L^{\phi_\alpha}(\mathbb{R})$ and $K_w^{\chi,\mathcal{G}} f \in L^{\phi_\alpha}(\mathbb{R})$ whenever $f \in L^{\phi_\alpha}(\mathbb{R})$.

Corollary 3.6 *For every $f \in L^{\phi_\alpha}(\mathbb{R})$ there exists a $\lambda > 0$ such that*

$$\lim_{w\to+\infty} \int_{\mathbb{R}} \left[\exp\left(\lambda \left|K_w^{\chi,\mathcal{G}} f(x) - f(x)\right|^\alpha\right) - 1\right] dx = 0.$$

4 Link to Classical Settings

As a first remark, we want to highlight that the operators $K_w^{\chi,\mathcal{G}}$ provide a generalization of the well-known sampling Kantorovich operators K_w^{χ} (see, e.g., [17, 19–21]).

Indeed, if we choose the family of noise sequences $\mathcal{G} = \mathcal{G}_L$, for a fixed value $L \in \mathbb{R}$, defined with the constant functions $g_{k,w}(u) = L$, $u \in \mathbb{R}$, for every $k \in \mathbb{Z}$ and $w > 0$, it is easy to see that:

$$(K_w^{\chi,\mathcal{G}_L} f)(x) = \sum_{k\in\mathbb{Z}} \chi(wx-k) \left[w \int_{k/w}^{(k+1)/w} f(u)\, du \right] = (K_w^{\chi} f)(x).$$

Further, in this section we also want to prove that, from Definition 2.2, we can get back also to the classical generalized sampling operators of P.L. Butzer. For showing this, we need the following definitions.

Definition 4.1 A sequence $(s_m(x))_{m\in\mathbb{N}} \subset L^1(\mathbb{R})$ is called a *delta convergent sequence* if

$$\lim_{m\to+\infty} \int_{\mathbb{R}} s_m(x-\xi)\, f(x)\, dx = f(\xi),$$

for any bounded function f that is continuous at $\xi \in \mathbb{R}$.

For more details and properties about delta sequences, see, e.g., [23]. We now recall some classical results.

Proposition 4.1 *Let $(s_m(x))_{m\in\mathbb{N}}$ be a sequence of non-negative functions such that*

$$\int_{-\infty}^{+\infty} s_m(x)\, dx = 1, \ \forall m \geq 1 \tag{7}$$

and, with the property that for any couple of parameters $a, b \in \mathbb{R}$, with $a < b$:

$$\lim_{m\to+\infty} \int_a^b s_m(x-\xi)\, dx = \begin{cases} 0, & a, b > \xi, \text{ or } a, b < \xi, \\ 1/2 & a = \xi \text{ or } b = \xi, \\ 1, & a < \xi < b, \end{cases} \tag{8}$$

for every fixed $\xi \in \mathbb{R}$. Then, it turns out that $(s_m)_m$ is a delta convergent sequence and further, we also have:

$$\lim_{m\to+\infty} \int_{\xi}^{b} s_m\left(\frac{x-\xi}{2}\right) f(x)\, dx = f(\xi), \quad b > \xi$$

for any bounded function f that is continuous at $\xi \in \mathbb{R}$.

Proof The proof that $(s_m(x))_{m\in\mathbb{N}}$ is a delta convergent sequence is a simple generalization of exercise 8 of p. 15 of [23], so we show only the second part of the statement.

Let $\xi \in \mathbb{R}$ be fixed and define

$$g(x) = f(x) - f(\xi), \quad x \in \mathbb{R}.$$

We can write the following equality:

$$\int_\xi^b s_m\left(\frac{x-\xi}{2}\right) f(x)\,dx = f(\xi)\int_\xi^b s_m\left(\frac{x-\xi}{2}\right) dx + \int_\xi^b s_m\left(\frac{x-\xi}{2}\right) g(x)\,dx.$$

Now, we note that

$$\lim_{m\to+\infty} f(\xi)\int_\xi^b s_m\left(\frac{x-\xi}{2}\right) dx = 2f(\xi)\lim_{m\to+\infty}\int_0^{\frac{b-\xi}{2}} s_m(u)\,du = f(\xi),$$

by (8) with $a = \xi = 0$. So it remains to prove that

$$\lim_{m\to+\infty}\int_\xi^b s_m\left(\frac{x-\xi}{2}\right) g(x)\,dx = 0.$$

It is sufficient to see that

$$\left|\int_\xi^b s_m\left(\frac{x-\xi}{2}\right) g(x)\,dx\right| \le 2\int_{\mathbb{R}} s_m(u-\xi)\,|g(2u-\xi)|\,du$$

and hence, putting $h(u) := |g(2u-\xi)|$, observing that $h(u)$ is continuous and bounded, and recalling that $s_m(x)$ is a delta convergent sequence, we finally have

$$\lim_{m\to+\infty}\int_{\mathbb{R}} s_m(u-\xi)\,h(u)\,du = h(\xi) = |g(\xi)| = 0$$

and so the claim follows. □

Some examples of delta convergent sequences are:

1. $s_m^1(x) := \dfrac{m}{\pi\left(1+m^2x^2\right)}$;
2. $s_m^2(x) := \sqrt{\dfrac{m}{\pi}}\,e^{-mx^2}$;

3. $s_m^3(x) := mF(mx)$, where $F(x)$ is the Fejer's kernel, i.e.,

$$F(x) := \frac{1}{2}\text{sinc}^2(x/2), \quad x \in \mathbb{R},$$

where the sinc function is defined as $\text{sinc}(x) = \sin(\pi x)/\pi x$, $x \neq 0$, and $\text{sinc}(0) = 1$.

To see this fact, just note that all the above examples represent sequences of non-negative functions satisfying (7). Moreover, the following equalities hold:

$$\int s_m^1(x-\xi)\,dx = \frac{m}{\pi}\int \frac{1}{1+m^2(x-\xi)^2}dx = \frac{\arctan(m(x-\xi))}{\pi} + C;$$

$$\int s_m^2(x-\xi)\,dx = \sqrt{\frac{m}{\pi}}\int e^{-m(x-\xi)^2}dx = \frac{1}{2}\text{erf}\left(\sqrt{m}(x-\xi)\right) + C;$$

$$\int s_m^3(x-\xi)\,dx = m\int F(m(x-\xi))\,dx$$

$$= \frac{m\pi(x-\xi)\,\text{Si}(m\pi(x-\xi)) + \cos(m\pi(x-\xi)) - 1}{m\pi^2(x-\xi)} + C$$

for $C \in \mathbb{R}$, and where

$$\text{erf}(x) := \frac{2}{\sqrt{\pi}}\int_0^x e^{-t^2}dt, \ x \in \mathbb{R}$$

is the error function and

$$\text{Si}(x) := \int_0^x \frac{\sin(t)}{t}dt, \ x \in \mathbb{R}$$

is the sine integral function. Now, since these functions are odd, and it is well-known that:

$$\lim_{x\to+\infty}\text{erf}(x) = 1, \qquad \lim_{x\to+\infty}\text{Si}(x) = \frac{\pi}{2},$$

(for further details see, e.g., [28]) it is easy to conclude that these examples verify the condition (8).

Now we can prove the main theorem of this section.

Theorem 4.1 *Let $\mathcal{S} := (s_m)_{m\in\mathbb{N}}$ be a sequence of non-negative functions $s_m : \mathbb{R} \to \mathbb{R}^+$ satisfying (7) and (8) and $f : \mathbb{R} \to \mathbb{R}$ be a bounded, and continuous* *function at* $\xi \in \mathbb{R}$. *Setting:*

$$\mathcal{G}_{\mathcal{S}}^{m} := (\mathcal{G}_{w,m})_{w>0},$$

$m \in \mathbb{N}$, *with*

$$\mathcal{G}_{w,m} := (g_{k,w,m})_{k\in\mathbb{Z}}, \qquad g_{k,w,m}(u) := s_m\left(\frac{u-k/w}{2}\right), \quad u \in \mathbb{R},$$

we obtain, for every fixed $x \in \mathbb{R}$:

$$\lim_{m\to+\infty}\left(K_w^{\chi,\mathcal{G}_{\mathcal{S}}^{m}} f\right)(x) = \sum_{k\in\mathbb{Z}} \chi(wx-k)\, f\left(\frac{k}{w}\right) = (S_w^{\chi} f)(x).$$

Proof Let $x \in \mathbb{R}$ be fixed. We consider, tacking $m \in \mathbb{N}$, the family:

$$\left(K_w^{\chi,\mathcal{G}_{\mathcal{S}}^{m}} f\right)(x) = \sum_{k\in\mathbb{Z}} \chi(wx-k)\, \frac{\displaystyle\int_{k/w}^{(k+1)/w} s_m\left(\frac{u-k/w}{2}\right) f(u)\,du}{\displaystyle\int_{k/w}^{(k+1)/w} s_m\left(\frac{u-k/w}{2}\right) du}.$$

Since f is bounded, we can note that

$$\left|\chi(wx-k)\, \frac{\int_{k/w}^{(k+1)/w} s_m\left(\frac{u-k/w}{2}\right) f(u)du}{\int_{k/w}^{(k+1)/w} s_m\left(\frac{u-k/w}{2}\right) du}\right| \leq \|f\|_\infty\, |\chi(wx-k)|,$$

for every $m \in \mathbb{N}$, and since

$$\|f\|_\infty \sum_{k\in\mathbb{Z}} |\chi(wx-k)| \leq \|f\|_\infty\, m_0(\chi) < +\infty,$$

by the dominated convergence theorem and Proposition 4.1, we have

$$\begin{aligned}\lim_{m\to+\infty}\left(K_w^{\chi,\mathcal{G}_{\mathcal{S}}^{m}} f\right)(x) &= \sum_{k\in\mathbb{Z}} \chi(wx-k) \lim_{m\to+\infty} \frac{\displaystyle\int_{k/w}^{(k+1)/w} s_m\left(\frac{u-k/w}{2}\right) f(u)du}{\displaystyle\int_{k/w}^{(k+1)/w} s_m\left(\frac{u-k/w}{2}\right) du} \\ &= \sum_{k\in\mathbb{Z}} \chi(wx-k)\, f\left(\frac{k}{w}\right) = (S_w^{\chi} f)(x).\end{aligned}$$

This completes the proof. □

Acknowledgments The authors are members of the Gruppo Nazionale per l'Analisi Matematica, la Probabilità e le loro Applicazioni (GNAMPA) of the Istituto Nazionale di Alta Matematica (INdAM), of the network RITA (Research ITalian network on Approximation), and of the UMI group "Teoria dell'Approssimazione e Applicazioni".

The third author has been partially supported within the projects: (1) Ricerca di Base 2019 dell'Università degli Studi di Perugia—"Integrazione, Approssimazione, Analisi Nonlineare e loro Applicazioni", (2) "Metodi e processi innovativi per lo sviluppo di una banca di immagini mediche per fini diagnostici" funded by the Fondazione Cassa di Risparmio di Perugia, (FCRP), 2018, (3) "Metodiche di Imaging non invasivo mediante angiografia OCT sequenziale per lo studio delle Retinopatie degenerative dell'Anziano (M.I.R.A.)", funded by FCRP, 2019, and (4) "CARE: A regional information system for Heart Failure and Vascular Disorder", PRJ Project—1507 Action 2.3.1 POR FESR 2014–2020, 2020.

References

1. Acar, T., Costarelli, D., Vinti, G.: Linear prediction and simultaneous approximation by m-th order Kantorovich type sampling serie. Banach J. Math. Anal. **14**(4), 1481–1508 (2020)
2. Altomare, F.: On some convergence criteria for nets of positive operators on continuous function spaces. J. Math. Anal. Appl. **398**(2), 542–552 (2013)
3. Altomare, F.: On the convergence of sequences of positive linear operators and functionals on bounded function spaces. Proc. Am. Math. Soc. **149**, 3837–3848 (2021)
4. Altomare, F., Campiti, M.: Korovkin-Type Approximation Theory and Its Applications. De Gruyter Studies in Mathematics, vol. 17. De Gruyter, Berlin (1994)
5. Altomare, F., Leonessa, V.: On a sequence of positive linear operators associated with a continuous selection of Borel measures. Mediter. J. Math. **3**, 363–382 (2006)
6. Altomare, F., Cappelletti Montano, M., Leonessa, V.: On a generalization of Szász-Mirakjan-Kantorovich operators. Res. Math. **63**, 837–863 (2013)
7. Angeloni, L., Costarelli, D., Vinti, G.: A characterization of the convergence in variation for the generalized sampling series. Ann. Acad. Sci. Fennicae Math. **43**, 755–767 (2018)
8. Bardaro, C., Mantellini, I.: Asymptotic expansion of generalized Durrmeyer sampling type series. Jean J. Approx. **6**(2), 143–165 (2014)
9. Bardaro, C., Musielak, J., Vinti, G.: Nonlinear Integral Operators and Applications. De Gruyter Series in Nonlinear Analysis and Applications, vol. 9. De Gruyter, Berlin (2003)
10. Bardaro, C., Butzer, P.L., Stens R.L., Vinti, G.: Kantorovich-type generalized sampling series in the setting of Orlicz spaces. Samp. Theory Sign. Image Proc. **6**, 29–52 (2007)
11. Bardaro, C., Faina, L., Mantellini, I.: Quantitative Voronovskaja formulae for generalized Durrmeyer sampling type series. Math. Nachr. **289**(14–15), 1702–1720 (2016)
12. Butzer, P.L., Nessel, R.J.: Fourier Analysis and Approximation I. Academic Press, New York (1971)
13. Butzer, P.L., Stens, R.L.: Linear prediction by samples from past. In: Advanced Topics in Shannon Sampling and Interpolation Theory R. J. Marks II. Springer Texts Electrical Engineering, pp. 157–183. Springer, New York (1993)
14. Butzer, P.L., Fisher, A., Stens, R.L.: Approximation of continuous and discontinuous functions by generalized sampling series. J. Approx. Theory **50**, 25–39 (1987)
15. Butzer, P.L., Fisher, A., Stens, R.L.: Generalized sampling aproximation of multivariate signals. Atti Sem. Mat. Fis. Univ. Modena **41**, 17–37 (1993)
16. Cantarini, M., Costarelli, D., Vinti, G.: A solution of the problem of inverse approximation for the sampling Kantorovich operators in case of Lipschitz functions. Dolomites Res. Notes Approx. **13**, 30–35 (2020)

17. Cantarini, M., Costarelli, D., Vinti, G.: Approximation of differentiable and not differentiable signals by the first derivative of sampling Kantorovich operators. J. Math. Anal. Appl. **509**, 125913 (2022)
18. Costarelli, D., Spigler, R.: How sharp is the Jensen inequality? J. Inequalities Appl. **2015**, 1–10 (2015)
19. Costarelli, D., Vinti, G.: An inverse result of approximation by sampling Kantorovich series. Proc. Edinburgh Math. Soc. **62**(1), 265–280 (2019)
20. Costarelli, D., Vinti, G.: Inverse results of approximation and the saturation order for the sampling Kantorovich series. J. Approx. Theory **242**, 64–82 (2019)
21. Costarelli, D., Vinti, G.: Saturation by the Fourier transform method for the sampling Kantorovich series based on bandlimited kernels. Anal. Math. Phys. **9**, 2263–2280 (2019)
22. Costarelli, D., Piconi, M., Vinti, G.: On the convergence properties of Durrmeyer-Sampling type operators in Orlicz spaces. Math. Nachrichten (2021). https://doi.org/10.1002/mana.202100117
23. Kanwal, R.P.: Generalized Functions: Theory and Applications, 3rd edn. Springer, New York (2004)
24. Karsli, H.: On Urysohn type generalized sampling operators. Dolomites Res. Notes Approx. **14**(2), 58–67 (2021)
25. Karsli, H.: On multidimensional Urysohn type generalized sampling operators. Math. Found. Comput. **4**(4), 271–280 (2021). Special issue on: Approximation by linear and nonlinear operators with applications
26. Kivinukk, A., Tamberg, G.: On window methods in generalized Shannon sampling operators. In: New Perspectives on Approximation and Sampling Theory. Birkhäuser, Cham, pp. 63–85 (2014)
27. Musielak, J., Orlicz, W.: Spaces and Modular Spaces. Lecture Notes in Mathematics, vol. 1034. Springer, Berlin (1983)
28. Olver, F.W.J., Lozier, D.W., Boisfert, R.F., Clark, C.W.: NIST Handbook of Mathematical Functions. Cambridge University Press, Cambridge (2010)
29. Orlova, O., Tamberg, G.: On approximation properties of generalized Kantorovich-type sampling operators. J. Approx. Theory **201**, 73–86 (2016)
30. Rao, M.M., Ren, Z.D.: Theory of Orlicz Spaces. Monographs and Textbooks in Pure and Applied Mathematics, vol. 146. Marcel Dekker, New York (1991)
31. Vinti, G.: A general approximation result for nonlinear integral operators and applications to signal processing. Appl. Anal. **79**, 217–238 (2001)

A Modification of Bernstein-Durrmeyer Operators with Jacobi Weights on the Unit Interval

Mirella Cappelletti Montano and Vita Leonessa

Dedicated to Francesco Altomare with great affection, gratitude and esteem

Abstract The present paper is devoted to the study of a sequence of positive linear operators, acting on the space of all continuous functions on [0, 1] as well as on some weighted spaces of integrable functions on [0, 1]. These operators are, as a matter of fact, a generalization of the Bernstein-Durrmeyer operators with Jacobi weights. In particular, we present qualitative and approximation properties of these operators, also providing estimates of the rate of convergence. Moreover, by means of their asymptotic formula, we compare our operators with the Bernstein-Durrmeyer ones and a suitable modification of theirs, showing that, in suitable intervals, they provide a lower approximating error estimate.

Keywords Bernstein-Durrmeyer-type operators · Jacobi weights · Positive approximation processes · Rate of convergence · Generalized convexity

1 Introduction

In [17] a modification of the classical Bernstein operators B_n on [0, 1] that fixes the constants and the function x^2 (instead of x) was introduced; the author in particular showed that this modification provides an error of approximation that is as least as good as the one of the Bernstein operators on certain subintervals of [0, 1].

M. Cappelletti Montano
Dipartimento di Matematica, Università degli Studi di Bari Aldo Moro, Bari, Italy
e-mail: mirella.cappellettimontano@uniba.it

V. Leonessa (✉)
Department of Mathematics, Computer Science and Economics, University of Basilicata, Potenza, Italy
e-mail: vita.leonessa@unibas.it

A. M. Candela et al. (eds.), *Recent Advances in Mathematical Analysis*,
Trends in Mathematics, https://doi.org/10.1007/978-3-031-20021-2_14

Since then, many other mathematicians have undertaken the task to construct other modifications of well-known approximation processes in the same spirit of [17], in order to get better approximation results. For a survey on such type of operators we refer the interested readers to [4].

In particular, in [13], the authors introduced a modification of the Bernstein operators that fixes the constants and a given strictly increasing function, as follows: for any $n \geq 1$ and $f \in C([0,1])$,

$$B_n^\tau(f) = B_n(f \circ \tau^{-1}) \circ \tau,$$

where τ is a suitable strictly increasing C^∞-function on $[0,1]$. The authors studied qualitative and quantitative properties of such operators and compared their approximation error estimate with the one of the B_n's.

Subsequently (see [3]) this idea was applied in order to define a modification M_n^τ of the Bernstein-Durrmeyer operators on $[0, 1]$ introduced in [15], and independently in [20], which are a useful tool to study the approximation properties also of integrable functions.

During the years, Bernstein-Durrmeyer operators have been object of investigations by many authors (see, e.g. [12, 14]); in particular, in [21] the author studied a generalization $M_{n,a,b}$ of Bernstein-Durrmeyer operators acting on weighted spaces of integrable functions, where the considered one is the classical Jacobi weight $w_{a,b}$ on $[0, 1]$. Those operators have been intensely studied during the years in the one-dimensional and in multidimensional setting (see, e.g., [1, 23, 26]), also in connection with certain partial differential problems (see [5, 9]).

In this paper, we present a modification of the Bernstein-Durrmeyer operators with Jacobi weights $M_{n,a,b}^\tau$ in the same spirit of [3, 13].

We establish some qualitative properties of the operators $M_{n,a,b}^\tau$, such as their behaviour with respect to Lipschitz-continuous functions; moreover, we prove that they preserve some forms of convexity. We also prove that the sequence $(M_{n,a,b}^\tau)_{n\geq 1}$ is an approximation process in $C([0,1])$, as well in suitable spaces of integrable functions, and we evaluate the rate of convergence by means of appropriate moduli of smoothness.

Finally, we use an asymptotic formula for the operators $M_{n,a,b}^\tau$ in order to compare them with the $M_{n,a,b}$'s and the M_n^τ's, showing under which conditions the operators introduced in the present paper provide a lower approximating error estimate at least on certain subintervals of $[0, 1]$.

2 Preliminaries

From now on fix $a, b \in]-1, +\infty[$ and consider the normalized Jacobi weight

$$w_{a,b}(x) := \frac{x^a(1-x)^b}{\int_0^1 y^a(1-y)^b\,dy} \quad (0 < x < 1). \tag{1}$$

Moreover, let us denote by $\mu_{a,b} \in M_1^+([0,1])$ the absolutely continuous measure with respect to the Borel-Lebesgue measure λ_1 on $[0,1]$ with density $w_{a,b}$. Obviously, if $a = b = 0$ then $\mu_{0,0} = \lambda_1$.

By means of such a measure, it is possible to define the so called Bernstein-Durrmeyer operators with Jacobi weights (see [21, 22]). More precisely, for every $f \in L^1([0,1], \mu_{a,b})$, $n \geq 1$ and $x \in [0,1]$, set

$$M_{n,a,b}(f)(x) := \sum_{h=0}^{n} \omega_{n,h}(f) \binom{n}{h} x^h (1-x)^{n-h}, \tag{2}$$

where

$$\begin{aligned} \omega_{n,h}(f) : &= \frac{1}{\int_0^1 t^h (1-t)^{n-h} \, d\mu_{a,b}} \int_0^1 t^h (1-t)^{n-h} f(t) \, d\mu_{a,b} \\ &= \frac{\Gamma(n+a+b+2)}{\Gamma(h+a+1)\Gamma(n-h+b+1)} \int_0^1 t^{h+a} (1-t)^{n-h+b} f(t) \, dt, \end{aligned} \tag{3}$$

Γ being the classical Euler Gamma function.

We also recall that in [9] it has been noted that

$$M_{n,a,b}(f) = B_n(D_{n,a,b}(f)) \quad (f \in L^1([0,1], \mu_{a,b})), \tag{4}$$

where B_n stand for the classical Bernstein operators on $[0,1]$ and $D_{n,a,b} : L^1([0,1], \mu_{a,b}) \longrightarrow L^1([0,1], \mu_{a,b})$ are the positive linear operators defined in [10, formula (4.6)] by

$$D_{n,a,b}(f)(x) = \frac{\Gamma(n+a+b+2)}{\Gamma(nx+a+1)\Gamma(n-nx+b+1)} \int_0^1 t^{nx+a} (1-t)^{n-nx+b} f(t) \, dt. \tag{5}$$

Given (4) and denoted by $e_m(x) = x^m$, $m \in \mathbb{N}$, it is possible to evaluate $M_n(e_m)$, $m \in \mathbb{N}$, since

$$D_{n,a,b}(e_m) = \frac{\Gamma(n+a+b+2)}{\Gamma(m+n+a+b+2)} (a+1+ne_1) \cdots (a+m+ne_1). \tag{6}$$

In particular (see [22, Section 5.2]),

$$M_{n,a,b}(e_0) = e_0 \tag{7}$$

$$M_{n,a,b}(e_1) = \frac{a+1+ne_1}{n+a+b+2}, \tag{8}$$

and

$$M_{n,a,b}(e_2) = \frac{(a+1)(a+2) + 2n(a+2)e_1 + n(n-1)e_2}{(n+a+b+2)(n+a+b+3)}. \tag{9}$$

Moreover, if, for a given $x \in [0, 1]$ we denote by $\psi_x^i(t) = (t-x)^i$, $i \geq 1$, we have that

$$M_{n,a,b}(\psi_x)(x) = \frac{a+1-(a+b+2)x}{n+a+b+2} \tag{10}$$

and

$$\begin{aligned} M_{n,a,b}(\psi_x^2)(x) &= \frac{2nx(1-x) + x^2(a+b+2)(a+b+3)}{(n+a+b+2)(n+a+b+3)} \\ &- \frac{2x(a+1)(a+b+3)}{(n+a+b+2)(n+a+b+3)} + \frac{(a+1)(a+2)}{(n+a+b+2)(n+a+b+3)}. \end{aligned} \tag{11}$$

3 Modified Bernstein-Durrmeyer Operators with Jacobi Weights

In what follows, τ will be an infinitely differentiable function on $[0, 1]$ such that $\tau(0) = 0$, $\tau(1) = 1$, and $\tau'(x) > 0$ for $x \in [0, 1]$.

Consider now the image measure $\mu_{a,b}^{\tau}$ of $\mu_{a,b}$ by means of τ and the corresponding Lebesgue space $L^p([0, 1], \mu_{a,b}^{\tau})$, with $1 \leq p < +\infty$. Namely, a function f belongs to $L^p([0, 1], \mu_{a,b}^{\tau})$ if

$$\int_0^1 |f|^p \, d\mu_{a,b}^{\tau} = \int_0^1 |f \circ \tau^{-1}|^p \, d\mu_{a,b} < +\infty.$$

Such a space is equipped with the norm $\|\cdot\|_{L^p([0,1],\mu_{a,b}^{\tau})}$. If $\tau = e_1$, $\mu_{a,b}^{e_1} = \mu_{a,b}$ and, if this is the case, we will omit the superscript e_1. Note that $f \in L^p([0, 1], \mu_{a,b})$ if and only if $f \in L^p([0, 1], \mu_{a,b}^{\tau})$. Moreover, also $\mu_{a,b}^{\tau} \in M_1^+([0, 1])$. Finally, whenever $a = b = 0$ and $\tau = e_1$, the corresponding space is indeed $L^p([0, 1])$ endowed with the usual norm $\|\cdot\|_p$.

For every $n \geq 1$, the positive linear operator

$$M_{n,a,b}^{\tau} : L^1([0, 1], \mu_{a,b}^{\tau}) \longrightarrow L^1([0, 1], \mu_{a,b}^{\tau})$$

that we are going to consider is defined by setting, for every $f \in L^1([0,1], \mu_{a,b}^\tau)$, $0 \le x \le 1$,

$$M_{n,a,b}^\tau(f)(x) := \sum_{h=0}^{n} \binom{n}{h} \tau^h(x)(1-\tau(x))^{n-h} \omega_{n,h}(f \circ \tau^{-1}) \tag{12}$$

(see (3)). More precisely

$$M_{n,a,b}^\tau(f) = M_{n,a,b}(f \circ \tau^{-1}) \circ \tau. \tag{13}$$

If we choose $\tau = e_1$ we get the original Bernstein-Durrmeyer operators with Jacobi weights $M_{n,a,b}$ [21].

Further, in the particular case of $a = b = 0$, we get the modified Bernstein-Durrmeyer operators introduced and studied in [3]. Observe that, if in addition $\tau = e_1$, then those operators turn into the classical Bernstein-Durrmeyer operators [15, 20].

If $f = \tau^m$ ($m \in \mathbb{N}$), then $f \circ \tau^{-1} = e_m$. Hence we have

$$M_{n,a,b}^\tau(\tau^m) = (M_{n,a,b}(e_m)) \circ \tau. \tag{14}$$

In particular, from (7)–(9), we get

$$M_{n,a,b}^\tau(e_0) = e_0 \tag{15}$$

$$M_{n,a,b}^\tau(\tau) = \frac{a+1+n\tau}{n+a+b+2}, \tag{16}$$

and

$$M_{n,a,b}^\tau(\tau^2) = \frac{(a+1)(a+2)+2n(a+2)\tau+n(n-1)\tau^2}{(n+a+b+2)(n+a+b+3)}. \tag{17}$$

For the operators $M_{n,a,b}^\tau$, a formula similar to (4) can be obtained considering the following modification B_n^τ of the Bernstein operators introduced in [13]:

$$B_n^\tau(f)(x) = \sum_{h=0}^{n} \binom{n}{h} \tau^h(x)(1-\tau(x))^{n-h} (f \circ \tau^{-1}) \left(\frac{h}{n}\right)$$

($n \ge 1$, $f \in C([0,1])$, $0 \le x \le 1$). In fact, on account of (5),

$$M_{n,a,b}^\tau(f) = B_n^\tau(D_{n,a,b}(f \circ \tau^{-1}) \circ \tau) \quad (f \in L^1([0,1], \mu_{a,b}^\tau)). \tag{18}$$

Theorem 3.1 *For every* $f \in C([0,1])$ *we have*

$$\lim_{n\to\infty} M^{\tau}_{n,a,b}(f) = f \quad \textit{uniformly on } [0,1]\,.$$

Proof It is sufficient to note that $\{e_0, \tau, \tau^2\}$ is an extended complete Tchebychev system on $[0,1]$ and that, thanks to (15)-(17), $\lim_{n\to\infty} M^{\tau}_{n,a,b}(e_0) = e_0$, $\lim_{n\to\infty} M^{\tau}_{n,a,b}(\tau) = \tau$, and $\lim_{n\to\infty} M^{\tau}_{n,a,b}(\tau^2) = \tau^2$, uniformly on $[0,1]$. □

In order to get some estimates of the rate of convergence in Theorem 3.1, we use a general result (see [8, 22]) which involves the usual first modulus of continuity $\omega(f,\delta)$ and the second modulus of smoothness $\omega_2(f,\delta)$. To this end, we need some further notations.

For $x \in [0,1]$, let $\psi^x_{\tau,i}$ be the function defined by

$$\psi^x_{\tau,i}(t) = (\tau(t) - \tau(x))^i \quad (i = 0, 1, 2, \ldots)\,.$$

If $\tau = e_1$ we shall simply write $\psi^i_x(t) = (t-x)^i$.

For any $n \geq 1$ and $x \in [0,1]$ (see (11)), we have

$$\begin{aligned} M^{\tau}_{n,a,b}(\psi^x_{\tau,2})(x) &= \frac{2n\tau(x)(1-\tau(x)) + \tau(x)^2(a+b+2)(a+b+3)}{(n+a+b+2)(n+a+b+3)} \\ &- \frac{2\tau(x)(a+1)(a+b+3)}{(n+a+b+2)(n+a+b+3)} + \frac{(a+1)(a+2)}{(n+a+b+2)(n+a+b+3)}. \end{aligned} \tag{19}$$

Moreover, by using a result due to Freud (see [16]), we get that there exists a constant $K > 0$ such that

$$K\psi^2_x(t) \leq \tau'(x)\psi^x_{\tau,2}(t) \quad \text{for every } x, t \in [0,1]\,. \tag{20}$$

Obviously, $K = 1$ if $\tau = e_1$.

We can now state the following result.

Proposition 3.1 *Consider* $n \geq 1$, $f \in C([0,1])$ *and* $0 \leq x \leq 1$. *Then*

$$|M^{\tau}_{n,a,b}(f)(x) - f(x)| \leq \omega(f, \sigma^{\tau}_n(x)) + \frac{3}{2}\omega_2(f, \sigma^{\tau}_n(x))\,, \tag{21}$$

where

$$\sigma^{\tau}_n(x) = \frac{\sqrt{\tau'(x)}}{\sqrt{K}} \times$$
$$\sqrt{\frac{2n\tau(x)(1-\tau(x)) + \tau(x)^2(a+b+2)(a+b+3) - 2\tau(x)(a+1)(a+b+3) + (a+1)(a+2)}{(n+a+b+2)(n+a+b+3)}}.$$

Moreover,

$$\|M^{\tau}_{n,a,b}(f) - f\|_{\infty} \le \omega(f, \delta^{\tau}_{n}) + \frac{3}{2}\omega_2(f, \delta^{\tau}_{n}), \tag{22}$$

where

$$\delta^{\tau}_{n} = \frac{\sqrt{\|\tau'\|_{\infty}}}{\sqrt{K}} \sqrt{\frac{n/2 + \max\{a^2 + 3a + 2, b^2 + 3b + 2\}}{(n + a + b + 2)(n + a + b + 3)}}.$$

Proof Let $n \ge 1$, $f \in C([0, 1])$, $0 \le x \le 1$ and $\delta > 0$. Using [22, Theorem 2.2.1] (see also [8, Theorem 1.6.2]), we have that

$$|M^{\tau}_{n,a,b}(f)(x) - f(x)| \le |f(x)||M^{\tau}_{n,a,b}(e_0)(x) - 1| + \frac{1}{\delta}|M^{\tau}_{n,a,b}(\psi_x)(x)|\omega(f, \delta)$$
$$+ \left\{M^{\tau}_{n,a,b}(e_0)(x) + \frac{1}{2\delta^2}M^{\tau}_{n,a,b}(\psi_x^2)(x)\right\}\omega_2(f, \delta)$$
$$= \frac{1}{\delta}|M^{\tau}_{n,a,b}(\psi_x)(x)|\omega(f, \delta) + \left\{1 + \frac{1}{2\delta^2}M^{\tau}_{n,a,b}(\psi_x^2)(x)\right\}\omega_2(f, \delta).$$

By Cauchy-Schwarz inequality we get

$$|M^{\tau}_{n,a,b}(\psi_x)| \le \sqrt{M^{\tau}_{n,a,b}(\psi_x^2)},$$

therefore

$$|M^{\tau}_{n,a,b}(f)(x) - f(x)| \le$$
$$\le \frac{1}{\delta}\sqrt{M^{\tau}_{n,a,b}(\psi_x^2)(x)}\omega(f, \delta) + \left\{1 + \frac{1}{2\delta^2}M^{\tau}_{n,a,b}(\psi_x^2)(x)\right\}\omega_2(f, \delta).$$

From (20) and the positivity of the $M^{\tau}_{n,a,b}$'s, we get

$$M^{\tau}_{n,a,b}(\psi_x^2) \le \frac{\tau'(x)}{K}M^{\tau}_{n,a,b}(\psi^x_{\tau,2}).$$

Taking (19) into account and setting $\delta = \sigma^{\tau}_{n}(x)$, we get (21).

To get (22), note that, for every $x \in [0, 1]$, $2n\tau(x)(1-\tau(x)) \le n/2$ and that the function

$$g(x) = \tau(x)^2(a + b + 2)(a + b + 3) - 2\tau(x)(a + 1)(a + b + 3) + (a + 1)(a + 2)$$

has a unique critical point at $x = \tau^{-1}\left(\dfrac{a+1}{a+b+2}\right)$, which is a local minimum point, so that its global maximum has to be $\max\{g(0), g(1)\}$. □

We pass now to discuss approximation properties of $(M^{\tau}_{n,a,b}(f))_{n\geq 1}$ also in the space of $L^p([0,1], \mu^{\tau}_{a,b})$, $p \geq 1$. We note that these results seem to be new also in the context $a = b = 0$.

First we recall that a measure μ on $[0, 1]$ is said to be *invariant* for an operator B of domain $D(B)$ if

$$\int_0^1 B(f)\, d\mu = \int_0^1 f\, d\mu \quad \text{for every } f \in D(B)$$

(see [18, Section 5.1, p. 178]).

Lemma 3.1 *The measure $\mu^{\tau}_{a,b}$ is an invariant measure for the operators $M^{\tau}_{n,a,b}$ on $L^1([0,1], \mu^{\tau}_{a,b})$, and in particular for their restrictions to $C([0,1])$. Moreover, each $M^{\tau}_{n,a,b}$ is a contraction from $L^p([0,1], \mu^{\tau}_{a,b})$ into itself.*

Proof Fix $f \in L^1([0,1], \mu^{\tau}_{a,b})$; then

$$\int_0^1 M^{\tau}_{n,a,b}(f)\, d\mu^{\tau}_{a,b} = \sum_{h=0}^{n} \binom{n}{k} \omega_{n,h}(f \circ \tau^{-1}) \int_0^1 \tau^h (1-\tau)^{n-h}\, d\mu^{\tau}_{a,b}$$

$$= \frac{1}{\int_0^1 y^a (1-y)^b\, dy} \int_0^1 \left[\sum_{h=0}^{n} \binom{n}{k} y^h (1-y)^{n-h}\right] y^a (1-y)^b (f \circ \tau^{-1})(y)\, dy$$

$$= \frac{1}{\int_0^1 y^a (1-y)^b\, dy} \int_0^1 y^a (1-y)^b (f \circ \tau^{-1})(y)\, dy = \int_0^1 f\, d\mu^{\tau}_{a,b},$$

hence the first part of the claim is proven.

In order to prove the second part, first note that from Jensen's inequality it follows that, if $f \in L^p([0,1], \mu^{\tau}_{a,b})$, $|M^{\tau}_{n,a,b}(f)|^p \leq M^{\tau}_{n,a,b}(|f|^p)$. Then

$$\int_0^1 |M^{\tau}_{n,a,b}(f)|^p\, d\mu^{\tau}_{a,b} \leq \int_0^1 M^{\tau}_{n,a,b}(|f|^p)\, d\mu^{\tau}_{a,b} = \int_0^1 |f|^p\, d\mu^{\tau}_{a,b},$$

that is $\|M^{\tau}_{n,a,b}\|_{L^p([0,1],\mu^{\tau}_{a,b})} \leq 1$. □

Theorem 3.2 *For every $f \in L^p([0,1], \mu^{\tau}_{a,b})$,*

$$\lim_{n\to\infty} M^{\tau}_{n,a,b}(f) = f \quad \text{in } L^p([0,1], \mu^{\tau}_{a,b}).$$

Proof As a consequence of the previous lemma, the sequence $(M^{\tau}_{n,a,b})_{n\geq 1}$ is equibounded from $L^p([0,1], \mu^{\tau}_{a,b})$ into $L^p([0,1], \mu^{\tau}_{a,b})$. On account of Theorem 3.1 and recalling that $C([0,1])$ is dense in $L^p([0,1], \mu^{\tau}_{a,b})$ (see [11, Lemma 26.2 and Theorem 29.14]) the proof is given. □

We proceed by obtaining an estimate of the convergence in Theorem 3.2 in the particular case $a = b = 0$ for which, as quoted before, our operators turn into those ones considered in [3]. Let us denote by M^{τ}_n the operators $M^{\tau}_{n,0,0}$ and by μ^{τ} the measure $\mu^{\tau}_{0,0}$. We shall use a result due to Swetits and Wood [25, Theorem 1] which involves the second-order integral modulus of smoothness defined, for $f \in L^p([0,1])$, $1 \leq p < +\infty$, as

$$\omega_{2,p}(f,\delta) := \sup_{0<t\leq\delta} \|f(\cdot+t) - 2f(\cdot) + f(\cdot-t)\|_p \quad (\delta > 0).$$

In particular, recalling that $f \circ \tau^{-1} \in L^p([0,1])$, [25, Theorem 1] states that there exists a constant $C_p > 0$ such that

$$\begin{aligned}\|M^{\tau}_n(f) - f\|_{L^p([0,1],\mu^{\tau})} &= \|M_n(f\circ\tau^{-1}) - f\circ\tau^{-1}\|_p \\ &\leq C_p\{\rho^2_{n,p}\|f\circ\tau^{-1}\|_p + \omega_{2,p}(f\circ\tau^{-1}, \rho_{n,p})\},\end{aligned}$$

where the sequence $\rho_{n,p} \to 0$ as $n \to \infty$ and it is defined as follows:

$$\rho_{n,p} := \max\left\{\|M_n(\psi_x)\|_p^{1/2}, \|M_n(\psi_x^2)\|_p^{p/(2p+1)}\right\}.$$

From (10) we get $\|M_n(\psi_x)\|_p^{1/2} \leq \dfrac{1}{\sqrt{n+2}(p+1)^{1/(2p)}} =: \beta_{n,p} \to 0.$ Moreover,

$$\|M_n(\psi_x^2)\|_p^{p/(2p+1)} = \frac{\|2ne_1(1-e_1) - 6e_1(1-e_1) + 2\|_p^{p/(2p+1)}}{(n+2)^{p/(2p+1)}(n+3)^{p/(2p+1)}} =: \gamma_{n,p}\,.$$

Note that

$$0 \leq \gamma_{n,p} \leq \left(\frac{n+4}{2(n+2)(n+3)}\right)^{p/(2p+1)} \to 0\,.$$

By setting

$$\alpha_{n,p} := \max\{\beta_{n,p}, \gamma_{n,p}\}$$

we obtain that

$$\|M^{\tau}_n(f) - f\|_{L^p([0,1],\mu^{\tau})} \leq C_p(\alpha^2_{n,p}\|f\circ\tau^{-1}\|_p + \omega_{2,p}(f\circ\tau^{-1}, \alpha_{n,p})). \quad (23)$$

We now present some shape preserving properties of the operators $M^{\tau}_{n,a,b}$.

For every $k \in \mathbb{N}$, consider the linear subspace $\mathbb{P}_{\tau,k}$ generated by the set $\{\tau^i : i = 0, \dots, k\}$. This space is invariant under our operators, i.e.

$$M^{\tau}_{n,a,b}(\mathbb{P}_{\tau,k}) \subset \mathbb{P}_{\tau,k} \qquad (k \in \mathbb{N}, n \geq 1).$$

Indeed, as shown in (6), $D_{n,a,b}$ maps polynomials on [0, 1] into polynomials on [0, 1] of the same degree; by this, (18), (14) and the fact that $B^{\tau}_n(\mathbb{P}_{\tau,k}) \subset \mathbb{P}_{\tau,k}$ (see [13, Section 2]), the statement easily follows.

We also prove that the $M^{\tau}_{n,a,b}$'s preserve some forms of convexity and we investigate their behaviour with respect to Lipschitz-continuous functions.

First of all, we point out that the operators $M^{\tau}_{n,a,b}$ do not preserve the usual convexity. For instance, if $\tau(x) = 4/\pi \arctan(x)$ $(0 \leq x \leq 1)$, then $M^{\tau}_{n,0,0}(e_1)$ is not convex for low values of n.

Anyway, the operators $M^{\tau}_{n,a,b}$ preserve other forms of convexity.

We recall (see [27]) that a function $f \in C([0, 1])$ is said to be *convex with respect to* τ if, whenever $0 \leq x_0 < x_1 < x_2 \leq 1$,

$$\begin{vmatrix} 1 & 1 & 1 \\ \tau(x_0) & \tau(x_1) & \tau(x_2) \\ f(x_0) & f(x_1) & f(x_2) \end{vmatrix} \geq 0.$$

In particular, f is convex with respect to τ if and only if $f \circ \tau^{-1}$ is convex.

We can state the following result.

Proposition 3.2 *Let* $f \in C([0, 1])$ *be convex with respect to* τ. *Then* $M^{\tau}_{n,a,b}(f)$ *is convex with respect to* τ *for any* $n \geq 1$.

Proof Since, for every $n \geq 1$, the operators $M_{n,a,b}$ map continuous convex functions into (continuous) convex functions (see, for example, [9, Proposition 2]), if $f \in C([0, 1])$ is continuous with respect to τ, then $M_{n,a,b}(f \circ \tau^{-1})$ is convex and, hence, $M^{\tau}_{n,a,b}(f)$ is convex with respect to τ by means of (13). □

Another form of convexity can be considered.

Let us fix $k \geq 1$ and $a_0 < a_1 < \dots < a_k \in \mathbb{R}$; moreover, for $x \in \mathbb{R}$ set $u(x) := (x - a_0) \cdots (x - a_k)$. If $f : [a_0, a_k] \to \mathbb{R}$, the *divided difference* of f with respect to $a_0, \dots, a_k$ is defined by

$$[a_0, \dots, a_k; f] := \sum_{h=0}^{k} \frac{f(a_h)}{u'(a_h)}.$$

A function $f : I \to \mathbb{R}$ is said to be *k-convex* (see, e.g., [8, Appendix 2]) on the interval I if for all $a_0 < a_1 < \dots < a_k$ in I one has $[a_0, \dots, a_k; f] \geq 0$.

On the other hand, if $a \in I$ and $h > 0$ are such that $a, a+h, a+2h, \ldots, a+kh \in I$, then

$$[a, a+h, a+2h, \ldots, a+kh;\, f] = \frac{1}{k!h^k}\Delta_h^k f(a),$$

where $\Delta_h^k f(a)$ is the classical k-th difference of f with step h at point a.

Hence, if f is k-convex, then $\Delta_h^k f(a) \geq 0$ for all $a \in I$. Moreover, if f is continuous, it is always possible to choose the $a_0, \ldots, a_k$ in the definition of k-convex functions in such a way they are equally spaced (see [24]), so a continuous function is k-convex if and only if $\Delta_h^k f(a) \geq 0$ for all $a \in I$.

We remark that a function $f \in C^k([0,1])$ is k-convex if $f^{(k)} \geq 0$.

Obviously, 1-convex functions are just the increasing ones, while 2-convex functions are the usual convex ones.

It is possible to further extend the definition of k-convex functions following [19]. If $f \in C(I)$, set, for all $a \in I$,

$$\Delta_{h,\varphi}^k f(a) := \Delta_h^k (f \circ \varphi^{-1})(\varphi(a)),$$

φ being a C^∞-function on I such that $\varphi'(x) \neq 0$ for all $x \in I$ and that $\lim_{x\to 0}\varphi(x) = 0$, provided that 0 is a cluster point for I. f is said *φ-convex of order k* (see [19]) if $\Delta_{h,\varphi}^k f(a) \geq 0$.

If $f \in C^k([0,1])$, then f is φ-convex of order k if

$$D_\varphi^{(k)}(f)(x) := (f \circ \varphi^{-1})^{(k)}(\varphi(x)) \geq 0 \quad (x \in I).$$

It is easy to show that, given our assumptions on τ, a function $f \in C^k([0,1])$ is τ-convex of order k if and only if

$$(f \circ \tau^{-1})^{(k)} \geq 0;$$

in other words, f is τ-convex of order k if $f \circ \tau^{-1}$ is k-convex.

Many classical approximation processes preserve k-convex functions, like for example Bernstein operators (see [8, Prop. A.2.5]) or the classical Bernstein-Durrmeyer operators (see [2]). Also Berstein-Durrmeyer operators with Jacobi weights preserve k-convexity.

First off, given $k \in \mathbb{N}$ and $h = 0, \ldots, n$, we set

$$\Delta_1^k \omega_{n,h}(f) = \sum_{l=0}^{k} (-1)^{k-l}\binom{k}{l}\omega_{n,h+l}(f).$$

By induction, it is easy to prove that, for every $n \geq 1$, $k \in \mathbb{N}$, $f \in C([0,1])$, and $x \in [0,1]$,

$$M_{n,a,b}^{(k)}(f)(x) = n(n-1)\cdots(n-k+1)\sum_{h=0}^{n-k}\binom{n-k}{h}\Delta_1^k\omega_{n,h}(f)x^h(1-x)^{n-h-k}. \tag{24}$$

Following the same reasoning as in [2] (see also [9, Proposition 2.7]), we can prove the following result.

Proposition 3.3 *If $f \in C([0,1])$ is k-convex, then $M_{n,a,b}(f)$ is k-convex.*
As a consequence, the operators $M_{n,a,b}^{\tau}$ preserve τ-convexity of order k.

Proof Let us fix a k-convex function $f \in C([0,1])$. It is enough to assume $f \in C^k([0,1])$; in fact every continuous k-convex function is the uniform limit a sequence $(f_m)_{m\geq 1}$ of k-convex and C^k functions (take, for example $f_m = B_m(f)$ for all $m \geq 1$).

To show that $M_{n,a,b}(f)$ is k-convex, taking (24) into account, we have to prove that

$$\Delta_1^k\omega_{n,h}(f) \geq 0.$$

Indeed,

$$\Delta_1^k\omega_{n,h}(f) = (-1)^k\frac{\Gamma(n+a+b+2)}{\Gamma(h+a+1+k)\Gamma(n-h+b+1)}\int_0^1 F^{(k)}(x)f(x)\,dx,$$

where $F(x) = x^{h+a+k}(1-x)^{n-h+b}$. From this, integrating by parts,

$$(-1)^k\int_0^1 F^{(k)}(x)f(x)\,dx, = \int_0^1 F(x)f^{(k)}(x)\,dx \geq 0$$

and this completes the proof.

Consider now a τ-convex function f of order k; we have to show that $M_{n,a,b}^{\tau}(f)\circ\tau^{-1}$ is k-convex but this is a straightforward consequence of the previous considerations, (13) and the fact that $f \circ \tau^{-1}$ is k-convex. □

We pass now to investigate the behavior of the operators $M_{n,a,b}^{\tau}$ on Lipschitz-continuous functions. We first recall that we denote by Lip([0, 1]) the space consisting of those $f \in C([0,1])$ such that

$$|f|_{\mathrm{Lip}} := \sup_{\substack{x,y\in[0,1]\\ x\neq y}} \frac{|f(x)-f(y)|}{|x-y|} < +\infty.$$

Moreover, for $M > 0$, $f \in \mathrm{Lip}_M 1$ if

$$|f(x) - f(y)| \le M|x - y| \quad \text{for every } 0 \le x, y \le 1.$$

$\mathrm{Lip}_M 1$ is said to be the space of all Lipschitz continuous functions with Lipschitz constant M.

Finally, for $0 \le \alpha \le 1$, we shall write $f \in \mathrm{Lip}_M \alpha$ if

$$|f(x) - f(y)| \le M|x - y|^\alpha \quad \text{for every } 0 \le x, y \le 1.$$

Observe that, both τ and τ^{-1} are Lipschitz continuous functions. More precisely, $\tau \in \mathrm{Lip}_L 1$ with $L := \|\tau'\|_\infty$ and $\tau^{-1} \in \mathrm{Lip}_N 1$ with $N := (\min_{[0,1]} \tau')^{-1}$.

Proposition 3.4 *$M^\tau_{n,a,b}(f) \in \mathrm{Lip}([0, 1])$ for every $n \ge 1$ and $f \in \mathrm{Lip}([0, 1])$; moreover*

$$|M^\tau_{n,a,b}(f)|_{\mathrm{Lip}} \le \left(1 + \frac{\omega}{n}\right) LN|f|_{\mathrm{Lip}}. \tag{25}$$

where

$$\omega := -\frac{a + b + 2}{a + b + 3} < 0. \tag{26}$$

As a consequence

$$M^\tau_{n,a,b}(\mathrm{Lip}_M 1) \subset \mathrm{Lip}_{MLN} 1 \quad \textit{for every } n \ge 1\,. \tag{27}$$

Further, for every $n \ge 1$, $f \in C([0, 1])$, $\delta > 0$, $M > 0$ and $0 < \alpha \le 1$,

$$\omega(M^\tau_{n,a,b}(f), \delta) \le (1 + LN)\omega(f, \delta) \quad \textit{and} \quad M^\tau_{n,a,b}(\mathrm{Lip}_M \alpha) \subset \mathrm{Lip}_{(LN)^\alpha M} \alpha\,. \tag{28}$$

Proof By recalling [9, Theorem 3.2], $M_{n,a,b}(\mathrm{Lip}([0, 1])) \subset \mathrm{Lip}([0, 1])$ and

$$|M_{n,a,b}(f)|_{\mathrm{Lip}} \le \left(1 + \frac{\omega}{n}\right) |f|_{\mathrm{Lip}} \le |f|_{\mathrm{Lip}},$$

hence we get $M^\tau_{n,a,b}(\mathrm{Lip}([0, 1])) \subset \mathrm{Lip}([0, 1])$ and (25) easily follows from

$$|M^\tau_{n,a,b}(f)|_{\mathrm{Lip}} \le \left(1 + \frac{\omega}{n}\right) |f|_{\mathrm{Lip}} |\tau|_{\mathrm{Lip}} |\tau^{-1}|_{\mathrm{Lip}}\,.$$

As a consequence (27) is fulfilled.

Finally, taking [6, Cor. 6.1.20] into account and since $\|M^\tau_{n,a,b}\| = 1$ and property (27) holds, for every $n \ge 1$, $f \in C([0, 1])$, $\delta > 0$, $M > 0$ and $0 < \alpha \le 1$, (28) is proven. □

4 Asymptotic Formula and Its Consequences

In this section we want to find a tool to compare the operators $M_{n,a,b}$ and M_n^τ with the operators $M_{n,a,b}^\tau$, showing under which conditions the latter perform better in order to approximate certain functions. A way to do that consists in comparing the corresponding asymptotic formulae.

We recall that (see [9]), for every $u \in C^2([0,1])$,

$$\lim_{n\to\infty} n(M_{n,a,b}(u) - u) = A_{a,b}(u) \tag{29}$$

uniformly in $[0, 1]$, where, for every $u \in C^2([0,1])$,

$$A_{a,b}(u)(x) = x(1-x)u''(x) + (a+1-(a+b+2)x)u'(x).$$

On the other hand, also in view of [3], it is easy to obtain the following result.

Proposition 4.1 *For every $u \in C^2([0,1])$,*

$$\lim_{n\to\infty} n(M_{n,a,b}^\tau(u)(x) - u(x)) = (A_{a,b}(u\circ\tau^{-1})\circ\tau)(x) = \frac{\tau(x)(1-\tau(x))}{\tau'(x)^2}u''(x)$$
$$+\frac{1}{\tau'(x)}\left((a+1)-(a+b+2)\tau(x) - \frac{\tau(x)(1-\tau(x))\tau''(x)}{\tau'(x)^2}\right)u'(x) \tag{30}$$

uniformly w.r.t. $x \in [0, 1]$.

By comparing (29) and (30), we can infer the next theorem.

Theorem 4.1 *If $f \in C^2([0,1])$ and there exists $n_0 \in \mathbb{N}$ such that, for all $n \geq n_0$ and $x \in]0,1[$,*

$$f(x) \leq M_{n,a,b}^\tau(f)(x) \leq M_{n,a,b}(f)(x),$$

then, for $x \in]0,1[$,

$$f''(x) \geq \frac{\tau''(x)}{\tau'(x)}f'(x) - \frac{\tau'(x)}{\tau(x)(1-\tau(x))}\left((a+1)-(a+b+2)\tau(x)\right)f'(x)$$
$$\geq \left(1 - \frac{x(1-x)\tau'(x)^2}{\tau(x)(1-\tau(x))}\right)f''(x) - \frac{(a+1)-(a+b+2)x}{\tau(x)(1-\tau(x))}\tau'(x)^2 f'(x) \tag{31}$$

Conversely, if there exists $x_0 \in]0,1[$, in which (31) holds with strict inequalities, then there exists $n_0 \in \mathbb{N}$ such that, for all $n \geq n_0$,

$$f(x_0) < M^{\tau}_{n,a,b}(f)(x_0) < M_{n,a,b}(f)(x_0)\,.$$

Example 4.1 Consider

$$\tau = \frac{e_2 + \alpha e_1}{1+\alpha} \quad (\alpha > 0)\,.$$

Moreover, for the sake of simplicity, let us suppose that $a = 1/2$, $b = -1/2$, and $f = e_2$.

We prove that, for a fixed $\alpha > 0$, there exist a subinterval J_α of $]0, 5/6[$ and $n_0 \in \mathbb{N}$ such that, for each $x \in J_\alpha$ and $n \geq n_0$,

$$x^2 < M^{\tau}_{n,1/2,-1/2}(e_2)(x) < M_{n,1/2,-1/2}(e_2)(x)\,.$$

Taking Theorem 4.1 into account, we have to show that, in a suitable interval of $]0, 5/6[$,

$$1 > \frac{\tau''(x)}{\tau'(x)}x - \frac{\tau'(x)\,(3-4\tau(x))}{2\tau(x)(1-\tau(x))}x > 1 - \frac{x(5-6x)\tau'(x)^2}{2\tau(x)(1-\tau(x))}$$

or, equivalently,

$$\begin{cases} F^{\tau}(x) = 1 - \dfrac{\tau''(x)}{\tau'(x)}x + \dfrac{\tau'(x)}{\tau(x)} \cdot \dfrac{3-4\tau(x)}{2(1-\tau(x))}x > 0\,, \\ G^{\tau}(x) = -F^{\tau}(x) + \dfrac{x(5-6x)\tau'(x)^2}{2\tau(x)(1-\tau(x))} > 0\,. \end{cases} \tag{32}$$

Direct calculations show that, for all $x \in]0, 5/6[$,

$$F^{\tau}(x) = \frac{\alpha}{2x+\alpha} + \frac{2x+\alpha}{x+\alpha} \cdot \frac{3(\alpha+1) - 4(x^2+\alpha x)}{2(1+\alpha-(x^2+\alpha x))} > 0\,.$$

Observe that $\lim_{x\to 0^+} F^{\tau}(x) = 5/2$. Moreover

$$\lim_{x\to 0^+} G^{\tau}(x) = -\frac{5}{2(1+\alpha)}\,, \quad \lim_{x\to 5/6^-} G^{\tau}(x) = -\frac{105\alpha+200}{(5+3\alpha)(5+6\alpha)(11+6\alpha)}\,.$$

Finally $G^{\tau}(2/3) > 0$.

In what follows we continue to denote by F^{τ} and G^{τ} respectively the extensions by continuity in 0 and in 5/6 of the functions in (32).

First we observe that from $G^{\tau}(0) < 0 < G^{\tau}(2/3)$ if follows that there exists $y_\alpha \in]0, 2/3[$ such that $G^{\tau}(y_\alpha) = 0$. Analogously, from $G^{\tau}(5/6) < 0 < G^{\tau}(2/3)$ it follows that there exists $z_\alpha \in]2/3, 5/6[$ such that $G^{\tau}(z_\alpha) = 0$.

It can be proven that G^{τ} is positive in $]y_\alpha, z_\alpha[$.

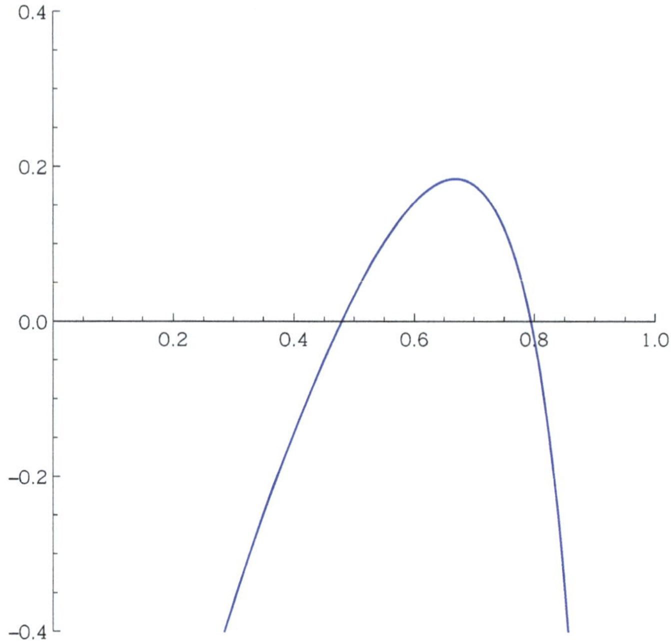

Fig. 1 The plot of G^τ for $\alpha = 1$

Such a neighborhood is contained in $]0, 5/6[$ and it is the desired subinterval in which (32) holds.

As an example we provide the plot of G^τ for $\alpha = 1$ (Fig. 1).

Under suitable assumptions, the operators $M^\tau_{n,a,b}$ also perform better than the operators M^τ_n considered in [3], as showed in the following result. Note that the asymptotic formula for the operators M^τ_n is (30) for $a = b = 0$.

Theorem 4.2 *If* $f \in C^2([0, 1])$ *and there exists* $n_0 \in \mathbb{N}$ *such that, for all* $n \geq n_0$ *and* $x \in]0, 1[$,

$$f(x) \leq M^\tau_{n,a,b}(f)(x) \leq M^\tau_n(f)(x),$$

then, for $x \in]0, 1[$,

$$\begin{aligned} f''(x) &\geq \frac{\tau''(x)}{\tau'(x)} f'(x) - \frac{\tau'(x)}{\tau(x)(1-\tau(x))}((a+1) - (a+b+2)\tau(x)) f'(x) \\ &\geq \frac{\tau''(x)}{\tau'(x)} f'(x) - \frac{\tau'(x)}{\tau(x)(1-\tau(x))}(1 - 2\tau(x)) f'(x). \end{aligned} \tag{33}$$

Conversely, if there exists $x_0 \in]0,1[$*, in which* (33) *holds with strict inequalities, then there exists* $n_0 \in \mathbb{N}$ *such that, for all* $n \geq n_0$*,*

$$f(x_0) < M_{n,a,b}^{\tau}(f)(x_0) < M_n^{\tau}(f)(x_0)\,.$$

Example 4.2 Fix $\tau = \frac{e_2+\alpha e_1}{1+\alpha}$ $(\alpha > 0)$, $-1 < a = b < 0$ and let $f = e_2$. Then, (33) with strict inequalities becomes

$$1 > \frac{\tau''(x)}{\tau'(x)}x - \frac{\tau'(x)(a+1)(1-2\tau(x))}{\tau(x)(1-\tau(x))}x > \frac{\tau''(x)}{\tau'(x)}x - \frac{\tau'(x)(1-2\tau(x))}{\tau(x)(1-\tau(x))}x$$

and it is easy to see that it holds whenever $1 - 2\tau(x) > 0$, that is for every $0 < x < (\sqrt{\alpha^2+2\alpha+2}-\alpha)/2 < 1$.

A further consequence of the asymptotic formula (30) consists in finding a representation in terms of the operators $M_{n,a,b}^{\tau}$ of suitable semigroups acting on spaces of continuous as well as integrable functions. For similar results see, e.g. [8], where the reader can also find more details about semigroup theory.

Corollary 4.1 *There exists a Markov semigroup* $(T(t))_{t\geq 0}$ *such that for every* $f \in C([0,1])$, $t \geq 0$ *and for every sequence* $(k_n)_{n\geq 1}$ *of positive integers such that* $\lim_{n\to\infty} k_n/n = t$*,*

$$\lim_{n\to\infty} (M_{n,a,b}^{\tau})^{k_n}(f) = T(t)(f\circ\tau^{-1})\circ\tau \quad \textit{uniformly on } [0,1]. \tag{34}$$

Moreover, for every $f \in C([0,1])$*,*

$$\lim_{m\to\infty} (M_{n,a,b}^{\tau})^{m}(f) = \int_0^1 f\,d\mu_{a,b}^{\tau} = \lim_{t\to\infty} T(t)(f\circ\tau^{-1})\circ\tau \tag{35}$$

uniformly on $[0,1]$.

Further, for every $p \geq 1$, $(T(t))_{t\geq 0}$ *has a unique extention* $(T_p(t))_{t\geq 0}$ *which is a positive contraction semigroup on* $L^p([0,1],\mu_{a,b})$ *and, if* $t \geq 0$ *and* $(k_n)_{n\geq 1}$ *is a sequence of positive integers satisfying* $\lim_{n\to\infty} k_n/n = t$*, then for every* $f \in L^p([0,1],\mu_{a,b}^{\tau})$*,*

$$\lim_{n\to\infty} (M_{n,a,b}^{\tau})^{k_n}(f) = T_p(t)(f\circ\tau^{-1})\circ\tau \quad \textit{in } L^p([0,1],\mu_{a,b}^{\tau}). \tag{36}$$

Finally, if $f \in L^p([0,1],\mu_{a,b}^{\tau})$ *and* $n \geq 1$*,*

$$\lim_{m\to\infty} (M_{n,a,b}^{\tau})^{m}(f) = \int_0^1 f\,d\mu_{a,b}^{\tau} = \lim_{t\to\infty} T_p(t)(f\circ\tau^{-1})\circ\tau \tag{37}$$

in $L^p([0,1],\mu_{a,b}^{\tau})$.

Proof First note that, for every $k, n \geq 1$ and $f \in C([0,1])$ we put

$$(M^{\tau}_{n,a,b})^k(f) := M^k_{n,a,b}(f \circ \tau^{-1}) \circ \tau.$$

Formula (34) follows from [7, Theorem 3.3]. Formula (35) is a consequence of [3, formula (8)] and [9, Theorem 4.2].

On the other hand, (36) derives directly from [9, Theorem 4.4], since $f \in L^p([0,1], \mu^{\tau}_{a,b})$ if and only if $f \in L^p([0,1], \mu_{a,b})$. Finally, formula (37) can be obtained from (35), since $C([0,1])$ is dense in $L^p([0,1], \mu^{\tau}_{a,b})$. □

Remark 4.1 We notice that the generator of the semigroup $(T(t))_{t\geq 0}$ in Corollary 4.1 is the closure of the differential operator $A_{a,b}(u \circ \tau^{-1}) \circ \tau$ on $C^2([0,1])$ (see (30)).

Acknowledgments The paper has been performed within the activities of GNAMPA-INdAM (Gruppo Nazionale per l'Analisi Matematica, la Probabilità e le loro Applicazioni dell' Istituto Nazionale di Alta Matematica), of the network RITA (Research ITalian network on Approximation), and of the UMI Group "Teoria dell'Approssimazione e Applicazioni".

References

1. Abel, U., Heilmann, M.: The complete asymptotic expansion for Bernstein-Durrmeyer operators with Jacobi weights. Mediterr. J. Math. **1**(4), 487–499 (2004)
2. Abel, U., Leviatan, D., Raşa, I.: On the q-monotonicity preservation of Durrmeyer-type operators. Mediterr. J. Math. **18**(4), 173, 15pp. (2021)
3. Acar, T., Aral, A., Raşa, I.: Modified Bernstein-Durrmeyer operators. Gen. Math. **22**(1), 27–41 (2014)
4. Acar, T., Cappelletti Montano, M., Garrancho, P., Leonessa, V.: On sequences of J. P. King-type operators. J. Funct. Spaces **2019**, 2329060, 12pp. (2019)
5. Albanese, A., Campiti, M., Mangino, E.: Approximation formulas for C_0-semigroups and their resolvent operators. J. Appl. Funct. Anal. **1**(3), 343–358 (2006)
6. Altomare, F., Campiti, M.: Korovkin-Type Approximation Theory and its Applications. De Gruyter Studies in Mathematics, vol. 17. Walter de Gruyter, Berlin (1994)
7. Altomare, F., Raşa, I.: Lipschitz Contractions, unique ergodicity and asymptotics of Markov semigroups. Boll. U. M. I Serie 9 **5**(1), 1–17 (2012)
8. Altomare, F., Cappelletti Montano, M., Leonessa, V., Raşa, I.: Markov Operators, Positive Semigroups and Approximation Processes. De Gruyter Studies in Mathematics, vol. 61. Walter de Gruyter GmbH, Berlin (2014)
9. Altomare, F., Cappelletti Montano, M., Leonessa, V.: On the positive semigroups generated by Fleming-Viot type differential operators. Commun. Pure Appl. Anal. **18**(1), 323–340 (2019)
10. Attalienti, A., Campiti, M.: Degenerate evolution problems and Beta-type operators. Stud. Math. **140**, 117–139 (2000)
11. Bauer, H.: Measure and Integration Theory. De Gruyter Studies in Mathematics, vol. 26. Walter de Gruyter, Berlin (2001)
12. Berens, H., Xu, Y.: On Bernstein-Durrmeyer polynomials with Jacobi weights. In: Chui, C.K. (ed.) Approximation Theory and Functional Analysis, pp. 25–46. Academic Press, Boston (1991)

13. Cárdenas-Morales, D., Garrancho, P., Raşa, I.: Bernstein-type operators which preserve polynomials. Comput. Math. Appl. **62**(1), 158–163 (2011)
14. Derriennic, M.M.: Sur l'approximation des fonctions integrables sur [0, 1] par des polynomes de Bernstein modifiés. J. Approx. Theory **31**, 325–343 (1981)
15. Durrmeyer, J.L.: Une formule d'inversion de la transformée de Laplace: application a la theorie des moments. These de 3e cycle, Faculté des Sciences de l'Université de Paris (1967)
16. Freud, G.: On approximation by positive linear methods I, II. Stud. Sci. Math. Hungar. **2**, 63–66 (1967). **3**, 365–370 (1968)
17. King, J.P.: Positive linear operators which preserve x^2. Acta Math. Hungar. **99**(3), 203–208 (2003)
18. Krengel, U.: Ergodic Theorems. De Gruyter Studies in Mathematics, vol. 6. Walter de Guyter, Berlin (1985)
19. López-Moreno, A.-J., Muñoz-Delgado, F.-J.: Asymptotic expansion of multivariate conservative linear operators. J. Comput. Appl. Math. **150**, 219–251 (2003)
20. Lupaş, A.: Die Folge der Betaoperatoren. Dissertation, Universität Stuttgart (1972)
21. Păltănea, R.: Sur un opérateur polynomial défini sur l'ensemble des fonctions intégrables. (French) [A polynomial operator defined on the set of integrable functions] Itinerant seminar on functional equations, approximation and convexity (Cluj-Napoca, 1983), 101–106, Preprint, 83–2, University of Babeş-Bolyai, Cluj-Napoca (1983)
22. Păltănea, R.: Approximation Theory Using Positive Linear Operators. Birkhäuser, Boston (2004)
23. Păltănea, R.: Durrmeyer type operators on a simplex. Constr. Math. Anal. **4**(2), 215–228 (2021)
24. Popoviciu, T.: Sur le reste dans certaines formules linéaires d'approximation de l'analyse. Mathematica (Cluj) **1**(24), 95–142 (1959)
25. Swetits, J.J., Wood, B.: Quantitative estimates for L_p approximation with positive linear operators. J. Approx. Theory **38**, 81–89 (1983)
26. Waldron, S.: A generalized beta integral and the limit of Bernstein-Durrmeyer operator with Jacobi weights. J. Approx. Theory, **122**, 141–150 (2003)
27. Ziegler, Z.: Linear approximation and generalized convexity. J. Approx. Theory **1**, 420–433 (1968)

On a Particular Scaling for the Prototype Anisotropic p-Laplacian

Simone Ciani, Umberto Guarnotta, and Vincenzo Vespri

To celebrate Francesco Altomare's 70th genethliac

Abstract In this brief note we show that under a volume non-preserving scaling it is possible to recover the basics for a regularity theory regarding local weak solutions to the fully anisotropic equation

$$\partial_t u = \sum_{i=1}^{N} \partial_i(|\partial_i u|^{p_i-2}\partial_i u) \quad \text{in} \quad \Omega_T = \Omega \times (-T, T), \quad \text{with} \quad \Omega \subset\subset \mathbb{R}^N. \tag{1}$$

We characterize self-similar solutions regarding this particular scaling and we show that semi-continuity for solutions to this equation is a consequence of a simple property that is itself invariant under scaling.

Keywords Anisotropic p-laplacian · Critical mass lemma · Intrinsic scaling · Lower Semi-Continuity

---

S. Ciani
Department of Mathematics, Technische Universität Darmstadt, Darmstadt, Germany
e-mail: ciani@mathematik.tu-darmstadt.de

U. Guarnotta
Dipartimento di Matematica e Informatica, Università degli Studi di Palermo, Palermo, Italy
e-mail: umberto.guarnotta@unipa.it

V. Vespri (✉)
Dipartimento DIMAI, Università degli Studi di Firenze, Firenze, Italy
e-mail: vincenzo.vespri@unifi.it

A. M. Candela et al. (eds.), *Recent Advances in Mathematical Analysis*,
Trends in Mathematics, https://doi.org/10.1007/978-3-031-20021-2_15

1 Introduction to the Problem

Equation (1) is a parabolic anisotropic equation with non-standard growth. We refer to the introduction of [5, 24] and the surveys [21, 22] for a non-exhaustive introduction to the origin of the problem, and to the introduction of [8] and the book [1] for a more general account to the parabolic problem. At a first glance Eq. (1) may look similar to the equation

$$u_t - \operatorname{div}(|\nabla u|^{p-2}\nabla u) = 0 \quad \text{locally weakly in } \Omega_T. \tag{2}$$

Literature on this topic is very developed, and even if the problem of regularity of solutions to (1) is old more than fifty years, still very much is unknown from the point of view of basic regularity, as local Hölder continuity or Harnack inequality. The principal motivation is that the techniques usually employed for nonlinear equations (as p-Laplacian equations, porous medium equations, doubly nonlinear equations, and so on) are not directly applicable to it. Let us explain this point in detail.

Up to our knowledge, in the setting of evolutionary nonlinear operators of p-growth (whose prototype is (2) with $p \neq 2$), the main technique to prove a Harnack inequality is exploiting a parabolic continuous transformation having the general form

$$w(x, t) = e^{t/(p-2)}u(x, e^t), \qquad x \in \Omega,\ t > 0. \tag{3}$$

This transformation maps super-solutions to (2) to super-solutions to a similar equation, that has an exponential dependence on time only on the non-homogeneous terms. Along this strategy, the possibility to stretch time and control the non-homogeneous terms is crucial, in order to employ a technique originally conceived by E. DeGiorgi for solutions to elliptic partial differential equations (see, e.g., [10], [13]), based on a version of the isoperimetric inequality (cf. [12, Lemma 2.2., page 5]). This argument allows to prove an expansion of positivity for the transformed super-solutions that, if carried back to solutions to (2), provides the expansion of positivity necessary for an intrinsic Harnack inequality to hold true.

The main issue dealing with (1) is that, in general, a continuous transformation with an exponential-type dependence on time necessarily affects the space variables. Taking into account also the strong nonlinear behavior of the equation along the space variables, the control of the non-homogeneous terms in the transformed equation is encumbered. More precisely, from the energetic point of view, the new equation is no more of the same kind of (1), and this leads the whole machinery to fail.

On the other hand, in [6] the authors proved that an intrinsic Harnack type inequality is valid for local weak solutions to (1), by adapting a classic idea of E. DiBenedetto (see [11]) consisting in a comparison between the solution and a particular one, called Barenblatt solution in honor to its discoverer (see the original in [2, 3] and [7] for an overview on the anisotropic case). However, the generalization of this inequality to a wider class of parabolic operators patterned after (1) is still an open and challenging problem. The purpose of the present work is to investigate a particular scaling of the equation: it would permit to free the time variable from the space ones, opening the way to an application of a transformation similar to (3). This homogenization seems to unveil a new insight on the anisotropic behavior of these operators. See also [15] for a detailed analysis of self-similarity in the case of fast anisotropic diffusion. From the energetic point of view, serious difficulties appear even with the stationary counterpart of (1), because the competition among different directional p_i-diffusions encodes both singular and degenerate behavior. Roughly speaking, this can be illustrated within the scaling of [9], looking at the kind of degeneration that the set

$$\prod_{i=1}^{N} \left\{ |x_i| < \rho^{\frac{\bar{p}}{p_i}} M^{\frac{p_i-\bar{p}}{p_i}} \right\}, \qquad M, \rho > 0$$

exhibits as M vanishes. This is a volume-preserving set of self-similar geometry where the equation evolves, and the parameter M is usually chosen to be a multiple of the oscillation of u, in order to restore the homogeneity of the energy. The problem is that, depending on the sign of $(p_i - \bar{p})$, the set stretches or vanishes along the respective coordinates.

The different scaling that we propose in this note (see (6) and (7)) possesses the following properties: the intrinsic geometry associated with it degenerates monotonically with M, so we say that the geometry is only *degenerate*, not singular; it does not affect anyhow the time variables from the intrinsic point of view. From this perspective, this particular scaling seems promising; see for instance, the energy $\mathcal{E}_n$ in Lemma 3.2. The crucial point is that we can identify the self-similar solutions to (1), namely, the solutions that coincide with their scaled functions; this is done via correspondence with a Fokker-Planck equation (cf. Proposition 2.3). As a consequence, all the properties of solutions to (1) proved in [6] hold true, in a re-interpreted formulation, also for solutions of a 'wild' Fokker-Planck equation (see (15)). The existence of a Barenblatt solution is of fundamental importance to understand the behaviour of solutions.

Moreover, we show that this special scaling preserves the energy of the solutions, as well as other properties, that will be called for this reason *invariants* (standing for scale-invariants). An example is furnished by the Critical Mass Lemma, that can be regarded as a measure-theoretical maximum principle (see Lemma 3.2 for details; see also [27, p. 8] and Proposition 2 of [18] in the context of anisotropic porous medium).

Dimensional analysis is a simple consequence of the well-known covariance principle of Physics: all physical laws can be represented in a form which is equally valid for all observers. The very idea of self-similarity is connected with the group of transformations of solutions: see, e.g., [4]. These groups are inborn in the differential equations governing the process, and are determined by the physical dimensions of the variables appearing in them; transformations of units of time, length, mass, etc. are the simplest examples. This kind of self-similarity is obtained by power laws with exponents that are simple fractions defined in an elementary way from dimensional considerations. These arguments led to an interpretation of nonlinear parabolic theory, developed by DiBenedetto ([12]), Vazquez ([26]), and many others, which is nowadays known as method of intrinsic scaling (cf. also [25]). The key feature of the argument of intrinsic scaling is that, by appropriately scaling the geometry, the energy of solutions enjoys a homogeneous form that is easier to manipulate. This idea can be used in turn to interpret the energy of solutions to anisotropic equations like (1) in a homogeneous fashion. This is the purpose of the present scaling, whose side-effect on energy $\mathcal{E}_n$ is here shown by the non-scaled version of Lemma 3.3. We present here a general version of this lemma, that we could not find in literature for the full parabolic anisotropic equation and that is propaedeutic to the study of further properties.

Indeed, as a byproduct of our analysis, by applying the ideas of [20] to the parabolic setting, we show that lower semi-continuity of super-solutions is a sole consequence of these general invariants.

The existence of a lower semi-continuous representative for local weak super-solutions has already been obtained in [14] by using an idea of [19]. The authors observe that a proper $L^r - L^\infty$ estimate for weak super-solutions suffices to obtain a lower semi-continuous representative. This technique is however linked to the particular structure of the equation, that allowed them to add a constant to the solution to generate another solution. The new approach of [20] is more general, since the existence of a lower semi-continuous representative is linked only to a more general property, that is the analogue of Lemma 3.2. In this way the authors of the aforementioned [14, 19, 20] proved that weak solutions are p-super-harmonic solutions. The latter ones are, on an appropriate setting, proper lower semi-continuous functions, that can be compared with any sufficiently regular solution to the same equation. Since the comparison principle for equations driven by monotone operators holds true, the main step consists in proving semi-continuity.

It would be an interesting subject to determine whether p-super-harmonic functions, whose derivatives a priori may be even unbounded, can satisfy a Critical Mass Lemma as Lemma 3.2.

1.1 Structure of the Paper

Section 2 is devoted to set up the functional framework and propose the particular scaling. In Sect. 3 we show that energetic properties of the equation are invariant under this scaling. Finally, in Sect. 4, we furnish a new proof of semi-continuity for super-solutions to (1).

Notation

- Let $N \in \mathbb{N}$, $N > 1$, and let $\Omega \subset \mathbb{R}^N$ be a bounded open set. Given $T > 0$, we set $\Omega_T = \Omega \times (-T, T)$.
  The symbol $A \subset\subset B$ means 'A is compactly contained in B'.
- For any $\varphi \in W^{1,1}(\Omega)$, we denote by $\partial_i \varphi$ the i-th directional weak derivative of φ.
  If moreover $\varphi \in W^{1,2}([s,t], L^2(\Omega))$ for some $s, t \in \mathbb{R}$, $s < t$, then $\partial_t \varphi$ stands for the weak time-derivative of φ.
- We denote the cube of side $2\rho > 0$ and center $x \in \mathbb{R}^N$ with $x + K_\rho$, while

$$\begin{cases} (x,t) + Q^-_{\rho,\tau} = (x + K_\rho) \times (-\tau, 0], \\ (x,t) + Q^+_{\rho,\tau} = (x + K_\rho) \times [0, \tau), \\ (x,t) + Q_{\rho,\tau} = (x + K_\rho) \times (-\tau, \tau], \end{cases}$$

  stand for, respectively, the backward, forward and full cylinders centered at $(x,t) \in \mathbb{R}^{N+1}$.
  When $\tau = 1$ we simply write $Q^-_\rho, Q_\rho, Q^+_\rho$ instead of $Q^-_{\rho,1}$, $Q_{\rho,1}$, $Q^+_{\rho,1}$.
- We fix a vector of N numbers $\mathbf{p} = (p_1, \dots, p_N)$; the index i will run through $1, \dots, N$. We define the harmonic mean of p_is as $\bar{p} = N(\sum_{i=1}^N 1/p_i)^{-1}$, and for $\bar{p} < N$ the Sobolev exponent of the harmonic mean by $\bar{p}^* = N\bar{p}/(N - \bar{p})$. Hereafter we suppose

$$2 < p_1 \leq p_2 \leq \dots \leq p_N < \bar{p}^*.$$

- In the sequel we will make use of the following numbers:

$$\lambda = N(\bar{p} - 2) + \bar{p}, \quad \alpha = \frac{N}{\lambda}, \quad \alpha_i = \frac{N(\bar{p} - p_i) + \bar{p}}{\lambda p_i}. \tag{4}$$

- For any $M, \rho > 0$, the *intrinsic cube* and the *backward intrinsic cylinder* are defined respectively as

$$\mathcal{K}_\rho(M) = \prod_{i=1}^N \left\{ |x_i| < M^{\frac{p_i - 2}{p_i}} \rho^{\frac{\bar{p}}{p_i}} \right\},$$

$$Q^-_\rho(M) = \prod_{i=1}^N \left\{ |x_i| < M^{\frac{p_i - 2}{p_i}} \rho^{\frac{\bar{p}}{p_i}} \right\} \times \left(-\rho^{\bar{p}}, 0 \right].$$

  The notation of forward and full intrinsic cylinders is analogous to the one above.
- The function $\pi_i : \mathbb{R}^N \to \mathbb{R}$, $\pi_i(x) = x_i$, $i = 1, \dots, N$, will denote the projection with respect to the i-th space variable. Moreover, $\pi : \mathbb{R}^N \times \mathbb{R} \to \mathbb{R}^N$, $\pi(x,t) = x$, stands for the projection in the space variables.

– We denote by γ a positive constant (depending only on the data, i.e., N and p_is) that may vary from line to line.

2 Preliminaries

We introduce the parabolic anisotropic spaces, which are the natural setting to work within. We define

$$W_o^{1,\mathbf{p}}(\Omega) := \{u \in W_o^{1,1}(\Omega) |\, \partial_i u \in L^{p_i}(\Omega)\},$$

$$W_{loc}^{1,\mathbf{p}}(\Omega) := \{u \in W_{loc}^{1,1}(\Omega) |\, \partial_i u \in L^{p_i}(\Omega)\},$$

$$L_{loc}^{\mathbf{p}}(0,T; W_o^{1,\mathbf{p}}(\Omega)) := \{u \in L_{loc}^{1}(0,T; W_o^{1,1}(\Omega)) |\, \partial_i u \in L_{loc}^{p_i}(0,T; L_{loc}^{p_i}(\Omega))\}.$$

A function

$$u \in C_{loc}^0(0,T; L_{loc}^2(\Omega)) \cap L_{loc}^{\mathbf{p}}(0,T; W_{loc}^{1,\mathbf{p}}(\Omega))$$

is called a *local weak solution* of (1) if, for any $0 < t_1 < t_2 < T$ and any compact set $K \subset\subset \Omega$, it satisfies

$$\int_K u\varphi\, dx \Big|_{t_1}^{t_2} + \int_{t_1}^{t_2} \int_K (-u\, \partial_t \varphi + \sum_{i=1}^N |\partial_i u|^{p_i-2} \partial_i u\, \partial_i \varphi)\, dxdt = 0, \tag{5}$$

for all $\varphi \in C_{loc}^\infty(0,T; C_o^\infty(\Omega))$. By a density and approximation argument, we can consider test functions in (5) in the bigger space

$$\varphi \in W_{loc}^{1,2}(0,T; L_{loc}^2(\Omega)) \cap L_{loc}^{\mathbf{p}}(0,T; W_o^{1,\mathbf{p}}(\Omega)),$$

provided $\Omega \subset\subset \mathbb{R}^N$ is a rectangular domain (see [16] for an extension to more general domains).

2.1 *Scaling Properties of Solutions*

In the present subsection we show some important scaling properties of solutions to (1) and their correspondence with stationary solutions to a Fokker-Planck-type equation.

Proposition 2.1 *Let u be a local weak solution to the Eq. (1) in Ω_T. For any $M, \rho > 0$ appropriate for the inclusion $Q_\rho(M) \subset \Omega_T$, we define the parametric* *transformation*

$$\tilde{T}_{\rho,M}(x,t) = \left(M^{\frac{p_i-2}{p_i}} \rho^{\frac{\bar{p}}{p_i}} x_i, \rho^{\bar{p}} t \right). \tag{6}$$

Then the transformed function

$$\mathcal{T}(u)(x,t) = M^{-1} u\left(\tilde{T}_{\rho,M}(x,t) \right) = M^{-1} u\left(M^{\frac{p_i-2}{p_i}} \rho^{\frac{\bar{p}}{p_i}} x_i, \rho^{\bar{p}} t \right) \tag{7}$$

is a solution to (1) *in* $\tilde{T}^{-1}_{\rho,M}(\Omega_T)$.

Proof We perform some formal algebraic computations, representing change of variables in the integrals of definition (5).
If we generally suppose

$$\mathcal{T}(u) = M^{-1} u\left(L_i x_i, Tt \right)$$

for some $M, L_i, T > 0$, then

$$\partial_t \left(\mathcal{T} u \right) = M^{-1} T \partial_t u(L_i x_i, Tt) \quad \text{and} \quad \partial_i \left(\mathcal{T} u \right) = M^{-1} L_i \partial_i u(L_i x_i, Tt).$$

Thus, imposing the equation for $\mathcal{T}u$, namely,

$$\partial_t \left(\mathcal{T} u \right) = \sum_{i=1}^{N} \partial_i \left(|\partial_i (\mathcal{T}u)|^{p_i-2} \partial_i (\mathcal{T}u) \right),$$

we find

$$\begin{aligned} M^{-1} T \partial_t u(L_i x_i, Tt) = \partial_t \left((\mathcal{T}u)(x,t) \right) &= \sum_{i=1}^{N} \partial_i \left(|\partial_i (\mathcal{T}u)(x,t)|^{p_i-2} \partial_i (\mathcal{T}u)(x,t) \right) \\ &= \sum_{i=1}^{N} L_i^{p_i} M^{1-p_i} \partial_i \left(|\partial_i u|^{p_i-2} \partial_i u \right)(L_i x_i, Tt). \end{aligned}$$

Furthermore, we impose

$$M^{-1} T = L_i^{p_i} M^{1-p_i} \quad \forall i = 1, \ldots, N$$

to restore the homogeneity in the equation. We find $L_i = [M^{p_i-2}T]^{\frac{1}{p_i}}$, whence

$$\mathcal{T}u = M^{-1}u\left(\left[M^{p_i-2}T\right]^{\frac{1}{p_i}} x_i, Tt\right). \tag{8}$$

Taking $T = \rho^{\bar{p}}$ concludes the proof. □

Remark 2.1 The peculiarity of the scaling (7) is that it does not alter the time variable, from the point of view of intrinsic geometry. Indeed, the parameter M is usually chosen to be a suitable multiple of either the oscillation or the L^∞ norm of the solution itself, therefore leading to a geometry within the equation evolves in an intrinsic fashion (see [12, 25]).

Moreover, the proof of Proposition 2.1 reveals that (7) is not the only invariant: we may consider, for instance, also the transformation

$$\mathcal{T}_{\rho,M}u = M^{-1}u\left(M^{\frac{p_i-\bar{p}}{p_i}}\rho^{\frac{\bar{p}}{p_i}}x_i,\ M^{2-\bar{p}}\rho^{\bar{p}}t\right), \tag{9}$$

corresponding to $T = M^{2-\bar{p}}\rho^{\bar{p}}$ in (8). This transformation has been used extensively in [6], with the aim of obtaining a Harnack inequality which intrinsically scales within the particular geometry dictated by the transformation.

Definition 2.1 We define the intrinsic anisotropic cube by transformation (6) on the space variables,

$$\mathcal{K}_\rho(M) = \prod_{i=1}^{N}\left\{|x_i| < M^{\frac{p_i-2}{p_i}}\rho^{\frac{\bar{p}}{p_i}}\right\}, \tag{10}$$

and the intrinsic anisotropic cylinders

$$Q_\rho^-(M) := \tilde{T}_{\rho,M}(Q_1^-) = \prod_{i=1}^{N}\left\{|x_i| < M^{\frac{p_i-2}{p_i}}\rho^{\frac{\bar{p}}{p_i}}\right\} \times \left(-\rho^{\bar{p}}, 0\right]. \tag{11}$$

Similarly we define forward and full intrinsic cylinders.

Remark 2.2 Definition 2.1 is motivated by Proposition 2.1 and leads to the following consequence.

If u solves (1) in $Q_\rho^-(M)$, then $\mathcal{T}u$ solves (1) in Q_1^-. Vice-versa if u solves (1) in Q_1^- then $\mathcal{T}^{-1}(u) = Mu\left(M^{\frac{2-p_i}{p_i}}\rho^{-\frac{\bar{p}}{p_i}}x_i, \rho^{-\bar{p}}t\right)$ solves (1) in $Q_\rho^-(M)$.

Proposition 2.2 *Let u be a local weak solution to Eq.* (1). *Then the parametric transformations preserving the L^1 norm of u correspond to* (7) *for $M = \rho^{-\alpha\bar{p}}$, that is,*

$$\mathcal{T}_\rho u = \rho^{\alpha\bar{p}} u\left(\rho^{\alpha_i \bar{p}} x_i, \rho^{\bar{p}} t\right), \tag{12}$$

where α, α_i were defined in (4).

Proof Performing a change of variables, besides recalling (4), we get

$$\begin{aligned}\int_{\pi(\tilde{T}_{\rho,M}(Q_1))} \mathcal{T}u(x,t)dx &= M^{-1}\prod_{i=1}^{N}\left(M^{\frac{p_i-2}{p_i}}\rho^{\frac{\bar{p}}{p_i}}\right)^{-1}\int_{K_1} u(y,s)dy \\ &= \left(M^{-\frac{\lambda}{\bar{p}}}\rho^{-N}\right)\int_{K_1} u(y,s)dy.\end{aligned} \tag{13}$$

Hence, imposing $M^{-\frac{\lambda}{\bar{p}}}\rho^{-N} = 1$, we find $M = \rho^{-\alpha\bar{p}}$, as desired. □

Remark 2.3 It is worth noticing the following important geometric property, used also in the proof of Proposition 2.2: for any $M, \rho > 0$, the total volumes of the anisotropic cube and the anisotropic cylinder depend on p_is, i.e.,

$$|\mathcal{K}_\rho(M)| = 2^N \rho^N M^{\frac{N(\bar{p}-2)}{\bar{p}}} = \rho^N M^{\frac{N(\bar{p}-2)}{\bar{p}}} |K_1|,$$

$$|Q_\rho^-(M)| = 2^N \rho^{N+\bar{p}} M^{\frac{N(\bar{p}-2)}{\bar{p}}} = \rho^{N+\bar{p}} M^{\frac{N(\bar{p}-2)}{\bar{p}}} |Q_1^-|.$$

Definition 2.2 A solution u to (1) in $\mathbb{R}^{N+1}$ is said to be a *self-similar solution* if it satisfies $\mathcal{T}_\rho u = u$ for all $\rho > 0$, where T_ρ was defined in (12).

Now we consider the continuous transformation Φ and its inverse Φ^{-1} defined as

$$\Phi(u)(x,t) = w(x,t) = e^{\alpha t} u(e^{\alpha_i t} x_i, e^t), \quad \Phi^{-1}(w)(y,s) = u(y,s) = s^{-\alpha} w(s^{-\alpha_i} y_i, \log s). \tag{14}$$

This map formally sends solutions to (1) in $\Sigma^+ := \mathbb{R}^N \times (0, +\infty)$ into solutions of the anisotropic Fokker-Planck-type equation

$$\partial_t w = \sum_{i=1}^{N} \partial_i[(|\partial_i w|^{p_i-2}\partial_i w) + \alpha_i y_i w] \quad \text{in} \quad \Sigma := \mathbb{R}^N \times \mathbb{R}. \tag{15}$$

For each fixed time $t = \log(\rho^{-\lambda})$, $\rho > 0$, Φ corresponds to a parametric transformation of type (12), thus preserving the L^1 norm; indeed, it is readily seen that

$$\Phi(u)(x, \log(\rho^{-\lambda})) = \mathcal{T}_{\rho^{-\lambda/\bar{p}}} u(x, 1). \tag{16}$$

Now we present a characterization of the self-similar solutions.

Proposition 2.3 *Self-similar solutions to* (1) *in* Σ^+ *correspond to stationary solutions to the Fokker-Planck equation* (15) *and vice-versa.*

Proof Let us consider a self-similar solution u to (1) in Σ^+. We already know that $w = \Phi u$ is a solution to (15). It remains to show that w is stationary. By (16) and the self-similarity of u, for all $(x, t) \in \Sigma$ we get

$$w(x, t) = w(x, \log(\rho^{-\lambda})) = \mathcal{T}_{\rho^{-\lambda/\bar{p}}} u(x, 1) = u(x, 1) = w(x, 0),$$

being $t = \log(\rho^{-\lambda})$ for some $\rho > 0$.

Vice-versa, let w be a stationary solution to (15). We already know that u solves (1) in Σ^+, so it suffices to show that u is self-similar. For any $\rho > 0$, we choose $t = \log(\rho^{\bar{p}} l)$, $l > 0$, in (14) and use the fact that w is stationary to obtain

$$l^{\alpha} \rho^{\alpha \bar{p}} u(\rho^{\alpha_i \bar{p}} l^{\alpha_i} x_i, \rho^{\bar{p}} l) = w(x, t) = w(x, \log l) = l^{\alpha} u(l^{\alpha_i} x_i, l).$$

Dividing by l^{α}, besides performing the change of variables $y_i = l^{\alpha_i} x_i$, leads to

$$\mathcal{T}_{\rho} u = u \quad \forall \rho > 0,$$

which is the self-similarity of u. □

Definition 2.3 A self-similar solution to (1) in Σ^+ (or, equivalently, a solution to (1) corresponding to a stationary solution to the Fokker-Planck equation (15)) is said to be a *Barenblatt Fundamental solution*; it is denoted by $\mathcal{B}$, in analogy with the literature regarding the p-Laplacian.[1]

3 Scaling Invariants

Definition 2.3 is invariant under the scalings (7) and (9). In this section we show that also the energy of solutions is invariant, and the same holds for a particular energetic property of solutions, that can be regarded as a measure-theoretical maximum principle.

[1] Indeed, the epithet *Fundamental* does not mean that solutions are represented by an integral convolution with kernel $\mathcal{B}$, but that the classic $\mathcal{B}$ function approaches to the heat kernel as $p \to 2$. The Barenblatt solution for the p-Laplacian equation can be found in [2].

Lemma 3.1 (Energy Estimates) *Let u be a local weak solution to Eq. (1) in Ω_T. Let $(x_o, t_o) \in \Omega_T$ and $\rho, M > 0$ be such that $(x_o, t_o) + Q_\rho^-(M) \subset \Omega_T$. Then, for each function of the form*

$$C_o^\infty((x_o, t_o) + Q_\rho^-(M)) \ni \eta = \prod_{i=1}^N \eta_i^{p_i}(x_i, t) \text{ with } \eta_i \in C_o^\infty(\pi_i(x_o + \mathcal{K}_\rho(M)) \times (t_o - \rho^{\bar{p}}, t_o]),$$

we have the following estimates, valid for all $t_o - \rho^{\bar{p}} < s < t < t_o$ and $k \in \mathbb{R}$:

$$\begin{aligned}
&\int_{\mathcal{K}_\rho(M)} (u-k)_\pm^2 \eta(x,\tau)dx\Big|_{\tau=s}^{\tau=t} + \sum_{i=1}^N \int_s^t \int_{\mathcal{K}_\rho(M)} |\partial_i[\eta(u-k)_\pm]|^{p_i}\, dxd\tau \\
&\quad \le \gamma \Big\{ \int_s^t \int_{\mathcal{K}_\rho(M)} (u-k)_\pm^2 \partial_\tau \eta(x,\tau)\, dxd\tau \\
&\quad + \sum_{i=1}^N \int_s^t \int_{\mathcal{K}_\rho(M)} |(u-k)_\pm|^{p_i} \hat{\eta}_i |\partial_i \eta_i|^{p_i}\, dxd\tau \Big\},
\end{aligned} \tag{17}$$

where $\hat{\eta}_i := \eta/\eta_i^{p_i}$ and $\gamma > 0$ is a suitable constant (depending only on N and p_is).

Proof The function u solves Eq. (1) in $(x_o, t_o) + Q_\rho^-(M)$, so $\mathcal{T}(u)$ (defined in (7)) solves (1) in Q_1^-, according to Remark 2.2. Now, Lemma 3.1 of [14] on unitary cylinders ensures that for each function of the form

$$C_o^\infty(Q_1) \ni \eta = \prod_{i=1}^N \eta_i^{p_i}(x_i, t) \quad \text{with} \quad \eta_i \in C_o^\infty(\pi_i(K_1) \times (-1, 0]), \tag{18}$$

we have, for all $-1 < s_1 < s_2 < 0$ and $\bar{k} \in \mathbb{R}$,

$$\begin{aligned}
&\int_{K_1} (\mathcal{T}u - \bar{k})_\pm^2 \eta\, dy\Big|_{s_1}^{s_2} + \sum_{i=1}^N \int_{s_1}^{s_2} \int_{K_1} |\partial_i[\eta(\mathcal{T}u - \bar{k})_\pm]|^{p_i}\, dyds \\
&\quad \le \gamma \int_{s_1}^{s_2} \int_{K_1} |(\mathcal{T}u - \bar{k})_\pm|^2 \partial_s \eta\, dyds \\
&\quad + \sum_{i=1}^N \int_{s_1}^{s_2} \int_{K_1} |(\mathcal{T}u - \bar{k})_\pm|^{p_i} \hat{\eta}_i |\partial_i \eta_i|^{p_i}\, dyds.
\end{aligned} \tag{19}$$

Now we show that (17) comes from (19) by performing the change of variables (6) and (7), besides observing that $[\mathcal{T}u > \bar{k}] \cap K_1 = [u > k] \cap \mathcal{K}_\rho(M)$ provided

$\bar{k} = k/M$. Indeed, let us consider the change of variables

$$u(x,t) = u(M^{(p_i-2)/p_i}\rho^{\bar{p}/p_i}y_i, \rho^{\bar{p}}s) = M\mathcal{T}(u)(y,s),$$

with the stipulations

$$\begin{cases} x_i = M^{\frac{p_i-2}{p_i}}\rho^{\frac{\bar{p}}{p_i}}y_i, \\ t = \rho^{\bar{p}}s. \end{cases}$$

We observe that

$$\begin{cases} \partial_{x_i}u(x,t) = M^{2/p_i}\rho^{-\bar{p}/p_i}\partial_{y_i}\mathcal{T}u(y,s), \\ \partial_t u(x,t) = M\rho^{-\bar{p}}\partial_s\mathcal{T}u(y,s), \end{cases}$$

and $dx(y) = \left(\prod_i |dx_i/dy_i|\right)dy = \rho^N M^{\frac{N(\bar{p}-2)}{\bar{p}}}dy$. Hence the first integral in (19) becomes

$$\begin{aligned} \int_{K_1}(\mathcal{T}u-\bar{k})_\pm^2\eta\,dy\Big|_{s_1}^{s_2} &= \int_{K_\rho(M)}(M^{-1}(u(x,t)-k))_\pm^2\eta(y(x),s(t))\,(\rho^{-N}M^{\frac{-N(\bar{p}-2)}{\bar{p}}}dx)\Big|_{t_1}^{t_2} \\ &= \rho^{-N}M^{-\frac{[N(\bar{p}-2)+2\bar{p}]}{\bar{p}}}\int_{K_\rho(M)}(u(y,t)-k)_\pm^2\eta\,dx\Big|_{t_1}^{t_2}, \end{aligned}$$

being $t_1 := \rho^{\bar{p}}s_1 < \rho^{\bar{p}}s_2 =: t_2$. Similarly we evaluate the other integrals of (19), obtaining

$$\int_{s_1}^{s_2}\int_{K_1}|\partial_{y_i}[\eta(\mathcal{T}u-\bar{k})_\pm]|^{p_i}\,dyds =$$
$$\rho^{-N}M^{-\frac{[N(\bar{p}-2)+2\bar{p}]}{\bar{p}}}\int_{t_1}^{t_2}\int_{K_\rho(M)}|\partial_{x_i}[\eta(u-k)_\pm]|^{p_i}\,dxdt.$$

$$\int_{s_1}^{s_2}\int_{K_1}|(\mathcal{T}u-\bar{k})_\pm|^2\partial_s\eta\,dyds =$$
$$\rho^{-N}M^{-\frac{[N(\bar{p}-2)+2\bar{p}]}{\bar{p}}}\int_{t_1}^{t_2}\int_{K_\rho(M)}|(u(y,t)-k)_\pm^2\partial_t\eta\,dxdt.$$

$$\int_{s_1}^{s_2}\int_{K_1}|(\mathcal{T}u-\bar{k})_\pm|^{p_i}\hat{\eta}_i|\partial_{y_i}\eta_i|^{p_i}\,dyds =$$
$$\rho^{-N}M^{-\frac{[N(\bar{p}-2)+2\bar{p}]}{\bar{p}}}\int_{t_1}^{t_2}\int_{K_\rho(M)}|u(x,t)-k)_\pm|^{p_i}\hat{\eta}_i|\partial_{x_i}\eta_i|^{p_i}\,dxdt.$$

Collecting the terms $\rho^{-N} M^{-\frac{[N(\bar{p}-2)+2\bar{p}]}{\bar{p}}}$ we get (17). Hence, energy estimates are invariant under the scaling transformation (7). □

Remark 3.1 Clearly, the energy estimates above are valid also in forward and full cylinders $(x_o, t_o) + Q_\rho^+(M)$, $(x_o, t_o) + Q_\rho(M)$, provided they are contained in Ω_T.

The next Lemma is a sort of measure-theoretical maximum principle, popular amongst nonlinear analysts as Critical Mass Lemma (following Caffarelli), or De Giorgi-type Lemma (following DiBenedetto). It may be proven at ease for unitary cylinders, and then re-interpreted in the intrinsic geometry dictated by the scaling (7). To show the convenience of using (7), first we prove the lemma in its general form, and then we discuss its invariance with respect to the scaling.

We recall that local weak sub-solutions (resp, super-solutions) to (1) are locally bounded from above (resp., below) in Ω_T (see, e.g., [14, 23]), provided an additional condition constraining the spareness of p_is is ensured. Let us fix a cylinder $(y, s) + Q_{2\rho}(\theta) \subset\subset \Omega_T$, being $(y, s) \in \Omega_T$ and $\rho, \theta > 0$ appropriate. Let μ^+, μ^- be such that

$$\mu^- \leq \operatorname*{ess\,inf}_{(y,s)+Q_{2\rho}(\theta)} u \leq \operatorname*{ess\,sup}_{(y,s)+Q_{2\rho}(\theta)} u \leq \mu^+.$$

We also fix $\omega > 0$, $\xi \in (0, 1]$, and $a \in (0, 1)$.

Lemma 3.2 (De Giorgi-Type/Critical Mass) *Let u be a local weak super-solution to* (1) *in Ω_T locally bounded from below, and let $\rho, \theta, \mu^\pm, \omega, \xi, a$ be defined as above. Then there exists $\nu^- \in (0, 1)$, depending on the data N, p_is and on the parameters θ, ω, ξ, a but not on the radius ρ, such that if*

$$|[u \leq \mu^- + \xi\omega] \cap [(y, s) + Q_{2\rho}^-(\theta)]| \leq \nu^- |Q_{2\rho}^-(\theta)| \tag{20}$$

then

$$u \geq \mu^- + a\xi\omega \quad \text{a.e. in } Q_\rho^-(\theta). \tag{21}$$

Likewise, if u is a local weak sub-solution to (1) *in Ω_T which is locally bounded from above, then there exists $\nu^+ \in (0, 1)$, depending on the data N, p_is and on the parameters θ, ω, ξ, a but not on the radius ρ, such that if*

$$|[u \geq \mu^+ - \xi\omega] \cap [(y, s) + Q_{2\rho}^-(\theta)]| \leq \nu^+ |Q_{2\rho}^-(\theta)| \tag{22}$$

then

$$u \leq \mu^+ - a\xi\omega \quad \text{a.e. in } Q_\rho^-(\theta). \tag{23}$$

Proof We prove (21), since the proof of (23) is analogous. Without loss of generality we assume $(y, s) = (0, 0)$, just to ease the notation. Let us set, for any $n, \bar{n} \in \mathbb{N} \cup \{0\}$,

$$\rho_n = \rho + \frac{\rho}{2^n}, \qquad \mathcal{K}_n = \prod_{i=1}^{N} \left\{ |x_i| < \theta^{\frac{p_i-2}{p_i}} \rho^{\frac{\bar{p}}{p_i}} \left(1 + \frac{1}{2^{n+\bar{n}}}\right) \right\}, \qquad Q_n = \mathcal{K}_n \times (-\rho_n^{\bar{p}}, 0].$$

Since $K_0 \to \mathcal{K}_\rho(\theta)$ as $\bar{n} \to \infty$, we fix $\bar{n}$ such that $\mathcal{K}_0 \subset \mathcal{K}_{2\rho}(\theta)$. Notice that $\bar{n}$ can be chosen in such a way that it depends only on N and p_is. We apply energy estimates (17) over $\mathcal{K}_n, Q_n$ to the truncations $(u - k_n)_-$ at the levels

$$k_n = \mu^- + \xi_n \omega, \quad \text{where} \quad \xi_n = a\xi + \frac{(1-a)\xi}{2^n}.$$

Incidentally, notice that

$$|(u - k_n)_-| \leq \xi_n \omega \leq \xi \omega.$$

For any n, we pick a cut-off function η_n of the form $\eta_n = \bar{\eta}(t) \prod_{i=1}^{N} \eta_i^{p_i}(x_i)$, where

$$\eta_i(x_i) = \begin{cases} 1, & \text{in} \quad \pi_i(\mathcal{K}_{n+1}), \\ 0, & \text{in} \quad \mathbb{R} \setminus \pi_i(\mathcal{K}_n), \end{cases} \qquad |\partial_i \eta_i| \leq \frac{\gamma 2^n}{\theta^{\frac{p_i-2}{p_i}} \rho^{\frac{\bar{p}}{p_i}}}.$$

$$\bar{\eta}(t) = \begin{cases} 1, & \text{when} \quad t \geq -\rho_{n+1}^{\bar{p}}, \\ 0, & \text{when} \quad t < -\rho_n^{\bar{p}}, \end{cases} \qquad |\partial_t \bar{\eta}| \leq \frac{\gamma}{(\rho_n^{\bar{p}} - \rho_{n+1}^{\bar{p}})} \leq \frac{\gamma 2^{(n+1)\bar{p}}}{\rho^{\bar{p}}},$$

The energy estimate (17), applied with these choices, yields

$$\begin{aligned} \mathcal{E}_n &:= \sup_{(-\rho_n^{\bar{p}}, 0]} \int_{\mathcal{K}_n} (u - k_n)_-^2 \eta_n \, dx + \sum_{i=1}^{N} \int\int_{Q_n} |\partial_i [\eta_n (u - k_n)_-]|^{p_i} \, dxdt \\ &\leq \gamma \int\int_{Q_n} \left\{ (u - k_n)_-^2 |\partial_t \bar{\eta}| + \sum_{i=1}^{N} |(u - k_n)_-|^{p_i} |\partial_i \eta_i|^{p_i} \right\} dxdt \\ &\leq \frac{\gamma 2^{\bar{p}(n+1)}}{\rho^{\bar{p}}} (\xi \omega)^2 \left(1 + \sum_{i=1}^{N} \left(\frac{\xi \omega}{\theta}\right)^{p_i - 2}\right) |[u < k_n] \cap Q_n|. \end{aligned}$$

Now we combine these estimates of the energies $\mathcal{E}_n$ with the embedding inequality (see [14])

$$\begin{aligned}&\int\int_{Q_n} |(u-k_n)_-\eta_n|^{\bar{p}(\frac{N+2}{N})}dxdt\\&\leq \gamma\Big(\sup_{(-\rho_n^{\bar{p}},0]}\int_{\mathcal{K}_n}(u-k_n)_-^2\eta_n^2\,dx+\sum_{i=1}^N\int\int_{Q_n}|\partial_i[\eta_n[(u-k_n)_-]|^{p_i}\,dxdt\Big)^{\frac{N+\bar{p}}{N}}\\&=\gamma\mathcal{E}_n^{\frac{N+\bar{p}}{N}}.\end{aligned}$$

Observing that $(u-k_n)_-\geq(k_n-k_{n+1})=(1-a)\xi\omega/2^{n+1}$ in $[u<k_{n+1}]\cap Q_{n+1}$, we get the chain

$$\begin{aligned}\Big(\frac{(1-a)\xi\omega}{2^{n+1}}\Big)^{\bar{p}(\frac{N+2}{N})}|[u<k_{n+1}]\cap Q_{n+1}|&\leq\int\int_{Q_{n+1}}|(u-k_n)_-|^{\bar{p}(\frac{N+2}{N})}dxdt\\&\leq\int\int_{Q_n}|(u-k_n)_-\eta_n|^{\bar{p}(\frac{N+2}{N})}dxdt\leq\gamma\mathcal{E}_n^{\frac{N+\bar{p}}{N}}\\&\leq\gamma\Big[\frac{2^{\bar{p}(n+1)}}{\rho^{\bar{p}}}(\xi\omega)^2\Big(1+\sum_{i=1}^N\Big(\frac{\xi\omega}{\theta}\Big)^{p_i-2}\Big)\Big]^{\frac{N+\bar{p}}{N}}|[u<k_n]\cap Q_n|^{\frac{N+\bar{p}}{N}}.\end{aligned}\tag{24}$$

We set $Y_n=\frac{|[u<k_n]\cap Q_n|}{|Q_n|}$ and recall that up to a constant we have $|Q_n|^{\frac{\bar{p}}{N}}\leq\theta^{\bar{p}-2}\rho^{\frac{\bar{p}}{N}(\bar{p}+N)}$. Dividing (24) for $|Q_{n+1}|$, as well as noticing that $|Q_{n+1}|\geq 2^{-(N+1)}|Q_n|$, we obtain

$$\begin{aligned}Y_{n+1}\leq&\gamma\Big(\frac{2^{n+1}}{(1-a)\xi\omega}\Big)^{\bar{p}(\frac{N+2}{N})}\Big[\frac{2^{\bar{p}(n+1)}}{\rho^{\bar{p}}}(\xi\omega)^2\Big(1+\sum_{i=1}^N\Big(\frac{\xi\omega}{\theta}\Big)^{p_i-2}\Big)\Big]^{\frac{N+\bar{p}}{N}}Y_n^{1+\frac{\bar{p}}{N}}|Q_n|^{\frac{\bar{p}}{N}}\\&\leq 2^{n(2N+\bar{p}+2)\frac{\bar{p}}{N}}\Big[\frac{\gamma(\xi\omega/\theta)^{2-\bar{p}}}{(1-a)^{\frac{\bar{p}}{N}(N+2)}}\Big(1+\sum_{i=1}^N(\xi\omega/\theta)^{p_i-2}\Big)^{\frac{N+\bar{p}}{N}}\Big]Y_n^{1+\frac{\bar{p}}{N}}.\end{aligned}\tag{25}$$

According to [12, Lemma 4.1, page 12], (25) produces $Y_\infty:=\lim_{n\to\infty}Y_n=0$ provided

$$Y_0\leq 2^{-\frac{N}{\bar{p}}(2N+\bar{p}+2)}\Big[\frac{\gamma(\xi\omega/\theta)^{2-\bar{p}}}{(1-a)^{\frac{\bar{p}}{N}(N+2)}}\Big(1+\sum_{i=1}^N(\xi\omega/\theta)^{p_i-2}\Big)^{\frac{N+\bar{p}}{N}}\Big]^{-\frac{N}{\bar{p}}}:=\nu^*.\tag{26}$$

Then $Y_\infty = 0$ implies (21), concluding the proof. It remains to ensure (26).

We set $\nu^- := \nu^*/\gamma$, where ν^* stems from (26) and γ is such that $|Q^-_{2\rho}(\theta)| \leq \gamma |Q_0|$. Therefore

$$|[u < \mu^- + \xi\omega] \cap Q_0| \leq |[u < \mu^- + \xi\omega] \cap Q^-_{2\rho}(\theta)| \leq \nu^-\gamma|Q_0| = \nu^*|Q_0|,$$

which guarantees (26).

In order to prove (23), we may proceed as above, considering energy estimates (17) in the same iterative geometry, but this time using the truncations $(u - k_n)_+$, being $k_n = \mu^+ - \xi_n\omega$.

□

Remark 3.2 We observe that setting $\theta = \xi\omega$ frees ν^- from any dependence on θ, ω, ξ. When $\omega \geq \mu^+ - \mu^-$, this choice of θ represents the intrinsic geometry, since ω is related to the oscillation of the solution u in $Q^-_{2\rho}(\theta)$.

Simple adjustments on Q_n guarantee the validity of Lemma 3.2 also for forward cylinders and full cylinders.

Lemma 3.2 is invariant under the particular scalings (7) and (9). To show this, one can use the change of variables employed in the proof of Lemma 17 to prove the following lemma, and then one can go back to Lemma 3.2. Accordingly, the argument sketched here shows the equivalence of Lemmas 3.2 and 3.3.

Lemma 3.3 *Let u be a local weak super-solution to* (1) *in Q_1 such that* $\operatorname{ess\,inf}_{Q_1^-} u \geq 0$. *Then for any $a \in (0, 1)$ there exists $\nu_a^+ > 0$, depending on a, p_i, N but neither on u nor on ρ, such that*

$$|[u \leq 1] \cap Q_1^-(1)| \leq \nu_a^+ |Q_1^-(1)| \quad \Rightarrow \quad \operatorname*{ess\,inf}_{Q^-_{1/2}(1)} u \geq a\,. \tag{27}$$

Let u be a local weak sub-solution to (1) *in Q_1 such that* $\operatorname{ess\,sup}_{Q_1^-} u \leq 1$. *Then for any $a \in (0, 1)$ there exists $\nu_a^- > 0$, depending on a, p_i, N but neither on u nor on ρ, such that*

$$|[u \geq 1/2] \cap Q_1^-(1/2)| \leq \nu_a^- |Q_1^-(1/2)| \quad \Rightarrow \quad \operatorname*{ess\,sup}_{Q^-_{1/2}(1/2)} u \leq (1 - a/2)\,. \tag{28}$$

Proof It suffices to apply Lemma 3.2 to the function $(u - \mu^-)/(\xi\omega)$ (resp., $(\mu^+ - u)/(\xi\omega)$) with the choices $\mu^- = 0, \theta = \xi\omega = 1$ (resp., $\mu^+ = 1, \theta = \xi\omega = 1/2$), and $\rho = 1$ and in the first (resp., second) case. □

4 A Topological Consequence of Energy Invariants: Lower Semi-Continuity of Super-Solutions

Theorem 4.1 *Let u be a weak local super-solution to* (1) *in Ω_T locally bounded from below. Then u is lower semi-continuous.*

Proof We proceed in a way reminiscent of [20]. Set $\mathcal{Q}_\rho := \mathcal{Q}_\rho(1)$ for all $\rho > 0$, and consider the lower semi-continuous regularization of u, defined as

$$u_*(x,t) = \lim_{\rho\to 0^+} \operatorname*{ess\,inf}_{(x,t)+\mathcal{Q}_\rho} u \quad \forall (x,t) \in \Omega_T. \tag{29}$$

We observe that this function is well defined, since $(x,t)+\mathcal{Q}_\rho \subset \Omega_T$ for small values of ρ. It is a well-known fact that u_* is lower semi-continuous. Accordingly, proving that $u_* = u$ almost everywhere in Ω_T furnishes the lower semi-continuity of u. In order to show this equality, we also define the set

$$\mathcal{L} = \left\{(x,t) \in \Omega_T : \; |u(x,t)| < \infty, \text{ and } \lim_{\rho\to 0^+} \fint_{(x,t)+\mathcal{Q}_\rho} |u(x,t)-u(y,t)|\, dydt = 0\right\}. \tag{30}$$

This set is well defined, since $u \in L^1_{loc}(0,T; L^1_{loc}(\Omega))$. Moreover,

$$|\mathcal{L}| = |\Omega_T|. \tag{31}$$

As we will see, this is a consequence of the fact that $X := (\Omega_T, \mathscr{L}^{N+1}, d)$, being $\mathscr{L}^{N+1}$ the $(N+1)$-Lebesgue measure and d a particular distance to be introduced, is a doubling space. We consider the following distance d: for any (x,t), $(y,s) \in \Omega_T$ we define

$$d((x,t),(y,s)) := \max\{|x_i - y_i|^{\frac{p_i}{\bar{p}}}, |t-s|^{\frac{1}{\bar{p}}}\},$$

and we denote by $\mathbb{B}_\rho(x,t)$ the balls with respect to distance d. It turns out that $\mathbb{B}_\rho(x,t) = (x,t)+\mathcal{Q}_\rho$. The doubling property follows from

$$\mathscr{L}^{N+1}(\mathbb{B}_{2\rho}) = \mathscr{L}^{N+1}(\mathcal{Q}_{2\rho}) = (2\rho)^{N+\bar{p}} = 2^{N+\bar{p}}\mathscr{L}^{N+1}(\mathcal{Q}_\rho) = 2^{N+\bar{p}}\mathscr{L}^{N+1}(\mathbb{B}_\rho).$$

Accordingly, [17, p. 12] provides (31).

Taking (31) into account, it is sufficient to prove $u_* = u$ in $\mathcal{L}$. For all $(x,t) \in \mathcal{L}$ we have

$$u_*(x,t) = \lim_{\rho\to 0} \operatorname*{ess\,inf}_{(x,t)+\mathcal{Q}_\rho} u \le \lim_{\rho\to 0} \fint_{(x,t)+\mathcal{Q}_\rho} u\, dydt = u(x,t).$$

To show the opposite inequality, let us pick $(x_o,t_o) \in \mathcal{L}$ and suppose by contradiction that $u_*(x_o,t_o) < u(x_o,t_o)$. Let $r, b > 0$ be small enough such that

$(x_o, t_o) + Q_r \subset \Omega_T$ and

$$\operatorname*{ess\,inf}_{(x_o,t_o)+Q_r} u := \mu_- \leq u_*(x_o, t_o) < \mu_- + b < u(x_o, t_o).$$

This choice is possible, since u_* is close to μ^- for small values of ρ, as well as Q_ρ shrinks to (x_o, t_o) as $\rho \to 0^+$. Let us introduce $a \in (0, 1)$ such that

$$\mu_- + ab > u_*(x_o, t_o), \qquad \text{i.e.,} \qquad \frac{u_*(x_o, t_o) - \mu_-}{b} < a < 1.$$

Then there exists $\nu_a^- > 0$, depending only on a, b, p_i, N, such that for some $\rho \in (0, r)$ we have

$$|[u \leq \mu_- + b] \cap (x_o, t_o) + Q_\rho| \leq \nu_a^- |Q_\rho|,$$

since otherwise we have, for all $\rho \in (0, r)$,

$$\begin{aligned}\int_{(x_o,t_o)+Q_\rho} |u(x_o, t_o) - u(x, t)|\, dxdt &\geq \int_{[u\leq\mu_-+b]\cap(x_o,t_o)+Q_\rho} [u(x_o, t_o) - (\mu_- + b)]\, dxdt \\ &\geq \nu_a^- [u(x_o, t_o) - (\mu_- + b)]|Q_\rho|,\end{aligned}$$

contradicting $(x_o, t_o) \in \mathcal{L}$. Now we are in the position to apply Lemma 3.2 and reach

$$u(x, t) \geq \mu_- + ab > u_*(x_o, t_o), \qquad \text{for a.a. } (x, t) \in (x_o, t_o) + Q_{\rho/2}.$$

This contradicts the definition of $u_*(x_o, t_o)$, since

$$u_*(x_o, t_o) < \operatorname*{ess\,inf}_{(x_o,t_o)+Q_{\rho/2}} u \leq \lim_{\rho\to 0^+} \operatorname*{ess\,inf}_{(x_o,t_o)+Q_\rho} u = u_*(x_o, t_o).$$

Therefore, we obtain $u_* = u$ in $\mathcal{L}$, concluding the proof. □

Remark 4.1 Semi-continuity of solutions to (1), proved in Theorem 4.1, jointly with the structure of the equation, ensures that the estimates (21) and (23) of Lemma 3.2 hold true everywhere (and not merely 'almost everywhere') in $Q_\rho^-(\theta)$.

Acknowledgments The authors are grateful to Sunra J.N. Mosconi, for interesting conversations on the subject. S. Ciani acknowledges the support of the Department of Mathematics of T.U. Darmstadt. U. Guarnotta was supported by the following research projects: (1) PRIN 2017 'Nonlinear Differential Problems via Variational, Topological and Set-valued Methods' (Grant No. 2017AYM8XW) of MIUR; (2) PRA 2020–2022 Linea 3 'MO.S.A.I.C.' of the University of Catania. All the authors acknowledge the support of GNAMPA (INdAM).

References

1. Antontsev, S., Shmarev, S.: Evolution PDEs with Nonstandard Growth Conditions. Existence, Uniqueness, Localization, Blow-Up. Atlantis Studies in Differential Equations, vol. 4. Atlantis Press, Paris (2015)
2. Barenblatt, G.I.: On some unsteady motions of a fluid and a gas in a porous medium (in Russian). Akad. Nauk SSSR. Prikl. Mat. Meh. **16**, 67–78 (1952)
3. Barenblatt, G.I.: Scaling, Self-Similarity, and Intermediate Asymptotics. Cambridge Texts in Applied Mathematics, vol. 14. Cambridge University Press, Cambridge (1996)
4. Barenblatt, G.I.: Scaling. Cambridge Texts in Applied Mathematics, vol. 34. Cambridge University Press, Cambridge (2003)
5. Bousquet, P., Brasco, L.: Lipschitz regularity for orthotropic functionals with nonstandard growth conditions. Rev. Mat. Iberoam. **36**(7), 1989–2032 (2020)
6. Ciani, S., Mosconi, S., Vespri, V.: Parabolic Harnack estimates for anisotropic slow diffusion, J. Anal. Math. (in press, 2022). http://math.huji.ac.il/~w-jam/articles.html
7. Ciani, S., Vespri, V.: An introduction to barenblatt solutions for anisotropic p-laplace equations. In: Anomalies in Partial Differential Equations. Springer INdAM Series, vol. 43, pp. 99–125. Springer, Cham (2021)
8. Bousquet, P., Brasco, L., Leone, C., Verde, A.: Gradient estimates for an orthotropic nonlinear diffusion equation. Adv. Calc. Var. (2021). https://doi.org/10.1515/acv-2021-0052
9. Ciani, S., Skrypnik, I.I., Vespri, V.: On the local behavior of local weak solutions to some singular anisotropic elliptic equations. Adv. Nonlinear Anal. **12**(1), 237–265, (2023)
10. De Giorgi, E.: Sulla differenziabilità e l'analiticità delle estremali degli integrali multipli regolari (in Italian). Mem. Accad. Sci. Torino. Cl. Sci. Fis. Mat. Nat. **3**, 25–43 (1957)
11. DiBenedetto, E.: Intrinsic Harnack type inequalities for solutions of certain degenerate parabolic equations. Arch. Rational Mech. Anal. **100**(2), 129–147 (1988)
12. DiBenedetto, E.: Degenerate Parabolic Equations. Universitext Springer, New York (1993)
13. DiBenedetto, E., Gianazza, U., Vespri, V.: Harnack estimates for quasi-linear degenerate parabolic differential equations. Acta Math. **200**(2), 181–209 (2008)
14. Düzgün, F.G., Mosconi, S., Vespri, V.: Anisotropic Sobolev embeddings and the speed of propagation for parabolic equations. J. Evol. Equ. **19**(3), 845–882 (2019)
15. Feo, F., Vázquez, J.L., Volzone, B.: Anisotropic p-laplacian evolution of fast diffusion type. Adv. Nonlinear Stud. **21**(3), 523–555 (2021)
16. Haškovec, J., Schmeiser, C.: A note on the anisotropic generalizations of the Sobolev and Morrey embedding theorems. Monatsh. Math. **158**(1), 71–79 (2009)
17. Heinonen, J.: Lectures on Analysis on Metric Spaces. Universitext. Springer, New York (2001)
18. Henriques, E.: Concerning the regularity of the anisotropic porous medium equation. J. Math. Anal. Appl. **377**(2), 710–731 (2011)
19. Kuusi, T.: Lower semicontinuity of weak supersolutions to nonlinear parabolic equations. Differ. Integral Equ. **22**(11–12), 1211–1222 (2009)
20. Liao, N.: Regularity of weak supersolutions to elliptic and parabolic equations: lower semicontinuity and pointwise behavior. J. Math. Pures Appl. (9) **147**, 179–204 (2021)
21. Marcellini, P.: Regularity under general and p, q-growth conditions. Discrete Contin. Dyn. Syst. Ser. S **13**(7), 2009–2031 (2020)
22. Mingione, G.: Regularity of minima: an invitation to the dark side of the calculus of variations. Appl. Math. **51**(4), 355–426 (2006)
23. Mingqi, Y., Xiting, L.: Boundedness of solutions of parabolic equations with anisotropic growth conditions. Canad. J. Math. **49**(4), 798–809 (1997)
24. Ural'tseva, N.N., Urdaletova, A.B.: Boundedness of gradients of generalized solutions of degenerate nonuniformly elliptic quasilinear equations (in Russian). Vestnik Leningrad. Univ. Mat. Mekh. Astronom. **4**, 50–56 (1983)

25. Urbano, J.M.: The method of intrinsic scaling. In: A Systematic Approach to Regularity for Degenerate and Singular PDEs. Lecture Notes in Mathematics, vol. 1930. Springer, Berlin (2008)
26. Vázquez, J.L.: The porous medium equation. In: Mathematical Theory. Oxford Mathematical Monographs, The Clarendon Press/Oxford University Press, Oxford (2007)
27. Vespri, V.: What I learnt from Emmanuele DiBenedetto. In: *Harnack Inequalities and Nonlinear Operators*. Springer INdAM Series, vol. 46, pp. 1–27. Springer, Cham (2021)

A Deformation Theory in Augmented Spaces and Concentration Results for NLS Equations Around Local Maxima

Silvia Cingolani and Kazunaga Tanaka

Dedicated to Professor Francesco Altomare with admiration and friendship

Abstract We present some new existence results of semiclassical bound states for a nonlinear Schrödinger equation concentrating at local maxima of an external potential, by means of deformation arguments in augmented spaces.

Keywords Nonlinear Schrödinger equation · Singular perturbation · Local maxima · Deformation theory

1 Introduction

In recent years a large amount of papers has been devoted to investigate concentration phenomena of solutions of nonlinear Schrödinger equations of the form

$$-\varepsilon^2\Delta v + V(x)v = g(v) \quad \text{in } \mathbb{R}^N, \qquad v \in H^1(\mathbb{R}^N) \tag{1}$$

where $N \geq 3$, $\varepsilon > 0$, V is an external potential and $g \in C(\mathbb{R}, \mathbb{R})$ is a nonlinear term.

Starting to the celebrated works by Floer and Weinstein [22] and Rabinowitz [29], variational approaches were developed extensively. We refer to [1, 2, 22, 28] for some first existence and concentration results under nondegeneracy conditions

S. Cingolani (✉)
Dipartimento di Matematica, Università degli Studi di Bari Aldo Moro, Bari, Italy
e-mail: silvia.cingolani@uniba.it

K. Tanaka
Department of Mathematics, School of Science and Engineering, Waseda University, Tokyo, Japan
e-mail: kazunaga@waseda.jp

A. M. Candela et al. (eds.), *Recent Advances in Mathematical Analysis*,
Trends in Mathematics, https://doi.org/10.1007/978-3-031-20021-2_16

both on the critical points of the potential V (local minima or maxima) and on the solutions of some limiting problems, which are crucial to apply finite dimensional reduction arguments. For studies without nondegeneracy conditions we refer to [4, 7–10, 15, 16, 20, 23] for concentration at local minima of $V(x)$ and [5, 19, 21] for concentration at local maxima and saddle points of $V(x)$. More precisely for concentration at local maxima and saddle points, del Pino and Felmer [21] introduced a new reduction method and showed the existence under Ambrosetti-Rabinowitz type condition and monotonicity condition on $g(\xi)$. Their approach takes advantage of the Nehari manifold. In [19], D'Avenia, Pomponio and Ruiz improved the result in [21] removing the monotonicity assumption. They used a different minimax argument in certain cones in the variational space. More general results were given in [5] with a different approach. See also [6], where the existence of clustering solutions at local maxima or saddle points of $V(x)$.

In this paper we give a slight generalization of the result in [5] and furnish a simplified proof. Precisely, for the potential $V(x)$, we assume

(V1) $V(x) \in C^N(\mathbb{R}^N, \mathbb{R})$, $\nabla V(x) \in L^{N/2}(\mathbb{R}^N) + L^\infty(\mathbb{R}^N)$.
(V2) $\inf_{x\in\mathbb{R}^N} V(x) \equiv \underline{V} > 0$, $\sup_{x\in\mathbb{R}^N} V(x) \equiv \overline{V} < \infty$.
(V3) There exists a bounded open set $\Omega \subset \mathbb{R}^N$ such that

$$\nabla V(x) \neq 0 \quad \text{for all } x \in \partial\Omega.$$

For simplicity, we focus on the case that $V(x)$ takes its local maximum in Ω. That is, we assume

(LM) $V_0 \equiv \sup_{x\in\Omega} V(x) > \sup_{x\in\partial\Omega} V(x)$

and we consider concentration of solutions to maxima of $V(x)$ in Ω.

For $g(\xi)$ we assume the following conditions:

(g1) $g(0) = 0$, $\lim_{\xi\to 0} \frac{g(\xi)}{\xi} = 0$.
(g2) $\lim_{\xi\to\infty} \frac{g(\xi)}{\xi^{\frac{N+2}{N-2}}} = 0$.
(g3) There exists $\xi_0 > 0$ such that

$$\frac{1}{2} V_0 \xi_0^2 < G(\xi_0),$$

where $G(\xi) = \int_0^\xi g(\tau)\, d\tau$ and $V_0 > 0$ is the constant appeared in (LM).

Our main result is

Theorem 1.1 *Assume (V1)–(V3), (LM) and (g1)–(g3). Then* (1) *has a family of positive solutions which concentrates in* Ω. *More precisely, there exist* $\varepsilon_0 > 0$ *and a family* $(v_\varepsilon(x))_{\varepsilon\in(0,\varepsilon_0]}$ *of positive solutions of* (1) *with the following property: for any sequence* $(\varepsilon_j)_{j=1}^\infty \subset (0, \varepsilon_0]$ *with* $\varepsilon_j \to 0$ *after extracting a subsequence—we denote it by* (ε_j) *for simplicity of notation —, there exist* $(x_j)_{j=1}^\infty \subset \mathbb{R}^N$, $x_0 \in \Omega$

with $V(x_0) = \max_{x \in \overline{\Omega}} V(x)$ *and a positive least energy solution* $\omega_0 \in H^1(\mathbb{R}^N)$ *of the limit problem* $-\Delta u + V(x_0)u = g(u)$ *in* $\mathbb{R}^N$ *such that*

$$\varepsilon_j x_j \to x_0,$$
$$v_{\varepsilon_j}(\varepsilon_j(x - x_j)) \to \omega_0(x) \quad \textit{strongly in } H^1(\mathbb{R}^N).$$

Remark 1.1 We can deal with more general potential $V(x)$, for example, where $V(x)$ has a MP geometry in Ω, and we can show concentration at MP critical point of $V(x)$. See [5] and [14].

We note that slightly different result is given in [5], where $g(\xi)$ is assumed to be of class C^1. In [5], an iteration of 2 flows, the standard deformation flow in $H^1(\mathbb{R}^N)$ and the flow in $\mathbb{R}^N$ associated to $-\nabla V(x)$, are used together with the tail minimizing operator $\tau_\varepsilon(v)$, which is defined as $\tau_\varepsilon(v) = w$, where w is a solution of an exterior problem:

$$\begin{cases} -\varepsilon^2 \Delta w + V(x)w = g(w) \text{ in } |x - p| > \varepsilon L, \\ w(x) = v(x) \qquad\qquad\quad \text{on } |x - p| = \varepsilon L. \end{cases} \tag{2}$$

Here v is a given function in a neighborhood of expected solutions, $L \gg 1$ is a big constant and $p = \beta(v)$ is the center of mass. See Sect. 2 below. The procedure in [5] is rather involved and in the present note we give a simplified deformation approach through a construction of a flow in an augmented space $\mathbb{R}^N \times H^1(\mathbb{R}^N)$.

Denoting $u(x) = v(\varepsilon x)$, we introduce an equivalent problem to (1):

$$-\Delta u + V(\varepsilon x)u = g(u) \quad \text{in } \mathbb{R}^N, \qquad u \in H^1(\mathbb{R}^N) \tag{3}$$

and we try to find critical points of the corresponding functional:

$$I_\varepsilon(u) = \frac{1}{2}\int_{\mathbb{R}^N} |\nabla u|^2 + V(\varepsilon x)u^2 - \int_{\mathbb{R}^N} G(u) : \; H^1(\mathbb{R}^N) \to \mathbb{R}.$$

To construct our deformation flow, we adapt our idea in recent paper [14] in which we study singular perturbation problem for the nonlinear Choquard equations. We consider the following functional which is defined in the augmented space $\mathbb{R}^N \times H^1(\mathbb{R}^N)$:

$$J_\varepsilon(z, u) = \frac{1}{2}\int_{\mathbb{R}^N} |\nabla u|^2 + V(\varepsilon x + z)u^2 - \int_{\mathbb{R}^N} G(u);\; \mathbb{R}^N \times H^1(\mathbb{R}^N) \to \mathbb{R}. \tag{4}$$

We note that

$$J_\varepsilon(z, u) = I_\varepsilon(u(x - \frac{z}{\varepsilon})) \quad \text{for } (z, u) \in \mathbb{R}^N \times H^1(\mathbb{R}^N)$$

and we will construct a deformation flow $\widetilde{\eta}_\varepsilon$ for $I_\varepsilon(u)$ as a composition of a deformation flow for $J_\varepsilon(z,u)$ in $\mathbb{R}^N \times H^1(\mathbb{R}^N)$ and a projection $\pi : \mathbb{R}^N \times H^1(\mathbb{R}^N) \to H^1(\mathbb{R}^N)$ defined by

$$\pi(z,u)(x) = u(x - \frac{z}{\varepsilon}).$$

An advantage of this construction is that for $h \in \mathbb{R}^N$ a flow $\widetilde{\eta}_\varepsilon(t,0,u) = (ht,u)$ in $\mathbb{R}^N \times H^1(\mathbb{R}^N)$ corresponds to the translation flow $t \mapsto u(x - \frac{h}{\varepsilon}t)$ in $H^1(\mathbb{R}^N)$. We can include naturally the translation flow in our deformation flow. We note that the translation is continuous in $H^1(\mathbb{R}^N)$ but not of class C^1 in $H^1(\mathbb{R}^N)$. Thus it can not be obtained via the standard deformation argument. Similar construction of deformation flows by means of augmented functionals are studied in [13, 17, 18, 24, 26] for scalar field equations, where scaling $\lambda \mapsto u(x/\lambda)$ is used instead of translation.

In following sections, we give an outline of construction of our deformation flow and a proof of our Theorem 1.1.

2 Limit Problem and Neighborhood of Expected Solutions

In what follows we assume (V1)–(V3), (LM), (g1)–(g3). Since we look for a positive solution, we may assume that $g(\xi)$ is odd in ξ without loss of generality.

We use the following notation: for $u \in H^1(\mathbb{R}^N)$

$$\|u\|_{H^1} = \left(\int_{\mathbb{R}^N} |\nabla u|^2 + u^2\right)^{1/2},$$

$$\|u\|_r = \left(\int_{\mathbb{R}^N} |u|^r\right)^{1/r} \quad \text{for } r \in [1,\infty),$$

$$\|u\|_\infty = \text{ess sup}_{x\in\mathbb{R}^N} |u(x)|.$$

We also use notation: $B(p,r) = \{x \in \mathbb{R}^N : |x-p| < r\}$ for $p \in \mathbb{R}^N, r > 0$.

2.1 Limit Problems

For $a > 0$ we define

$$L_a(u) = \frac{1}{2}\|\nabla u\|_2^2 + \frac{a}{2}\|u\|_2^2 - \int_{\mathbb{R}^N} G(u) : H^1(\mathbb{R}^N) \to \mathbb{R}.$$

Critical points of $L_a(u)$ is a solution of

$$-\Delta u + au = g(u) \quad \text{in } \mathbb{R}^N. \tag{5}$$

If a family $(u_\varepsilon(x))$ of solutions of (3) and $(x_\varepsilon) \subset \mathbb{R}^N$ satisfy as $\varepsilon \to 0$

$$\varepsilon x_\varepsilon \to x_0, \tag{6}$$

$$u_\varepsilon(x - x_\varepsilon) \to u_0(x) \neq 0 \quad \text{in } H^1(\mathbb{R}^N), \tag{7}$$

then u_0 is a critical point of $L_{V(x_0)}(u)$, that is, a solution of (5) with $a = V(x_0)$.

The scalar field equation (5) has been well-studied since the celebrated work [3] of Berestycki and Lions, where the following classical result has been established. We also refer to [27].

Proposition 2.1 *We assume that* $a > 0$ *satisfies*

$$\frac{1}{2}a\xi_0^2 < G(\xi_0) \quad \textit{for some } \xi_0 > 0.$$

Then we have

(i) (5) *has a weak solution. Moreover solution* $u(x)$ *satisfies Pohozaev identity:* $P_a(u) = 0$, *where*

$$P_a(u) = \frac{N-2}{2}\|\nabla u\|_2^2 + \frac{N}{2}a\,\|u\|_2^2 - N\int_{\mathbb{R}^N} G(u).$$

(ii) We define the least energy level E_a *by*

$$E_a = \inf\{L_a(u);\ u \neq 0,\ L_a'(u) = 0\}.$$

Then $E_a > 0$ *and* E_a *is attained by a positive solution of* (5).

(iii) E_a *is characterized as the infinimum on Pohozaev manifold:*

$$E_a = \inf\{L_a(u);\ u \neq 0,\ P_a(u) = 0\}.$$

Moreover it is also characterized by the mountain pass minimax method:

$$E_a = \inf_{\gamma\in\Gamma_a}\max_{t\in[0,1]} L_a(\gamma(t)),$$

where $\Gamma_a = \{\gamma(t) \in C([0,1], H^1(\mathbb{R}^N));\ \gamma(0) = 0,\ L_a(\gamma(1)) < 0\}$.

We write $Q = [0,1]^N$ and $n+Q = [n_1, n_1+1]\times[n_2, n_2+1]\times\cdots\times[n_N, n_N+1]$ for $n = (n_1, \cdots, n_N) \in \mathbb{Z}^N$. For $c > 0$ we set

$$C_a^c = \{u \in H^1(\mathbb{R}^N)\setminus\{0\} :\ L_a'(u) = 0,\ L_a(u) \le c,\ \|u\|_{L^2(Q)} = \max_{n\in\mathbb{Z}^N}\|u\|_{L^2(n+Q)}\}.$$

We have

Lemma 2.1 *Suppose that* $c < 2E_a$. *Then* C_a^c *is compact in* $H^1(\mathbb{R}^N)$ *and any* $u \in C_a^c$ *has constant sign.*

For the proof we refer to [4]. We also set

$$C_{a,+}^c = \{u \in C_a^c : u(x) > 0 \text{ in } \mathbb{R}^N\}.$$

Since $g(\xi)$ is odd, we have

$$C_a = C_{a,+}^c \cup (-C_{a,+}^c), \qquad C_{a,+}^c \cap (-C_{a,+}^c) = \emptyset$$

and $C_{a,+}^c$ is compact for $c < 2E_a$.

From now on, we assume the following (V4) in addition to (V1)–(V3) and (LM).

(V4) For any $p \in \Omega$, $2E_{V(p)} > E_{V_0}$.

Replacing Ω with a smaller set, we can assume (V4) without loss of generality. In fact, noting that $a \mapsto E_a$ is continuous and $V(x)$ is of class C^N, by Sard theorem we can see that for almost all $\alpha > 0$ close to 0,

$$\Omega_\alpha = \{x \in \Omega : V(x) > V_0 - \alpha\}$$

satisfies (V1)–(V4) and (LM).

We note that if a family $(u_\varepsilon) \subset H^1(\mathbb{R}^N)$ of solutions of (3) and $(u_\varepsilon) \subset H^1(\mathbb{R}^N)$ and $(x_\varepsilon) \subset \mathbb{R}^N$ satisfy (6) and (7), then x_0 is a critical point of $V(x)$. This fact is observed by X. Wang [30]. Thus the limit point (x_0, u_0) of (u_ε) is characterized as a critical point of the following limit functional:

$$L(z,u) = \frac{1}{2}\|\nabla u\|_2^2 + \frac{1}{2}V(z)\,\|u\|_2^2 - \int_{\mathbb{R}^N} G(u) : \mathbb{R}^N \times H^1(\mathbb{R}^N) \to \mathbb{R}.$$

That is, (x_0, u_0) satisfies $DL(x_0, u_0) = 0$, where $D = (\partial_z, \partial_u)$. Moreover we have

$$I_\varepsilon'(u_\varepsilon) \to L(x_0, u_0) \quad \text{as } \varepsilon \to 0. \tag{8}$$

We set

$$b = E_{V_0} \tag{9}$$

and

$$\begin{aligned}\mathcal{K}_b = \{(\xi,\omega) \in \Omega \times H^1(\mathbb{R}^N) :\ & DL(\xi,\omega) = 0,\ L(\xi,\omega) = b, \\ & \|\omega\|_{L^2(Q)} = \max_{n\in\mathbb{Z}^N} \|\omega\|_{L^2(n+Q)}\}, \\ = \{(\xi,\omega) \in \Omega \times H^1(\mathbb{R}^N) :\ & \nabla V(\xi) = 0,\ L_{V(\xi)}'(\omega) = 0,\ L_{V(\xi)}(\omega) = b,\end{aligned}$$

$$\|\omega\|_{L^2(Q)} = \max_{n\in\mathbb{Z}^N} \|\omega\|_{L^2(n+Q)}\},$$

$$\mathcal{K}_{b,+} = \{(\xi,\omega) \in \mathcal{K}_b : \omega(x) > 0 \text{ in } \mathbb{R}^N\}.$$

Under the conditions (V1)–(V4), we deduce from Lemma 2.1 that $\mathcal{K}_b$ and $\mathcal{K}_{b,+}$ are compact in $\mathbb{R}^N \times H^1(\mathbb{R}^N)$.

3 ε-Dependent Distance and a Neighborhood of Expected Solutions

For $\varepsilon > 0$ small we try to find a solution of the rescaled problem (3) which concentrates to a point in $\mathcal{K}_{b,+}$ in the sense (6)–(8). Precisely for $\varepsilon > 0$ small we aim to find a critical point of $I_\varepsilon(u)$ is a "suitable" neighborhood of

$$K_{b,+}^{(\varepsilon)} = \{\omega(x - \frac{\xi}{\varepsilon}) : (\xi,\omega) \in \mathcal{K}_{b,+}\}.$$

To define suitable neighborhoods of $K_{b,+}^{(\varepsilon)}$, we introduce the following ε-dependent distance $\operatorname{dist}_\varepsilon(\cdot,\cdot)$ in $H^1(\mathbb{R}^N)$. For $\varepsilon > 0$ and $u, v \in H^1(\mathbb{R}^N)$ we set

$$\operatorname{dist}_\varepsilon(u,v) = \inf_{h\in\mathbb{R}^N} \left(|h|^2 + \left\|u(x) - v(x - \frac{h}{\varepsilon})\right\|_{H^1}^2\right)^{1/2}.$$

We also introduce $H_\varepsilon(u) : H^1(\mathbb{R}^N) \to \mathbb{R}^N$ by

$$H_\varepsilon(u) = \int_{\mathbb{R}^N} \nabla V(\varepsilon x) u(x)^2.$$

The following lemma shows that $\operatorname{dist}_\varepsilon(\cdot,\cdot)$ is a natural distance to consider concentration of a sequence $(u_{\varepsilon_j}) \subset H^1(\mathbb{R}^N)$, $\varepsilon_j \to 0$ to a limit $(\xi,\omega) \in \mathcal{K}_{b,+}$.

Lemma 3.1 *For $(\xi,\omega) \in \mathbb{R}^N \times H^1(\mathbb{R}^N)$, if $(u_{\varepsilon_j}) \subset H^1(\mathbb{R}^N)$, $\varepsilon_j \to 0$ satisfies*

$$\operatorname{dist}_{\varepsilon_j}(u_{\varepsilon_j}, \omega(x - \frac{\xi}{\varepsilon_j})) \to 0,$$

then

$$I_{\varepsilon_j}(u_{\varepsilon_j}) \to L(\xi,\omega),$$

$$\left\|I'_{\varepsilon_j}(u_{\varepsilon_j})\right\|_{(H^1(\mathbb{R}^N))^*} \to \|\partial_u L(\xi,\omega)\|_{(H^1(\mathbb{R}^N))^*},$$

$$H_{\varepsilon_j}(u_{\varepsilon_j}) \to \nabla V(\xi)\,\|\omega\|_2^2 = \partial_z L(\xi,\omega).$$

We set for $\rho > 0$

$$
\begin{aligned}
N_\rho^{(\varepsilon)} &= \{u \in H^1(\mathbb{R}^N) : \mathrm{dist}_\varepsilon(u, K_{b,+}^{(\varepsilon)}) < \rho\} \\
&= \{\omega(x - \frac{\xi + h}{\varepsilon}) + \varphi(x - \frac{\xi + h}{\varepsilon}) : (\xi, \omega) \in \mathcal{K}_{b,+},\ |h|^2 + \|\varphi\|_{H^1}^2 < \rho^2\}.
\end{aligned}
$$

For technical reasons, we introduce another sets of neighborhoods $A_\rho^{(\varepsilon)}$ as follows; we set

$$
S_{b,+} = \{\omega(\frac{x}{s}) : \omega \in P_2\mathcal{K}_{b,+},\ s \in [\frac{1}{2}, \frac{3}{2}]\}, \tag{10}
$$

where $P_2 : \mathbb{R}^N \times H^1(\mathbb{R}^N) \to H^1(\mathbb{R}^N)$ is the standard projecton. We introduce

$$
\begin{aligned}
\mathcal{Z}_{b,+} &= \{(\xi, \omega) : \xi \in \overline{\Omega},\ \omega \in S_{b,+}\}, \\
Z_{b,+}^{(\varepsilon)} &= \{\omega(x - \frac{\xi}{\varepsilon}) : (\xi, \omega) \in \mathcal{Z}_{b,+}\} \\
&= \{\omega(\frac{x - \xi/\varepsilon}{s}) : \xi \in \overline{\Omega},\ \omega \in P_2\mathcal{K}_{b,+},\ s \in [\frac{1}{2}, \frac{3}{2}]\}.
\end{aligned}
$$

We remark that $S_{b,+}$, $\mathcal{Z}_{b,+}$, $Z_{b,+}^{(\varepsilon)}$ are compact by Lemma 2.1. Since

$$
\begin{aligned}
\mathcal{K}_{b,+} &\subset \{(\xi, \omega(x)) : \xi \in \overline{\Omega},\ \omega \in P_2\mathcal{K}_{b,+}\} \\
&\subset \{(\xi, \omega(\frac{x}{s})) : \xi \in \overline{\Omega},\ \omega \in P_2\mathcal{K}_{b,+}, s \in [\frac{1}{2}, \frac{3}{2}]\},
\end{aligned}
$$

we have $\mathcal{K}_{b,+} \subset \mathcal{Z}_{b,+}$ and $K_{b,+}^{(\varepsilon)} \subset Z_{b,+}^{(\varepsilon)}$. We also set for $\rho > 0$

$$
A_\rho^{(\varepsilon)} = \{u \in H^1(\mathbb{R}^N) : \mathrm{dist}_\varepsilon(u, Z_{b,+}^{(\varepsilon)}) < \rho\}.
$$

Then we have $N_\rho^{(\varepsilon)} \subset A_\rho^{(\varepsilon)}$.

4 Concentration-Compactness Type Result

In this section we give an ε-dependent concentration-compactness result, which will be useful to develop deformation argument.

Proposition 4.1 *There exists* $\rho_0 > 0$ *such that if* $(\varepsilon_j)_{j=1}^\infty \subset (0, 1]$ *and* $(u_j)_{j=1}^\infty \subset H^1(\mathbb{R}^N)$ *satisfy*

$$u_j \in A_{\rho_0}^{(\varepsilon_j)}, \tag{11}$$

$$\varepsilon_j \to 0, \tag{12}$$

$$I_{\varepsilon_j}(u_j) \to b, \tag{13}$$

$$I'_{\varepsilon_j}(u_j) \to 0 \quad \text{strongly in } (H^1(\mathbb{R}^N))^*, \tag{14}$$

$$H_{\varepsilon_j}(u_j) \to 0 \quad \text{in } \mathbb{R}^N \tag{15}$$

as $j \to \infty$, *where* b *is defined in* (9), *then*

$$\operatorname{dist}_{\varepsilon_j}(u_j, K_{b,+}^{(\varepsilon_j)}) \to 0 \quad \text{as } j \to \infty.$$

In particular, for any $\rho > 0$ *there exists* $j_\rho \in \mathbb{N}$ *such that*

$$u_j \in \overline{N_\rho^{(\varepsilon_j)}} \quad \text{for } j \geq j_\rho.$$

Proof Argument for Proposition 4.5 in [14] works after slight modification. □

Remark 4.1 Assume (V1)–(V3) and suppose that V_0 is a critical value of $V|_\Omega$. Then, defining $b = E_{V_0}$, we can consider the convergence of sequence $(u_j)_{j=1}^\infty \subset H^1(\mathbb{R}^N)$ with (11)–(15). When V_0 is a critical value, which is not a minimum, the condition (15) is important to show convergence of (u_{ε_j}) to $\mathcal{K}_{b,+}$. When V_0 is a minimum of $V(x)$ in Ω, we can show the result of Proposition 4.1 without the condition (15).

Remark 4.2 Conditions (13) and (14) are related to the Palais-Smale condition. Proposition 4.1 ensures compactness of ε-dependent Palais-Smale sequences with an additional condition (15), which is related to $\mathbb{R}^N$-action on $H^1(\mathbb{R}^N)$: $h \to u(x-h)$. Similar condition, which is related to Pohozaev identity, was considered in [12, 13, 24–26] and called (PSP) condition.

As a corollary to Proposition 4.1, we have

Proposition 4.2 *For sufficiently small* $0 < \rho_* < \rho_{**}$, *we have*

(i) There exist $\varepsilon_0 > 0$, $\nu_0 > 0$ *and* $\delta_0 > 0$ *with the following properties: For* $\varepsilon \in (0, \varepsilon_0]$ *it holds that*

$$\|(H_\varepsilon(u), I'_\varepsilon(u))\|_{(\mathbb{R}^N \times H^1(\mathbb{R}^N))^*} \equiv \left(|H_\varepsilon(u)|^2 + \|I'_\varepsilon(u)\|^2_{(H^1(\mathbb{R}^N))^*}\right)^{1/2} \geq \nu_0$$

for all $u \in A_{\rho_{**}}^{(\varepsilon)} \setminus \overline{N_{\rho_*}^{(\varepsilon)}}$ *with* $I_\varepsilon(u) \in [b - \delta_0, b + \delta_0]$.

(ii) Suppose that for some $\varepsilon \in (0, \varepsilon_0]$

$$(H_\varepsilon(u), I'_\varepsilon(u)) \neq 0 \quad \text{for all } u \in \overline{N^{(\varepsilon)}_{\rho_*}} \text{ with } I_\varepsilon(u) \in [b - \delta_0, b + \delta_0].$$

Then there exists $\nu_\varepsilon > 0$ *such that*

$$\left\| (H_\varepsilon(u), I'_\varepsilon(u)) \right\|_{(\mathbb{R}^N \times H^1(\mathbb{R}^N))^*} \geq \nu_\varepsilon$$

for $u \in A^{(\varepsilon)}_{\rho_{**}}$ *with* $I_\varepsilon(u) \in [b - \delta_0, b + \delta_0]$.

Proof (i) follows from Proposition 4.1. (ii) follows from the fact that the following Palais-Smale type condition holds. For fixed $\varepsilon \in (0, 1]$, if $(u_j)_{j=1}^\infty \subset A^{(\varepsilon)}_{\rho_{**}}$ satisfies

$$(H_\varepsilon(u_j), I'_\varepsilon(u_j)) \to 0 \quad \text{strongly in } (\mathbb{R}^N \times H^1(\mathbb{R}^N))^*,$$

then $(u_j)_{j=1}^\infty$ has a strongly convergent subsequence in $H^1(\mathbb{R}^N)$. □

5 Tail Minimizing Vector Field

To find critical points in a neighborhood $N^{(\varepsilon)}_\rho$ of expected solutions, it is important to control the size of $u \in N^{(\varepsilon)}_\rho$ outside a large ball $B(\beta(u), L)$, where $\beta(u)$ is a center of mass of u and $L \gg 1$.

To control the size of u outside of a large ball—we call it tail of function u—in many works minimizing methods in a large ball is used. That is,

$$\text{minimize}\{I_\varepsilon(w) : w \in H^1(\mathbb{R}^N),\ w = u \text{ in } B(\beta(u), L),\ \|w\|_{H^1(|x-\beta(u)| \geq L)} \leq r\}, \tag{16}$$

where $r > 0$ is a small constant. We note that the minimizer is a solution of (2) after rescaling. In this approach, we need to have uniqueness of the minimizer and usually g is assumed to be of class C^1. We remark that such an approach is difficult to adapt to non-local problems, e.g. nonlinear Choquard equations.

In this note, we take another approach developed in [11] for nonlinear Choquard equations. Instead of minimizing problem (16), we use a special deformation flow for the tail of functions, which works not only for local problems but also in non-local contexts. In [11], a deformation flow for the tail of functions and the standard deformation flow are used separately. Here we use them in a unified way.

To define the tail of functions, we use a special center of mass

$$\beta(u) : \widetilde{S}_{b,+,\rho_1} \to \mathbb{R}^N,$$

which is introduced in Section 3.3 of [14] (see also [5, 6]). Here $\rho_1 > 0$ is a small constant and

$$\widetilde{S}_{b,+,\rho_1} = \{\omega(x-p) + \varphi(x) : \omega \in S_{b,+},\ p \in \mathbb{R}^N,\ \|\varphi\|_{H^1} < \rho_1\}.$$

We recall that $S_{b,+}$ is defined in (10). Our center of mass enjoys properties of the following lemma.

Lemma 5.1 *There exist $\rho_1 > 0$ and $R_0 > 0$ such that*

(i) For $u(x) = \omega(x-p) + \varphi(x) \in \widetilde{S}_{b,+,\rho_1}$ with $\omega \in S_{b,+}$, $p \in \mathbb{R}^N$, $\|\varphi\|_{H^1} < \rho_1$

$$|\beta(u) - p| \le R_0.$$

(ii) $\beta(u)$ is shift-invariant, that is, $\beta(u(x-p)) = \beta(u) + p$ for all $u \in \widetilde{S}_{b,+,\rho_1}$ and $p \in \mathbb{R}^N$.
(iii) $\beta(u)$ is of class C^1 and there exists $C > 0$ independent of u such that $\|\beta'(u)\|_{(H^1(\mathbb{R}^N))^*} \le C$ for all u.
(iv) If $u, v \in \widetilde{S}_{b,+,\rho_1}$ satisfy $u(x) = v(x)$ in $B(\beta(u), 4R_0)$, then $\beta(u) = \beta(v)$.

We will study tail minimizing property in $B(\beta(u), \frac{4}{\sqrt{\varepsilon}})$.

We note that the standard center of mass like $\beta_0(u) = \frac{\int_{\mathbb{R}^N} x(|\nabla u|^2 + u^2)\,dx}{\int_{\mathbb{R}^N} |\nabla u|^2 + u^2\,dx}$ is not well-defined in a H^1-neighborhoods of expected solutions and it does not have properties in Lemma 2.1. Here we recall the definition of our $\beta(u)$ briefly.

For $R > 1$ we choose functions $\zeta_R(x), \widetilde{\zeta}_R(x) \in C^\infty(\mathbb{R}^N, \mathbb{R})$ such that

$$\zeta_R(x) = \begin{cases} 1 \text{ for } |x| \le R, \\ 0 \text{ for } |x| \ge R+1, \end{cases} \qquad \widetilde{\zeta}_R(x) = \begin{cases} 0 \text{ for } |x| \le R, \\ 1 \text{ for } |x| \ge R+1, \end{cases}$$
$$\zeta_R(x),\ \widetilde{\zeta}_R(x) \in [0,1],\ |\nabla\zeta_R(x)|,\ \left|\nabla\widetilde{\zeta}_R(x)\right| \le 2 \quad \text{for all } x \in \mathbb{R}^N.$$

We set

$$r_* = \inf_{\omega \in S_{b,+}} \|\omega\|_{H^1} > 0.$$

Since $S_{b,+}$ is compact, there exists $R_* > 0$ such that

$$\|\omega\|_{H^1(|x| \le R_*)} \ge \frac{2}{3} r_*, \quad \|\omega\|_{H^1(|x| \ge R_*)} \le \frac{1}{6} r_* \quad \text{for all } \omega \in S_{b,+}.$$

For $u = \omega(x-p) + \varphi(x) \in \widetilde{S}_{b,+,r_*/6}$ with $p \in \mathbb{R}^N$, $\omega \in S_{b,+}$ and $\|\varphi\|_{H^1} < \frac{1}{6} r_*$, we have

$$\|u(x)\|_{H^1(|x-p| \le R_*)} \ge \frac{1}{2} r_*, \tag{17}$$

$$\|u(x)\|_{H^1(|x-p|\geq R_*)} \leq \frac{1}{3}r_*. \tag{18}$$

We set for $q \in \mathbb{R}^N$ and $u \in \widetilde{S}_{b,+,r_*/6}$

$$\Phi(q,u) = \int_{\mathbb{R}^N} \zeta_{R_*}(x-q)(|\nabla u|^2 + u^2)\,dx.$$

By (17) and (18), we have for $u(x) = \omega(x-p) + \varphi(x) \in \widetilde{S}_{b,+,r_*/6}$

$$\Phi(p,u) \geq \left(\frac{1}{2}r_*\right)^2,$$
$$\Phi(q,u) \leq \left(\frac{1}{3}r_*\right)^2 \quad \text{for } |q-p| \geq 2R_* + 1,$$
$$\Phi(q, u(x-q')) = \Phi(q-q', u(x)) \quad \text{for all } q, q' \in \mathbb{R}^N.$$

We choose and fix a function $\psi(s) \in C^\infty([0,\infty),\mathbb{R})$ such that

$$\psi(s) = \begin{cases} 1 \text{ for } s \in [(\frac{1}{2}r_*)^2, \infty), \\ 0 \text{ for } s \in [0, (\frac{1}{3}r_*)^2], \end{cases} \qquad \psi(s) \in [0,1] \quad \text{for all } s \in \mathbb{R}.$$

Then we have for $u = \omega(x-p) + \varphi(x) \in \widetilde{S}_{b,+,r_*/6}$

$$\psi(\Phi(p,u)) = 1,$$
$$\psi(\Phi(q,u)) = 0 \quad \text{for } |q-p| \geq 2R_* + 1.$$

We set

$$\beta(u) = \frac{\int_{\mathbb{R}^N} q\psi(\Phi(q,u))\,dq}{\int_{\mathbb{R}^N} \psi(\Phi(q,u))\,dq} : \widetilde{S}_{b,+,r_*/6} \to \mathbb{R}^N.$$

Thus we can show the desired properties in Lemma 2.1 for $\rho_1 = r_*/6$ and $R_0 = 2R_* + 1$.

To study tail minimizing property in $B(\beta(u), \frac{4}{\sqrt{\varepsilon}})$, we define $T_\varepsilon(u) : \widetilde{S}_{b,+,\rho_1} \to \mathbb{R}$ by

$$T_\varepsilon(u) = \int_{\mathbb{R}^N} \widetilde{\zeta}_{4/\sqrt{\varepsilon}}(x-\beta(u))(|\nabla u|^2 + |u|^2). \tag{19}$$

We use $T_\varepsilon(u)$ to estimate the size of the tail of u

We also define $M_1(u)$, $M_2(u)$; $\widetilde{S}_{b,+,\rho_1} \to H^1(\mathbb{R}^N)$ by

$$M_1(u) = \zeta_{1/\sqrt{\varepsilon}}(x - \beta(u))u,$$
$$M_2(u) = (1 - \zeta_{1/\sqrt{\varepsilon}}(x - \beta(u)))u.$$

We note that $u = M_1(u) + M_2(u)$ and $\|M_1(u)\|_{H^1}, \|M_2(u)\|_{H^1} \le 3\,\|u\|_{H^1}$.

Choosing $0 < \rho_* < \rho_{**}$ small, we may assume $A^{(\varepsilon)}_{\rho_{**}} \subset \widetilde{S}_{b,+,\rho_1}$.

Proposition 5.1 *There exists $c_\varepsilon > 0$ independent of $u \in \widetilde{S}_{b,+,\rho_1}$ which satisfies $c_\varepsilon \to 0$ as $\varepsilon \to 0$ and the following properties: There exist $u^{(1)}, u^{(2)} \in H^1(\mathbb{R}^N)$ depending on u such that for $\varepsilon > 0$ small*

(i)

$$\operatorname{supp} u^{(1)} \subset B(\beta(u), \frac{3}{\sqrt{\varepsilon}})$$
$$\operatorname{supp} u^{(2)} \subset \mathbb{R}^N \setminus B(\beta(u), \frac{2}{\sqrt{\varepsilon}}),$$
$$\operatorname{supp} u^{(1)} \cap \operatorname{supp} u^{(2)} = \emptyset,$$
$$\left\| u - u^{(1)} - u^{(2)} \right\|_{H^1} < c_\varepsilon,$$
$$\left| (u - u^{(2)}, u^{(2)})_{H^1} \right| < c_\varepsilon,$$
$$\left| (I'_\varepsilon(u) - I'_\varepsilon(u^{(2)}))u^{(2)} \right| < c_\varepsilon. \tag{20}$$

(ii)

$$\beta'(u)u^{(2)} = 0, \tag{21}$$
$$M'_1(u)u^{(2)} = 0,$$
$$\partial_u(\|M_2(u)\|^2_{H^1})u^{(2)} \ge -c_\varepsilon. \tag{22}$$

(iii) For $T_\varepsilon(u)$ defined in (19),

$$T_\varepsilon(u) \le \left\| u^{(2)} \right\|^2_{H^1},$$
$$T'_\varepsilon(u)u^{(2)} = 2T_\varepsilon(u).$$

(iv) For $c_0 > 0$ independent of ε, we have

$$I'_\varepsilon(u)u^{(2)} \ge c_0 T_\varepsilon(u) - c_\varepsilon. \tag{23}$$

Proof First we note that there exists $c > 0$ independent of $\varepsilon \in (0, 1]$ and $u \in H^1(\mathbb{R}^N)$ such that

$$I_\varepsilon(u),\ I_\varepsilon'(u)u \geq c\,\|u\|_{H^1}^2 \quad \text{for } \varepsilon \in (0,1] \text{ and for all } \|u\|_{H^1} \text{ small.} \tag{24}$$

We may assume that (24) holds for $\|u\|_{H^1} \leq \rho_1$.

We define $u^{(1)}$ and $u^{(2)}$ as follows:

$$u^{(1)}(x) = \zeta_{\frac{2}{\sqrt{\varepsilon}}+\frac{k}{\varepsilon^{1/4}}}(x - \beta(u))u(x),$$
$$u^{(2)}(x) = \widetilde{\zeta}_{\frac{2}{\sqrt{\varepsilon}}+\frac{k+1}{\varepsilon^{1/4}}}(x - \beta(u))u(x).$$

Here $k \in \{1, 2, \cdots, [\varepsilon^{-1/4}] - 1\}$ is chosen so that

$$\|u\|^2_{H^1(|x-\beta(u)|\in[\frac{2}{\sqrt{\varepsilon}}+\frac{k}{\varepsilon^{1/4}}, \frac{2}{\sqrt{\varepsilon}}+\frac{k+1}{\varepsilon^{1/4}}])} \leq \frac{C}{[\varepsilon^{-1/4}]} \to 0 \quad \text{as } \varepsilon \to 0. \tag{25}$$

Since $\widetilde{S}_{b,+,\rho_1}$ is bounded and

$$\sum_{k=0}^{[\varepsilon^{-1/4}]-1} \|u\|^2_{H^1(|x-\beta(u)|\in[\frac{2}{\sqrt{\varepsilon}}+\frac{k}{\varepsilon^{1/4}}, \frac{2}{\sqrt{\varepsilon}}+\frac{k+1}{\varepsilon^{1/4}}])} \leq \|u\|^2_{H^1(|x-\beta(u)|\in[\frac{2}{\sqrt{\varepsilon}}, \frac{3}{\sqrt{\varepsilon}}])}$$
$$\leq \|u\|^2_{H^1} \leq C,$$

we can find k with (25).

The properties (i)–(iv) follow from the choice of k. We show just (20)–(23). First we show (21). Since $\mathrm{supp} u^{(2)} \subset \mathbb{R}^N \setminus B(\beta(u), \frac{2}{\sqrt{\varepsilon}})$, we may assume that $u^{(2)} = 0$ on $B(\beta(u), 4R_0)$. Thus $\beta(u + su^{(2)}) = \beta(u)$ for small s by Lemma 5.1 (iv) and we have (21).

For (20), we compute

$$(I_\varepsilon'(u) - I_\varepsilon'(u^{(2)}))u^{(2)} = \int_{K_k} (\nabla u - \nabla u^{(2)})\nabla u^{(2)} + \int_{K_k} V(\varepsilon x)(u - u^{(2)})u^{(2)}$$
$$- \int_{K_k} (g(u) - g(u^{(2)}))u^{(2)}.$$

Here we denote $K_k = \{x : |x - \beta(u)| \in [\frac{2}{\sqrt{\varepsilon}} + \frac{k}{\varepsilon^{1/4}}, \frac{2}{\sqrt{\varepsilon}} + \frac{k+1}{\varepsilon^{1/4}}]\}$. Thus (20) follows from (25).

For (23), it follows from (20) that

$$I_\varepsilon'(u)u^{(2)} \geq I_\varepsilon'(u^{(2)})u^{(2)} - c_\varepsilon \geq c\left\|u^{(2)}\right\|^2_{H^1} - c_\varepsilon$$
$$\geq c_0 T_\varepsilon(u) - c_\varepsilon.$$

Finally for (22) we note that

$$\begin{aligned}
\partial_u M_2(u)u^{(2)} &= \partial_u[(1-\zeta_{\frac{1}{\sqrt{\varepsilon}}}(x-\beta(u)))u]u^{(2)} \\
&= (1-\zeta_{\frac{1}{\sqrt{\varepsilon}}}(x-\beta(u)))u^{(2)} + \zeta'_{\frac{1}{\sqrt{\varepsilon}}}(x-\beta(u)))(\beta'(u)u^{(2)})u \\
&= (1-\zeta_{\frac{1}{\sqrt{\varepsilon}}}(x-\beta(u)))u^{(2)} = u^{(2)}.
\end{aligned}$$

Here we used (21). Thus

$$\begin{aligned}
\frac{1}{2}\partial_u(\|M_2(u)\|_{H^1})u^{(2)} &= (M_2(u), \partial_u M_2(u)u^{(2)})_{H^1} \\
&= ((1-\zeta_{1/\sqrt{\varepsilon}}(x-\beta(u)))u, u^{(2)})_{H^1} = (u, u^{(2)})_{H^1} \\
&= \left\|u^{(2)}\right\|^2_{H^1} + (u-u^{(2)}, u^{(2)})_{H^1} \\
&\geq (u-u^{(2)}, u^{(2)})_{H^1} = -c_\varepsilon.
\end{aligned}$$

□

From Proposition 4.1, we find a vector field $u \mapsto -u^{(2)}$ which has good properties for deformation. By (ii), (iii), $-u^{(2)}$ does not effect the center of mass $\beta(u)$ and the center part $M_1(u)$ of u. Moreover $-u^{(2)}$ gives a direction which decreases both of $I_\varepsilon(u)$ and $T_\varepsilon(u)$ provided $T_\varepsilon(u) \geq \frac{c_\varepsilon}{c_0}$.

We note that by the compactness of $S_{b,+}$

$$\sup_{\omega\in S_{b,+}} T_\varepsilon(\omega) \to 0 \quad \text{as } \varepsilon \to 0.$$

We set

$$\kappa_\varepsilon = \max\{2\max_{\omega\in S_{b,+}} T_\varepsilon(\omega), \frac{2c_\varepsilon}{c_0}\}. \tag{26}$$

We use notation:

$$\begin{aligned}
[I_\varepsilon \leq c] &= \{u \in H^1(\mathbb{R}^N) : I_\varepsilon(u) \leq c\}, \\
[T_\varepsilon \geq c] &= \{u \in H^1(\mathbb{R}^N) : T_\varepsilon(u) \geq c\}.
\end{aligned}$$

As a corollary to Proposition 5.1 (iii), (iv) we have

Corollary 5.1 *For $u \in A^{(\varepsilon)}_{\rho_{**}} \cap [T_\varepsilon \geq \kappa_\varepsilon]$,*

$$I'_\varepsilon(u)u^{(2)} \geq c_\varepsilon, \quad T'_\varepsilon(u)u^{(2)} \geq c_\varepsilon.$$

Conversely, for $u \in A^{(\varepsilon)}_{\rho_{**}} \cap [T_\varepsilon \le \kappa_\varepsilon]$ we can say that $u(x)$ concentrates near $\beta(u)$. In particular we have

Proposition 5.2 *Assume $u \in A^{(\varepsilon)}_{\rho_{**}} \cap [T_\varepsilon \le \kappa_\varepsilon]$. Then we have*

$$I_\varepsilon(u) \ge L(\varepsilon\beta(u), u) - d_\varepsilon.$$

Here $d_\varepsilon > 0$ is independent of u and satisfies $d_\varepsilon \to 0$ as $\varepsilon \to 0$.

Proof For $u \in A^{(\varepsilon)}_{\rho_{**}} \cap [T_\varepsilon \le \kappa_\varepsilon]$ we compute

$$\begin{aligned}
I_\varepsilon(u) &= L(\varepsilon\beta(u), u) + \frac{1}{2}\int_{\mathbb{R}^N} (V(\varepsilon x) - V(\varepsilon\beta(u)))u^2 \\
&= L(\varepsilon\beta(u), u) + \frac{1}{2}\int_{|x-\beta(u)|\le\frac{4}{\sqrt{\varepsilon}}} (V(\varepsilon x) - V(\varepsilon\beta(u)))u^2 \\
&\quad + \frac{1}{2}\int_{|x-\beta(u)|\ge\frac{4}{\sqrt{\varepsilon}}} (V(\varepsilon x) - V(\varepsilon\beta(u)))u^2 \\
&\ge L(\varepsilon\beta(u), u) - \frac{1}{2}\|V(y) - V(\varepsilon\beta(u))\|_{L^\infty(|y-\varepsilon\beta(u)|\le 4\sqrt{\varepsilon})} \|u\|_2^2 \\
&\quad - \frac{1}{2}\overline{V}\,\|u\|^2_{H^1(|x-\beta(u)|\ge\frac{4}{\sqrt{\varepsilon}})} \\
&\ge L(\varepsilon\beta(u), u) - \frac{1}{2}\|V(y) - V(\varepsilon\beta(u))\|_{L^\infty(|y-\varepsilon\beta(u)|\le 4\sqrt{\varepsilon})} \|u\|_2^2 - \frac{1}{2}\overline{V}T_\varepsilon(u) \\
&\ge L(\varepsilon\beta(u), u) - \frac{1}{2}\|V(y) - V(\varepsilon\beta(u))\|_{L^\infty(|y-\varepsilon\beta(u)|\le 4\sqrt{\varepsilon})} \|u\|_2^2 - \frac{1}{2}\overline{V}\kappa_\varepsilon.
\end{aligned}$$

Since $\|V(y) - V(\varepsilon\beta(u))\|_{L^\infty(|y-\varepsilon\beta(u)|\le 4\sqrt{\varepsilon})}$, $\kappa_\varepsilon \to 0$ as $\varepsilon \to 0$ uniformly in u. We have the conclusion. □

6 Deformation Argument

6.1 Deformation Result

We have the following deformation result.

Proposition 6.1 *For sufficiently small $0 < \rho_* < \rho_{**}$, let $\varepsilon_0, \nu_0, \delta_0 > 0$ be numbers given in Proposition 4.2 and let $\kappa_\varepsilon > 0$ be a number given in (26), which satisfies $\kappa_\varepsilon \to 0$ as $\varepsilon \to 0$. Moreover suppose for some $\varepsilon \in (0, \varepsilon_0]$*

$$(H_\varepsilon(u), I'_\varepsilon(u)) \ne 0 \quad \textit{for } u \in \overline{N^{(\varepsilon)}_{\rho_*}} \textit{ with } I_\varepsilon(u) \in [b - \delta_0, b + \delta_0]. \tag{27}$$

Then for any $\delta_1 \in (0, \delta_0)$ *there exist* $\delta \in (0, \delta_1)$ *and a continuous map* $\eta(t, u)$: $[0, 1] \times A^{(\varepsilon)}_{\rho_{**}} \to A^{(\varepsilon)}_{\rho_{**}}$ *such that*

(i) $\eta(0, u) = u$ *for all* $u \in A^{(\varepsilon)}_{\rho_{**}}$,
(ii) $\eta(t, u) = u$ *for all* $t \in [0, 1]$ *if* $I_\varepsilon(u) \notin [b - \delta_1, b + \delta_1]$ *or* $u \notin A^{(\varepsilon)}_{\frac{3\rho_{**}+\rho_*}{4}}$,
(iii) $t \mapsto I_\varepsilon(\eta(t, u))$ *is a non-increasing function of t for all* $u \in A^{(\varepsilon)}_{\rho_{**}}$,
(iv) $\eta(1, u) \in [I_\varepsilon \le b - \delta]$ *if* $u \in A^{(\varepsilon)}_{\rho_*} \cap [I_\varepsilon \le b + \delta]$.
(v) $\eta(t, u) \in [T_\varepsilon \le \kappa_\varepsilon]$ *for all* $t \in [0, 1]$ *if* $u \in [T_\varepsilon \le \kappa_\varepsilon]$.

To show Proposition 6.1, we use an augmented functional $J_\varepsilon(z, u)$ defined in (4). We note that

(i) $J_\varepsilon(z, u) = I_\varepsilon(u(x - \frac{z}{\varepsilon}))$.
(ii) $\partial_u J_\varepsilon(z, u)\varphi = I'_\varepsilon(u(x - \frac{z}{\varepsilon}))\varphi(x - \frac{z}{\varepsilon})$.
(iii) $\partial_z J_\varepsilon(z, u) = H_\varepsilon(u(x - \frac{z}{\varepsilon}))$.

In particular, for $D = (\partial_z, \partial_u)$ we have

$$\|DJ_\varepsilon(z, u)\|^2_{(\mathbb{R}^N \times H^1(\mathbb{R}^N))^*} = \left|H_\varepsilon(u(x - \frac{z}{\varepsilon})\right|^2 + \left\|I'_\varepsilon(u(x - \frac{z}{\varepsilon}))\right\|^2_{(H^1(\mathbb{R}^N))^*}. \tag{28}$$

We set

$$\mathcal{N}^{(\varepsilon)}_\rho = \{(z, u) \in \mathbb{R}^N \times H^1(\mathbb{R}^N) : u(x - \frac{z}{\varepsilon}) \in N^{(\varepsilon)}_\rho\},$$
$$\mathcal{A}^{(\varepsilon)}_\rho = \{(z, u) \in \mathbb{R}^N \times H^1(\mathbb{R}^N) : u(x - \frac{z}{\varepsilon}) \in A^{(\varepsilon)}_\rho\}.$$

By (28) and Proposition 4.2, we have

Proposition 6.2 *Let* $0 < \rho_* < \rho_{**}$ *be small constants. Then we have*

(i) There exist $\nu_0 > 0$ *and* $\delta_0 > 0$ *independent of* ε *such that for* $\varepsilon > 0$ *small*

$$\|DJ_\varepsilon(z, u)\|_{(\mathbb{R}^N \times H^1(\mathbb{R}^N))^*} \ge \nu_0$$

for all $(z, u) \in \mathcal{A}^{(\varepsilon)}_{\rho_{**}} \setminus \overline{\mathcal{N}^{(\varepsilon)}_{\rho_*}}$ *with* $J_\varepsilon(z, u) \in [b - \delta_0, b + \delta_0]$.
(ii) Suppose that (27) *holds. Then there exists* $\nu_\varepsilon > 0$ *such that*

$$\|DJ_\varepsilon(z, u)\|_{(\mathbb{R}^N \times H^1(\mathbb{R}^N))^*} \ge \nu_\varepsilon \quad \textit{for all } (z, u) \in \mathcal{A}^{(\varepsilon)}_{\rho_{**}} \textit{ with } J_\varepsilon(z, u) \in [b - \delta_0, b + \delta_0].$$

We may assume $\nu_\varepsilon < \nu_0$.

6.2 A Special Vector Field for Deformation

To show Proposition 6.1, it is important to find a vector field $W(z, u)$ such that both of I_ε and T_ε decrease along the corresponding flow provided $T_\varepsilon(u) \geq \kappa_\varepsilon$.

Using $u^{(2)}$ given in Proposition 4.1, we have the following

Proposition 6.3 *Suppose that* (27) *holds. Then for* $\varepsilon > 0$ *small, there exists a locally Lipschitz vector field* $W(z, u) : \mathcal{A}^{(\varepsilon)}_{\rho_{**}} \cap \{(z, u) : J_\varepsilon(z, u) \in [b - \delta_0, b + \delta_0]\} \to \mathbb{R}^N \times H^1(\mathbb{R}^N)$ *with the following properties.*

(i) $DT_\varepsilon(u)W(z, u) > 0$ *if* $T_\varepsilon(u) > \kappa_\varepsilon$.
(ii) $DJ_\varepsilon(z, u)W(z, u) > \nu_\varepsilon$ *if* $(z, u) \in \mathcal{A}^{(\varepsilon)}_{\rho_{**}}$ *and* $J_\varepsilon(z, u) \in [b - \delta_0, b + \delta_0]$.
(iii) $DJ_\varepsilon(z, u)W(z, u) > \nu_0$ *if* $(z, u) \in \mathcal{A}^{(\varepsilon)}_{\rho_{**}} \setminus \overline{\mathcal{N}^{(\varepsilon)}_{\rho_*}}$ *and* $J_\varepsilon(z, u) \in [b - \delta_0, b + \delta_0]$.
(iv) There exist $C, C' > 0$ *independent of* ε *and* u *such that*

$$\|DM_1(u)W(z, u)\|_{H^1} \leq C, \quad D(\|M_2(u)\|^2_{H^1})W(z, u) \geq -C'.$$

Proof By Proposition 6.2, for any $(z, u) \in \mathcal{A}^{(\varepsilon)}_{\rho_{**}}$ with $J_\varepsilon(u) \in [b - \delta_0, b + \delta_0]$, we can find $(\xi, w) \in \mathbb{R}^N \times H^1(\mathbb{R}^N)$ such that

$$\begin{aligned} &|\xi|^2 + \|w\|^2_{H^1} \leq 1, \\ &DJ_\varepsilon(z, u)(\xi, w) > \nu_0 \quad \text{if } (z, u) \in \mathcal{A}^{(\varepsilon)}_{\rho_{**}} \setminus \overline{\mathcal{N}^{(\varepsilon)}_{\rho_*}}, \\ &DJ_\varepsilon(z, u)(\xi, w) > \nu_\varepsilon \quad \text{if } (z, u) \in \overline{\mathcal{N}^{(\varepsilon)}_{\rho_*}}. \end{aligned}$$

We compute for $(z, u) \in \mathcal{A}^{(\varepsilon)}_{\rho_{**}}$ and $\ell > 0$

$$\begin{aligned} \partial_u T_\varepsilon(u)(w + \ell u^{(2)}) &= \partial_u T_\varepsilon(u)w + \ell \partial_u T_\varepsilon(u)u^{(2)} \\ &\geq -C_1 + 2\ell T_\varepsilon(u). \end{aligned}$$

where $C_1 > 0$ is independent of ε and u. For $\kappa_\varepsilon > 0$ defined in (26), we set

$$\ell_\varepsilon \equiv \frac{C_1}{\kappa_\varepsilon} \to \infty \quad \text{as } \varepsilon \to 0. \tag{29}$$

Finally we define $V_{z,u} \in \mathbb{R}^N \times H^1(\mathbb{R}^N)$ for $(z, u) \in \mathcal{A}^{(\varepsilon)}_{\rho_{**}}$ with $J_\varepsilon(z, u) \in [b - \delta_0, b + \delta_0]$ by

$$V_{z,u} = \begin{cases} (\xi, w + \ell_\varepsilon u^{(2)}) & \text{if } T_\varepsilon(u) \geq \kappa_\varepsilon, \\ (\xi, w) & \text{if } T_\varepsilon(u) < \kappa_\varepsilon. \end{cases}$$

Then, replacing $W(z,u)$ with $V_{z,u}$, we have the statements (i)–(iv) of Proposition 6.3.

Using a partition of unity in the standard way, we can find the desired locally Lipschitz vector field $W(z,u)$. □

We consider the following ODE:

$$\frac{d\widetilde{\eta}}{dt} = -\varphi_1(J_\varepsilon(\widetilde{\eta}))\varphi_2(\widetilde{\eta})W(\widetilde{\eta}), \qquad \widetilde{\eta}(0,z,u) = (z,u), \tag{30}$$

where $\varphi_1(s) : \mathbb{R} \to [0,1]$ and $\varphi_2(u) : H^1(\mathbb{R}^N) \to [0,1]$ are suitable cut-off functions. We can deduce the following proposition.

Proposition 6.4 *For $\varepsilon > 0$ small, suppose that* (27) *holds. Then for any given $\delta_1 \in (0,\delta_0)$ there exist $\delta \in (0,\delta_1)$ and a continuous map $\widetilde{\eta}(t,z,u) : [0,1]\times\mathcal{A}^{(\varepsilon)}_{\rho_{**}} \to \mathcal{A}^{(\varepsilon)}_{\rho_{**}}$ such that*

(i) $\widetilde{\eta}(0,z,u) = (z,u)$ *for all* $(z,u) \in \mathcal{A}^{(\varepsilon)}_{\rho_*}$.
(ii) $\widetilde{\eta}(t,z,u) = (z,u)$ *for all* $t \in [0,1]$ *if* $J_\varepsilon(z,u) \notin [b-\delta_1, b+\delta_1]$ *or* $(z,u) \notin \mathcal{A}^{(\varepsilon)}_{\frac{3\rho_{**}+\rho_*}{4}}$.
(iii) $t \mapsto J_\varepsilon(\widetilde{\eta}(t,z,u))$ *is non-increasing on* $[0,1]$ *for all* $(z,u) \in \mathcal{A}^{(\varepsilon)}_{\rho_{**}}$.
(iv) $J_\varepsilon(\widetilde{\eta}(1,z,u)) \le b-\delta$ *if* $J_\varepsilon(z,u) \le b+\delta$.

We note that $W(z,u)$ is not uniformly bounded with respect to ε. The property (iv) in Proposition 6.3 is important to ensure the global solvability of (30). In fact, along the flow $\widetilde{\eta}(t) = \widetilde{\eta}(t,z,u)$, we have

$$\begin{aligned}\frac{d}{dt}(\|M_2(\widetilde{\eta}(t))\|^2_{H^1}) &= D(\|M_2(\widetilde{\eta}(t))\|^2_{H^1})\frac{d\widetilde{\eta}}{dt}(t)\\ &= -\varphi_1(J_\varepsilon(\widetilde{\eta}(t)))\varphi_2(\widetilde{\eta}(t))D(\|M_2(\widetilde{\eta}(t))\|^2_{H^1})W(\widetilde{\eta}(t))\\ &\le C'.\end{aligned}$$

We also have $\frac{d}{dt}(\|M_1(\widetilde{\eta}(t))\|_{H^1}) \le C_1$. Thus $\widetilde{\eta}(t) = M_1(\widetilde{\eta}(t)) + M_2(\widetilde{\eta}(t))$ stays bounded uniformly in ε.

The flow $\eta(t,u)$ in Proposition 6.1 is obtained as

$$\eta(t,u) = \pi(\widetilde{\eta}(t,0,u)), \quad \text{where } \pi(z,u)(x) = u(x-\frac{z}{\varepsilon}).$$

7 Proof of Theorem 1.1

To prove Theorem 1.1, it suffices to show the following proposition.

Proposition 7.1 *Assume (f1)–(f4), (V1)–(V4) and (LM) and let* $b = E_{V_0}$. *For any* $\rho_* > 0$ *and* $\overline{\delta} > 0$ *there exists* $\varepsilon_0 = \varepsilon_0(\rho_*, \overline{\delta}) > 0$ *such that for* $\varepsilon \in (0, \varepsilon_0]$, $I_\varepsilon(u)$ *has a critical point u in* $N_{\rho_*}^{(\varepsilon)} \cap [T_\varepsilon \le \kappa_\varepsilon]$ *satisfying* $I_\varepsilon(u) \in [b - \overline{\delta}, b + \overline{\delta}]$.

Proof of Proposition 7.1 Let $\omega_0(x)$ be a positive least energy solution of $L'_{V_0}(u) = 0$ corresponding to $b = E_{V_0}$. For $s_0 \in (0, \frac{1}{2})$ small, set $\gamma_{0\varepsilon}(s,\xi) : [1-s_0, 1+s_0] \times \overline{\Omega} \to H^1(\mathbb{R}^N)$ by

$$\gamma_{0\varepsilon}(s,\xi)(x) = \omega_0(\frac{x - \xi/\varepsilon}{s}).$$

We note that

$$\begin{aligned} I_\varepsilon(\gamma_{0\varepsilon}(s,\xi)) &= \frac{1}{2}\left\|\nabla(\omega_0(\frac{x}{s}))\right\|_2^2 + \frac{1}{2}\int_{\mathbb{R}^N} V(\varepsilon x + \xi)\omega_0(\frac{x}{s})^2 - \frac{1}{2}\mathcal{D}(\omega_0(\frac{x}{s})) \\ &\to L(\xi, \omega_0(\frac{x}{s})) = L_{V_0}(\omega_0(\frac{x}{s})) - \frac{1}{2}(V_0 - V(\xi))\left\|\omega_0(\frac{x}{s})\right\|_2^2 \end{aligned}$$

as $\varepsilon \to 0$ uniformly in $(s,\xi) \in [1-s_0, 1+s_0] \times \overline{\Omega}$. We also have

$$\begin{aligned} &\varepsilon\beta(\gamma_{0\varepsilon}(s,\xi)) \to \xi \quad \text{as } \varepsilon \to 0 \text{ uniformly in } (s,\xi) \in [1-s_0, 1+s_0] \times \overline{\Omega}, \\ &P_{V_0}(\gamma_{0\varepsilon}(1-s_0,\xi)) > 0, \\ &P_{V_0}(\gamma_{0\varepsilon}(1+s_0,\xi)) < 0 \quad \text{for all } \xi \in \overline{\Omega}. \end{aligned}$$

Choosing small $s_0 \in (0, \frac{1}{2})$, we have $\gamma_{0\varepsilon}(s,\xi) \in A_{\rho_{**}}^{(\varepsilon)}$ for all $(s,\xi) \in [1-s_0, 1+s_0] \times \overline{\Omega}$.

Since there exists $\overline{\delta} > 0$ such that

$$\max_{(s,\xi)\in[1-s_0,1+s_0]\times\overline{\Omega}} L(\xi, \omega_0(\frac{x}{s})) \le b,$$

$$\max_{(s,\xi)\in\partial([1-s_0,1+s_0]\times\overline{\Omega})} L(\xi, \omega_0(\frac{x}{s})) \le b - 2\overline{\delta},$$

for any $\delta \in (0, \overline{\delta})$ we have for sufficiently small $\varepsilon > 0$

$$\max_{(s,\xi)\in[1-s_0,1+s_0]\times\overline{\Omega}} I_\varepsilon(\gamma_{0\varepsilon}(s,\xi)) \le b + \delta,$$

$$\max_{(s,\xi)\in\partial([1-s_0,1+s_0]\times\overline{\Omega})} I_\varepsilon(\gamma_{0\varepsilon}(s,\xi)) \le b - \overline{\delta}.$$

We also note that

$$\gamma_{0\varepsilon}(s,\xi) \in [T_\varepsilon \le \kappa_\varepsilon] \quad \text{for all } (s,\xi) \in [1-s_0, 1+s_0] \times \overline{\Omega}.$$

We define $F_\varepsilon : A^{(\varepsilon)}_{\rho_{**}} \to \mathbb{R} \times \mathbb{R}^N$ by

$$F_\varepsilon(u) = (P_{V_0}(u), \varepsilon\beta(u)).$$

Let x_0 be a maximum point in Ω, that is, $V(x_0) = V_0$. Using the degree theory, we can see for any continuous deformation $\gamma_\varepsilon(s, \xi)$ of $\gamma_{0\varepsilon}(s, \xi)$, which keeps the boundary $\partial([1-s_0, 1+s_0] \times \overline{\Omega})$, there exists $(s_0, \xi_0) \in [1-s_0, 1+s_0] \times \overline{\Omega}$ such that

$$F_\varepsilon(\gamma_\varepsilon(s_0, \xi_0)) = (0, x_0).$$

Thus if $\gamma_\varepsilon(s, \xi) \in [T_\varepsilon \le \kappa_\varepsilon]$ for all (s, ξ), we have for $u_\varepsilon = \gamma_\varepsilon(s_0, \xi_0)$

$$\begin{aligned} I_\varepsilon(\gamma_\varepsilon(s_0, \xi_0)) &\ge L(x_0, u_\varepsilon) - d_\varepsilon \\ &\ge E_{V_0} - d_\varepsilon = b - d_\varepsilon. \end{aligned} \tag{31}$$

Here we used Propositions 5.2 and 2.1 (iii).

Now we argue indirectly and assume that there are no critical points in $N^{(\varepsilon)}_{\rho_*} \cap [T_\varepsilon \le \kappa_\varepsilon]$. Then by Proposition 6.1, there exists $\delta \in (0, \overline{\delta}]$ and a continuous map $\eta(t, u) : [0, 1] \times A^{(\varepsilon)}_{\rho_{**}} \to A^{(\varepsilon)}_{\rho_{**}}$ with the properties (i)–(v) in Proposition 6.1.

We set $\gamma_\varepsilon(s, \xi) = \eta(1, \gamma_{0\varepsilon}(s, \xi))$. We can see

$$\gamma_\varepsilon(s, \xi) = \gamma_{0\varepsilon}(s, \xi) \quad \text{for } (s, \xi) \in \partial([1-s_0, 1+s_0] \times \overline{\Omega}),$$

$$\gamma_\varepsilon(s, \xi) \in [T_\varepsilon \le \kappa_\varepsilon] \quad \text{for all } (s, \xi) \in [1-s_0, 1+s_0] \times \overline{\Omega}$$

and

$$I_\varepsilon(\gamma_\varepsilon(s, \xi)) \le b - \delta \quad \text{for all } (s, \xi) \in [1-s_0, 1+s_0] \times \overline{\Omega}.$$

This contradicts with (31) for $\varepsilon > 0$ small. Therefore Proposition 7.1 holds. □

Acknowledgments The first author is supported by PRIN 2017JPCAPN "Qualitative and quantitative aspects of nonlinear PDEs" and by INdAM-GNAMPA. The second author is supported in part by Grant-in-Aid for Scientific Research (19H00644, 18KK0073, 17H02855, 16K13771) of Japan Society for the Promotion of Science.

References

1. Ambrosetti, A., Badiale, M., Cingolani, S: Semiclassical states of nonlinear Schrödinger equations. Arch. Rational Mech. Anal. **140**(3), 285–300 (1997)
2. Ambrosetti, A., Malchiodi, A., Secchi, S.: Multiplicity results for some nonlinear Schrödinger equations with potentials. Arch. Ration. Mech. Anal. **159**(3), 253–271 (2001)

3. Berestycki, H., Lions, P.-L.: Nonlinear scalar field equations. I. Existence of a ground state. Arch. Rational Mech. Anal. **82**, 313–345 (1983)
4. Byeon, J., Jeanjean, L.: Standing waves for nonlinear Schrödinger equations with a general nonlinearity. Arch. Ration. Mech. Anal. **185**(2), 185–200 (2007)
5. Byeon, J., Tanaka, K.: Semi-classical standing waves for nonlinear Schrödinger equations at structurally stable critical points of the potential. J. Eur. Math. Soc. **15**, 1859–1899 (2013)
6. Byeon, J., Tanaka, K.: Semiclassical standing waves with clustering peaks for nonlinear Schrödinger equations. Memoir Am. Math. Soc. **229**, 1–87 (2014)
7. Byeon, J., Tanaka, K.: Multi-bump positive solutions for a nonlinear elliptic problem in expanding tubular domains. Calc. Var. Partial Differ. Equ. **50**(1–2), 365–397 (2014)
8. Cingolani, S., Gallo, M.: On the fractional NLS equation and the effects of the potential well's topology. Adv. Nonlinear Stud. **21**(1), 1–40 (2021)
9. Cingolani, S., Lazzo, M.: Multiple semiclassical standing waves for a class of nonlinear Schrödinger equations. Topol. Methods Nonlinear Anal. **10**(1), 1–13 (1997)
10. Cingolani, S., Lazzo, M.: Multiple positive solutions to nonlinear Schrödinger equations with competing potential functions. J. Differ. Equ. **160**(1), 118–138 (2000)
11. Cingolani, S., Tanaka, K.: Semi-classical states for the nonlinear Choquard equations: existence, multiplicity and concentration at a potential well. Rev. Mat. Iberoam. **35**(6), 1885–1924 (2019)
12. Cingolani, S., Tanaka, K.: Ground state solutions for the nonlinear Choquard equation with prescribed mass. In: Geometric Properties for Parabolic and Elliptic PDE's. Springer INdAM Series, vol. 47, pp. 23–41. Springer, Cham (2021)
13. Cingolani, S., Tanaka, K.: Deformation argument under PSP condition and applications. Anal. Theory Appl. **37**(2), 191–208 (2021)
14. Cingolani, S., Tanaka, K.: Semi-classical states for nonlinear Choquard equations: concentration at saddle points of potentials (2022, preprint)
15. Cingolani, S., Jeanjean, L., Tanaka, K.: Multiplicity of positive solutions of nonlinear Schrödinger equations concentrating at a potential well. Calc. Var. **53**, 413–439 (2015)
16. Cingolani, S., Jeanjean, L., Tanaka, K.: Multiple complex-valued solutions for nonlinear magnetic Schrödinger equations. J. Fixed Point Theory Appl. **19**(1), 37–66 (2017)
17. Cingolani, S., Gallo, M., Tanaka, K.: Normalized solutions for fractional nonlinear scalar field equations via Lagrangian formulation. Nonlinearity **34**, 4017–4056 (2021)
18. Cingolani, S., Gallo, M., Tanaka, K.: Normalized solutions for fractional nonlinear scalar field equation via Lagrangian formulation. Calc. Var. Partial Differ. Equ. **61**(2), 68, 34 pp. 35B38 (2022)
19. d'Avenia, P., Pomponio, A., Ruiz, D.: Semiclassical states for the nonlinear Schrödinger equation on saddle points of the potential via variational methods. J. Funct. Anal. **262**(10), 4600–4633 (2012)
20. del Pino, M., Felmer, P.: Local mountain passes for semilinear elliptic problems in unbounded domains. Cal. Var. Partial Differ. Equ. **4**, 121–137 (1996)
21. del Pino, M., Felmer, P.: Semi-classical states of nonlinear Schrödinger equations: a variational reduction method. Math. Ann. **324**, 1–32 (2002)
22. Floer, A., Weinstein, A.: Nonspreading wave packets for the cubic Schrödinger equation with a bounded potential. J. Funct. Anal. **69**(3), 397–408 (1986)
23. Gallo, M.: Multiplicity and concentration results for local and fractional NLS equations with critical growth. Adv. Differential Equations. **26**(9–10), 397–424 (2021).
24. Hirata, J., Tanaka, K.: Nonlinear scalar field equations with L^2 constraint: mountain pass and symmetric mountain pass approaches. Adv. Nonlinear Stud. **9**(2), 263–290 (2019)
25. Hirata, J., Ikoma, N., Tanaka, K.: Nonlinear scalar field equations in $\mathbb{R}^N$: mountain pass and symmetric mountain pass approaches. Topol. Methods Nonlinear Anal. **35**(2), 253–276 (2010)
26. Ikoma, N., Tanaka, K.: A note on deformation argument for L^2 normalized solutions of nonlinear Schrödinger equations and systems. Adv. Differ. Equ. **24**(11–12), 609–646 (2019)
27. Jeanjean, L., Tanaka, K.: A remark on least energy solutions in $\mathbb{R}^N$. Proc. Am. Math. Soc. **131**(8), 2399–2408 (2003)

28. Oh, Y.-G.: Existence of semiclassical bound states of nonlinear Schrödinger equations with potentials of the class $(V)_a$. Commun. Partial Differ. Equ. **13**(12), 1499–1519 (1988)
29. Rabinowitz, P.H.: On a class of nonlinear Schrödinger equations. Z. Angew. Math. Phys. **43**(2), 270–291 (1992)
30. Wang, X.: On concentration of positive bounded states of nonlinear Schrödinger equations. Commun. Math. Phys. **153**, 229–244 (1993)

Some Geometric Observations on Heat Kernels of Markov Semigroups with Non-local Generators

Kristian P. Evans and Niels Jacob

Dedicated to Professor Francesco Altomare

Abstract In this paper we study the possibility to represent the heat kernel associated with a pseudo differential operator with negative definite symbol with the help of a combination of two metrics associated with the symbol. We discuss in great detail two examples and we outline the consequences of this approach.

Keywords Non-local generators · Fractional Laplacians · Potential kernels · Heat kernel decays · Metric measure spaces

1 Introduction

The heat kernel of a "nice" uniformly elliptic differential operator in divergence form acting on functions defined on a Riemannian manifold is well understood in geometric terms. Moreover, using expansions of the heat kernel enables us to determine geometric properties of the manifold using analytic data. The monographs of Davies [6] and that of Varopoulos et. al. [30] may serve as classical texts. Clearly these results stimulated a lot of generalisations, maybe the most far reaching is that on the geometry induced on rather general spaces with the help of local regular Dirichlet forms, we refer to the monograph of Bakry, Gentil and Ledoux [1], but also to the work of Sturm [29]. For the Laplacian in $\mathbb{R}^n$ we obtain the heat kernel, i.e., the Gauss kernel, as the inverse Fourier transform of $(2\pi)^{-\frac{n}{2}}e^{-t|\xi|^2}$ and when replacing $\xi \mapsto |\xi|^2$ by a continuous negative definite function $\psi : \mathbb{R}^n \to \mathbb{R}$ the inverse Fourier transform of $(2\pi)^{-\frac{n}{2}}e^{-t\psi(\xi)}$ leads to a convolution semigroup $(\mu_t)_{t\geq 0}$ of sub-probability measures, and hence to a Lévy process. The reader should note that

K. P. Evans · N. Jacob (✉)
Department of Mathematics, Swansea University, Swansea, Wales, UK
e-mail: K.Evans@Swansea.ac.uk; N.Jacob@swansea.ac.uk

A. M. Candela et al. (eds.), *Recent Advances in Mathematical Analysis*,
Trends in Mathematics, https://doi.org/10.1007/978-3-031-20021-2_17

the constant $(2\pi)^{-\frac{n}{2}}$ is due to our normalisation of the Fourier transform in (10). In fact, all symmetric Lévy processes are obtained in this way. The corresponding Feller semigroup $(T_t^\psi)_{t\geq 0}$ has the generator $-\psi(D)$, $\psi(D)u = F^{-1}(\psi(\cdot)\hat{u})$, and if $e^{-t\psi} \in L^1$ for all $t > 0$, then

$$(T_t^\psi u)(x) = (2\pi)^{-\frac{n}{2}} \int_{\mathbb{R}^n} e^{ix\cdot\xi} e^{-t\psi(\xi)} \hat{u}(\xi) \mathrm{d}\xi \tag{1}$$
$$= \int_{\mathbb{R}^n} u(x-y)\mu_t(\mathrm{d}y)$$
$$= \int_{\mathbb{R}^n} p_t^\psi(x-y)u(y)\mathrm{d}y.$$

We call $(p_t^\psi)_{t>0}$ the heat kernel corresponding to ψ and our aim is to study p_t^ψ for real-valued negative definite $\psi : \mathbb{R}^n \to \mathbb{R}$ such that $e^{-t\psi} \in L^1(\mathbb{R}^n)$ and $\psi(\xi) = 0$ if and only if $\xi = 0$. In this case $d_\psi(\xi, \eta) = \psi^{\frac{1}{2}}(\xi - \eta)$ is a metric on $\mathbb{R}^n$ and we assume that d_ψ generates on $\mathbb{R}^n$ the Euclidean topology, see below. With such a function ψ we can associate the symmetric Dirichlet form

$$\mathcal{E}^\psi(u, v) = \int_{\mathbb{R}^n} \psi(\xi)\hat{u}(\xi)\overline{\hat{v}(\xi)}\mathrm{d}\xi, \tag{2}$$

which is a closed form on $L^2(\mathbb{R}^n)$ when taking as domain the Hilbert space $H^{\psi,1}(\mathbb{R}^n)$ where with

$$\|u\|_{\psi,s} := \left(\int_{\mathbb{R}^n} (1 + \psi(\xi))^s |\hat{u}(\xi)|^2 \mathrm{d}\xi, \right)^{\frac{1}{2}}, \ s \geq 0, \tag{3}$$

we have

$$H^{\psi,s}(\mathbb{R}^n) := \left\{ u \in L^2(\mathbb{R}^n) | \|u\|_{\psi,s} < \infty \right\}. \tag{4}$$

Note that the classical Bessel potential space $H^t(\mathbb{R}^n)$ is $H^{|\cdot|,t}(\mathbb{R}^n)$ with $\|u\|_t^2 = \int_{\mathbb{R}^n} (1 + |\xi|^2)^t |\hat{u}(\xi)|^2 \mathrm{d}\xi$. To study p_t^ψ we study $F^{-1}((2\pi)^{-\frac{n}{2}} e^{-t\psi})$ and certain results are easy to obtain, e.g. scaling results if $\psi(\xi) = |\xi|^s, 0 < s < 2$, or the fact that p_t^ψ is rotational invariant if ψ is.
A theorem due to I. Schoenberg [27] states (in a formulation suitable for our purpose) that a metric space $(\mathbb{R}^n, d)$ is isometrically embedded into a Hilbert space if and only if d^2 is a continuous negative function. Based on this result and P.-A. Meyer's [23] on the carré du champs for non-local Dirichlet forms the second named author introduced the idea to try to identify p_t^ψ with certain geometric objects. First ideas were given during the Conference on Lévy Processes in Paris, the paper [15] contains first rigorous results which were further investigated in [4, 7] and [16]. The

aim of this paper is to continue these studies, partly by some surprising structural results, partly by providing a lot of new and rather interesting examples.

In studying p_t^ψ many authors try to carry over techniques successfully when dealing with local operators, and this works to a certain point. Our approach is, however, fundamentally different by assuming that p_t^ψ has the structure

$$p_t^\psi(x-y) = p_t^\psi(0)\frac{p_t^\psi(x-y)}{p_t^\psi(0)} \tag{5}$$

and $p_t^\psi(0)$ is determined by the metric $d_\psi(\xi,\eta) = \psi^{\frac{1}{2}}(\xi-\eta)$ and

$$\frac{p_t^\psi(x-y)}{p_t^\psi(0)} = e^{-\delta_t^2(x,y)}$$

where δ_t^2 is a further metric. While d_ψ is always under some minimal assumptions at our disposal and we may find

$$p_t^\psi(0) = t\mathcal{L}(\lambda^{(n)}(B^{d_\psi}(0,\sqrt{\cdot})))(t), \tag{6}$$

where $\mathcal{L}$ denotes the Laplace transform and $\lambda^{(n)}$ the n-dimensional Lebesgue measure. In the case where the doubling property holds for d_ψ we even get

$$p_t^\psi(0) \asymp \lambda^{(n)}\left(B^{d_\psi}(0,\sqrt{\tfrac{1}{t}})\right), \tag{7}$$

the existence of $\delta_{\psi,t}$ is by no means trivial and so far in general not known.
One of the major objectives of this paper is to come closer to an understanding as to what happens in the case where a semigroup $(T_t)_{t\geq 0}$ is generated by a pseudo differential operator $-q(x,D)$ with $q(x,\xi)$ depending on x. We discuss in Sect. 3 how a type of perturbation argument can be used to study $T_t u$ in a neighbourhood of a point x_0 with the help of $S_t^{q(x_0,\cdot)}$ where $(S_t^{q(x_0,\cdot)})_{t\geq 0}$ is the semigroup generated by the constant coefficient operator $-q(x_0,D)$ with symbol $-q(x_0,\xi)$ and by assumption $\xi \mapsto q(x_0,\xi)$ is a continuous negative definite function. This suggests to try to understand $(T_t)_{t\geq 0}$ either by the family $(S_t^{q(x_0,\cdot)})_{t\geq 0, x_0\in\mathbb{R}^n}$, or by the two families of metrics $d_{q(x_0,\cdot)}$ and $\delta_{q(x_0,\cdot),t}$, $x_0 \in \mathbb{R}^n$ and $t > 0$. We do not yet have a complete general theory, however we can support our ideas by a lot of examples. Section 4 is entirely devoted to the two examples

$$L(x,\xi) = \sum_{k=1}^n a_k(x)|\xi_k|, \quad \xi_k \in \mathbb{R}, \tag{8}$$

and

$$\Lambda(x,\xi) = \left(\sum_{k=1}^{n} b_k(x)\xi_k^2\right)^{\frac{1}{2}}, \quad \xi_k \in \mathbb{R}, \tag{9}$$

which we discuss in our context in great detail. Section 2 is on the one hand meant to set the scene and give background material which is, however, most likely not standard. On the other hand, it contains some observations which seems to us completely new and may hint to further connections. We show that for certain symbols of the type $\sum |\xi_k|^{2r}, 0 < r \leq 1$, we may use fractional derivatives to represent the corresponding Dirichlet form as finite sum of squares of operators. This is picked up in Sect. 5 where the result is extended to symbols of the type $\sum |\xi_k|^{2r_k}, 0 < r_k \leq 1$. From here the way to treat symbols which have a certain x-dependence is open.

Our notations are standard, we refer to [13], see also [2] for some notions. For the reader's convenience we recall that $\psi : \mathbb{R}^n \to \mathbb{C}$ is called a continuous negative definite function if it is continuous, $\psi(0) \geq 0$ and $\xi \mapsto e^{-t\psi(\xi)}$ is for all $t > 0$ positive definite in the sense of Bochner, which is equivalent to ψ admitting a Lévy-Khinchin representation. Moreover, $f : (0,\infty) \to \mathbb{R}$ is called a Bernstein function if f is arbitrarily often differentiable, $f \geq 0$ and $(-1)^k \frac{\mathrm{d}^k f}{\mathrm{d}t^k} \leq 0, k \in \mathbb{N}$.

2 Setting the Scene: The Operators $(-\Delta)^r$ and $\Delta_{(r)}$

For $u \in \mathcal{S}(\mathbb{R}^n)$ we define its Fourier transform $\hat{u} = Fu$ by

$$\hat{u}(\xi) = (Fu)(\xi) = (2\pi)^{-\frac{n}{2}} \int_{\mathbb{R}^n} e^{-ix\cdot\xi} u(x)\mathrm{d}x \tag{10}$$

implying for the inverse Fourier transform F^{-1} the formula

$$(F^{-1}u)(x) = (2\pi)^{-\frac{n}{2}} \int_{\mathbb{R}^n} e^{ix\cdot\xi} u(\xi)\mathrm{d}\xi. \tag{11}$$

With this normalisation, Plancherel's theorem holds with the constant 1, i.e., the Fourier transform is an isometry on $L^2(\mathbb{R}^n)$ and the convolution theorem reads as

$$(u \cdot v)^\wedge(\xi) = (2\pi)^{-\frac{n}{2}}(\hat{u} * \hat{v})(\xi) \text{ and } (u * v)^\wedge(\xi) = (2\pi)^{\frac{n}{2}}\hat{u}(\xi)\hat{v}(\xi). \tag{12}$$

For $1 \leq k \leq n$ we have the identity

$$\left(\frac{\partial u}{\partial x_k}\right)^\wedge(\xi) = i\xi_k\hat{u}(\xi) \tag{13}$$

and therefore

$$\left(\frac{\partial}{\partial x_k}\right) u(x) = \left(F^{-1}(i\xi_k \hat{u})\right)(x). \tag{14}$$

The formal L^2-adjoint of $\frac{\partial}{\partial x_k}$ is given by

$$\left(\frac{\partial}{\partial x_k}\right)^* v(x) = -\frac{\partial}{\partial x_k} v(x) \tag{15}$$

implying that

$$\left(\frac{\partial}{\partial x_k}\right)^* v(x) = \left(F^{-1}(-i\xi_k \hat{v})\right)(x) \tag{16}$$

and on the level of symbols we have as expected

$$\sigma\left(\left(\frac{\partial}{\partial x_k}\right)^*\right)(\xi) = \overline{\sigma\left(\frac{\partial}{\partial x_k}\right)(\xi)} \tag{17}$$

For $u \in \mathcal{S}(\mathbb{R}^n)$ it follows that

$$\begin{aligned}\left(\frac{\partial}{\partial x_k}\right) \circ \left(\frac{\partial}{\partial x_k}\right)^* u(x) &= F^{-1}(\xi_k^2 \hat{u})(x) \\ &= -\frac{\partial^2}{\partial x_k^2} u(x) = \left(\frac{\partial}{\partial x_k}\right)^* \circ \left(\frac{\partial}{\partial x_k}\right) u(x).\end{aligned}$$

Therefore we find for the Laplace operator Δ on $\mathbb{R}^n$

$$\Delta = \sum_{k=1}^{n} \frac{\partial^2}{\partial x_k^2} = -\sum_{k=1}^{n} \left(\frac{\partial}{\partial x_k}\right)^* \circ \left(\frac{\partial}{\partial x_k}\right) \tag{18}$$

and $\sigma(\Delta) = -|\xi|^2 = -\sum_{k=1}^{n} \xi_k^2$.
Let $\xi_k \in \mathbb{R}$ and $0 < r < 1$. The fractional power (with exponent r) of $\pm i\xi_k$ is given by

$$\begin{aligned}(\pm i\xi_k)^r &= |\xi_k|^r e^{\pm i \frac{r\pi \operatorname{sgn}\xi_k}{2}} \\ &= |\xi_k|^r \left(\cos\frac{r\pi \operatorname{sgn}\xi_k}{2} \pm i \sin\frac{r\pi \operatorname{sgn}\xi_k}{2}\right) \\ &= |\xi_k|^r \left(\cos\frac{r\pi}{2} \pm i (\operatorname{sgn}\xi_k) \sin\frac{r\pi}{2}\right)\end{aligned} \tag{19}$$

and we have

$$\overline{(\pm i\xi_k)}^r = (\mp i\xi_k)^r \text{ as well as } (\pm i\xi_k)^r\overline{(\pm i\xi_k)^r} = |\xi_k|^{2r}. \tag{20}$$

This allows us to define for $u \in \mathcal{S}(\mathbb{R}^n)$, $1 \leq k \leq n$ and $0 < r < 1$, the fractional partial derivative

$$\begin{aligned}\left(\frac{\partial}{\partial x_k}\right)^r u(x) &:= F^{-1}((i\xi_k)^r\hat{u})(\xi) \\ &= (2\pi)^{-\frac{n}{2}}\int_{\mathbb{R}^n} e^{ix\cdot\xi}(i\xi_k)^r\hat{u}(\xi)\mathrm{d}\xi.\end{aligned} \tag{21}$$

(The reader may note that different definitions of the Fourier transforms may lead to a different sign in (13) and (21).)
The formal L^2-adjoint $\left(\left(\frac{\partial}{\partial x_k}\right)^r\right)^*$ of $\left(\frac{\partial}{\partial x_k}\right)^r$ is determined on $\mathcal{S}(\mathbb{R}^n)$ by

$$\begin{aligned}\int_{\mathbb{R}^n}\left(\frac{\partial}{\partial x_k}\right)^r u(x)\overline{v(x)}\mathrm{d}x &= \int_{\mathbb{R}^n} F\left(\left(\frac{\partial}{\partial x_k}\right)^r u\right)(\xi)\overline{(Fv)}(\xi)\mathrm{d}\xi \\ &= \int_{\mathbb{R}^n}(i\xi_k)^r\hat{u}(\xi)\overline{\hat{v}(\xi)}\mathrm{d}\xi \\ &= \int_{\mathbb{R}^n}\hat{u}(\xi)\overline{(-i\xi_k)^r\hat{v}(\xi))}\mathrm{d}\xi,\end{aligned}$$

i.e., we have

$$\left(\left(\frac{\partial}{\partial x_k}\right)^r\right)^* v(x) = F^{-1}((-i\xi_k)^r\hat{v})(\xi). \tag{22}$$

For the corresponding symbols we find

$$\sigma\left(\left(\frac{\partial}{\partial x_k}\right)^r\right)(\xi) = (i\xi_k)^r \text{ and } \sigma\left(\left(\frac{\partial}{\partial x_k}\right)^r\right)^* = (-i\xi_k)^r = \overline{\sigma\left(\left(\frac{\partial}{\partial x_k}\right)^r\right)(\xi)}. \tag{23}$$

For $u \in \mathcal{S}(\mathbb{R}^n)$ we can calculate

$$\begin{aligned}\left(\frac{\partial}{\partial x_k}\right)^r\circ\left(\left(\frac{\partial}{\partial x_k}\right)^r\right)^* u(x) &= F^{-1}\left((i\xi_k)^r F\left(\left(\left(\frac{\partial}{\partial x_k}\right)^r\right)^* u\right)\right)(x) \\ &= F^{-1}((i\xi_k)^r(-i\xi_k)^r\hat{u}))(x) \\ &= F^{-1}(|\xi_k|^{2r}\hat{u})(x) \\ &= \left(\left(\frac{\partial}{\partial x_k}\right)^r\right)^*\circ\left(\frac{\partial}{\partial x_k}\right)^r u(x).\end{aligned} \tag{24}$$

It is obvious that both $\left(\frac{\partial}{\partial x_k}\right)^r$ and $\left(\left(\frac{\partial}{\partial x_k}\right)^r\right)^*$ have continuous extensions from $H^{rs}(\mathbb{R}^n)$ to $H^{r(s-1)}(\mathbb{R}^n)$, $s \geq 1$. Both operators also admit a representation as a fractional Liouville derivative, see [18], namely

$$\left(\frac{\partial}{\partial x_k}\right)^r u(x) = -\frac{1}{\Gamma(1-r)}\frac{\partial}{\partial x_k}\int_{x_k}^{\infty}\frac{u(t)}{(t-x_k)^r}\mathrm{d}t \tag{25}$$

and

$$\left(\left(\frac{\partial}{\partial x_k}\right)^r\right)^* u(x) = \frac{1}{\Gamma(1-r)}\frac{\partial}{\partial x_k}\int_{\infty}^{x_k}\frac{u(t)}{(x_k-t)^r}\mathrm{d}t. \tag{26}$$

These representations admit further extensions of $\left(\frac{\partial}{\partial x_k}\right)^r$ and $\left(\left(\frac{\partial}{\partial x_k}\right)^r\right)^*$, respectively. Further, using the fact that $\left(\frac{\partial}{\partial x_k}\right)^r$ and $\left(\left(\frac{\partial}{\partial x_k}\right)^r\right)^*$ are pseudo-differential operators with constant coefficients, hence they are translation invariant, and using the fact that their symbols vanish at $\xi = 0$, we can extend both operators in a consistent way to the constants by

$$\left(\frac{\partial}{\partial x_k}\right)^r c = \left(\left(\frac{\partial}{\partial x_k}\right)^r\right)^* c = 0, \tag{27}$$

and now we can extend both operators to functions $a = u + c$, where u belongs, for example, to $H^r(\mathbb{R}^n)$. The point is that $a = u + c$ need not be integrable over $\mathbb{R}^n$ or vanish at infinity. We also would like to remark that identifying $\pm i\xi_k$ as a continuous negative definite function and $-\left(\frac{\partial}{\partial x_k}\right)^r$ as well as $-\left(\left(\frac{\partial}{\partial x_k}\right)^r\right)^*$ as generators of Markovian semigroups on $L^p(\mathbb{R}^n)$, $1 \leq p < \infty$, or $C_\infty(\mathbb{R}^n)$ obtained by subordination in the sense of Bochner allows further extensions using the functional calculus developed in [24], see also [26].
Let $\varphi : \mathbb{R}^n \to \mathbb{R}$ be a scalar function in $\mathcal{S}(\mathbb{R}^n)$ and let $\Phi : \mathbb{R}^n \to \mathbb{R}^n$ be a vector field, $\Phi = (\Phi_1, \ldots, \Phi_n)$, with $\Phi_k \in \mathcal{S}(\mathbb{R}^n)$. For $0 < r < 1$ we introduce the r**-gradient** $\operatorname{grad}_{(r)}\varphi$ of φ and the r**-divergence** $\operatorname{div}_{(r)}\Phi$ of Φ by

$$\operatorname{grad}_{(r)}\varphi = \sum_{k=1}^{n}\left(\frac{\partial}{\partial x_k}\right)^r \varphi e_k \tag{28}$$

where e_k is the kth unit vector in $\mathbb{R}^n$, and

$$\operatorname{div}_{(r)}\Phi = \sum_{k=1}^{n}\left(\left(\frac{\partial}{\partial x_k}\right)^r\right)^*\Phi_k. \tag{29}$$

Now we define the operator $\Delta_{(r)}$ by

$$\Delta_{(r)}\varphi := \mathrm{div}_{(r)}\mathrm{grad}_{(r)}\varphi. \tag{30}$$

Since

$$\mathrm{div}_{(r)}\mathrm{grad}_{(r)}\varphi = \sum_{k=1}^{n}\left(\left(\frac{\partial}{\partial x_k}\right)^r\right)^* \circ \left(\frac{\partial}{\partial x_k}\right)^r \varphi$$

it follows that

$$\Delta_{(r)}\varphi(\xi) = F^{-1}\left(\sum_{k=1}^{n} -|\xi_k|^{2r}\hat{\varphi}\right)(\xi). \tag{31}$$

It is convenient to introduce the continuous negative definite function

$$\psi_r(\xi) := \sum_{k=1}^{n} |\xi_k|^{2r}, \quad 0 < r < 1, \tag{32}$$

which is the symbol of the operator $\Delta_{(r)}$, $0 < r < 1$. Thus we may write

$$-\Delta_{(r)}\varphi = \psi_r(D)\varphi. \tag{33}$$

Starting with the Laplacian Δ on $\mathbb{R}^n$ and $0 < r < 1$, we have now two ways to generalise this operator. With the help of Bernstein functions $f_{(r)}$, $f_{(r)}(s) = s^r$, $s \geq 0$, we can form the fractional power Δ^r defined on $\mathcal{S}(\mathbb{R}^n)$ by

$$\Delta^r u(x) := -f_{(r)}(-\Delta)u(x) := -(2\pi)^{-\frac{n}{2}} \int_{\mathbb{R}^n} e^{ix\xi}|\xi|^{2r}\hat{u}(\xi)\mathrm{d}\xi, \tag{34}$$

and we refer to [22] for several other ways to define Δ^r, and we may consider the operator

$$\Delta_{(r)}u = -\psi_r(D)u. \tag{35}$$

Of course, for $n = 1$ we have $\Delta_{(r)} = -\psi_r(D)$, but for $n > 1$ it follows for the symbol $\sigma(-\Delta^r)$ of $-\Delta^r$ that

$$\sigma(-\Delta^r)(\xi) = |\xi|^{2r} = \left(\sum_{k=1}^{n} |\xi_k|^2\right)^r$$

and

$$\sigma(-\Delta_{(r)})(\xi) = \sum_{k=1}^{n} |\xi_k|^{2r}$$

and hence $\sigma(-\Delta^r)(\xi) \neq \sigma(-\Delta_{(r)})(\xi)$. Note that we have with two constants $0 < \gamma_{1,n,r} < \gamma_{2,n,r}$ the estimates

$$\gamma_{1,n,r}|\xi|^{2r} \leq \sum_{k=1}^{n} |\xi_k|^{2r} = \psi_r(\xi) \leq \gamma_{2,n,r}|\xi|^{2r} \tag{36}$$

as well as

$$\tilde{\gamma}_{1,n,r}(1+|\xi|^2)^{\frac{r}{2}} \leq (1+\sum_{k=1}^{n} |\xi_k|^{2r})^{\frac{1}{2}} = (1+\psi_r(\xi))^{\frac{1}{2}} \leq \tilde{\gamma}_{2,n,r}(1+|\xi|^2)^{\frac{r}{2}}. \tag{37}$$

The first set of estimates means that the pseudo-differential operator $-\Delta^r$ and $-\Delta_{(r)}$ are comparable elliptic pseudo-differential operators of order $2r$, while the second pair of estimates yield that the space $H^{rs}(\mathbb{R}^n)$ and $H^{\psi_r,s}(\mathbb{R}^n)$ with norms $\|\cdot\|_{rs}$ and $\|\cdot\|_{\psi_r,s}$ are equivalent Banach spaces. However, these two operators inherit different properties from the Laplacian, for example Δ^r is again rotational invariant, but for $n > 1$ not of divergence form while $\Delta_{(r)}$ is of divergence form, but of course it is not rotational invariant. Looking at the corresponding Dirichlet forms (when dealing with Dirichlet forms we always assume the functions in their domain to be real-valued) we find for $\mathcal{E}^{(r)}$ corresponding to Δ^r and for $\mathcal{E}_{\psi_r}$ corresponding to $-\Delta_{(r)}$ that they have the common domain $H^r(\mathbb{R}^n) = H^{\psi_r,1}(\mathbb{R}^n)$. Further we have on $H^r(\mathbb{R}^n)$ (obtained by extending from $\mathcal{S}(\mathbb{R}^n)$)

$$\begin{aligned}\mathcal{E}^{(r)}(u,v) &= \int_{\mathbb{R}^n} |\xi|^{2r}\hat{u}(\xi)\overline{\hat{v}(\xi)}\mathrm{d}\xi \\ &= c_{n,r}\int_{\mathbb{R}^n}\left(\int_{\mathbb{R}^n\setminus\{0\}} \frac{(u(x+y)-u(x))(v(x+y)-v(x))}{|y|^{n+r}}\mathrm{d}y\right)\mathrm{d}x \\ &= \int_{\mathbb{R}^n}(-\Delta)^{\frac{r}{2}}u(x)(-\Delta)^{\frac{r}{2}}v(x)\mathrm{d}x\end{aligned} \tag{38}$$

while for $\mathcal{E}_{\psi_r}$ we find

$$\begin{aligned}\mathcal{E}_{\psi_r}(u,v) &= \int_{\mathbb{R}^n} \psi_r(\xi)|\hat{u}(\xi)\overline{\hat{v}(\xi)}\mathrm{d}\xi \\ &= \int_{\mathbb{R}^n}\left(\sum_{k=1}^{n} c_{1,r}\int_{\mathbb{R}} \frac{(u(x+y_ke_k)-u(x))(v(x+y_ke_k)-v(x))}{|y_k|^{1+r}}\mathrm{d}y_k\right)\mathrm{d}x\end{aligned} \tag{39}$$

$$= \int_{\mathbb{R}^n} (\psi_r(D)^{\frac{1}{2}} u(x) (\psi_r(D))^{\frac{1}{2}} v(x) \mathrm{d}x$$

$$= \int_{\mathbb{R}^n} \sum_{k=1}^{n} \left(\frac{\partial}{\partial x_k}\right)^r u(x) \left(\frac{\partial}{\partial x_k}\right)^r v(x) \mathrm{d}x. \tag{40}$$

The important difference is that $\mathcal{E}_{\psi_r}$ admits the representation (40) with a finite number of operators $\left(\frac{\partial}{\partial x_k}\right)^r$ and it allows us to interpret $L(D)$ with symbol (8) as a type of fractional derivative Hörmander operator, which clearly is different to a fractional (power of a) Hörmander type operator. This is not a decomposition of $L(D) = \psi_r(D)$ or of $\mathcal{E}_{\psi_r}$ into a sum of compositions or products, respectively, of derivations, which only holds for Δ. However, we want to refer to the old paper of P.-A. Meyer [23], where a decomposition for generators of symmetric Lévy processes into an infinite sum of squares of operators was discussed and we will return to this later in our paper. For the readers convenience we would like to provide the calculation leading to (40). For $u, v \in \mathcal{S}(\mathbb{R}^n)$ we find

$$\begin{aligned}
\int_{\mathbb{R}^n} \sum_{k=1}^{n} |\xi_k|^{2r} \hat{u}(\xi) \overline{\hat{v}(\xi)} \mathrm{d}\xi &= \sum_{k=1}^{n} \int_{\mathbb{R}^n} (i\xi_k)^r \overline{(i\xi_k)^r} \hat{u}(\xi) \overline{\hat{v}(\xi)} \mathrm{d}\xi \\
&= \sum_{k=1}^{n} \int_{\mathbb{R}^n} (i\xi_k)^r \hat{u}(\xi) \overline{(i\xi_k)^r \hat{v}(\xi)} \mathrm{d}\xi \\
&= \sum_{k=1}^{n} \int_{\mathbb{R}^n} \left(\frac{\partial}{\partial x_k}\right)^r u(x) \left(\frac{\partial}{\partial x_k}\right)^r v(x) \mathrm{d}x
\end{aligned}$$

where we have used in the last step that for v real-valued we have that $\overline{\hat{v}(-\xi)} = \hat{v}(\xi)$. There is a third natural representation of $\mathcal{E}_{\psi_r}$ which we obtain by factorising each term $|\xi_k|^{2r}$, i.e., we have

$$\begin{aligned}
\mathcal{E}_{\psi_r}(u, v) &= \int_{\mathbb{R}^n} \sum_{k=1}^{n} |\xi_k|^r \hat{u}(\xi) |\xi_k|^r \overline{\hat{v}(\xi)} \mathrm{d}\xi \qquad (41) \\
&= \int_{\mathbb{R}^n} \sum_{k=1}^{n} \left(-\frac{\partial^2}{\partial x_k^2}\right)^{\frac{r}{2}} u(x) \left(-\frac{\partial^2}{\partial x_k^2}\right)^{\frac{r}{2}} v(x) \mathrm{d}x,
\end{aligned}$$

however, we will discuss this representation in a slightly different context in a forthcoming paper.

We want to summarise certain estimates for $\mathcal{E}^{(r)}$ and $\mathcal{E}_{\psi_r}$. In each of the following estimates we can replace $\|\cdot\|_{rs}$ by a constant times $\|\cdot\|_{\psi_r,1}$ and vice versa. For $u \in H^r(\mathbb{R}^n) = H^{\psi_r,1}(\mathbb{R}^n)$ we have

$$|\mathcal{E}^{(r)}(u,v)| \leq \|u\|_r \|v\|_r, \tag{42}$$

$$|\mathcal{E}_{\psi_r}(u,v)| \leq \|u\|_{\psi_r,1} \|v\|_{\psi_r,1} \tag{43}$$

as well as

$$\mathcal{E}^{(r)}(u,u) \geq \|u\|_r^2 - \|u\|_{L^2}^2 \tag{44}$$

and

$$\mathcal{E}_{\psi_r}(u,u) \geq \|u\|_{\psi,r}^2 - \|u\|_{L^2}, \tag{45}$$

in fact, the last two estimates we have equality. Moreover, we have Sobolev's inequality

$$\|u\|_{L^p}^2 \leq c_{n,r,p} \mathcal{E}^{(r)}(u,u) \tag{46}$$

with $p = \frac{2n}{n-2r}$, $n \geq 2$, and hence we have

$$\|u\|_{L^p}^2 \leq \tilde{c}_{n,r,p} \mathcal{E}_{\psi_r}(u,u). \tag{47}$$

For us, most important is estimate (46) and we want to explain this in a wider context. Thus taking the equivalence of the norms $\|\cdot\|_r$ and $\|\cdot\|_{\psi_r,1}$ into account the two forms satisfy equivalent estimates, but we will see that this does not hold for the corresponding heat kernels. Suppose that $(B, D(B))$ is a symmetric Dirichlet form on $L^2(\mathbb{R}^n)$ associated with the Markovian semigroup $(T_t)_{t\geq 0}$. Further suppose that

$$\|u\|_{L^p}^2 \leq c_0 B(u,u) \tag{48}$$

holds for all $u \in D(B)$ for some $p > 2$ and set $N := \frac{2p}{p-2}$. Then the operators T_t have a kernel $p_t(x,y)$ which is symmetric, i.e., $p_t(x,y) = p_t(y,x)$, and

$$(T_t u)(x) = \int_{\mathbb{R}^n} p_t(x,y) u(y) \mathrm{d}y, \quad u \in L^2(\mathbb{R}^n), \tag{49}$$

and we have

$$0 \leq p_t(x,y) \leq \|T_t\|_{L^1-L^\infty} = \operatorname{ess\,sup}_{x\in\mathbb{R}^n} p_t(x,x) \leq c_1 t^{-\frac{N}{2}}. \tag{50}$$

In the case that $(B, D(B))$ is translation invariant then there exists a continuous negative definite function $\psi : \mathbb{R}^n \to \mathbb{R}$ such that $D(B) = H^{\psi,1}(\mathbb{R}^n)$ and

$$B(u, v) = \int_{\mathbb{R}^n} \psi(\xi)\hat{u}(\xi)\overline{\hat{v}(\xi)}\mathrm{d}\xi \tag{51}$$

for all $u, v \in H^{\psi,1}(\mathbb{R}^n)$. In this case we write $\mathcal{E}_\psi$ for B. The corresponding semigroup is associated with a convolution semigroup $(\mu_t)_{t\geq 0}$, i.e.,

$$\hat{\mu}_t(\xi) = (2\pi)^{-\frac{n}{2}} e^{-t\psi(\xi)}, \tag{52}$$

and $T_t^\psi u = \mu_t * u$. Since we assume $(T_t^\psi)_{t\geq 0}$ to be a Markovian semigroup we must have $\psi(0) = 0$ and estimate (46) implies now that μ_t has a density p_t^ψ with respect to $\lambda^{(n)}$ given by

$$p_t^\psi(x) = (2\pi)^{-n} \int_{\mathbb{R}^n} e^{ix\cdot\xi} e^{-t\psi(\xi)}\mathrm{d}\xi. \tag{53}$$

It follows further from (50) that

$$0 \leq p_t^\psi(x) \leq \|T_t^\psi\|_{L^1-L^\infty} = p_t(0) \leq ct^{-\frac{N}{2}}. \tag{54}$$

For example, for $\tilde{\psi}_r(\xi) = |\xi|^{2r}$ or $\psi_r(\xi) = \sum_{k=1}^n |\xi_k|^{2r}$, $n \geq 2$, we find $p = \frac{2n}{n-2r}$, $n \geq 2$, and therefore $\frac{N}{2} = \frac{n}{r}$ implying for both semigroups $(T_t^{\tilde{\psi}_r})_{t\geq 0}$ and $(T_t^{\psi_r})_{t\geq 0}$ the estimate

$$p_t^{\tilde{\psi}_r}(0) \leq \tilde{c}_{n,r} t^{-\frac{n}{2r}} \text{ and } p_t^{\psi_r}(0) \leq c_{n,r} t^{-\frac{n}{2r}}. \tag{55}$$

Recall that if $\psi : \mathbb{R}^n \to \mathbb{R}$ is a continuous negative definite function such that $\psi(\xi) = 0$ if and only if $\xi = 0$ then

$$d_\psi(\xi, \eta) := \psi^{\frac{1}{2}}(\xi - \eta) \tag{56}$$

is a translation invariant metric on $\mathbb{R}^n$ and the condition

$$\liminf_{|\xi|\to\infty} \psi(\xi) > 0 \tag{57}$$

is equivalent to the fact that the topology generated by d_ψ is the Euclidean one, see [15], and in the following we always assume that d_ψ generates the Euclidean topology. We introduce the volume function $V_\psi(\cdot)$ by

$$V_\psi(\rho) := \lambda^{(n)}(B^{d_\psi}(0, \rho)), \tag{58}$$

where $B^{d_\psi}(x,\rho)$ is the ball with respect to d_ψ having centre $x \in \mathbb{R}^n$ and radius $\rho > 0$. A result from [15] states that with $V_\psi^*(\rho) := V_\psi(\sqrt{\rho})$ we have

$$p_t^\psi(0) = (2\pi)^{-n} t\mathcal{L}(V_\psi^*)(t), \tag{59}$$

where $\mathcal{L}$ denotes the Laplace transform, here we assume that $e^{-t\psi} \in L^1(\mathbb{R}^n)$ for $t > 0$. If we assume that d_ψ has the doubling property, i.e.,

$$V_\psi(\rho) \le \gamma V_\psi(2\rho), \quad \rho > 0, \tag{60}$$

then $e^{-t\psi} \in L^1(\mathbb{R}^n)$ and further we have, see [15]

$$p_t^{\psi(0)} \asymp V_\psi\left(\frac{1}{\sqrt{t}}\right), \tag{61}$$

where $a \asymp b$, $0 < a < b$, means that with two constants $\kappa_1, \kappa_2 > 0$ we have $0 < \kappa_1 \le \frac{a}{b} \le \kappa_2$. It is not difficult to see that both metrics $d_{\tilde{\psi}_r}$ and d_{ψ_r} have the doubling property and that

$$V_{\tilde{\psi}_r}(\rho) \asymp V_{\psi_r}(\rho). \tag{62}$$

The term $p_t^\psi(0)$ determines the diagonal behaviour of $p_t^\psi(x,y) = p_t^\psi(x-y)$. In [15] it was suggested that in many cases $p_t^\psi(x-y)$ has the form

$$p_t^\psi(x-y) = c_\psi p_t^\psi(0) e^{-\delta_{\psi,t}^2(x,y)} \tag{63}$$

where $\delta_{\psi,t}$ is a further metric on $\mathbb{R}^n$. We refer to [15, 16] or [4] and [7] where many examples are given, however, no general result is so far known. In this section we now return to our previous discussion of the semigroups $(T_t^{\tilde{\psi}_{\frac{1}{2}}})_{t\ge0}$ and $(T_t^{\psi_{\frac{1}{2}}})_{t\ge0}$. The continuous negative definite function $\tilde{\psi}_{\frac{1}{2},n}(\xi) = |\xi|$, $\xi \in \mathbb{R}^n$, corresponds to the n-dimensional Cauchy process and we know explicitly the associated (heat) kernels $p_t^{C,n}(x)$ and they are given by

$$p_t^{C,n}(x-y) = \pi^{-\frac{n+1}{2}}\Gamma\left(\frac{n+1}{2}\right)\frac{t}{(t^2+(x-y)^2)^{\frac{n-1}{2}}} \tag{64}$$

$$= p_t^{C,n}(0)e^{-\delta_{C,t}^2(x,y)} \tag{65}$$

where

$$\delta_{C,n,t}(x,y) = \sqrt{\frac{n+1}{2}}\left(\ln\left(\frac{(|x-y|^2+t^2)}{t^2}\right)\right)^{\frac{1}{2}}. \tag{66}$$

The function $s \mapsto \ln(1+\frac{s}{a}), a>0$, is a Bernstein function and $\xi \mapsto t^{-2}(|\xi|^2+t^2), t>0$ is a continuous negative definite function, hence $\ln\left(\frac{|\xi|^2+t^2}{t^2}\right)$ is a continuous negative definite function vanishing only for $\xi=0$. Thus its square root is a metric on $\mathbb{R}^n$. It follows that for $p_t^{C,n}$ we have the representation (63), i.e.,

$$\begin{aligned} p_t^{C,n}(x-y) &= t\mathcal{L}(\lambda^{(n)}(B^{d_{\tilde{\psi}_{\frac{1}{2},n}}}(0,\sqrt{\cdot})))(t)e^{-\delta^2_{C,n,t}(x,y)} \\ &= t\mathcal{L}(\lambda^{(n)}(V^*_{\tilde{\psi}_{\frac{1}{2},n}}))(t)e^{-\delta^2_{C,n,t}(x,y)}. \end{aligned} \tag{67}$$

For $\psi_{\frac{1}{2},n}=\sum_{k=1}^n|\xi_k|$ we note that

$$\begin{aligned} p_t^{\psi_{\frac{1}{2},n}}(x-y) &= (2\pi)^{-n}\int_{\mathbb{R}^n} e^{i(x-y)\xi}e^{-t\psi_{\frac{1}{2},n}(\xi)}\,\mathrm{d}\xi \\ &= \prod_{k=1}^n (2\pi)^{-1}\int_{\mathbb{R}^n} e^{i(x_k-y_k)\xi_k}e^{-t\tilde{\psi}_{\frac{1}{2},1}(\xi_k)}\,\mathrm{d}\xi_k \\ &= \prod_{k=1}^n p_t^{C,1}(x_k-y_k) \\ &= \prod_{k=1}^n t\mathcal{L}(\lambda^{(1)}(V^*_{\tilde{\psi},\frac{1}{2},1}))(t)e^{-\delta^2_{C,1,t}(x_k,y_k)}. \end{aligned}$$

In particular we find

$$p_t^{\psi_{\frac{1}{2},n}}(x-y) = p_t^{\psi_{\frac{1}{2},n}}(0)e^{-\delta^2_{\psi_{\frac{1}{2},t}}(x,y)} \tag{68}$$

with

$$\begin{aligned} \delta^2_{\psi_{\frac{1}{2},t}}(x,y) &= \sum_{k=1}^n \delta^2_{C,1,t}(x_k,y_k) \\ &= \sum_{k=1}^n \ln\left(\frac{|x_k-y_k|^2+t^2}{t^2}\right). \end{aligned} \tag{69}$$

Since $d_{\tilde{\psi}_{\frac{1}{2},n}}(\xi,\eta)=\left(\sum_{k=1}^n|\xi_k-\eta_k|^2\right)^{\frac{1}{4}}$ and $d_{\psi_{\frac{1}{2},n}}(\xi,\eta)=\left(\sum_{k=1}^n|\xi_k-\eta_k|\right)^{\frac{1}{2}}$, it is clear that $V_{\tilde{\psi}_{\frac{1}{2},n}}(\rho)$ and $V_{\psi_{\frac{1}{2},n}}(\rho)$ are comparable, implying that $p_t^{\tilde{\psi}_{\frac{1}{2},n}}(0)$ and

$p_t^{\psi_{\frac{1}{2},n}}(0)$ are comparable. However, for $t > 0$ fixed, $p_t^{C,n}(x)$ and $p_t^{\psi_{\frac{1}{2},n}}(x)$ have a different (asymptotic) behaviour. For this, note that

$$p_t^{C,n}(x) = \pi^{-\frac{n+1}{2}}\Gamma\left(\frac{n+1}{2}\right)\frac{t}{(t^2+|x|^2)^{\frac{n+1}{2}}} \tag{70}$$

and

$$p_t^{\psi_{\frac{1}{2},n}}(x) = \frac{1}{\pi^n}\frac{t^n}{\prod_{k=1}^n(t^2+|x_k|^2)}. \tag{71}$$

We want to remark that our findings are rather natural. For two continuous negative definite functions ψ_1 and ψ_2 such that $e^{-t\psi_j} \in L^1(\mathbb{R}^n)$ and $\psi_1 \asymp \psi_2$ we have of course $\mathcal{E}^{\psi_1}(u,u) \asymp \mathcal{E}^{\psi_2}(u,u)$ as well as $e^{-t\psi_1} \asymp e^{-t\psi_2}$ implying $p_t^{\psi_1(0)} \asymp p_t^{\psi_2}(0)$, and obviously $d_{\psi_1} \asymp d_{\psi_2}$. However, the off-diagonal behaviour of $p_t^{\psi_1}(x)$ and $p_t^{\psi_2}(x)$, respectively, depends on the smoothness of ψ_1 and ψ_2, respectively, and if $\psi_1 \asymp \psi_2$ but ψ_1 and ψ_2 have different smoothness properties, then we shall not expect a comparison of $p_t^{\psi_1}(x)$ and $p_t^{\psi_2}(x)$. Consequently we can extend our previous consideration for $p_t^{\tilde{\psi}_r}(0)$ and $p_t^{\psi_r}(0)$.

3 Some Comparisons of Semigroups

In this chapter we want to summarise some results and ideas on pseudo-differential operators generating Feller and/or sub-Markovian semigroups on $C_\infty(\mathbb{R}^n)$, the space of all continuous functions vanishing at infinity, or $L^2(\mathbb{R}^n)$. The generator of a Feller semigroup has to satisfy the positive maximum principle and if an operator mapping $C_0^\infty(\mathbb{R}^n)$ into $C(\mathbb{R}^n)$ satisfies the positive maximum principle, by a result of Courrège [5], this operator must be a pseudo-differential operator $-q(x,D)$ where

$$q(x,D)u(x) = (2\pi)^{-\frac{n}{2}}\int_{\mathbb{R}^n} e^{ix\cdot\xi}q(x,\xi)\hat{u}(\xi)d\xi \tag{72}$$

has a symbol $q : \mathbb{R}^n \times \mathbb{R}^n \to \mathbb{C}$ which is continuous and for $x \in \mathbb{R}^n$ fixed $\xi \mapsto q(x,\xi)$ is a continuous negative definite function. Such symbols have been later called by W. Hoh negative definite symbols. The key question is of course when $-q(x,D)$ with a negative definite symbol extends from $C_0^\infty(\mathbb{R}^n)$ (or $\mathcal{S}(\mathbb{R}^n)$) to a generator of a Feller semigroup or to a sub-Markovian semigroup.

Of course, for certain special classes of symbols such as x-independent symbols or when $q(x,\xi)$ is the symbol of a uniformly elliptic differential operator of second order in divergence form, such results have been known for a long time. Using the Lévy-Khinchin representation of $q(x,\cdot)$ it seems that T. Komatsu [19–21] was the first who could for certain $q(x,\xi)$ construct corresponding processes and hence semigroups.

In [12] a perturbation method in the spirit of pseudo-differential operator theory was suggested to construct with the help of $-q(x,D)$ a Feller semigroup provided that $q(x,\xi)$ is bounded from above and below by a fixed continuous negative definite function ψ and that $q(x,\xi)$ is in a rather precise sense close to $\psi(\cdot)$. We may think that $\psi(\xi) := q(x_0,\xi)$ for some $x_0 \in \mathbb{R}^n$ and then to decompose $q(x,\xi) = q(x_0,\xi) + (q(x,\xi) - q(x_0,\xi))$. The smallness condition becomes now a condition on $q(x,\xi) - q(x_0,\xi)$ as well as a condition on certain partial derivatives of this term. While such results are looking rather restrictive, combined with techniques related to the martingale problem, Hoh in [8–10], see also [11], could get rid of the smallness assumptions, however, certain regularity conditions are still needed.

In [17], under assumptions posed in [12] some interesting estimates for corresponding semigroups have been proved. Suppose that $(T_t)_{t\geq 0}$ is a Feller semigroup which extends to a Markovian semigroup on $L^2(\mathbb{R}^n)$ and is generated by $-q(x,D)$, and denote by $(S_t^{\psi})_{t\geq 0}$ the Feller and Markovian semigroup generated by $-\psi(D)$. (We may think that $\psi(\xi) = q(x_0,\xi)$ for some fixed $x_0 \in \mathbb{R}^n$.) In this case it follows for all $u \geq 0$ that $0 \leq T_t u \leq \|T_t\|_{L^1 - L^\infty}\|u\|_{L^1}$. We assume that $q(x,D)$ as well as $\psi(D)$ have both the L^2-domain $H^{\psi,2}(\mathbb{R}^n)$. In this case we find for all $u \in H^{\psi,2}(\mathbb{R}^n)$

$$T_t u - S_t^{\psi} u = \int_0^t T_s \circ (-q(x,D) + \psi(D)) \circ S_{t-s}^{\psi} u \, ds. \tag{73}$$

Under suitable assumptions (see also below) it was proved in [17] that

$$|T_t u(x) - S_t^{\psi} u(x)| \leq d_1 t \|(1+\psi)\hat{u}\|_{L^1} + d_2 t^2 \|(1+\psi)^2 \hat{u}\|_{L^1} \tag{74}$$

holds for all $u \in C_0^\infty(\mathbb{R}^n)$, $t > 0$. It is easy to derive from these estimates

$$|T_t u(x) - S_t^{\psi} u(x)| \leq \tilde{d}_1 t \|u\|_{n+3} + \tilde{d}_2 t^2 \|u\|_{n+5}. \tag{75}$$

We now add the assumptions that $d_\psi(\xi,\eta) = \psi^{\frac{1}{2}}(\xi - \eta)$ is a metric on $\mathbb{R}^n$ generating the Euclidean topology, compare with (56) and (57), and having the doubling property, we refer to [15] for a discussion of the doubling property and many examples. Furthermore, we assume that

$$\begin{aligned} p_t^{\psi}(x-y) &= (2\pi)^{-n} \int_{\mathbb{R}^n} e^{i(x-y)\cdot\xi} e^{-t\psi(\xi)} d\xi \\ &= p_t^{\psi}(0) e^{-\delta^2_{\psi,t}(x,y)} && (76) \\ &= (2\pi)^{-n} t \mathcal{L}(V_\psi^*)(t) e^{-\delta^2_{\psi,t}(x,y)}, && (77) \end{aligned}$$

recall that the existence of $\delta_{\psi,t}$ is the crucial point of this assumption. Our aim is to combine (74) and (77) to study $(T_t)_{t\geq 0}$, more precisely we want to understand T_t for small $t > 0$ in a neighbourhood of a point $x \in \mathbb{R}^n$. For this we pick two bounded measurable sets $G_1 \subset G_2$ and a function $\varphi \in C_0^\infty(\mathbb{R}^n)$ such that for all $x \in \mathbb{R}^n$ we have

$$\chi_{G_1}(x) \leq \varphi(x) \leq \chi_{G_2}(x), \tag{78}$$

where χ_G is the characteristic function of the set G. Since we will fix φ and are interested in $t > 0$ to be small we may replace (74) by

$$|T_t\varphi(x) - S_t^\psi \varphi(x)| \leq \kappa_0(\varphi)t \quad \text{as } t \to 0, \tag{79}$$

or

$$-\kappa_0(\varphi)t + S_t^\psi \varphi(x) \leq T_t\varphi(x) \leq S_t^\psi \varphi(x) + \kappa_0(\varphi)t \tag{80}$$

and the monotonicity of S_t^ψ yields

$$-\kappa_0(\varphi)t + \left(S_t^\psi \chi_{G_1}\right)(x) \leq (T_t\varphi)(x) \leq (S_t^\psi \chi_{G_2})(x) + \kappa_0(\varphi)t. \tag{81}$$

Since we know the estimate $0 \leq T_t\varphi \leq \|T_t\|_{L^1 - L^\infty} \|\varphi\|_{L^1}$, in (80) as well as in the following estimate we can replace the left hand side by $\max(0, -\kappa_0(\varphi)t + S_t^\psi \varphi(x))$ and the right hand side by $\min(\|T_t\|_{L^1 - L^\infty}\|\varphi\|_{L^1}, \kappa_0(\varphi)t + (S_t^\psi \varphi)(x))$. However, for any set $G \subset \mathbb{R}^n$ we have by our assumptions

$$\begin{aligned}(S_t^\psi \chi_G)(x) &= (2\pi)^{-n} t\mathcal{L}(V_\psi^*)(t) \int_{\mathbb{R}^n} \chi_G(y) e^{-\delta_{\psi,t}(x,y)} \mathrm{d}y \\ &= (2\pi)^{-n} t\mathcal{L}(V_\psi^*)(t) \int_G e^{-\delta^2_{\psi,t}(x,y)} \mathrm{d}y\end{aligned}$$

which yields

$$(\inf_{y\in G} e^{-\delta^2_{\psi,t}(x,y)})\lambda^{(n)}(G) \leq \int_G e^{-\delta^2_{\psi,t}(x,y)} \mathrm{d}y \leq (\sup_{y\in G} e^{-\delta^2_{\psi,t}(x,y)})\lambda^{(n)}(G).$$

Now taking the doubling property of d_ψ into account and apply (61) we arrive with suitable constants γ_0 and γ_1 at

$$\begin{aligned}-\kappa_0(\varphi)t + \gamma_0 V_\psi\left(\frac{1}{\sqrt{t}}\right) \inf_{y\in G_1} e^{-\delta^2_{\psi,t}(x,y)} \lambda^{(n)}(G_1) &\leq T_t\varphi(x) \\ \leq \kappa_0(\varphi)t + \gamma_1 V_\psi\left(\frac{1}{\sqrt{t}}\right) \sup_{y\in G_2} e^{-\delta^2_{\psi,t}(x,y)} \lambda^{(n)}(G_2)\end{aligned} \tag{82}$$

as $t \to 0$. In the case where the inf and the sup are attained, say at y_{min} and y_{max} respectively we obtain

$$-\kappa_0(\varphi)t+\gamma_0 V_\psi\left(\frac{1}{\sqrt{t}}\right)e^{-\delta^2_{\psi,t}(x,y_{min})}\lambda^{(n)}(G_1)\leq T_t\varphi(x) \tag{83}$$

$$\leq \kappa_0(\varphi)t+\gamma_1 V_\psi\left(\frac{1}{\sqrt{t}}\right)e^{-\delta^2_{\psi,t}(x,y_{max})}\lambda^{(n)}(G_2)$$

as $t \to 0$.We remind the reader that in (83) we can again use $0 \leq T_t\varphi(x) \leq \|T_t\|_{L^1-L^\infty}\|\varphi\|_{L^1}$ to improve the estimates as before. Of course, we can now use properties of $\delta_{\psi,t}$ and properties of the sets G_1 and G_2 to derive more special results. We want to end these considerations with a few remarks to the type of smallness condition imposed in [12] or [17]. Let $q(x,\xi) = q(x_0,\xi)+q(x,\xi)-q(x_0,\xi)$ be a decomposition of the negative definite symbol $q(x,\xi)$ and for simplicity set $q(x_0,\xi) = q_1(\xi)$ and write $q(x,\xi) = q_1(\xi)+q_2(x,\xi)$. Further let $\psi(\xi)$ be a fixed continuous negative definite function, it could be $q_1(\xi)$. Standard assumptions are

$$\eta_1\psi(\xi)\leq q_1(\xi)\leq \eta_2\psi(\xi), \quad |\xi|\geq \rho_0\geq 0, \tag{84}$$

as well as

$$|\partial_x^\alpha q_2(x,\xi)|\leq \varphi_\alpha(x)(1+\psi(\xi)) \tag{85}$$

for some $m \in \mathbb{N}$, $|\alpha| \leq m$, and $\varphi_\alpha \in L^1(\mathbb{R}^n)$. The smallness condition now becomes

$$\kappa\sum_{|\alpha|\leq m}\|\varphi_\alpha\|_{L^1}<\eta_1 \tag{86}$$

where κ is a constant controlling $\|q_1(D)u\|_{\psi,s}$ from below, the most detailed account is given in [17]. Thus under the condition of that type we may derive (74). As mentioned before, using stopping time techniques in the martingale problem, we can free ourselves from the smallness condition and still can identify $-q(x,D)$ as a generator of a Feller semigroup. If we do this, we may reverse the argument, i.e., we may consider now the process (or semigroup) obtained with the help of the martingale problem in a (small) neighbourhood of a point and then we may try to get estimates such as (74) which is often possible.

For pseudo-differential operators having a more special structure, estimates such as (75) may be derived much easier provided that we know already the operators involved generate a Feller semigroup. For continuous negative definite functions $\psi_k : \mathbb{R}^n \to \mathbb{R}$, $1 \leq k \leq N$, and bounded continuous functions $a_k \in C_b(\mathbb{R}^n)$ assume that

$$-q(x,D) = -\sum_{k=1}^N a_k(x)\psi_k(D), \quad a_k\geq 0, \tag{87}$$

generates a Feller semigroup $(T_t)_{t\geq 0}$ and for some $x_0 \in \mathbb{R}^n$ let $\psi(\xi) = q(x_0, \xi) = \sum_{k=1}^{N} a_k(x_0)\psi_k(\xi)$. Note that for $\psi_k(\xi) = |\xi_k|^{r_k}$, $1 \leq k \leq n$, hence for $q(x, \xi) = \sum_{k=1}^{n} |a_k(x)^{\frac{1}{r_k}} \xi_k|^{r_k}$, Corollary 4.1 in Schilling and Schnurr [25] yields that the corresponding process is the unique solution of a stochastic differential equation driven by the r_k-stable process. Since ψ is a continuous negative definite function, $-\psi(D)$ extends also to a generator of a Feller semigroup $(S_t^{\psi})_{t\geq 0}$. We assume that $C_0^{\infty}(\mathbb{R}^n)$ is a subset of $D(q(x, D)) \cap D(\psi(D))$ and both operators are considered on $C_{\infty}(\mathbb{R}^n)$ as a generator of a Feller semigroup. Then it follows that for $u \in C_0^{\infty}(\mathbb{R}^n)$ that

$$
\begin{aligned}
|T_t u(x) - S_t^{\psi} u(x)| &= \left| \int_0^t T_s \circ (-q(x, D) + q(x_0, D)) \circ S_{t-s}^{\psi} u(x)\, \mathrm{d}s \right| \\
&\leq \int_0^t \|T_s\|_{L^\infty - L^\infty} \|(-q(x, D) + q(x_0, D)) S_{t-s}^{\psi} u\|_\infty \mathrm{d}s \\
&\leq \int_0^t \|(-q(x, D) + q(x_0, D)) S_{t-s}^{\psi} u\|_\infty \mathrm{d}s \\
&= \int_0^t \|\sum_{k=1}^{N} ((a_k(x) - a_k(x_0))\psi_k(D)) S_{t-s}^{\psi} u\|_\infty \mathrm{d}s \\
&\leq \left(\max_{1\leq k\leq N} \sup_{x\in\mathbb{R}^n} |a_k(x) - a_k(x_0)| \right) \sum_{k=1}^{N} \int_0^t \|\psi_k(D) S_{t-s}^{\psi} u\|_\infty \mathrm{d}s \\
&= \left(\max_{1\leq k\leq N} \sup_{x\in\mathbb{R}^n} |a_k(x) - a_k(x_0)| \right) \sum_{k=1}^{N} \int_0^t \|S_{t-s}^{\psi} (\psi_k(D) u)\|_\infty \mathrm{d}s \\
&\leq t \left(\max_{1\leq k\leq N} \sup_{x\in\mathbb{R}^n} |a_k(x) - a_k(x_0)| \right) \sum_{k=1}^{N} \|\psi_k(D) u\|_\infty \\
&\leq t \left(\max_{1\leq k\leq N} \sup_{x\in\mathbb{R}^n} |a_k(x) - a_k(x_0)| \right) \sum_{k=1}^{N} c_k \|u\|_{2,\infty}.
\end{aligned}
$$

If we define for $a : \mathbb{R}^n \to \mathbb{R}$ its absolute oscillations as

$$
\operatorname{osc} a := \sup_{x,y\in\mathbb{R}^n} |a(x) - a(y)| \tag{88}
$$

we find

$$
|T_t u(x) - S_t^{\psi} u(x)| \leq t \left(\max_{1\leq k\leq N} \operatorname{osc} a_k \right) \sum_{k=1}^{N} c_k \|u\|_{2,\infty}. \tag{89}
$$

Let us note that as S_t^{ψ}, u and of course $t > 0$ are independent of $\max_{1\leq k\leq N} \operatorname{osc} a_k$ and therefore fixing x_0, t and ψ we can determine all semigroups $(T_t)_{t\geq 0}$ constructed as in [12] for which

$$\max_{1\leq k\leq N} \operatorname{osc} a_k \leq \frac{1}{2}t\sum_{k=1}^{N} c_k\|\varphi\|_{2,\infty} \tag{90}$$

implies

$$\frac{1}{2}S_t^{\psi}\varphi(x) \leq T_t\varphi(x) < \frac{3}{2}S_t^{\psi}\varphi(x), \tag{91}$$

and we have not used the estimate $0 \leq T_t\varphi \leq \|T_t\|_{L^1-L^\infty}\|\varphi\|_{L^1}$. Note that (91) can be seen as a first step to obtain Aronson-type estimates. Clearly a continuity argument will allow us to obtain (91) for all $\tilde{t}$ in a small interval around t.

4 Two Detailed Examples and a Problem of General Nature

Let us consider the two operators $L(D) := \psi_{\frac{1}{2}}(D)$ with symbol $L(\xi) = \psi_{\frac{1}{2}}(\xi) = \sum_{k=1}^{n}|\xi_k|$, and $\Lambda(D) := \tilde{\psi}_{\frac{1}{2}}(D)$ with symbol $\Lambda(\xi) = |\xi| = \left(\sum_{k=1}^{n}|\xi_k|^2\right)^{\frac{1}{2}}$. We also introduce the operators $L(x, D)$ and $\Lambda(x, D)$ with "variable coefficients" by

$$L(x, D)u(x) = (2\pi)^{-\frac{n}{2}}\int_{\mathbb{R}^n} e^{ix\cdot\xi}\left(\sum_{k=1}^{n} a_k(x)|\xi_k|\right)\hat{u}(\xi)\,\mathrm{d}\xi \tag{92}$$

and

$$\Lambda(x, D)u(x) = (2\pi)^{-\frac{n}{2}}\int_{\mathbb{R}^n} e^{ix\cdot\xi}\left(\sum_{k=1}^{n} b_k(x)\xi_k^2\right)^{\frac{1}{2}}\hat{u}(\xi)\,\mathrm{d}\xi \tag{93}$$

where we assume (at least) $a_k, b_k \in C_b(\mathbb{R}^n)$, $1 \leq k \leq n$, and

$$0 < \nu_0 \leq a_k(x) \leq \nu_1 \tag{94}$$

and

$$0 < \mu_0 \leq b_k(x) \leq \mu_1 \tag{95}$$

for all $x \in \mathbb{R}^n$. We add the assumption that $-L(x, D)$ as well as $-\Lambda(x, D)$ have extensions from $\mathcal{S}(\mathbb{R}^n)$ to generators of Feller semigroups $(T_t^L)_{t\geq 0}$ and $(T_t^\Lambda)_{t\geq 0}$.

Later we may add further assumptions. We refer to the results in [11, 14] or [3] where sufficient criteria on $L(x,\xi)$ and $\Lambda(x,\xi)$ are given to satisfy the assumptions. We will also consider the operators $L(x_0, D)$ and $\Lambda(x_0, D)$ where we freeze the "coefficient" at $x_0 \in \mathbb{R}^n$, i.e.,

$$L(x_0,\xi) = \sum_{k=1}^{n} a_k(x_0)|\xi_k|. \tag{96}$$

and

$$\Lambda(x_0,\xi) = \left(\sum_{k=1}^{n} b_k(x_0)\xi_k^2\right)^{\frac{1}{2}}, \tag{97}$$

with corresponding operators $L(x_0, D)$ and $\Lambda(x_0, D)$, respectively. We want to study $p_t^{L(x_0,D)}(x-y)$ and $p_t^{\Lambda(x_0,D)}(x-y)$.
The change of variable $\eta_k = a_k(x_0)\xi_k$ yields

$$\begin{aligned}
p_t^{L(x_0,\cdot)}(x-y) &= (2\pi)^{-n}\int_{\mathbb{R}^n} e^{i(x-y)\cdot\xi} e^{-t\sum_{k=1}^{n} a_k(x_0)|\xi_k|}\mathrm{d}\xi \\
&= (2\pi)^{-n}\prod_{k=1}^{n}\int_{\mathbb{R}^n} e^{i(x_k-y_k)\cdot\xi_k} e^{-ta_k(x_0)|\xi_k|}\mathrm{d}\xi_k \\
&= \prod_{k=1}^{n}\frac{1}{2\pi a_k(x_0)}\int_{\mathbb{R}^n} e^{i\left(\frac{x_k-y_k}{a_k(x_0)}\right)\eta_k} e^{-|\eta_k|t}\mathrm{d}\eta_k \\
&= \prod_{k=1}^{n}\frac{1}{\pi a_k(x_0)}\frac{a_k^2(x_0)t}{a_k^2(x_0)t^2+|x_k-y_k|^2} \\
&= p_t^{L(x_0,\cdot)}(0)\prod_{k=1}^{n}\frac{a_k^2(x_0)t^2}{a_k^2(x_0)t^2+|x_k-y_k|^2} \\
&= p_t^{L(x_0,\cdot)}(0)e^{-\sum_{k=1}^{n}\ln\left(\frac{|x_k-y_k|^2+a_k^2(x_0)t^2}{a_k^2(x_0)t^2}\right)},
\end{aligned}$$

and with

$$\delta_{L(x_0,\cdot),t}(x,y) = \left(\sum_{k=1}^{n}\ln\left(\frac{|x_k-y_k|^2+a_k^2(x_0)t^2}{a_k^2(x_0)t^2}\right)\right)^{\frac{1}{2}} \tag{98}$$

we find

$$p_t^{L(x_0,\cdot)}(x-y) = p_t^{L(x_0,\cdot)}(0)e^{-\delta^2_{L(x_0,\cdot),t}(x,y)} \tag{99}$$

and for $p_t^{L(x_0,\cdot)}(0)$ we have the expression

$$p_t^{L(x_0,\cdot)}(0) = \frac{1}{t^n} \prod_{k=1}^{n} \frac{1}{\pi a_k(x_0)}. \tag{100}$$

A similar calculation using the change of variable $\eta_k = \sqrt{b_k(x_0)}\xi_k$ gives for $z_k = \frac{(x_k - y_k)}{\sqrt{b_k(x_0)}}$

$$\begin{aligned}
p_t^{\Lambda(x_0,\cdot)}(x-y) &= (2\pi)^{-n} \int_{\mathbb{R}^n} e^{i(x-y)\cdot\xi} e^{-t\left(\sum_{k=1}^n b_k(x_0)\xi_k^2\right)^{\frac{1}{2}}} \mathrm{d}\xi \\
&= (2\pi)^{-n} \frac{1}{\left(\prod_{k=1}^n b_k(x_0)\right)^{\frac{1}{2}}} \int_{\mathbb{R}^n} e^{iz\cdot\eta} e^{-t|\eta|} \mathrm{d}\eta \\
&= \frac{\pi^{\frac{-n+1}{2}} \Gamma\left(\frac{n+1}{2}\right)}{\left(\prod_{k=1}^n b_k(x_0)\right)^{\frac{1}{2}}} \frac{t}{(t^2 + |z|^2)^{\frac{n+1}{2}}} \\
&= p_t^{\Lambda(x_0,\cdot)}(0) \left(\frac{t^2}{t^2 + |z|^2}\right)^{\frac{n+1}{2}} \\
&= p_t^{\Lambda(x_0,\cdot)}(0) e^{-\frac{n+1}{2} \ln\left(\frac{|z|^2 + t^2}{t^2}\right)}
\end{aligned}$$

and with

$$\delta_{\Lambda(x_0,\cdot),t}(x,y) = \sqrt{\frac{n+1}{2}} \left(\ln \left(\frac{\sum_{k=1}^n \frac{|x_k - y_k|^2}{b_k(x_0)} + t^2}{t^2} \right) \right)^{\frac{1}{2}} \tag{101}$$

we arrive at

$$p_t^{\Lambda(x_0,\cdot)}(x-y) = p_t^{\Lambda(x_0,\cdot)}(0) e^{-\delta^2_{\Lambda(x_0,\cdot),t}(x,y)} \tag{102}$$

where

$$p_t^{\Lambda(x_0,\cdot)}(0) = \frac{\pi^{\frac{-n+1}{2}} \Gamma\left(\frac{n+1}{2}\right)}{t^n \left(\prod_{k=1}^n b_k(x_0)\right)^{\frac{1}{2}}}. \tag{103}$$

We note that in (99) and (102) we can estimate $p_t^{L(x_0,\cdot)}(0)$ from above and from below by $V_{L(x_0,\cdot)}\left(\frac{1}{\sqrt{t}}\right)$ and $p_t^{L(x_0,\cdot)}(0)$ by $V_{\Lambda(x_0,\cdot)}\left(\frac{1}{\sqrt{t}}\right)$ and $V_{L(x_0,\cdot)}$ is determined by $d^{L(x_0,\cdot)}(\xi,\eta) = (L(x_0, \xi - \eta))^{\frac{1}{2}}$ whereas $V_{\Lambda(x_0,\cdot)}$ is determined by

$d^{\Lambda(x_0,\cdot)}(\xi,\eta) = (\Lambda(x_0,\xi-\eta))^{\frac{1}{2}}$. Moreover, using (94) and (95), respectively, we obtain with $p_t^{\psi_{\frac{1}{2}},n}$ as in (68) or (71) and with $p_t^{C,n}$ as in (64) that

$$\left(\frac{\nu_0}{\nu_1}\right)^n p_{\nu_0 t}^{\psi_{\frac{1}{2}},n}(x,y) \le p_t^{L(x_0,\cdot)}(x,y) \le \left(\frac{\nu_1}{\nu_0}\right)^n p_{\nu_1 t}^{\psi_{\frac{1}{2}},n}(x,y) \tag{104}$$

and

$$\left(\frac{\mu_0}{\mu_1}\right)^{\frac{n}{2}} p_{\sqrt{\mu_0}t}^{C,n}(x,y) \le p_t^{\Lambda(x_0,\cdot)}(x,y) \le \left(\frac{\mu_1}{\mu_0}\right)^{\frac{n}{2}} p_{\sqrt{\mu_1}t}^{C,n}(x,y) \tag{105}$$

where we used

$$\sum_{k=1}^n \ln\left(\frac{|x_k-y_k|^2+\nu_1^2 t^2}{\nu_1^2 t^2}\right) \le \delta^2_{L(x_0,\cdot),t}(x,y) \le \sum_{k=1}^n \ln\left(\frac{|x_k-y_k|^2+\nu_0^2 t^2}{\nu_0^2 t^2}\right) \tag{106}$$

and

$$\sqrt{\frac{n+1}{2}}\ln\left(\frac{\sum_{k=1}^n \frac{|x_k-y_k|^2}{\mu_1}+t^2}{t^2}\right) \le \delta^2_{\Lambda(x_0,\cdot),t}(x,y) \le \sqrt{\frac{n+1}{2}}\ln\left(\frac{\sum_{k=1}^n \frac{|x_k-y_k|^2}{\mu_0}+t^2}{t^2}\right). \tag{107}$$

We want to discuss our findings under a type of geometric idea. Our starting point shall be the two negative definite symbols $L(x,\xi)$ and $\Lambda(x,\xi)$ as in (92) and (93) subject to (94) and (95), respectively. Both symbols are comparable, i.e.,

$$L(x,\xi) \asymp \Lambda(x,\xi), \tag{108}$$

and in fact we have

$$L(x,\xi) \asymp \Lambda(x,\xi) \asymp |\xi|, \quad x\in\mathbb{R}^n, \xi\in\mathbb{R}^n. \tag{109}$$

This yields of course that the metrics $d^{L(x_0,\cdot)}$ and $d^{\Lambda(x_0,\cdot)}$ are for every $x_0\in\mathbb{R}^n$ comparable as they are comparable with

$$d(\xi,\eta) := |\xi-\eta|^{\frac{1}{2}} = \left(\sum_{k=1}^n |\xi_k-\eta_k|^2\right)^{\frac{1}{4}}, \tag{110}$$

which is the metric corresponding to $\tilde{\psi}_{\frac{1}{2}}(\xi) = |\xi| = (\xi_1^2+\cdots+\xi_n^2)^{\frac{1}{2}}$. Switching to the densities $p_t^{L(x_0,\cdot)}$ and $p_t^{\Lambda(x_0,\cdot)}$, this implies that in fact for all $x_1,x_2\in\mathbb{R}^n$

the terms $p_t^{L(x_0,\cdot)}(0)$ and $p_t^{\Lambda(x_0,\cdot)}(0)$ are comparable as they are comparable with $p_t^{C,n}(0) \asymp t^{-n}$. However, the two metrics $\delta_{L(x_0,\cdot),t}$ and $\delta_{\Lambda(x_0,\cdot),t}$ are not comparable, hence $p_t^{L(x_0,\cdot)}(x-y)$ and $p_t^{\Lambda(x_0,\cdot)}(x-y)$ are not comparable either. With each symbol $L(x,\xi)$ and $\Lambda(x,\xi)$ we can associate two families of metrics:

$$d_{L(x_0,\cdot)}(\xi,\eta) = (L(x_0,\xi-\eta))^{\frac{1}{2}} = \left(\sum_{k=1}^{n} a_k(x_0)|\xi_k-\eta_k|\right)^{\frac{1}{2}} \tag{111}$$

and

$$\delta_{L(x_0,\cdot)}(x,y) = \left(\sum_{k=1}^{n} \ln\left(\frac{|x_k-y_k|^2 + a_k(x_0)t^2}{a_k^2(x_0)t^2}\right)\right)^{\frac{1}{2}}, \tag{112}$$

as well as

$$d_{\Lambda(x_0,\cdot),t}(x,y) = (\Lambda(x_0,\xi-\eta))^{\frac{1}{2}} = \left(\sum_{k=1}^{n} b_k(x_0)|\xi_k-\eta_k|^2\right)^{\frac{1}{4}} \tag{113}$$

and

$$\delta_{\Lambda(x_0,\cdot),t}(x,y) = \sqrt{\frac{n+1}{2}}\left(\ln\left(\frac{\sum_{k=1}^{n}\frac{|x_k-y_k|^2}{b_k(x_0)} + t^2}{t^2}\right)\right)^{\frac{1}{2}}. \tag{114}$$

The metrics $d_{L(x_0,\cdot)}$, $d_{\Lambda(x_1,\cdot)}$ and d are all comparable, whereas for $\delta_{L(x_0,\cdot),t}(x,y)$ we have the estimate (104) and for $\delta_{\Lambda(x_0,\cdot),t}(x,y)$ we have the estimates (105).

Now let us assume that $-L(x,D)$ and $-\Lambda(x,D)$ generate Feller and Markov semigroups $(T_t^L)_{t\geq 0}$ and $(T_t^\Lambda)_{t\geq 0}$ each having densities $p_t^L(x,y)$ and $p_t^\Lambda(x,y)$, respectively, i.e., we have

$$(T_t^L u)(x) = \int_{\mathbb{R}^n} p_t^L(x,y)u(y)\,\mathrm{d}y \tag{115}$$

and

$$(T_t^\Lambda u)(x) = \int_{\mathbb{R}^n} p_t^\Lambda(x,y)u(y)\,\mathrm{d}y. \tag{116}$$

In each case we may try to compare $T_t^L u$ with $S_t^{L(x_0,\cdot)}$ and $T_t^\Lambda u$ with $S_t^{\Lambda(x_0,\cdot)}$, $t>0$ and $x_0 \in \mathbb{R}^n$ fixed. Here $(S_t^{L(x_0,\cdot)})_{t\geq 0}$ denotes the semigroup with generator $-L(x_0,D)$ and $(S_t^{\Lambda(x_0,\cdot)})_{t\geq 0}$ with generator $-\Lambda(x_0,D)$. Depending on assumptions on the variable coefficients a_k and b_k, $1 \leq k \leq n$, respectively, such a comparison is possible and we may apply results from Sect. 3, such as (82). However, we refer to the remark towards the end of this section. We may combine now these estimates and we obtain for example

$$\begin{aligned} - w_0(\varphi,t)+&\tilde{\gamma}_0 V_{\frac{1}{2},n}\left(\frac{1}{\sqrt{t}}\right)\inf_{y\in G_1} e^{-\delta^2_{\psi_{\frac{1}{2},n,t}}}(x,y)\lambda^{(n)}(G_1) \\ &\leq \int_{\mathbb{R}^n} p_t^L(x,y)\varphi(y)\,\mathrm{d}y = (T_t^L\varphi)(x) \\ &\leq w_0(\varphi,t)+\tilde{\gamma}_1 V_{\psi_{\frac{1}{2},n}}\left(\frac{1}{\sqrt{t}}\right)\sup_{y\in G_2} e^{-\delta^2_{\psi_{\frac{1}{2},n,t}}(x,y)}\lambda^{(n)}(G_2), \end{aligned} \tag{117}$$

with bounded measurable sets $G_1 \subset G_2$ such that $\chi_{G_1} \leq \varphi \leq \chi_{G_2}$ and $\lim_{t\to 0} w_0(\varphi,t) = 0$. An analogous result holds for $(T_t^{\Lambda})_{t\geq 0}$ and we may also use (104) or (105), respectively, to get further estimates. For the operator $L(x,D)$ we may also apply (91).

However, some caution is needed. While for second order uniformly elliptic differential operators with smooth, say C^K-coefficients, existence results for corresponding semigroups are by standard methods available as are perturbations and comparison results, in our case this is different. While there are several sets of conditions on the coefficients a_k or b_k to guarantee estimates such as (75), a simple unified approach is not yet at hand. In particular, this holds also for the more general examples in the next section, however, we want to mention already here that we can transfer the considerations of this section to symbols of the type

$$q(\xi) = \sum_{j=1}^{N}(\xi_{j1}^2+\cdots+\xi_{jm_j}^2)^{\frac{1}{2}} + \sum_{l=1}^{M}|\xi_l| \tag{118}$$

and consequently to

$$q(x,\xi) = \sum_{j=1}^{N} c_j(x)\left(\sum_{k=1}^{m_j} b_{jk}(x)\xi_{jk}^2\right)^{\frac{1}{2}} + \sum_{l=1}^{M} a_l(x)|\xi_l| \tag{119}$$

with bounded continuous non-negative functions being uniformly bounded away from 0.

These two examples treated here in detail together with the ones in the next section suggest a more fundamental question: given a negative definite symbol $q(x,\xi)$ such that for every $x \in \mathbb{R}^n$ by $d_{q(x,\cdot)}(\xi,\eta) := (q(x,\xi-\eta))^{\frac{1}{2}}$ a volume doubling metric is given on $\mathbb{R}^n$ generating the Euclidean topology and such that for each $x_0 \in \mathbb{R}^n$ fixed a metric $\delta_{q(x_0,\cdot),t}$ on $\mathbb{R}^n$ is given such that we have

$$p_t^{q(x_0,\cdot)}(x-y) = p_t^{q(x_0,\cdot)(0)} e^{-\delta^2_{q(x_0,\cdot),t}(x,y)}. \tag{120}$$

What geometric structures do the combined sections $x \mapsto d_{q(x,\cdot)}$ and $x \mapsto \delta_{q(x,\cdot),t}$ implement on $\mathbb{R}^n$? Note that $d_{q(x,\cdot)}$ acts on the co-variables ξ and η while $\delta_{q(x,\cdot),t}$ acts on the variable x.

5 More Examples and Observations

While we know that the metric $d_\psi(\xi, \eta) = \psi^{\frac{1}{2}}(\xi - \eta)$ exists whenever $\psi(0) = 0$ if and only if $\xi = 0$, and we have also a criterion for the doubling property to hold, the situation is different for $\delta_{\psi,t}$. Of course, the Gaussian case, i.e., $\psi(\xi) = |\xi|^2$, $\xi \in \mathbb{R}^n$, is well understood and the case $\psi(\xi) = |\xi|$ as well as $\psi(\xi) = \sum_{k=1}^n |\xi_k|$ is discussed in Sect. 4. A further example was given in [15], e.g. the continuous negative definite function $\psi(\xi) = \ln(\cosh \xi)$, $\xi \in \mathbb{R}$, with

$$\delta_{\psi,t}(x, y) = -\ln \left| \frac{\Gamma\left(\frac{t+i(x-y)}{2}\right)}{\Gamma\left(\frac{t}{2}\right)} \right| = \left(\sum_{j=1}^{\infty} \ln\left(1 + \frac{(x-y)^2}{(t+2j)^2}\right)\right)^{\frac{1}{2}}. \tag{121}$$

However, the general problem, when $\delta_{\psi,t}$ does exist is still open. So far we know only one general result, see below, but this result we can use to construct a lot of interesting examples. The following theorem is taken from [15, Theorem 7.1]:

Theorem 5.1 *If for the transition densities $(p_t(\cdot))_{t>0}$ of a Lévy process in $\mathbb{R}$ we have*

$$F^{-1}(p_t)(\xi) = e^{-tf(|\xi|)},\ \xi \in \mathbb{R}, \tag{122}$$

with a Bernstein function f such that $f(0) = 0$ and

$$C_t := \int_0^\infty e^{-tf(s)} \mathrm{d}s < \infty \textit{ for all } t > 0, \tag{123}$$

then there exists for each $t > 0$ a complete Bernstein function g_t such that

$$\frac{p_t(x)}{p_t(0)} = e^{-g_t(|x|^2)}. \tag{124}$$

Denote the continuous negative definite function associated with $(p_t)_{t\geq 0}$ by $\psi : \mathbb{R} \to \mathbb{R}$ we find

$$\delta_{\psi,t}(x, y) = g_t^{\frac{1}{2}}(|x-y|^2). \tag{125}$$

Since $s \mapsto s^{\frac{1}{2}}$ is a Bernstein function and $g_t^{\frac{1}{2}}$ is a Bernstein function and since $g_t(0) = 0$ only if $x = 0$ (by (124)), it follows indeed that $\delta_{\psi,t}$ is a metric on $\mathbb{R}$. As already remarked in [15] we have

Corollary 5.1 *The result of Theorem 5.1 holds for $\mathbb{R}^2$ and $\mathbb{R}^3$ too.*

Thus in light of Theorem 5.1 we need to find Bernstein functions f such that $f(0) = 0$ and

$$C_t = \int_0^\infty e^{-tf(s)} \mathrm{d}s < \infty.$$

Thus, if f is defined on $[0, \infty)$ as a continuous function with $f(0) = 0$, we need to verify that for every $t > 0$

$$\int_R^\infty e^{-tf(s)} \mathrm{d}s < \infty \tag{126}$$

for some $R > 0$. The condition $f(s) \geq cs^\epsilon$, $c > 0$, $\epsilon > 0$, yields

$$\int_R^\infty e^{-tf(s)} \mathrm{d}s \leq \int_R^\infty e^{-cts^\epsilon} \mathrm{d}s = \frac{1}{(ct)^{\frac{1}{\epsilon}}} \int_{(ct)^{\frac{1}{\epsilon}} R}^\infty e^{-r^\epsilon} \mathrm{d}r < \infty$$

which gives us now plenty of examples such as

$$|\xi|^\alpha, 0 < \alpha \leq 1; \quad |\xi|^{\frac{1}{2}} \arctan |\xi|^{\frac{1}{2}}; \quad \frac{|\xi|}{(|\xi| + a)^\alpha}, 0 < \alpha < 1; a \geq 0;$$

or

$$|\xi|^{\frac{1}{2}} (1 \pm e^{-2a|\xi|^{\frac{1}{2}}}), a > 0; \quad \frac{|\xi|(1 - e^{-2(|\xi|+a)^{\frac{1}{2}}})}{\sqrt{|\xi| + a}}, a > 0,$$

where for $\xi \in \mathbb{R}^n, n = 1, 2, 3$, we write $|\xi|$ for the Euclidean norm. We can also take sums

$$\psi(\xi) = \sum_{k=1}^N \psi_k(\tilde{\xi}_k), \tag{127}$$

where $\psi_k : \mathbb{R}^{\tilde{n}_k} \to \mathbb{R}$, $\tilde{n}_k = 1, 2, 3$. Applying Corollary 5.1 it follows in this case that

$$\delta_{\psi,t}(x, y) = \left(\sum_{k=1}^N \delta^2_{\psi_k,t}(\tilde{x}_k, \tilde{y}_k) \right)^{\frac{1}{2}}. \tag{128}$$

In particular, we may consider for $\xi_k \in \mathbb{R}$ and $0 < r_1 \leq \cdot \leq r_n < \frac{1}{2}$

$$\psi_{\vec{r}}(\xi) = \sum_{k=1}^{n} |\xi_k|^{2r_k} \tag{129}$$

and when looking at the corresponding Dirichlet form

$$\begin{aligned} \mathcal{E}^{\psi_{\vec{r}}}(u, v) = & \int_{\mathbb{R}^n} \sum_{k=1}^{n} |\xi_k|^{2r_k} \hat{u}(\xi)\overline{\hat{v}(\xi)} \, d\xi \\ & \int_{\mathbb{R}^n} \sum_{k=1}^{n} \left(\frac{\partial}{\partial x_k}\right)^{r_k} u(x) \left(\frac{\partial}{\partial x_k}\right)^{r_k} v(x) \, dx, \end{aligned} \tag{130}$$

where we used a calculation analogous to that leading to (40). The representation (40) and (130) should be studied in more detail. Clearly, the bilinear operator

$$\begin{aligned} \Gamma_{\psi_{\vec{r}}} &: H^{\psi_{\vec{r}},1} \times H^{\psi_{\vec{r}},1} \to L^1(\mathbb{R}^n) \\ & (u, v) \mapsto \Gamma_{\psi_{\vec{r}}}(u, v) := \sum_{k=1}^{n} \left(\frac{\partial}{\partial x_k}\right)^{r_k} u \left(\frac{\partial}{\partial x_k}\right)^{r_k} v \end{aligned}$$

is not the carré du champ associated with $\mathcal{E}^{\psi_{\vec{r}}}$, however, it gives a decomposition of $\mathcal{E}^{\psi_{\vec{r}}}$ into a finite sum of squares. Using the carré du champ, in [23], P.A. Meyer could give a decomposition of non-local Dirichlet forms of the type $\mathcal{E}^{\psi}$ into an infinite sum of squares. The proof of his results reminds us much about the proof of Schoenberg's theorem about the isometric embedding of metric spaces into Hilbert spaces, we refer to our discussions in [15] or [16]. The carré du champ and the second carré du champ are for **local** Dirichlet forms very closely connected to the underlying geometry (curvature considerations etc.) and allow us to define in a general setting the definition of a dimension with the help of the Bakry-Emery curvature-dimension inequality, see [1]. In a recent paper [28], Spener, Weber and Zacher have proved that in this sense the fractional Laplacian Δ^r has infinite dimension. In light of P.A. Meyer's result, this result might be less surprising and the considerations in Sect. 4 may give a hint that for non-local operators generating a Dirchlet form the classical carré du champ is not the natural object to introduce a type of associated geometry.

References

1. Bakry, D., Gentil, J., Ledoux, M.: Analysis and Geometry of Markov Diffusion Operators. Springer, Berlin (2014)
2. Berg, C., Forst, G.: Potential Theory on Locally Compact Abelian Groups. Springer, Berlin (1975)
3. Böttcher, B., Schilling, R., Wang, J.: Lévy-Type Processes: Construction, Approximation and Sample Path Properties. Springer, Berlin (2013)
4. Bray, L., Jacob, N.: Some considerations on the structure of transition densities of symmetric Lévy processes. Commun. Stoch. Anal. **10**, 405–420 (2016)
5. Courrège, P.: Sur la forme intégro-différentielle des opérateurs de C_k^∞ dans C satisfaiscent au principe du maximum. In: Sém. Theorie du Potential 1965/66. Exposé 2, 38pp., Paris (1966)
6. Davies, E.B.: Heat Kernels and Spectral Theory. Cambridge University Press, Cambridge (1989)
7. Evans, K., Jacob, N.: On adjoint additive processes. Probab. Math. Stat. **40**, 205–223 (2020)
8. Hoh, W.: The martingale problem for a class of pseudo-differential operators. Math. Ann. **300**, 121–147 (1994)
9. Hoh, W.: Pseudo differential operators with negative definite symbols and the martingale problem. Stoch. Stoch. Rys. **55**, 225–252 (1995)
10. Hoh, W.: A symbolic calculus for pseudo-differential operators generating Feller semigroups. Osaka J. Math. **35**, 798–820 (1998)
11. Hoh, W.: Pseudo Differential Operators Generating Markov Processes. Habilitationschrift, Bielefeld (1998)
12. Jacob, N.: A class of Feller semigroups generated by pseudo-differential operators. Math. Z. **215**, 151–166 (1994)
13. Jacob, N.: Pseudo Differential Operators and Markov Processes. Vol. I: Fourier Analysis and Semigroups. Imperial College Press, London (2001)
14. Jacob, N.: Pseudo Differential Operators and Markov Processes. Vol. II: Generators and Their Potential Theory. Imperial College Press, London (2002)
15. Jacob, N., Knopova, V., Landwehr, S., Schilling, R.: A geometric interpretation of the transition density of a symmetric Lévy process. Sci. China Math. **55**, 1099–1126 (2012)
16. Jacob, N., Rhind, E.O.T.: Aspects of micro-local analysis and geometry in the study of Lévy-type generators. In: Bahns, D., Pohl, A., Witt, I. (eds.) Open Quantum Systems, pp. 77–140. Springer, Berlin (2019)
17. Jacob, N., Schilling, R.: Estimates for Feller semigroups generated by pseudo-differential operators. In: J. Rakosnik (ed.) Function Spaces, Differential Operators and Nonlinear Analysis, pp. 27–49. Prometheus Publishing House, Buffalo (1996)
18. Kilbas, A., Srivastava, H., Trujello, J.: Theory and Applications of Fractional Differential Equations. Elsevier, Amsterdam (2006)
19. Komatsu, T.: Markov processes associated with certain intergo-differential operators. Osaka J. Math. **10**, 271–303 (1973)
20. Komatsu, T.: On the martingale problem for generators of stable processes with perturbations. Osaka J. Math. **21**, 113–132 (1984)
21. Komatsu, T.: Pseudo-differential operators and Markov processes. J. Math Soc. Japan **36**, 387–418 (1984)
22. Kwasnicki, M.: Ten equivalent definitions of the fractional Laplace operators. Fract Calc. Appl. Anal. **20**, 7–51 (2017)
23. Meyer, P.-A.: Démonstrations probabiliste de certaines inéqalités de Littlewood-Paley. Exposé 2: L'operateur carré du champ. In: Séminaire de Probabilités, vol. X. Springer, Berlin (1976)
24. Schilling, R.: Subordination in the sense of Bochner and a related functional calculus. J. Austr. Math. Soc. (Ser. A) **64**, 368–396 (1998)
25. Schilling, R., Schnurr, A.: The symbol associated with the solution of a stochastic differential equation. Electron. J. Probab. **15**, 1369–1393 (2010)

26. Schilling, R.L., Song, R., Vondraček, Z.: Bernstein Functions, 2nd edn. De Gruyter Verlag, Berlin (2012)
27. Schoenberg, I.J.: Metric spaces and positive definite functions. Trans. Amer. Math. Soc. **44**, 522–536 (1938)
28. Spener, A., Weber, I., Zacher, R.: The fractional Laplacian has infinite dimension. Commun. P.D.E. **45**, 57–75 (2020)
29. Sturm, K.T.: Diffusion processes and heat kernels on metric spaces. Ann. Probab. **26**, 1–55 (1998)
30. Varopoulos, N., Saloff-Coste, L., Coulhon, T.: Analysis and Geometry on Groups. Cambridge University Press, Cambridge (1992)

On Oscillatory Behavior of Third Order Half-Linear Difference Equations

Said R. Grace

Honoring the career of Professor F. Altomare on the occasion of his Seventieth birthday.

Abstract This paper deals with the oscillatory behavior of third order half-linear difference equations. We present new oscillation criteria, which improve, extend and simplify existing ones in the literature. The results are illustrated by some examples.

Keywords Oscillation · Asymptotic behavior · Third order · Difference equations

1 Introduction

This paper is concerned with oscillatory behavior of all solutions of the half-linear third order difference equations of the form

$$\Delta(a(t)(\Delta^2 x(t))^\alpha) + q(t)x^\alpha(t-m+1) = 0. \tag{1}$$

We shall assume that:

(i) $\{q(t)\}$ and $\{a(t)\}$ are positive real sequences,
(ii) $\alpha \geq 1$ is the ratio of positive odd integers,
(iii) $m \geq 1$ is a positive integer.

Moreover, it is assumed that

$$A(t, t_0) = \sum_{s=t_0}^{t-1} a^{-\frac{1}{\alpha}}(s) \to \infty \text{ as } t \to \infty. \tag{2}$$

S. R. Grace (✉)
Department of Engineering Mathematics, Faculty of Engineering, Cairo University, Giza, Egypt

A. M. Candela et al. (eds.), *Recent Advances in Mathematical Analysis*, Trends in Mathematics, https://doi.org/10.1007/978-3-031-20021-2_18

Recall that a solution of (1) is a nontrivial real-valued sequence $\{x(t)\}$ satisfying (1) for $t \geq t_0 - m + 1$. Solutions vanishing identically in some neighborhood of infinity will be excluded from our consideration. A solution x of (1) is said to be oscillatory if it is neither eventually positive nor eventually negative; otherwise, it is called nonoscillatory. An equation itself is said to be oscillatory if all its solutions are oscillatory.

The problem of investigating oscillation criteria for various types of difference equations has been a very active research area over the past several decades. A large number of papers and monographs have been devoted to this problem; for a few examples, see [1–14] and the reference contained therein.

Most of the known results for Eq. (1) deal with the case $\alpha = 1$ and it seems that there is no reported result for the difference equations of the form of (1). Motivated by the above literature, our aim is to establish few new oscillation criterions for Eq. (1) via a comparison with the first order delay difference equations whose oscillatory characters we know as well as by using the summing criterion.

2 Main Results

In this section we study the oscillatory behavior of solutions for Eq. (1).

To obtain out result, we need the following lemma:

Lemma 2.1 *Let $\{q(t)\}$ be a sequence of positive real numbers, m a positive real number and $f : \mathbb{R} \to \mathbb{R}$ a continuous nondecreasing function, and $xf(x) > 0$ for $x \neq 0$. If the first order delay differential inequality*

$$\Delta y(t) + q(t) f(y(t - m + 1)) \leq 0$$

has an eventually positive solution, so it does the delay equation

$$\Delta y(t) + q(t) f(y(t - m + 1)) = 0.$$

This lemma is an extension of the discrete analogue of well known results. See Lemma 6.2.2 in [2]; it also can be found in [11]. The proof is immediate.

Now, we present our first main result.

Theorem 2.1 *Let condition (i)–(iii) and* (2) *hold and assume that there exists a positive real number n such that*

$$m > 2n + 1. \tag{3}$$

If the first order delay differential equations

$$\Delta W(t) + \frac{1}{\alpha} q(t) \left(\sum_{s=t_1}^{t-m} A(s, t_1) \right)^{\alpha} W(t - m + 1) = 0 \tag{4}$$

$$\Delta Y(t) + \frac{1}{\alpha} q(t) \left(\sum_{s=t-m+1}^{t+n-m} A(s + n, s) \right)^{\alpha} Y(t + 2n - m + 1) = 0 \tag{5}$$

are oscillatory for $t \geq t_1 \geq t_0$, *then every solution of Eq.* (1) *is oscillatory.*

Proof Let $x(t)$ be a nonoscillatory solution of Eq. (1), say $x(t) > 0$ and $x(t - m + 1) > 0$ for $t \geq t_1$, for some $t_1 \geq t_0$. It follows from Eq. (1) that

$$\Delta(a(t)(\Delta^2 x(t))^{\alpha}) = -q(t) x^{\alpha}(t - m + 1) \leq 0 \tag{6}$$

Hence $a(t)(\Delta^2 x(t))^{\alpha}$ is of one sign. We shall distinguish the following two cases:

(I) $\Delta x(t) > 0$ or (II) $\Delta x(t) < 0$ for $t \geq t_1$.

From Eq. (1), we see that

$$\Delta(a(t)(\Delta^2 x(t))^{\alpha}) = \Delta \left(a^{\frac{1}{\alpha}}(t) \Delta^2 x(t) \right)^{\alpha}.$$

Thus we have

$$\Delta(a(t)(\Delta^2 x(t))^{\alpha}) \geq \alpha \left(a^{\frac{1}{\alpha}}(t) \Delta^2 x(t) \right)^{\alpha - 1} \Delta \left(a^{\frac{1}{\alpha}}(t) \Delta^2 x(t) \right)$$

and so, we get

$$\Delta \left(a^{\frac{1}{\alpha}}(t) \Delta^2 x(t) \right) + \frac{1}{\alpha} \left(a^{\frac{1}{\alpha}}(t) \Delta^2 x(t) \right)^{1 - \alpha} q(t) x^{\alpha}(t - m + 1) \leq 0. \tag{7}$$

First, we consider Case (I).

$$\begin{aligned} \Delta x(t) \geq \Delta x(t) - \Delta x(t_1) &= \sum_{s=t_1}^{t-1} a^{-1/\alpha}(s) \left(a^{1/\alpha}(s) \Delta^2 x(s) \right) \\ &\geq A(t, t_1) \left(a^{\frac{1}{\alpha}}(t) \Delta^2 x(t) \right) \end{aligned}$$

and so, we get

$$\Delta x(t) \geq A(t, t_1) \left(a^{\frac{1}{\alpha}}(t) \Delta^2 x(t) \right).$$

Summing this inequality from t_1 to t, one can easily get

$$x(t) \geq \sum_{s=t_1}^{t-1} A(s, t_1) \left(a^{\frac{1}{\alpha}}(t)\Delta^2 x(t)\right),$$

or

$$x(t-m+1) \geq \left(\sum_{s=t_1}^{t-m} A(s, t_1)\right) \left(a^{\frac{1}{\alpha}}(t-m+1)\Delta^2 x(t-m+1)\right). \tag{8}$$

Since the function $a^{\frac{1}{\alpha}}(t)\Delta^2 x(t)$ is nonincreasing and $\alpha \geq 1$, we have

$$\left(a^{\frac{1}{\alpha}}(t)\Delta^2 x(t)\right)^{1-\alpha} \geq \left(a^{\frac{1}{\alpha}}(t-m+1)\Delta^2 x(t-m+1)\right)^{1-\alpha}. \tag{9}$$

Using (8), (9) in (7) we have

$$\begin{aligned}
\Delta\left(a^{\frac{1}{\alpha}}(t)\Delta^2 x(t)\right) &\leq -\frac{1}{\alpha}\left(a^{\frac{1}{\alpha}}(t-m+1)\Delta^2 x(t-m+1)\right)^{1-\alpha} q(t) x^{\alpha}(t-m+1) \\
&\leq -\frac{1}{\alpha}\left(a^{\frac{1}{\alpha}}(t-m+1)\Delta^2 x(t-m+1)\right)^{1-\alpha} q(t) \\
&\quad \times \left(\sum_{s=t_1}^{t-m} A(s, t_1)\right)^{\alpha} \left(a^{\frac{1}{\alpha}}(t-m+1)\Delta^2 x(t-m+1)\right)^{\alpha}.
\end{aligned}$$

Setting $W(t) = a^{\frac{1}{\alpha}}(t)\Delta^2 x(t)$, we have

$$\Delta W(t) + \frac{1}{\alpha}\left(\sum_{s=t_1}^{t-m} A(s, t_1)\right)^{\alpha} W(t-m+1) \leq 0.$$

It follows from Lemma 2.1 that the corresponding differential equation (4) also has a positive solution, which is a contradiction.

Next, for the Case (II).

$$-\Delta x(t) \geq \sum_{t}^{t+n-1} a^{-\frac{1}{\alpha}}(s) a^{\frac{1}{\alpha}}(s)\Delta^2 x(s) \geq A(t+n, t)\left(a^{\frac{1}{\alpha}}(t+n)\Delta^2 x(t+n)\right).$$

Summing up this inequality from t to $t+n$ we get

$$x(t) \geq \sum_{t}^{t+n} A(t+n, s) a^{\frac{1}{\alpha}}(t+2n)\Delta^2 x(t+2n)$$

or,

$$x(t-m+1) \geq \left(\sum_{s=t-m+1}^{t+n-m} A(s+n,s) \right) a^{\frac{1}{\alpha}}(t+2n-m+1)\Delta^2 x(t+2n-m+1). \quad (10)$$

Using this inequality in (7) we obtain

$$\Delta\left(a^{\frac{1}{\alpha}}(t)\Delta^2 x(t)\right) + \frac{1}{\alpha}\left(a^{\frac{1}{\alpha}}(t+2n-m+1)\Delta^2 x(t+2n-m+1)\right)^{1-\alpha} q(t)$$
$$\times \left(\sum_{s=t-m+1}^{t+n-m} A(s+n,s) \right)^{\alpha} \left(a^{\frac{1}{\alpha}}(t+2n-m+1)\Delta^2 x(t+2n-m+1)\right)^{\alpha} \leq 0.$$

Setting $Y(t) = a^{\frac{1}{\alpha}}(t)\Delta^2 x(t)$ in the above inequality, we have

$$\Delta Y(t) + \frac{1}{\alpha} q(t) \left(\sum_{s=t-m+1}^{t+n-m} A(s+n,s) \right)^{\alpha} Y(t+2n-m+1) \leq 0.$$

The rest of the proof is similar to that of Case (I) and hence is omitted. □

The following corollary is immediate.

Corollary 2.1 *Let conditions (i)–(iii) hold and assume that there exists a positive number n such that condition* (3) *holds. If*

$$\liminf_{t\to\infty} \sum_{s=t-m+1}^{t-1} q(s) \left(\sum_{u=t_1}^{s-m} A(u,t_1) \right)^{\alpha} = \infty \quad (11)$$

for $t \geq t_1 \geq t_0$ *and*

$$\liminf_{t\to\infty} \sum_{s=t+2n-m+1}^{t-1} q(s) \left(\sum_{u=s-m+1}^{s+n-m} A(+n,u) \right)^{\alpha} > \alpha \left(\frac{m-2n-1}{m-2n} \right)^{m-2n} \quad (12)$$

then Eq. (1) *is oscillatory.*

The following example is illustrative:

Example 2.1 Consider the third order equation

$$\Delta\left(\frac{1}{t^3}(\Delta^2 x(t))^3\right) + q(t)x^3(t+m-1) = 0. \quad (13)$$

Here, $\alpha = 3$, $a(t) = \frac{1}{t^3}$ and $\{q(t)\}$ is a positive sequence. We let $m \geq 1$ and assume that there exists a positive number such that $m > 2n+1$.

Now, $A(t, t_0) = \sum_{s=t_0}^{t-1} s$. If the conditions of Corollary 2.1

$$\liminf_{t\to\infty} \sum_{s=t-m+1}^{t-1} q(s) \left(\sum_{u=t_1}^{s-m} \sum_{v=t_1}^{u-1} v \right)^3 = \infty$$

and

$$\liminf_{t\to\infty} \sum_{s=t+2n-m+1}^{t-1} q(s) \left(\sum_{u=s-m+1}^{s+n-m} \sum_{v=u}^{u+n-1} v \right)^3 > 3 \left(\frac{m-2n-1}{m-2n} \right)^{m-2n}$$

are satisfied, for certain appropriate sequence $\{q(t)\}$, then Eq. (13) is oscillatory.

Next, we have the following comparison result with third order linear differential inequalities.

Theorem 2.2 *Let conditions (i)–(iii) and* (2) *hold and assume that there exists a positive number n such that* (3) *holds. If the inequality*

$$\Delta \left(a^{\frac{1}{\alpha}} \Delta^2 y(t) \right) + \frac{1}{\alpha} \left(\sum_{s=t_1}^{t-m+1} A(s, t_1) \right)^{\alpha-1} q(t) y(t-m+1) \le 0 \tag{14}$$

has no eventually positive nondecreasing solution for $t \ge t_1 \le t_0$ and the inequality

$$\Delta \left(a^{\frac{1}{\alpha}} \Delta^2 w(t) \right) + \frac{1}{\alpha} \left(\sum_{t-m+1}^{t+n-m} A(s+n, s) \right)^{\alpha-1} q(t) w(t-m+1) \le 0 \tag{15}$$

has no eventually positive nonincreasing solution, then Eq. (1) *is oscillatory,*

Proof Let $x(t)$ be a nonoscillatory solution of equation (1), say $x(t) > 0$ and $x(t-m+1) > 0$ for $t > t_1$ for some $t_1 > t_0$, Proceeding as in the proof of Theorem 2.1, we obtain the two cases (I) and (II) and the inequalities (7) and (10). We consider Case (I).

From (8) we can easily see that

$$a^{\frac{1}{\alpha}}(t) \Delta^2 x(t) \le a^{\frac{1}{\alpha}}(t-m+1) \Delta^2 x(t-m+1) \le \left(\sum_{s=t_1}^{t-m+1} A(s, t_1) \right)^{-1} x(t-m+1)$$

and

$$\left(a^{\frac{1}{\alpha}}(t) \Delta^2 x(t) \right)^{1-\alpha} \ge \left(\sum_{s=t_1}^{t-m+1} A(s, t_1) \right)^{\alpha-1} (x(t-m+1))^{1-\alpha}.$$

Using this inequality in (7) we have

$$\Delta\left(a^{\frac{1}{\alpha}}(t)\Delta^2 x(t)\right)+\frac{1}{\alpha}\left(\sum_{s=t_1}^{t-m+1} A(s,t_1)\right)^{\alpha-1}(x(t-m+1))^{1-\alpha}q(t)x^{\alpha}(t-m+1)\le 0$$

or,

$$\Delta\left(a^{\frac{1}{\alpha}}(t)\Delta^2 x(t)\right)+\frac{1}{\alpha}\left(\sum_{s=t_1}^{t-m+1} A(s,t_1)\right)^{\alpha-1} q(t)x(t-m+1)\le 0.$$

By condition (14), we arrive at the desired contradiction.

Next, we consider Case (II). From (10) we find that

$$\left(\sum_{t-m+1}^{t+n-m} A(s+n,s)\right)^{-1} x(t-m+1)\ge\left((a^{\frac{1}{\alpha}}(t+2n-m+1)\Delta^2 x(t+2n-m+1)\right).$$

Using this inequality in (7) we have

$$\Delta\left(a^{\frac{1}{\alpha}}(t)\Delta^2 x(t)\right)+\frac{1}{\alpha}\left(\sum_{t-m+1}^{t+n-m} A(s+n,s)\right)^{\alpha-1} x(t-m+1)^{1-\alpha}q(t)x^{\alpha}(t-m+1)\le 0$$

or

$$\Delta\left(a^{\frac{1}{\alpha}}(t)\Delta^2 x(t)\right)+\frac{1}{\alpha}\left(\sum_{t-m+1}^{t+n-m} A(s+n,s)\right)^{\alpha-1} q(t)x(t-m+1)\le 0.$$

By (15) we arrive at the desired contradiction. This completes the proof. □

Example 2.2 Consider the third order equation

$$\Delta\left(\frac{1}{t^3}(\Delta^2 x(t))^3\right)+q(t)x^3(t+m-1)=0. \tag{16}$$

Here, $\alpha=3$, $a(t)=\frac{1}{t^3}$ and $\{q(t)\}$ is a positive sequence. We let $m\ge 1$ and assume that there exists a positive number such that $m>2n+1$.

Now, $A(t,t_0)=\sum_{s=t_0}^{t-1} s$. If the inequality

$$\Delta\left(\frac{1}{t}\Delta^2 y(t)\right)+\frac{1}{3}\left(\sum_{s=t_1}^{t-m+1}\sum_{u=t_1}^{s-1} u\right)^2 q(t)y(t-m+1)\le 0$$

has no eventually positive nondecreasing solution for $t \geq t_1 \geq t_0$ and

$$\Delta\left(\frac{1}{t}\Delta^2 w(t)\right) + \frac{1}{3}\left(\sum_{s=t-m+1}^{t+n-m}\sum_{u=s}^{s+n-1} u\right)^2 q(t)w(t-m+1) \leq 0$$

has no eventually positive nonincreasing solution, then the conditions of Theorem 2.2 are satisfied and hence Eq. (16) is oscillatory.

Remark 2.1

1. We would like to mention here that the result in the present paper are essentially new and can easily be extended to even more general class of difference equations with neutral terms.
2. The results of this paper can be extended to higher order half-linear difference equations of the form

$$\Delta(a(t)\Delta^{n-1}x^(t))^\alpha) + q(t)x^\alpha(t-m+1) = 0, \quad n \text{ is a positive integer.}$$

References

1. Agarwal, R.P.: Difference Equations and Inequalities. Dekker, New York (2000)
2. Agarwal, R.P., Bohner, M., Grace, S.R., O'Regan, D.: Discrete Oscillation Theory. Hindawi, New York (2005)
3. Agarwal, R.P., Grace, S.R., O'Regan, D.: Oscillation Theory for Difference and Functional Difference Equations. Kluwer, Dordrecht (2000)
4. Benekas, V., Kashkynbayev, A., Stavroulakis, I.P.: A sharp oscillation criterion for a difference equation with constant delay. Adv. Difference Equations **2020**, Article number: 566 (2020)
5. El-Morshedy, H.A.: Oscillation and nonoscillation criteria for half-linear second order difference equations. Dynam. Syst. Appl. **15**, 429–450 (2006)
6. El-Morshedy, H.A.: New oscillation criteria for second order linear difference equations with positive and negative coefficients. Comput. Math. Appl. **58**, 1988–1997 (2009)
7. Grace, S.R., Agarwal, R.P., Bohner, M., O'Regan, D.: Oscillation od second-order strongly superlinear and strongly sublinear dynamic equations. Commun. Nonlinear Sci. Numer. Simul. **14**, 3463–3471 (2009)
8. Grace, S.R., Bohner, M., Agarwal, R.P.: On the oscillation of second-order half-linear dynamic equations. J. Difference Equ. Appl. **15**, 451–460 (2009)
9. Grace, S.R., El-Morshedy, H.A.: Oscillation criteria of comparison type for second order difference equations. J. Appl. Anal. **6**, 87–103 (2000)
10. Grace, S.R., Graef, J.R.: Oscillatory behavior of second order nonlinear differential equations with a sublinear neutral term. Math. Model. Anal. **23**, 217–226 (2018)
11. Gyori, I., Ladas, G.: Oscillation Theory of Delay Differential Equations with Applications. Clarendon Press, Oxford (1991)
12. Thandapani, E., Ravi, K.: Oscillation of second order half-linear difference equations. Appl. Math. Lett. **13**, 43–49 (2000)
13. Thandapani, E., Ravi, K., Graef, J.R.: Oscillation and comparison theorem for halflinear second order difference equations. Comput. Math. Appl. **42**, 953–960 (2001)
14. Yan, J., Qian, C.: Oscillation and comparison results for delay difference equations. J. Math. Anal. Appl. **165**, 346–360 (1992)

Existence of Bounded Solutions for a Weighted Quasilinear Elliptic Equation in R^N

Federica Mennuni and Addolorata Salvatore

Dedicated to Francesco Altomare, with great esteem and gratitude.

Abstract In this paper we establish a new existence result for the quasilinear elliptic equation

$$-\operatorname{div}(A(x,u)|\nabla u|^{p-2}\nabla u)+\frac{1}{p}A_t(x,u)|\nabla u|^p+V(x)|u|^{p-2}u=\xi(x)|u|^{q-2}u \qquad \text{in } \mathbb{R}^N$$

with $N \geq 2$ and $1 < q < p$. Here, we suppose $A : \mathbb{R}^N \times \mathbb{R} \to \mathbb{R}$ is a C^1-Carathéodory function such that $A_t(x,t) = \frac{\partial A}{\partial t}(x,t)$ and $V : \mathbb{R}^N \to \mathbb{R}$, $\xi : \mathbb{R}^N \to \mathbb{R}$ are suitable measurable functions. Since the coefficient of the principal part depends on the solution itself, the study of the interaction of two different norms in a suitable Banach space is needed.

Thus, a variational approach and approximation arguments on bounded sets can be used to state the existence of a nontrivial weak bounded solution.

Keywords Modified Schrödinger equation · Quasilinear elliptic equation · Unbounded domain · Weak Cerami-Palais-Smale condition · Minimum Principle · Weak bounded nontrivial solution · "Sublinear" growth · Approximating problems

F. Mennuni (✉) · A. Salvatore
Dipartimento di Matematica, Università degli studi di Bari Aldo Moro, Bari, Italy
e-mail: federica.mennuni@uniba.it; addolorata.salvatore@uniba.it

A. M. Candela et al. (eds.), *Recent Advances in Mathematical Analysis*,
Trends in Mathematics, https://doi.org/10.1007/978-3-031-20021-2_19

1 Introduction

In this paper we look for weak bounded solutions for the generalized quasilinear Schrödinger equation

$$-\mathrm{div}(A(x,u)|\nabla u|^{p-2}\nabla u) + \frac{1}{p}A_t(x,u)|\nabla u|^p + V(x)|u|^{p-2}u = g(x,u) \quad \text{in } \mathbb{R}^N \tag{1}$$

with $p > 1$ and $N \geq 2$, where $A : \mathbb{R}^N \times \mathbb{R} \to \mathbb{R}$ is C^1-Carathéodory function with partial derivative $A_t(x,u) = \frac{\partial A}{\partial t}(x,u)$, potential $V : \mathbb{R}^N \to \mathbb{R}$ is a suitable measurable function and $g : \mathbb{R}^N \times \mathbb{R} \to \mathbb{R}$ is a given Carathéodory function.

Nonlinear elliptic problems in $\mathbb{R}^N$ like (1) have been extensively studied in the particular case $A(x,t) \equiv$ constant when Eq. (1) turns out to the p-Laplacian equation

$$-\Delta_p u + V(x)|u|^{p-2}u = g(x,u) \quad \text{in } \mathbb{R}^N. \tag{2}$$

If $p = 2$, Eq. (2) reduces to the following Schrödinger equation

$$-\Delta u + V(x)u = g(x,u) \quad \text{in } \mathbb{R}^N$$

which is a central topic in Nonlinear Analysis (see e.g., [4, 6, 18, 19, 21, 28, 29]).

In the more general case $p > 1$, the p-Laplacian equation (2) has been widely studied in many papers, as [3, 5, 24, 26].

More recently, many authors investigated Eq. (1) when the coefficient $A(x,t)$ of the principal part is not constant. In this case, under suitable assumption on A, V and g, the natural functional associated to the problem is

$$J(u) = \frac{1}{p}\int_{\mathbb{R}^N} A(x,u)|\nabla u|^p dx + \frac{1}{p}\int_{\mathbb{R}^N} V(x)|u|^p dx - \int_{\mathbb{R}^N} G(x,u)dx$$

with $G(x,t) = \displaystyle\int_0^t g(x,s)ds$, but, even if $A(x,t)$ is a smooth strictly positive bounded function, if $A_t(x,u) \not\equiv 0$ functional J is well defined in $W^{1,p}(\mathbb{R}^N)$, while it is Gâteaux differentiable only along directions in $W^{1,p}(\mathbb{R}^N) \cap L^\infty(\mathbb{R}^N)$.

Such loss of regularity, which occurs also in bounded domains, has been overcome in different ways: by using nonsmooth techniques as in [1, 2, 17] or by introducing a suitable change of variable but only if $A(x,t)$ has a particular form and is independent of x as in [20, 27, 30, 31].

Recently, following a different approach developed in [10–12], it has been proved that, if $g(x,\cdot)$ has a super-p-linear but subcritical growth, functional J is C^1 in the Banach space $X = W^{1,p}(\mathbb{R}^N) \cap L^\infty(\mathbb{R}^N)$ equipped with the intersection norm, then abstract results can be applied since a "weaker compactness" condition holds (see Definition 2.1). Hence, the existence of at least one bounded solution of (1) has been stated if V verifies suitable assumptions in [16] or if $V \equiv 1$ and all the involved functions are radially symmetric in [15] or 1-periodic in x in [13].

Using the same variational approach, in this paper we want to study Eq. (1) when $g(x, \cdot)$ is a particular function having a sub-p-linear growth, more precisely we consider the quasilinear elliptic equation

$$\begin{aligned} -\operatorname{div}(A(x,u)|\nabla u|^{p-2}\nabla u) + \frac{1}{p}A_t(x,u)|\nabla u|^p \\ +V(x)|u|^{p-2}u = \xi(x)|u|^{q-2}u \quad \text{in } \mathbb{R}^N \end{aligned} \tag{3}$$

with $\xi : \mathbb{R}^N \to \mathbb{R}$ measurable function and $1 < q < p$. We will prove that, if $A(x,t)$ and $V(x)$ verify the same assumptions made in [16] in the super-p-linear case, Eq. (3) admits at least one weak bounded solution.

The paper is organized as follows. In Sect. 2 we introduce the abstract framework and we recall a weaker version of the Cerami's variant of the Palais-Smale condition and the related Minimum Principle (see Proposition 2.1). In Sect. 3 we introduce some preliminary assumptions on the functions $A(x,t)$, $V(x)$ and $\xi(x)$ which allow to give a variational principle for Eq. (3). In Sect. 4 we consider some further hypotheses, we state our main result (see Theorem 4.1) and we prove some properties for the action functional associated to the problem. Then, in Sect. 5 we prove Theorem 4.1 by means of a method of approximation on bounded domains.

2 Abstract Setting

Throughout this section, we assume that:

- $(X, \|\cdot\|_X)$ is a Banach space with dual $(X', \|\cdot\|_{X'})$;
- $(W, \|\cdot\|_W)$ is a Banach space such that $X \hookrightarrow W$ continuously, i.e., $X \subset W$ and a constant $\sigma_0 > 0$ exists such that

$$\|u\|_W \le \sigma_0 \|u\|_X \qquad \text{for all } u \in X;$$

- $J : \mathcal{D} \subset W \to \mathbb{R}$ and $J \in C^1(X, \mathbb{R})$ with $X \subset \mathcal{D}$.

Anyway, in order to avoid any ambiguity and semplify, when possible, the notation, from now on by X we denote the space equipped with its given norm $\|\cdot\|_X$, while, if the norm $\|\cdot\|_W$ is involved, we write it explicitly.

For simplicity, taking $\beta \in \mathbb{R}$, we say that a sequence $(u_n)_n \subset X$ is a *Cerami–Palais–Smale sequence at level* β, briefly $(CPS)_\beta$*–sequence*, if

$$\lim_{n\to+\infty} J(u_n) = \beta \quad \text{and} \quad \lim_{n\to+\infty} \|dJ(u_n)\|_{X'}(1 + \|u_n\|_X) = 0.$$

Moreover, β is a *Cerami–Palais–Smale level*, briefly *(CPS)–level*, if there exists a $(CPS)_\beta$–sequence.

As $(CPS)_\beta$–sequences may exist which are unbounded in $\|\cdot\|_X$ but converge with respect to $\|\cdot\|_W$, we have to weaken the classical Cerami–Palais–Smale condition in a suitable way according to the ideas already developed in previous papers (see, e.g., [10–12]).

Definition 2.1 The functional J satisfies the *weak Cerami–Palais–Smale condition at level β ($\beta \in \mathbb{R}$), briefly $(wCPS)_\beta$ condition*, if for every $(CPS)_\beta$–sequence $(u_n)_n$, a point $u \in X$ exists such that

(i) $\lim_{n\to+\infty} \|u_n - u\|_W = 0$ (up to subsequences),
(ii) $J(u) = \beta,\ dJ(u) = 0$.

If J satisfies the $(wCPS)_\beta$ condition at each level $\beta \in I$, I real interval, we say that J satisfies the $(wCPS)$ condition in I.

Condition $(wCPS)_\beta$ implies that the set of critical points of J at level β is compact with respect to $\|\cdot\|_W$, hence a Deformation Lemma and some abstract critical point theorems can be stated (see [12]). In particular, the following Minimum Principle holds (for the proof, see [12, Theorem 1.6]).

Proposition 2.1 (Minimum Principle) *If $J \in C^1(X,\mathbb{R})$ is bounded from below in X and $(wCPS)_\beta$ holds at level $\beta = \inf_X J \in \mathbb{R}$, then J attains its infimum, i.e., $u_0 \in X$ exists such that $J(u_0) = \beta$.*

3 Variational Framework and Regularity Result

Let $\mathbb{N} = \{1, 2, \dots\}$ be the set of the strictly positive integers and, taking any Ω open subset of $\mathbb{R}^N$, $N \geq 2$, we denote by:

- $B_R(x) = \{y \in \mathbb{R}^N : |x - y| < R\}$, the open ball in $\mathbb{R}^N$ with center in $x \in \mathbb{R}^N$ and radius $R > 0$;
- $|D|$ the usual Lebesgue measure of a measurable set D in $\mathbb{R}^N$;
- $L^r(\Omega)$ the Lebesgue space with norm $|u|_{\Omega,r} = \left(\int_\Omega |u|^r dx\right)^{\frac{1}{r}}$ if $1 \leq r < +\infty$;
- $L^\infty(\Omega)$ the space of Lebesgue–measurable and essentially bounded functions $u : \Omega \to \mathbb{R}$ with norm $|u|_{\Omega,\infty} = \text{ess sup}_\Omega |u|$;
- $W^{1,p}(\Omega)$ and $W_0^{1,p}(\Omega)$ the classical Sobolev spaces both equipped with the standard norm $\|u\|_\Omega = (|\nabla u|_{\Omega,p}^p + |u|_{\Omega,p}^p)^{\frac{1}{p}}$ if $1 \leq p < +\infty$.

Moreover, if $V : \mathbb{R}^N \to \mathbb{R}$ is a measurable function such that

$$(V_1) \qquad V_0 = \text{ess inf}_{\mathbb{R}^N} V(x) > 0,$$

we denote by

- $L^r_V(\mathbb{R}^N)$ the weighted Lebesgue space

$$L^r_V(\Omega) = \left\{ u \in L^r(\Omega) : \int_\Omega V(x)|u|^r dx < +\infty \right\} \quad \text{for} \quad 1 \le r < +\infty,$$

  equipped with the norm

$$|u|_{\Omega,V,r} = \left(\int_\Omega V(x)|u|^r dx \right)^{\frac{1}{r}};$$

- $W^{1,p}_V(\Omega)$ and $W^{1,p}_{0,V}(\Omega)$ the weighted Sobolev spaces

$$W^{1,p}_V(\Omega) = \left\{ u \in W^{1,p}(\Omega) : \int_\Omega V(x)|u|^p dx < +\infty \right\},$$

$$W^{1,p}_{0,V}(\Omega) = \left\{ u \in W^{1,p}_0(\Omega) : \int_\Omega V(x)|u|^p dx < +\infty \right\},$$

  both equipped with the norm

$$\|u\|_{\Omega,V} = \left(\int_\Omega |\nabla u|^p + V(x)|u|^p dx \right)^{\frac{1}{p}} = (|\nabla u|^p_{\Omega,p} + |u|^p_{\Omega,V,p})^{\frac{1}{p}}. \tag{4}$$

For semplicity, we put $B_R = B_R(0)$ and, if $\Omega = \mathbb{R}^N$, we omit the subscript in the notation, i.e., we put

- $|\cdot|_r = |\cdot|_{\mathbb{R}^N,r}$ for the norm in $L^r(\mathbb{R}^N)$, for all $1 \le r < +\infty$;
- $|\cdot|_{V,r} = |\cdot|_{\mathbb{R}^N,V,r}$ for the norm in $L^r_V(\mathbb{R}^N)$, for all $1 \le r < +\infty$;
- $\|\cdot\| = \|\cdot\|_{\mathbb{R}^N}$ for the norm in $W^{1,p}(\mathbb{R}^N)$ and in $W^{1,p}_0(\mathbb{R}^N)$;
- $\|\cdot\|_V = \|\cdot\|_{\mathbb{R}^N,V}$ for the norm in $W^{1,p}_V(\mathbb{R}^N) = W^{1,p}_{0,V}(\mathbb{R}^N)$.

Remark 3.1 By assumption (V_1), the following continuous embeddings hold:

$$L^r_V(\mathbb{R}^N) \hookrightarrow L^r(\mathbb{R}^N) \quad \text{for all } 1 \le r < +\infty \tag{5}$$

and

$$W^{1,p}_V(\mathbb{R}^N) \hookrightarrow W^{1,p}(\mathbb{R}^N) \quad \text{for all } 1 \le p < +\infty. \tag{6}$$

From now on, we assume that V verifies also the following assumption:

$$(V_2) \qquad \int_{B_1(x)} \frac{1}{V(y)} dy \longrightarrow 0 \quad \text{as} \quad |x| \to +\infty.$$

From Remark 3.1 and Sobolev Embedding Theorems, we deduce the following result (for the compact embeddings, we refer to [7, Theorem 3.1]).

Proposition 3.1 *Let $V : \mathbb{R}^N \to \mathbb{R}$ be a Lebesgue measurable function satisfying assumption (V_1).*

Then, the following continuous embeddings hold:

- *if $p < N$, then*

$$W_V^{1,p}(\mathbb{R}^N) \hookrightarrow L^r(\mathbb{R}^N) \text{ for any } p \le r \le \frac{Np}{N-p}; \tag{7}$$

- *if $p = N$, then*

$$W_V^{1,p}(\mathbb{R}^N) \hookrightarrow L^r(\mathbb{R}^N) \text{ for any } p \le r < +\infty; \tag{8}$$

- *if $p > N$, then*

$$W_V^{1,p}(\mathbb{R}^N) \hookrightarrow L^r(\mathbb{R}^N) \text{ for any } p \le r \le +\infty. \tag{9}$$

Furthermore, if assumption (V_2) also occurs, the compact embedding

$$W_V^{1,p}(\mathbb{R}^N) \hookrightarrow\hookrightarrow L^r(\mathbb{R}^N) \text{ for any } p \le r < p^* \tag{10}$$

holds, with

$$p^* = \begin{cases} \frac{Np}{N-p} & \text{if } p < N, \\ +\infty & \text{if } p \ge N. \end{cases}$$

From Proposition 3.1 it follows that for any $r \ge p$ as in (7), respectively (8) or (9), a constant $\tau_r > 0$ exists such that

$$|u|_r \le \tau_r \|u\|_V \text{ for all } u \in W_V^{1,p}(\mathbb{R}^N). \tag{11}$$

On the other hand, if Ω is an open bounded domain of class C^1 in $\mathbb{R}^N$ with $\partial\Omega$ bounded and $p < N$, a classical embedding Theorem (see, e.g., [9, Corollary 9.14]), implies that a constant $\sigma^* > 0$, independent of Ω and depending only on p and N, exists such that

$$|u|_{\Omega,p^*} \le \sigma^* \|u\|_\Omega \text{ for all } u \in W_0^{1,p}(\Omega).$$

From now on, we assume that V is a measurable function verifying (V_1) and we set

$$X = W_V^{1,p}(\mathbb{R}^N) \cap L^\infty(\mathbb{R}^N), \text{ with } \|u\|_X = \|u\|_V + |u|_\infty \text{ for any } u \in X. \quad (12)$$

In the following we assume $p \leq N$ as, otherwise, from (9) it follows that $X = W_V^{1,p}(\mathbb{R}^N)$ and all the arguments and the proofs can be semplified.

The following lemmas hold.

Lemma 3.1 *For any $r \geq p$, the Banach space X is continuously embedded in $L_V^r(\mathbb{R}^N)$, i.e., $X \hookrightarrow L_V^r(\mathbb{R}^N)$. More precisely, it results*

$$|u|_{V,r} \leq \|u\|_X \qquad \text{for all } u \in X. \quad (13)$$

Proof For the proof, see [16, Lemma 3.3] □

Remark 3.2 Observe that, thanks to (5), Lemma 3.1 ensures that

$$X \hookrightarrow L^r(\mathbb{R}^N) \text{ for any } p \leq r \leq +\infty.$$

Lemma 3.2 *If $(u_n)_n \subset X$, $u \in X$ and $M > 0$ are such that*

$$\|u_n - u\|_V \to 0 \quad \text{as } n \to +\infty, \quad (14)$$

and

$$|u_n|_\infty \leq M \quad \text{for all } n \in \mathbb{N}, \quad (15)$$

then,

$$u_n \to u \quad \text{in } L_V^r(\mathbb{R}^N) \quad \text{for any } p \leq r < +\infty.$$

Proof For the proof, see [16, Lemma 3.5]. □

Let $A : \mathbb{R}^N \times \mathbb{R} \to \mathbb{R}$ be a given function such that the following conditions hold:

(h_0) A a is C^1–Carathéodory function, i.e., $A(\cdot, t)$ is measurable for all $t \in \mathbb{R}$, and $A(x, \cdot)$ is C^1 for a.e. $x \in \mathbb{R}^N$;

(h_1) for any $\rho > 0$ we have that

$$\sup_{|t|\leq\rho} |A(\cdot, t)| \in L^\infty\left(\mathbb{R}^N\right), \qquad \sup_{|t|\leq\rho} |A_t(\cdot, t)| \in L^\infty\left(\mathbb{R}^N\right).$$

Furthermore, we assume that

(ξ_1) $\xi \in L^{\frac{p}{p-q}}(\mathbb{R}^N) \cap L^\infty(\mathbb{R}^N)$.

Proposition 3.2 *The assumption* (ξ_1) *implies that*

$$\int_{\mathbb{R}^N} \xi(x)|u|^q dx \in \mathbb{R} \quad \textit{for all } u \in X \quad (\textit{ or better for all } u \in W^{1,p}(\mathbb{R}^N))$$

and

$$\int_{\mathbb{R}^N} \xi(x)|u|^{q-2}uvdx \in \mathbb{R} \quad \textit{for all } u, v \in X \quad (\textit{ or better for all } u, v \in W^{1,p}(\mathbb{R}^N)).$$

Proof Let $u \in W^{1,p}(\mathbb{R}^N)$. As $\xi \in L^{\frac{p}{p-q}}(\mathbb{R}^N)$ and $|u|^q \in L^{\frac{p}{q}}(\mathbb{R}^N)$ with $\frac{p}{p-q}$ and $\frac{p}{q}$ conjugate exponents, by Hölder's inequality we have that

$$\int_{\mathbb{R}^N} \xi(x)|u|^q dx \le |\xi|_{\frac{p}{p-q}} |u|_p^q \tag{16}$$

while for all $u, v \in W^{1,p}(\mathbb{R}^N)$ it is

$$\left| \int_{\mathbb{R}^N} \xi(x)|u|^{q-2}uvdx \right| \le |\xi|_{\frac{p}{p-q}} |u|_p^{q-1} |v|_p \tag{17}$$

by applying again the extention of Hölder's inequality with $\frac{p}{p-q}$, $\frac{p}{q-1}$ and p conjugate exponents. □

We note that, for all $u \in X$, from the assumptions $(h_0) - (h_1)$ it follows that $A(\cdot, u)|\nabla u(\cdot)|^p \in L^1(\mathbb{R}^N)$. Thus, (12) and (16) imply that the functional

$$\begin{aligned} \mathcal{J}(u) = &\frac{1}{p} \int_{\mathbb{R}^N} A(x, u)|\nabla u|^p dx + \frac{1}{p} \int_{\mathbb{R}^N} V(x)|u|^p dx \\ &- \frac{1}{q} \int_{\mathbb{R}^N} \xi(x)|u|^q dx \end{aligned} \tag{18}$$

is well defined for all $u \in X$. Moreover, taking $v \in X$, from (17), the Gâteaux differential of functional $\mathcal{J}$ in u along the direction v is given by

$$\begin{aligned} \langle d\mathcal{J}(u), v\rangle = &\int_{\mathbb{R}^N} A(x, u)|\nabla u|^{p-2}\nabla u \cdot \nabla v dx + \frac{1}{p} \int_{\mathbb{R}^N} A_t(x, u)v|\nabla u|^p dx \\ &+ \int_{\mathbb{R}^N} V(x)|u|^{p-2}uvdx - \int_{\mathbb{R}^N} \xi(x)|u|^{q-2}uvdx. \end{aligned} \tag{19}$$

As useful in the following, we recall this technical lemma (for the proof, see [22]).

Lemma 3.3 *A constant $C > 0$ exists such that for any $\eta_1, \eta_2 \in \mathbb{R}^N$, $N \geq 1$, it results*

$$||\eta_1|^{r-2}\eta_1 - |\eta_2|^{r-2}\eta_2| \leq C|\eta_1 - \eta_2|\,(|\eta_1| + |\eta_2|)^{r-2} \qquad \text{if } r > 2, \tag{20}$$

$$||\eta_1|^{r-2}\eta_1 - |\eta_2|^{r-2}\eta_2| \leq C|\eta_1 - \eta_2|^{r-1} \qquad \text{if } 1 < r \leq 2. \tag{21}$$

Now, we can state the following regularity result.

Proposition 3.3 *Let $p > 1$ and assume that conditions (V_1), (h_0)–(h_1) and (ξ_1) hold.*

If $(u_n)_n \subset X$ and $u \in X$ are such that

$$u_n \to u \text{ a.e. in } \mathbb{R}^N \tag{22}$$

and (14), (15) hold for a constant $M > 0$, then

$$\mathcal{J}(u_n) \to \mathcal{J}(u) \quad \text{and} \quad \|d\mathcal{J}(u_n) - d\mathcal{J}(u)\|_{X'} \to 0 \text{ as } n \to +\infty.$$

Hence, $\mathcal{J}$ is a C^1 functional on X with Fréchet differential as in (19).

Proof For the sake of convenience, we set $\mathcal{J}(u) = \mathcal{J}_1(u) - \mathcal{J}_2(u)$, where

$$\mathcal{J}_1(u) = \int_{\mathbb{R}^N} A(x,u)|\nabla u|^p dx + \frac{1}{p}\int_{\mathbb{R}^N} V(x)|u|^p dx \in \mathbb{R}$$
$$\mathcal{J}_2(u) = \frac{1}{q}\int_{\mathbb{R}^N} \xi(x)|u|^q dx$$

with related Gâteaux differentials

$$\langle d\mathcal{J}_1(u), v\rangle = \int_{\mathbb{R}^N} A(x,u)|\nabla u|^{p-2}\nabla u \cdot \nabla v dx + \frac{1}{p}\int_{\mathbb{R}^N} A_t(x,u)v|\nabla u|^p dx + \int_{\mathbb{R}^N} V(x)|u|^{p-2}uv dx$$
$$\langle d\mathcal{J}_2(u), v\rangle = \int_{\mathbb{R}^N} \xi(x)|u|^{q-2}uv dx.$$

Let $(u_n)_n \subset X$, $u \in X$ and $M > 0$ such that (14), (15) and (22) hold. Arguing as in [16, Proposition 3.10], we obtain that

$$\mathcal{J}_1(u_n) \to \mathcal{J}_1(u) \quad \text{and} \quad \|d\mathcal{J}_1(u_n) - d\mathcal{J}_1(u)\|_{X'} \to 0 \quad \text{as } n \to +\infty.$$

Now, we have to prove that

$$\mathcal{J}_2(u_n) \to \mathcal{J}_2(u) \quad \text{and} \quad \|d\mathcal{J}_2(u_n) - d\mathcal{J}_2(u)\|_{X'} \to 0 \quad \text{as } n \to +\infty.$$

From (22) we know that

$$\xi(x)|u_n(x)|^q - \xi(x)|u(x)|^q \to 0 \quad \text{a.e. in } \mathbb{R}^N \quad \text{as } n \to +\infty.$$

Moreover, from (16), (11) and (14) it follows that

$$\int_{\mathbb{R}^N} \xi(x)|u_n - u|^q dx \to 0 \quad \text{as } n \to +\infty$$

and then $\mathcal{J}_2(u_n)$ is bounded.

Arguing by contradiction, assume that the limit $\mathcal{J}_2(u_n) \to \mathcal{J}_2(u)$ as $n \to +\infty$ does not hold. Then, $\gamma \in \mathbb{R}$ and a subsequence $\mathcal{J}_2(u_{k_n})$ exist such that

$$\mathcal{J}_2(u_{k_n}) \to \gamma \quad \text{as } n \to +\infty \quad \text{and } \gamma \neq \mathcal{J}_2(u). \tag{23}$$

On the other hand, a function $h(x) \in L^1(\mathbb{R}^N)$ exists such that up to subsequence

$$\xi(x)|u_{k_n}(x) - u(x)|^q \leq h(x) \quad \text{a.e. in } \mathbb{R}, \quad \text{for all } n \in \mathbb{N}$$

(see [9, Theorem 4.9]). It follows that, for any $n \in \mathbb{N}$

$$\begin{aligned} |\xi(x)|(|u_{k_n}(x)|^q - |u(x)|^q)| &\leq 2^{q-1}|\xi(x)||u_{k_n}(x) - u(x)|^q \\ &\quad + 2^{q-1}\xi(x)|u(x)|^q + \xi(x)|u(x)|^q \\ &\leq 2^{q-1}h(x) + (2^{q-1}+1)\xi(x)|u(x)|^q \in L^1(\mathbb{R}^N). \end{aligned}$$

So, the Dominated Convergence Theorem implies that $\mathcal{J}_2(u_{k_n}) \to \mathcal{J}_2(u)$, in contradiction with (23).

Taking $v \in X$, $\|v\|_X = 1$, we have

$$|v|_\infty \leq 1, \quad \|v\|_V \leq 1 \tag{24}$$

and we observe that

$$\begin{aligned} |\langle d\mathcal{J}_2(u_n) - d\mathcal{J}_2(u), v\rangle| &= \left| \int_{\mathbb{R}^N} \xi(x)(|u_n|^{q-2}u_n - |u|^{q-2}u)v dx \right| \\ &\leq \int_{\mathbb{R}^N} |\xi(x)||\,|u_n|^{q-2}u_n - |u|^{q-2}u||v| dx. \end{aligned} \tag{25}$$

If $q \in (1, 2]$, from (21), (17), (11), (24) and (14) we have

$$\begin{aligned} &\int_{\mathbb{R}^N} |\xi(x)||\,|u_n|^{q-2}u_n - |u|^{q-2}u||v| dx \leq C \int_{\mathbb{R}^N} |\xi(x)||u_n - u|^{q-1}|v| dx \\ &\leq C|\xi|_{\frac{p}{p-q}} |u_n - u|_p^{q-1} |v|_p \\ &\leq C\tau_p |\xi|_{\frac{p}{p-q}} \|u_n - u\|_V^{q-1} \to 0 \quad \text{as } n \to +\infty. \end{aligned} \tag{26}$$

Now, if $2 < q < p$, from Hölder inequality with $\xi(x) = \xi(x)^{\frac{1}{q}}\xi(x)^{\frac{q-1}{q}}$, (20), (16) and (11) we have

$$\int_{\mathbb{R}^N} |\xi(x)|||u_n|^{q-2}u_n - |u|^{q-2}u||v|dx$$
$$\leq \left(\int_{\mathbb{R}^N} |\xi(x)|||u_n|^{q-2}u_n - |u|^{q-2}u|^{\frac{q}{q-1}}\right)^{\frac{q-1}{q}} \left(\int_{\mathbb{R}^N} |\xi(x)||v|^q dx\right)^{\frac{1}{q}} \tag{27}$$
$$\leq c_0 \left(\int_{\mathbb{R}^N} |\xi(x)||u_n - u|^{\frac{q}{q-1}} (|u_n| + |u|)^{\frac{q(q-2)}{q-1}} dx\right)^{\frac{q-1}{q}},$$

whence once again from Hölder inequality but with $\xi(x) = \xi(x)^{\frac{1}{q-1}}\xi(x)^{\frac{q-2}{q-1}}$, it results

$$\int_{\mathbb{R}^N} |\xi(x)||u_n - u|^{\frac{q}{q-1}} (|u_n| + |u|)^{\frac{q(q-2)}{q-1}} dx$$
$$\leq \left(\int_{\mathbb{R}^N} |\xi(x)||u_n - u|^q dx\right)^{\frac{1}{q-1}} \left(\int_{\mathbb{R}^N} |\xi(x)| (|u_n| + |u|)^q dx\right)^{\frac{q-2}{q-1}}.$$

Hence, (27), direct computations, (16), (11) and (14) imply that suitable constants c_0^* and c_0^{**} exist such that

$$\int_{\mathbb{R}^N} |\xi(x)|||u_n|^{q-2}u_n - |u|^{q-2}u)||v|dx$$
$$\leq c_0^* \left(\int_{\mathbb{R}^N} |\xi(x)||u_n - u|^q dx\right)^{\frac{1}{q}} \left(\int_{\mathbb{R}^N} |\xi(x)||u_n|^q dx\right)^{\frac{q-2}{q}}$$
$$+ c_0^* \left(\int_{\mathbb{R}^N} |\xi(x)||u_n - u|^q dx\right)^{\frac{1}{q}} \left(\int_{\mathbb{R}^N} |\xi(x)||u|^q dx\right)^{\frac{q-2}{q}} \tag{28}$$
$$\leq c_0^{**}|u_n - u|_p \leq c_0^{**}\tau_p\|u_n - u\|_V \to 0 \quad \text{as } n \to +\infty.$$

Thus, summing up, from (25), (26) and (28) it follows that

$$|\langle d\mathcal{J}_2(u_n) - d\mathcal{J}_2(u), v\rangle| \to 0 \quad \text{as } n \to +\infty,$$

uniformly with respect to $v \in X$, $\|v\|_X = 1$, i.e.,

$$\|d\mathcal{J}(u_n) - d\mathcal{J}(u)\|_{X'} \to 0 \quad \text{as } n \to +\infty.$$

□

4 The Main Theorem and Related Results

From now on, besides $(h_0) - (h_1)$, (ξ_1),$(V_1) - (V_2)$ we assume the following additional assumptions:

(h_2) a constant $\alpha_0 > 0$ exists such that

$$A(x,t) \geq \alpha_0 \quad \text{a.e. in } \mathbb{R}^N, \text{ for all } t \in \mathbb{R};$$

(h_3) some constants $\mu > p$ and $\alpha_1 > 0$ exist so that

$$(\mu - p)A(x,t) - A_t(x,t)t \geq \alpha_1 A(x,t) \quad \text{a.e. in } \mathbb{R}^N, \text{ for all } t \in \mathbb{R};$$

(h_4) a constant $\alpha_2 > 0$ exists such that

$$pA(x,t) + A_t(x,t)t \geq \alpha_2 A(x,t) \quad \text{a.e. in } \mathbb{R}^N, \text{ for all } t \in \mathbb{R};$$

(V_3) for any $\varrho > 0$, a constant $C_\varrho > 0$ exists such that

$$\text{ess sup}_{|x|\leq\varrho} V(x) \leq C_\varrho;$$

(ξ_2) $\xi \geq 0$ with $|\text{supp}(\xi)| > 0$.

Remark 4.1 We can always assume $\alpha_0 \leq 1$. Moreover, taking $t = 0$ in (h_3), from (h_2) it follows that $\mu - p \geq \alpha_1$.

Now, we are ready to state our main result.

Theorem 4.1 *Under assumptions* (h_0)–(h_4), (V_1)–(V_3) *and* (ξ_1)–(ξ_2), *then problem* (3) *admits at least one weak nontrivial bounded solution.*

Remark 4.2 If Ω is bounded, the existence of solutions for quasilinear elliptic problems like (1) has been stated in [14] when the principal part $A(x,t,\xi)$ and the sub-p-linear term $g(x,t)$ have a more general form. Here, we reduce to deal with the special case $g(x,t) = \xi(x)|t|^{q-2}t$, $1 < q < p$, for defining the C^1 action functional $\mathcal{J}$ on the space $X = W^{1,p}(\mathbb{R}^N) \cap L^\infty(\mathbb{R}^N)$. However, with small changes in the proof, it is possible to prove Theorem 4.1 even if $g(x,t)$ is a Carathéodory function such that $|g(x,t)| \leq \xi(x)|t|^{q-1}$, $1 < q < p$.

We start giving this convergence result.

Proposition 4.1 *Suppose that hypotheses* (h_0)–(h_3), (V_1)-(V_2) *and* (ξ_1) *hold. Then, taking any* $\beta \in \mathbb{R}$ *and a* $(CPS)_\beta$*–sequence* $(u_n)_n \subset X$, *it follows that* $(u_n)_n$ *is bounded in* $W_V^{1,p}(\mathbb{R}^N)$. *Furthermore,* $u \in W_V^{1,p}(\mathbb{R}^N)$ *exists such that, up to subsequence,*

$$u_n \rightharpoonup u \text{ weakly in } W_V^{1,p}(\mathbb{R}^N), \tag{29}$$

$$u_n \to u \text{ strongly in } L^r(\mathbb{R}^N) \text{ for each } r \in [p, p^*[, \tag{30}$$

$$u_n \to u \text{ a.e. in } \mathbb{R}^N \tag{31}$$

as $n \to +\infty$.

Proof Let $\beta \in \mathbb{R}$ be fixed and consider a $(CPS)_\beta$—sequence $(u_n)_n \subset X$, i.e.,

$$\mathcal{J}(u_n) \to \beta \quad \text{and} \quad \|d\mathcal{J}(u_n)\|_{X'}(1 + \|u_n\|_X) \to 0 \qquad \text{as } n \to +\infty. \tag{32}$$

Then, (32), (18), (19), (h_3), (16), (h_2), (4) and (11) give

$$\begin{aligned}
\mu\beta + \varepsilon_n &= \mu\mathcal{J}(u_n) - \langle d\mathcal{J}(u_n), u_n\rangle \\
&= \frac{\mu}{p}\int_{\mathbb{R}^N} A(x, u_n)|\nabla u_n|^p dx + \frac{\mu}{p}\int_{\mathbb{R}^N} V(x)|u_n|^p dx - \frac{\mu}{q}\int_{\mathbb{R}^N} \xi(x)|u_n|^q dx \\
&- \int_{\mathbb{R}^N} A(x, u_n)|\nabla u_n|^p dx - \frac{1}{p}\int_{\mathbb{R}^N} A_t(x, u_n)u_n|\nabla u_n|^p dx \\
&- \int_{\mathbb{R}^N} V(x)|u_n|^p dx + \int_{\mathbb{R}^N} \xi(x)|u_n|^q dx \\
&= \left(\frac{\mu}{p} - 1\right)\int_{\mathbb{R}^N} A(x, u_n)|\nabla u_n|^p dx + \left(\frac{\mu}{p} - 1\right)\int_{\mathbb{R}^N} V(x)|u_n|^p dx \\
&- \frac{1}{p}\int_{\mathbb{R}^N} A_t(x, u_n)u_n|\nabla u_n|^p dx - \left(\frac{\mu}{q} - 1\right)\int_{\mathbb{R}^N} \xi(x)|u_n|^q dx \\
&\geq \frac{\alpha_1}{p}\int_{\mathbb{R}^N} A(x, u_n)|\nabla u_n|^p dx + \frac{\mu - p}{p}\int_{\mathbb{R}^N} V(x)|u_n|^p dx - \frac{\mu - q}{q}|\xi|_{\frac{p}{p-q}}|u_n|_p^q \\
&\geq \frac{\alpha_0\alpha_1}{p}\int_{\mathbb{R}^N} |\nabla u_n|^p dx + \frac{\mu - p}{p}\int_{\mathbb{R}^N} V(x)|u_n|^p dx - \frac{\mu - q}{q}|\xi|_{\frac{p}{p-q}}|u_n|_p^q \\
&\geq c_1\|u_n\|_V^p - c_2\|u_n\|_V^q.
\end{aligned}$$

for suitable constants c_1, c_2.

Therefore, from $1 < q < p$ we have that $(u_n)_n$ is bounded in $W_V^{1,p}(\mathbb{R}^N)$. Hence, (29)–(31) follow by the reflexivity of that space and by (10) in Proposition 3.1. □

Proposition 4.2 *Assume that conditions* $(h_0)-(h_2)$, (ξ_1) *and* (V_1) *hold. Then, some positive constants* c_3 *and* c_4 *exist such that*

$$\mathcal{J}(u) \geq c_3\|u\|_V^p - c_4\|u\|_V^q \quad \textit{for any } u \in X.$$

Hence, functional $\mathcal{J}$ *is bounded from below, i.e.,* $\alpha \in \mathbb{R}$ *exists such that*

$$J(u) \geq \alpha \textit{ for any } u \in X, \ \textit{with } \alpha = \min_{s \geq 0}\left(c_3 s^p - c_4 s^q\right).$$

Proof From (h_2), (16) and Remark 4.1 we have

$$\mathcal{J}(u) = \frac{1}{p}\int_{\mathbb{R}^N} A(x,u)|\nabla u|^p dx + \frac{1}{p}\int_{\mathbb{R}^N} V(x)|u|^p dx - \frac{1}{q}\int_{\mathbb{R}^N} \xi(x)|u|^q dx$$
$$\geq \frac{\alpha_0}{p}\|u\|_V^p - \frac{1}{q}|\xi|_{\frac{p}{p-q}}|u|_p^q.$$

Hence, the conclusion follows from (11) with $c_3 = \frac{\alpha_0}{p}$ and $c_4 = \frac{1}{q}|\xi|_{\frac{p}{p-q}}\tau_p^q$. □

Remark 4.3 We note that the thesis holds even if we replace $\|\cdot\|_V$ with the usual norm $\|\cdot\|$ in $W^{1,p}(\mathbb{R}^N)$.

Remark 4.4 We are not able to prove that $\mathcal{J}$ satisfies the $(wCPS)$ condition in the whole space X. In particular, we are not able to prove that the weak limit $u \in L^\infty(\mathbb{R}^N)$ since the Ladyzhenskaya-Ural'tseva result (see [23, Theorem II 5.1]) holds for bounded domains. Hence, we use a method of approximation on bounded domains.

From now on, let Ω denote an open bounded domain in $\mathbb{R}^N$. Thus, we define

$$X_\Omega = W_{0,V}^{1,p}(\Omega) \cap L^\infty(\Omega)$$

endowed with the norm

$$\|u\|_{X_\Omega} = \|u\|_{\Omega,V} + |u|_{\Omega,\infty} \quad \text{for any } u \in X_\Omega \tag{33}$$

and we denote by X'_Ω its dual.

Remark 4.5 Since we set in a bounded domain, it follows that $\|u\|_\Omega$ and $|\nabla u|_{p,\Omega}$ are equivalent norms. Moreover, as (V_3) holds, a constant $c_\Omega \geq 1$ exists such that

$$\|u\|_{\Omega,V}^p = \int_\Omega (|\nabla u|^p + V(x)|u|^p)dx \leq \int_\Omega |\nabla u|^p dx + c_\Omega \int_\Omega |u|^p dx \leq c_\Omega \|u\|_\Omega^p,$$

which, together with (6), implies that the norms $\|\cdot\|_{\Omega,V}$ and $\|\cdot\|_\Omega$ are equivalent, too. We have, in particular, that the norm in (33) can be replaced with the equivalent one, still denoted $\|\cdot\|_{X_\Omega}$, given by

$$\|u\|_{X_\Omega} = \|u\|_\Omega + |u|_{\infty,\Omega} \quad \text{for any } u \in X_\Omega,$$

where

$$X_\Omega = W_0^{1,p}(\Omega) \cap L^\infty(\Omega). \tag{34}$$

Actually, since any function $u \in X_\Omega$ can be trivially extended to a function $\tilde{u} \in X$ just assuming $\tilde{u}(x) = 0$ for all $x \in \mathbb{R}^N \setminus \Omega$, then

$$\|\tilde{u}\| = \|u\|_\Omega, \quad \|\tilde{u}\|_V = \|u\|_{\Omega,V}, \quad |\tilde{u}|_\infty = |u|_{\Omega,\infty}, \quad \|\tilde{u}\|_X = \|u\|_{X_\Omega}.$$

Thus, we can consider $\mathcal{J}_\Omega = \mathcal{J}|_{X_\Omega}$ the restriction of the functional $\mathcal{J}$ to X_Ω.

Clearly, from (18) it results

$$\begin{aligned}\mathcal{J}_\Omega(u) = \mathcal{J}|_{X_\Omega}(u) &= \frac{1}{p}\int_\Omega A(x,u)|\nabla u|^p dx + \frac{1}{p}\int_\Omega V(x)|u|^p dx \\ &\quad - \int_\Omega \xi(x)|u|^q dx, \quad u \in X_\Omega.\end{aligned} \tag{35}$$

Hence, by simplifying the arguments in the proof in [11, Proposition 3.1], it follows that $\mathcal{J}_\Omega : X_\Omega \to \mathbb{R}$ is a C^1 functional, where, for any $u, v \in X_\Omega$, its Fréchet differential in u evaluated in v is given by

$$\begin{aligned}\langle d\mathcal{J}_\Omega(u), v\rangle &= \int_\Omega A(x,u)|\nabla u|^{p-2}\nabla u \cdot \nabla v dx + \frac{1}{p}\int_\Omega A_t(x,u)v|\nabla u|^p dx \\ &\quad + \int_\Omega V(x)|u|^{p-2}uv dx - \int_\Omega \xi(x)|u|^{q-2}uv dx.\end{aligned} \tag{36}$$

We need to state the following result.

Proposition 4.3 *Let $1 < q < p$ and suppose that hypotheses (h_0)–(h_4), (V_1)–(V_3) and (ξ_1) hold. Then, $\mathcal{J}_\Omega$ satisfies the $(wCPS)$ condition in $\mathbb{R}$.*

Proof Taking $\beta \in \mathbb{R}$, let $(u_n)_n \subset X_\Omega$ be a $(CPS)_\beta$–sequence, i.e.,

$$\mathcal{J}_\Omega(u_n) \to \beta \quad \text{and} \quad \|d\mathcal{J}_\Omega(u_n)\|_{X'_\Omega}(1 + \|u_n\|_{X_\Omega}) \to 0 \qquad \text{if } n \to +\infty.$$

We want to prove that $u \in X_\Omega$ exists such that

(i) $\|u_n - u\|_\Omega \to 0$ (up to subsequences),
(ii) $\mathcal{J}_\Omega(u) = \beta, d\mathcal{J}_\Omega(u) = 0$.

Reasoning as in Proposition 4.1 with $\mathcal{J}_\Omega$ instead of $\mathcal{J}$ and using the Sobolev Embedding Theorem for bounded domains, we have that $(u_n)_n$ is bounded in $W_0^{1,p}(\Omega)$. Hence, up to subsequences, there exists $u \in W_0^{1,p}(\Omega)$ such that if $n \to +\infty$, then

$$\begin{aligned}&u_n \rightharpoonup u \text{ weakly in } W_0^{1,p}(\Omega),\\ &u_n \to u \text{ strongly in } L^r(\Omega) \text{ for each } r \in [1, p^*[,\\ &u_n \to u \text{ a.e. in } \Omega.\end{aligned}$$

The remainder of this proof can be stated arguing as in [11, Proposition 4.6] taking the term $g(x,t) = \xi(x)|t|^{q-2}t - V(x)|t|^{p-2}t$ which satisfies the hypothesis (G_1) required in [11]. □

Proposition 4.4 *Assume conditions* $(h_0) - (h_4)$, $(V_1) - (V_3)$ *and* (ξ_1). *Then functional* $\mathcal{J}_\Omega$ *has at least a critical point* $u_\Omega \in X_\Omega$. *Moreover, if* (ξ_2) *holds and* Ω *is such that* $|\text{supp}\xi \cap \Omega| > 0$, *then* u_Ω *is not trivial.*

Proof The functional $\mathcal{J}_\Omega$ is bounded from below in X_Ω (see Proposition 4.2) and satisfies condition $(wCPS)$ in $\mathbb{R}$ (see Proposition 4.3), thus, from Proposition 2.1, $\mathcal{J}_\Omega$ admits a minimum point u_Ω in X_Ω. Clearly, it is

$$\mathcal{J}_\Omega(u_\Omega) = \min_{u \in X_\Omega} \mathcal{J}_\Omega(u) \leq \mathcal{J}_\Omega(0) = 0.$$

We want to prove that u_Ω is non trivial since $\mathcal{J}_\Omega(u_\Omega) < 0$.

We consider $\varphi_1 \in X_\Omega$ the unique eigenfunction associated to the first eigenvalue λ_1 of $-\Delta_p$ in Ω, such that

$$\varphi_1 > 0 \text{ a.e. in } \Omega, \ \varphi_1 \in L^\infty(\Omega), \ |\varphi_1|_{\Omega,p} = 1, \ \text{and so } |\nabla\varphi_1|^p_{\Omega,p} = \lambda_1 \ \text{(see [25])}$$

Then, taking $t \in (0,1)$, from (h_1) we have

$$\begin{aligned}
\mathcal{J}_\Omega(t\varphi_1) &= \frac{t^p}{p}\int_\Omega A(x,t\varphi_1))|\nabla\varphi_1|^p dx + \frac{t^p}{p}\int_\Omega V(x)|\varphi_1|^p dx - \frac{t^q}{q}\int_\Omega \xi(x)|\varphi_1|^p dx \\
&\leq \frac{t^p}{p}\sup_{0\leq\tau\leq|\varphi_1|_\infty} A(x,\tau)|\nabla\varphi_1|^p_{\Omega,p} + \frac{t^p}{p}\int_\Omega V(x)|\varphi_1|^p dx \\
&\quad - \frac{t^q}{q}\int_\Omega \xi(x)|\varphi_1|^q dx.
\end{aligned}$$

So, if (ξ_2) holds and Ω is an open bounded domain such that $|\text{supp}\xi \cap \Omega| > 0$, it follows that $\int_\Omega \xi(x)|\varphi_1|^q dx > 0$. Hence, $\mathcal{J}_\Omega(t\varphi_1) < 0$ for t small enough since $p > q > 1$. □

5 Proof of the Theorem 4.1

For the proof of our main result, we follow a similar approach to those in [13] and [16]. Anyway, the different assumptions for our problem requires to rehash the proofs and provide them with all the details.

Throughout this section, we suppose that hypotheses in Theorem 4.1 are satisfied. Thus, for any $k \in \mathbb{N}$, we can consider the spaces X_{B_k} as in (34) and the related functionals

$$\mathcal{J}_k(u) = \mathcal{J}_{B_k}(u) = \mathcal{J}|_{X_{B_k}}(u)$$

as in (35). For the sake of convenience, since any $u \in X_{B_k}$ can be trivially extended as $u = 0$ a.e. in $\mathbb{R}^N \setminus B_k$, we still denote by u such an extension.

Remark 5.1 From the expression of $\mathcal{J}_k$, it follows that if $u \in X_{B_k}$ then

$$\langle d\mathcal{J}_k(u), v\rangle = \langle d\mathcal{J}(u), v\rangle \text{ for all } v \in X_{B_k}.$$

By (ξ_2), an integer k_1 exists such that $|\text{supp}(\xi) \cap B_{k_1}| > 0$. Without loss of generality, assume that $k_1 = 1$. Thus, from Proposition 4.4, a sequence $(u_k)_k \subset X$ exists such that for every $k \in \mathbb{N}$ it results:

(i) $u_k|_{B_k} \in X_{B_k}$ with $u_k = 0$ a.e. in $\mathbb{R}^N \setminus B_k$,
(ii) $\alpha \leq \mathcal{J}(u_k) \leq \mathcal{J}(u_1) < 0$,
(iii) $\langle d\mathcal{J}(u_k), v\rangle = 0$ for all $v \in X_{B_k}$,

where, from Proposition 4.2, in (ii) we can choose α independent of k. Now, our aim is proving that sequence $(u_k)_k$ is bounded in X. First of all, we recall the following result which is a refinement of [23, Theorem II.51].

Lemma 5.1 *Let Ω be an open bounded domain in $\mathbb{R}^N$ and consider p, s so that $1 < p \leq N$ and $p \leq s < p^*$ (if $N = p$ we just require that p^* is any number larger than s) and take $u \in W_0^{1,p}(\Omega)$. If $a^* > 0$ and $m_0 \in \mathbb{N}$ exist such that*

$$\int_{\Omega_m^+} |\nabla u|^p dx \leq a^* \left(m^s\, |\Omega_m^+| + \int_{\Omega_m^+} |u|^s dx \right) \quad \textit{for all } m \geq m_0,$$

with $\Omega_m^+ = \{x \in \Omega : u(x) > m\}$, then ess $\sup_\Omega u$ *is bounded from above by a positive constant which can be chosen so that it depends only on $|\Omega_{m_0}^+|$, N, p, s, a^*, m_0, $\|u\|_\Omega$, or better by a positive constant which can be chosen so that it depends only on N, p, s, a^*, m_0 and a_0^* for any a_0^* such that* $\max\{|\Omega_{m_0}^+|, \|u\|_\Omega\} \leq a_0^*$.

Vice versa, if inequality

$$\int_{\Omega_m^-} |\nabla u|^p dx \leq a^* \left(m^s\, |\Omega_m^-| + \int_{\Omega_m^-} |u|^s dx \right) \quad \textit{for all } m \geq m_0,$$

holds, with $\Omega_m^- = \{x \in \Omega : u(x) < -m\}$, then ess $\sup_\Omega(-u)$ *is bounded from above by a positive constant which can be chosen so that it depends only on N, p, s, a^*, m_0 and any constant which is greater than both $|\Omega_{m_0}^-|$ and $\|u\|_\Omega$.*

Proof For the proof, see [16, Lemma 5.6]. □

We proceed stating a boundedness result in X.

Proposition 5.1 *A positive constant M_0 exists such that*

$$\|u_k\|_X \leq M_0 \quad \textit{for all } k \in \mathbb{N}; \tag{37}$$

then, for all $\tau \geq p$, it results

$$|u_k|_{V,\tau} \leq M_0 \quad \textit{for all } k \in \mathbb{N}. \tag{38}$$

Proof From (ii) and Proposition 4.2 it follows that

$$\alpha \leq c_3\|u_k\|_V^p - c_4\|u_k\|_V^q \leq \mathcal{J}(u_k) \leq \mathcal{J}(u_1) < 0.$$

So, since $q < p$, we have that $(\|u_k\|_V)_k$ is bounded. Furthermore, by (6) the sequence $(\|u_k\|)_k$ is bounded, too.

Now, we need to prove that $(|u_k|_\infty)_k$ is bounded.

We suppose that $|u_k|_\infty > 1$, i.e., ess $\sup_{\mathbb{R}^N} u_k > 1$ or ess $\sup_{\mathbb{R}^N}(-u_k) > 1$.

Assume that ess $\sup_{\mathbb{R}^N} u_k > 1$ and consider the set

$$B_{1,k}^+ = \{x \in \mathbb{R}^N : u_k(x) > 1\}.$$

Since condition (i) holds, clearly we have

$$B_{1,k}^+ \subset B_k.$$

In particular, it follows that $B_{1,k}^+$ is an open bounded domain such that $|B_{1,k}^+| > 0$ and also

$$|B_{1,k}^+| \leq \int_{B_{1,k}^+} |u_k|^p dx \leq \int_{\mathbb{R}^N} |u_k|^p dx \leq \|u_k\|^p \leq C,$$

since $(\|u_k\|)_k$ is bounded.

Now, for any $m \in \mathbb{N}$ we define the function $R_m^+ : \mathbb{R} \to \mathbb{R}$ such that

$$R_m^+ t = \begin{cases} 0 & \text{if } t \leq m \\ t - m & \text{if } t > m. \end{cases}$$

It follows that $R_m^+ u_k \in X_{B_k}$. Hence, from condition (iii), (36) and hypotheses (V_1), (h_2) and (h_4) we obtain

$$0 = \langle d\mathcal{J}(u_k), R_m^+ u_k\rangle = \int_{B_{m,k}^+} \left(1 - \frac{m}{u_k}\right)\left(A(x, u_k) + \frac{1}{p}A_t(x, u_k)u_k\right)|\nabla u_k|^p dx$$

$$+ \int_{B_{m,k}^+} \frac{m}{u_k} A(x, u_k)|\nabla u_k|^p dx + \int_{B_{m,k}^+} \left(1 - \frac{m}{u_k}\right) V(x)|u_k|^p dx$$

$$- \int_{B_{m,k}^+} \xi(x)|u_k|^{q-2} u_k R_m^+ u_k dx$$

$$\geq \frac{\alpha_0\alpha_2}{p}\int_{B_{m,k}^+} |\nabla u_k|^p dx - \int_{B_{m,k}^+} \xi(x)|u_k|^{q-2}u_k(u_k - m)dx$$

$$= \frac{\alpha_0\alpha_2}{p}\int_{B_{m,k}^+} |\nabla u_k|^p dx - \int_{B_{m,k}^+} \xi(x)|u_k|^q dx + m\int_{B_{m,k}^+} \xi(x)|u_k|^{q-2}u_k dx,$$

where $B_{m,k}^+ = \{x \in \mathbb{R}^N : u_k(x) > m\}$ and $\int_{B_{m,k}^+} \xi(x)|u_k|^{q-2}u_k dx \geq 0$ since $\xi \geq 0$ and $u_k \geq 0$ in $B_{m,k}^+$.

Thus, from (ξ_1), it results

$$\int_{B_{m,k}^+} |\nabla u_k|^p dx \leq \frac{p}{\alpha_0\alpha_2}\int_{B_{m,k}^+} \xi(x)|u_k|^q dx \leq \frac{p}{\alpha_0\alpha_2}\,|\xi|_\infty \int_{B_{m,k}^+} |u_k|^q dx.$$

Since $q < p$ and $u_k(x) > 1$ for any $x \in B_{m,k}^+$, it follows that

$$\int_{B_{m,k}^+} |\nabla u_k|^p dx \leq c^*\left(|B_{m,k}^+| + \int_{B_{m,k}^+} |u_k|^p dx\right) \quad \text{for all } m \geq 1,$$

where $c^* > 0$ is a constant independent of m and k.

From Lemma 5.1 with $\Omega = B_k$ and from the boundedness of $(\|u_k\|)_k$ and $\left(|B_{m,k}^+|\right)_k$, it follows that a constant $M > 0$, independent of $k \in \mathbb{N}$, exists such that

$$\text{ess sup}_{B_k} u_k \leq M.$$

Similar arguments apply if $\text{ess sup}_{\mathbb{R}^N}(-u_k) > 1$.

Thus, we have that

$$|u_k|_\infty \leq C \text{ for all } k \in \mathbb{N}$$

and the proof of (37) is complete.

At last, (38) follows from estimate (13). □

We observe that the estimate (37) ensures the existence of $u^* \in W_V^{1,p}(\mathbb{R}^N)$ such that, up to subsequences,

$$u_k \rightharpoonup u^* \quad \text{in } W_V^{1,p}(\mathbb{R}^N). \tag{39}$$

From (10) it results

$$u_k \to u^* \quad \text{strongly in } L^r(\mathbb{R}^N) \text{ for any } r \in [p, p^*), \tag{40}$$

and

$$u_k \to u^* \quad \text{a.e. in } \mathbb{R}^N. \tag{41}$$

Proposition 5.2 $u^* \in L^\infty(\mathbb{R}^N)$.

Proof For the proof, see [16, Proposition 6.5]. □

Corollary 5.1 *For all $p \le r < +\infty$ we have that*

$$u_k \to u^* \quad \text{in } L^r(\mathbb{R}^N) \tag{42}$$

while, fixing $R \ge 1$,

$$u_k \to u^* \quad \text{in } L_V^r(B_R). \tag{43}$$

Proof Taking $p \le r < p^*$, (42) follows from (40). If $r \ge p^*$, then Propositions 5.1 and 5.2 imply

$$\begin{aligned}\int_{\mathbb{R}^N} |u_k - u^*|^r dx &\le |u_k - u^*|_\infty^{r-p^*+\epsilon} \int_{\mathbb{R}^N} |u_k - u^*|^{p^*-\epsilon} dx \\ &\le 2^{r-p^*+\epsilon}\left(M_0^{r-p^*+\epsilon} + |u^*|_\infty^{r-p^*+\epsilon}\right) \int_{B_R} |u_k - u^*|^{p^*-\epsilon} dx,\end{aligned}$$

hence (42) follows again from (40).

Furthermore, (43) holds since from (V_1) and (V_3) it follows that the norms $|\cdot|_{B_R,p}$ and $|\cdot|_{B_R,V,p}$ are equivalent. □

Proposition 5.3 *We have that*

$$u_k \to u^* \quad \text{strongly in } W_V^{1,p}(B_R) \text{ for all } R \ge 1.$$

Proof From (43) it is enough to prove that

$$|\nabla u_k - \nabla u^*|_{B_R,p} \to 0.$$

To achieve this aim, following an idea introduced in [8], let us consider the real map $\psi(t) = te^{\eta t^2}$, where $\eta > \left(\frac{\beta_2}{2\beta_1}\right)^2$ will be fixed once $\beta_1, \beta_2 > 0$ are chosen in a suitable way later on.

By definition,

$$\beta_1\psi'(t) - \beta_2|\psi(t)| > \frac{\beta_1}{2} \qquad \text{for all } t \in \mathbb{R}.$$

Defining $v_k = u_k - u^*$, we have that (39) implies

$$v_k \rightharpoonup 0 \text{ weakly in } W_V^{1,p}(\mathbb{R}^N),$$

while from (42), respectively (41), it follows that

$$v_k \to 0 \quad \text{strongly in } L^p(\mathbb{R}^N), \tag{44}$$

respectively

$$v_k \to 0 \quad \text{a.e. in } \mathbb{R}^N. \tag{45}$$

Moreover, from (37) and Proposition 5.2 it is

$$v_k \in X \quad \text{and} \quad |v_k|_\infty \leq \bar{M}_0 \quad \text{for all } k \in \mathbb{N} \tag{46}$$

with $\bar{M}_0 = M_0 + |u^*|_\infty$. Now, let $\chi_R \in C^\infty(\mathbb{R}^N)$, $R \geq 1$, be a cut-off function such that

$$\chi_R(x) = \begin{cases} 1 \text{ if } |x| \leq R \\ 0 \text{ if } |x| \geq R+1 \end{cases}, \text{with} \quad 0 \leq \chi_R(x) \leq 1 \quad \text{for all } x \in \mathbb{R}^N, \tag{47}$$

and

$$|\nabla\chi_R(x)| \leq 2 \quad \text{for all } x \in \mathbb{R}^N. \tag{48}$$

Thus, for every $k \in \mathbb{N}$ we consider the new function

$$w_{R,k} : x \in \mathbb{R}^N \mapsto w_{R,k}(x) = \chi_R(x)\psi(v_k(x)) \in \mathbb{R}.$$

We note that (46) gives

$$|\psi(v_k)| \leq \psi(\bar{M}_0), \quad 0 < \psi'(v_k) \leq \psi'(\bar{M}_0) \qquad \text{a.e. in } \mathbb{R}^N, \tag{49}$$

while (45) implies

$$\psi(v_k) \to 0, \quad \psi'(v_k) \to 1 \qquad \text{a.e. in } \mathbb{R}^N. \tag{50}$$

Then, from (47) and (49) we have that $\operatorname{supp} w_{R,k} \subset \operatorname{supp} \chi_R \subset B_{R+1}$ and also

$$|w_{R,k}(x)| \leq \psi(\bar{M}_0) \qquad \text{a.e. in } \mathbb{R}^N. \tag{51}$$

Moreover (50) implies that

$$w_{R,k} \to 0 \qquad \text{a.e. in } \mathbb{R}^N,$$

while from

$$\nabla w_{R,k} = \psi(v_k)\nabla\chi_R + \chi_R\psi'(v_k)\nabla v_k \qquad \text{a.e. in } \mathbb{R}^N \tag{52}$$

and (46)–(49) and (51) it follows that $w_{R,k} \in X_{B_{R+1}}$. Hence, for all $k \geq R+1$ we have that

$$w_{R,k} \in X_{B_k}$$

so (iii), (19) and (52) imply that

$$0 = \langle d\mathcal{J}(u_k), w_{R,k}\rangle = \int_{B_{R+1}} \psi(v_k)A(x,u_k)|\nabla u_k|^{p-2}\nabla u_k \cdot \nabla\chi_R dx$$
$$+ \int_{B_{R+1}} \chi_R\psi'(v_k)A(x,u_k)|\nabla u_k|^{p-2}\nabla u_k \cdot \nabla v_k dx$$
$$+ \frac{1}{p}\int_{B_{R+1}} A_t(x,u_k)w_{R,k}|\nabla u_k|^p dx$$
$$+ \int_{B_{R+1}} V(x)|u_k|^{p-2}u_k w_{R,k} dx - \int_{B_{R+1}} \xi(x)|u_k|^{q-2}u_k w_{R,k} dx.$$

Now, from (46), (47), (17), (37) and (44) it results

$$\left|\int_{B_{R+1}} \xi(x)|u_k|^{q-2}u_k w_{R,k} dx\right| \leq \int_{B_{R+1}} \xi(x)|u_k|^{q-1}|v_k| e^{\eta v_k^2} dx$$
$$\leq e^{\eta\bar{M}_0^2}|\xi|_{\frac{p}{p-q}}|u_k|_p^{q-1}|v_k|_p \to 0.$$

The remainder of the proof follows as in [16, Proposition 6.8] since the term $g(x,u) = \xi(x)|u|^{q-2}u$ does not appear anymore, while the functions A, A_t and V satisfy the same hypotheses in [16]. □

Proposition 5.4 *We have that*

$$\langle d\mathcal{J}(u^*), \varphi\rangle = 0 \quad \text{for all } \varphi \in C_c^\infty(\mathbb{R}^N)$$

with $C_c^\infty(\mathbb{R}^N) = \{\varphi \in C^\infty(\mathbb{R}^N) : \text{supp}\varphi \subset\subset \mathbb{R}^N\}$. *Hence,* $d\mathcal{J}(u^*) = 0$ *in* X.

Proof The proof follows from Propositions 5.2 and 5.3 arguing as in [16, Proposition 6.9]. □

Proof of Theorem 4.1 From Proposition 5.4 we have that u^* is a critical point of $\mathcal{J}$, i.e., a weak bounded solution of (3). Suppose by contradiction that $u^* = 0$. From (40), it follows that $u_k \to 0$ in $L^p(\mathbb{R}^N)$. Hence, (16) implies that

$$\int_{\mathbb{R}^N} \xi(x)|u_k|^q dx \to 0. \tag{53}$$

Furthermore, from (iii), (19), (h_4) and (h_2) it follows that

$$0 = \langle d\mathcal{J}(u_k), u_k\rangle = \int_{\mathbb{R}^N} A(x, u_k)|\nabla u_k|^p dx + \frac{1}{p}\int_{\mathbb{R}^N} A_t(x, u_k)u_k|\nabla u_k|^p dx$$
$$+ \int_{\mathbb{R}^N} V(x)|u_k|^p dx - \int_{\mathbb{R}^N} \xi(x)|u_k|^q dx$$
$$\geq \frac{\alpha_0\alpha_2}{p}\int_{\mathbb{R}^N} |\nabla u_k|^p dx + \int_{\mathbb{R}^N} V(x)|u_k|^p dx - \int_{\mathbb{R}^N} \xi(x)|u_k|^q dx,$$

whence, by means of (53) we have that

$$\|u_k\|_V^p \to 0 \quad \text{as } k \to +\infty. \tag{54}$$

Hence, from (18), (37), (h_1), (53) and (54) a positive constant $c > 0$ exists such that

$$\mathcal{J}(u_k) \leq c\|u_k\|_V^p - \int_{\mathbb{R}^N} \xi(x)|u_k|^q dx \to 0$$

which contradicts (ii). Then, it must be $u^* \not\equiv 0$. □

Acknowledgments The research that led to the present paper was partially supported by MIUR–PRIN project "Qualitative and quantitative aspects of nonlinear PDEs" (2017JPCAPN_005), *Fondi di Ricerca di Ateneo* 2017/18 "Problemi differenziali non lineari".

Both the authors are members of the Research Group INdAM-GNAMPA.

References

1. Arcoya, D., Boccardo, L: Critical points for multiple integrals of the calculus of variations. Arch. Rational Mech. Anal. **134**, 249–274 (1996)
2. Arioli, G., Gazzola, F.: Existence and multiplicity results for quasilinear elliptic differential systems. Commub. Partial Differential Equations **25**, 125–153 (2000)
3. Badiale, M., Guida, M., Rolando, S.: Compactness and existence results for the p-Laplace equation. J. Math. Anal. Appl. **451**, 345–370 (2017)
4. Bartolo, R., Candela, A.M., Salvatore, A.: Infinitely many solutions for a perturbed Schrödinger equation. Discrete Contin. Dyn. Syst. Ser. S, 94–102 (2015)
5. Bartolo, R., Candela, A.M., Salvatore, A.: Multiplicity results for a class of asymptotically p–linear equation on $\mathbb{R}^N$. Commun. Contemp. Math. **18**, Article 1550031 (24 pp) (2016)
6. Bartsch, T., Wang, Z.Q.: Existence and multiplicity results for some superlinear elliptic problems on $\mathbb{R}^N$. Commun. Partial Differential Equations **20**, 1725–1741 (1995)
7. Benci, V., Fortunato, D.: Discreteness conditions of the spectrum of Schrödinger operators. J. Math. Anal. Appl. **64**, 695–700 (1978)
8. Boccardo, L., Murat, F., Puel, J.P.: Existence of bounded solutions for nonlinear elliptic unilateral problems. Ann. Mat. Pura Appl. IV Ser. **152**, 183–196 (1988)
9. Brezis, H.: Functional Analysis, Sobolev Spaces and Partial Differential Equations, Universitext, vol. XIV. Springer, New York (2011)
10. Candela, A.M., Palmieri, G.: Multiple solutions of some nonlinear variational problems. Adv. Nonlinear Stud. **6**, 269–286 (2006)
11. Candela, A.M., Palmieri, G.: Infinitely many solutions of some nonlinear variational equations. Calc. Var. Partial Differential Equations **34**, 495–530 (2009)
12. Candela, A.M., Palmieri, G.: Some abstract critical point theorems and applications. In: Hou, X., Lu, X., Miranville, A., Su, J., Zhu, J. (eds.) Dynamical Systems, Differential Equations and Applications. *Discrete Contin. Dynam. Syst.* **Suppl. 2009**, 133–142 (2009)
13. Candela, A.M., Palmieri, G., Salvatore, A.: Positive solutions of modified Schrödinger equations on unbounded domains. Preprint.
14. Candela, A.M., Salvatore, A.: Existence of minimizer for some quasilinear elliptic problems. Discrete Contin. Dynam. Syst. **Ser. S 13**, 3335–3345 (2020)
15. Candela, A.M., Salvatore, A.: Existence of radial bounded solutions for some quasilinear elliptic equations in $\mathbb{R}^N$. Nonlinear Anal. **191**, Article 111625 (26 pp) (2020)
16. Candela, A.M., Salvatore, A., Sportelli, C.: Bounded solutions for weighted quasilinear modified Schrödinger equations. Calc. Var. Partial Differential Equations **61**, 220 (2022)
17. Canino, A., Degiovanni, M.: Nonsmooth critical point theory and quasilinear elliptic equations. In: Granas, A., Frigon, M., Sabidussi, G. (eds.) Topological Methods in Differential Equations and Inclusions 1–50. NATO Adv. Sci. Inst. Ser. C Math. Phys. Sci., vol. 472. Kluwer Acad. Publ., Dordrecht (1995)
18. Cerami, G., De Villanova, G., Solimini, S.: Solutions for a quasilinear Schrödinger equations: a dual approach. Nonlinear Anal. TMA. **56**, 213–226 (2004)
19. Cerami, G., Passaseo, D., Solimini, S.: Nonlinear scalar field equations: existence of a positive solution with infinitely many bumps. Ann. Inst. H. Poincaré Anal. Non Linéaire **32**, 23–40 (2015)
20. Colin, M., Jeanjean, L.: Infinitely many bound states for some nonlinear field equations. Calc. Var. Partial Differential Equations **23**, 139–168 (2005)
21. Ding, Y., Szulkin, A.: Bound states for semilinear Schrödinger equations with sign-changing potential. Calc. Var. **29**, 397–419 (2007)
22. Glowinski, R., Marrocco, A.: Sur l'approximation, par éléments finis d'ordre un, et la résolution, par pénalisation-dualité, d'une classe de problémes de Dirichlet non linéaires, Rev. Française Automat. Informat. Recherche Opérationnelle Sér. Rouge Anal. Numér. **9**, 41–76 (1975)

23. Ladyzhenskaya, O.A., Ural'tseva, N.N.: Linear and Quasilinear Elliptic Equations. Academic Press, New York (1968)
24. Li, G., Wang, C.: The existence of a nontrivial solution to p-Laplacian equations in $\mathbb{R}^N$ with supercritical growth. Math. Methods Appl. Sci. **36**, 69–79 (2013)
25. Lindqvist, P.: On the equation $\text{div}(|\nabla u|^{p-2}\nabla u) + \lambda|u|^{p-2}u = 0$. Proc. Amer. Math. Soc. **109**, 157–164 (1990)
26. Liu, C., Zheng, Y.: Existence of nontrivial solutions for p–Laplacian equations in $\mathbb{R}^N$. J. Math. Anal. Appl. **380**, 669–679 (2011)
27. Liu, J.Q., Wang, Y.Q., Wang, Z.Q.: Soliton solutions for quasilinear Schrödinger equations, II. J. Differential Equations **187**, 473–493 (2003)
28. Rabinowitz, P.H.: On a class of nonlinear Schrödinger equations. Z. Angew. Math. Phys. **43**, 270–291 (1992)
29. Salvatore, A.: Multiple solutions for perturbed elliptic equations in unbounded domains. Adv. Nonlinear Stud. **3**, 1–23 (2003)
30. Shen, Y., Wang, Y.: Soliton solutions for generalized quasilinear Schrödinger equations. Nonlinear Anal. **80**, 194–201 (2013)
31. Shi, H., Chen, H.: Existence and multiplicity of solutions for a class of generalized quasilinear Schrödinger equations. J. Math. Anal. Appl. **452**, 578–594 (2017)

Elliptic and Parabolic Problems for a Bessel-Type Operator

Giorgio Metafune, Luigi Negro, and Chiara Spina

Abstract We study elliptic and parabolic problems governed by the singular elliptic operators

$$\mathcal{L} = \sum_{i,j=1}^{N} q_{ij} D_{x_i x_j} + D_{yy} + \frac{c}{y} D_y,$$

in the half-space $\mathbb{R}^{N+1}_+ = \{(x, y) : x \in \mathbb{R}^N, y > 0\}$ under Neumann boundary conditions at $y = 0$. Here $Q = \left(q_{ij}\right)_{i,j=1,\dots,N}$ is a uniformly elliptic symmetric matrix.

Keywords Degenerate elliptic operators · Boundary degeneracy · Harmonic analysis

1 Introduction

We study solvability and regularity of elliptic and parabolic problems associated to the degenerate operators

$$\mathcal{L} = \sum_{i,j=1}^{N} q_{ij} D_{x_i x_j} + D_{yy} + \frac{c}{y} D_y \tag{1}$$

G. Metafune (✉) · L. Negro · C. Spina
Dipartimento di Matematica e Fisica "Ennio De Giorgi", Università del Salento, Lecce, Italy
e-mail: giorgio.metafune@unisalento.it; luigi.negro@unisalento.it; chiara.spina@unisalento.it

A. M. Candela et al. (eds.), *Recent Advances in Mathematical Analysis*,
Trends in Mathematics, https://doi.org/10.1007/978-3-031-20021-2_20

in the half-space $\mathbb{R}^{N+1}_+ = \{(x, y) : x \in \mathbb{R}^N, y > 0\}$ under Neumann boundary conditions at $y = 0$. Here $c \in \mathbb{R}$ and $Q = (q_{ij})_{i,j=1,\dots,N}$ is a constant, real symmetric matrix such that for some $M, \nu > 0$

$$\nu|\zeta|^2 \le \sum_{i,j=1}^{N} q_{ij}\zeta_i\zeta_j \le M|\zeta|^2, \qquad \forall \zeta \in \mathbb{R}^N \tag{2}$$

and $B_y = D_{yy} + \frac{c}{y}D_y$ is a singular operator of Bessel type.

These problems have been extensively studied in [9] and in [1–4] with different methods. In particular, in [9], methods from operator valued harmonic analysis have been used.

In this paper we prove solvability and regularity in L^p spaces with respect to the Lebesgue measure (see Sect. 3.1), using classical harmonic analysis, therefore simplifying the above approaches which, however, can be applied to more general underlying measures. Moreover, we show how to obtain higher order estimates up to the boundary for this kind of operators.

The case of variable (bounded and uniformly continuous) coefficients $q_{ij} = q_{ij}(x, y)$ can be treated without difficulties by standard perturbation arguments and Korn artifice and it is not discussed here, see also [3, 4] for more irregular coefficients. Results in spaces of continuous functions and Schauder estimates will appear elsewhere.

Let us explain in more detail the main differences between the approach in [9] and that of the present paper, assuming $q_{ij} = \delta_{ij}$ for simplicity.

If $\lambda u - \Delta_x u - B_y u = f$, in [9] we take the Fourier transform $\hat{u}$ only with respect to x with covariable ξ, thus obtaining

$$(\lambda + |\xi|^2)\hat{u}(\xi, y) - B_y\hat{u}(\xi, y) = \hat{f}(\xi, y).$$

Then $\hat{u} = (\lambda + |\xi|^2 - B_y)^{-1}\hat{f}$ which leads to the study of the operator valued multiplier

$$M : \mathbb{R}^N \setminus \{0\} \to \mathcal{B}(L^p(\mathbb{R}^+)), \qquad M(\xi) = (\lambda + |\xi|^2 - B_y)^{-1}$$

which allows to write $u = \mathcal{F}^{-1}M(\xi)\mathcal{F}f$. At this point, tools from operator valued harmonic analysis relying on the $\mathcal{R}$-boundedness of the heat kernel of B_y, which we prove through heat kernel estimates, allow to estimate the Sobolev norm of u with the L^p norm of f. This method has also the advantage to work in weighted spaces (the measure we consider is $y^m dxdy$) and allow to add an inverse square potential to $\mathcal{L}$ or to deal with Dirichlet boundary conditions. Estimates for the mixed derivatives are then deduced by classical localization techniques and Rellich inequalities.

Here instead, we take the Fourier transform with respect to all variables in the equation $\lambda u - \mathcal{L}u = f$ thus obtaining, see Lemma 3.1,

$$\hat{u}(\xi, \eta) = \frac{\hat{f}(\xi, \eta)}{\lambda + Q(\xi, \eta)} - \frac{c}{\lambda + Q(\xi, \eta)} \int_1^{+\infty} \hat{f}(\xi, \eta\sigma) \frac{\eta^2 \sigma}{\lambda + Q(\xi, \eta\sigma)} N_{\lambda,\sigma}(\xi, \eta\sigma)\, d\sigma,$$

where $Q(\xi, \eta) = \sum_{i,j=1}^{N} q_{ij}\xi_i\xi_j + \eta^2$ and

$$N_{\lambda,\sigma}(\xi, s) = e^{-cM_{\lambda,\sigma}(\xi,s)}, \qquad M_{\lambda,\sigma}(\xi, s) = \int_{\frac{s}{\sigma}}^{s} \frac{r}{\lambda + Q(\xi, r)}\, dr.$$

While the multiplier $(\lambda + Q(\xi, \eta))^{-1}$ and its variants $\frac{\xi_i\xi_j}{\lambda+Q(\xi,\eta)}$ are well-known, being connected with the resolvent of $\mathcal{L}_0 = \sum_{i,j=1}^{N+1} q_{ij} D_{x_i x_j}$, $N_{\lambda,\sigma}(\xi, \eta)$ is new and we prove in Sect. 2 that the Marcinkiewicz condition holds. Elliptic solvability is proved in Sect. 3 and regularity results are treated in Sect. 4. In Sect. 5 we show how to extend our results to parabolic equations. Further results and a comparison with the existing literature are treated in Sect. 6. Well known results on multipliers are collected in Sect. 7 where we give complete though elementary proofs to check the dependence on parameters, since we need in Sect. 2.

Notation For $N \geq 0$, $\mathbb{R}^{N+1}_+ = \{(x, y) : x \in \mathbb{R}^N, y > 0\}$. For $m \in \mathbb{R}$ we consider the measure $y^m dxdy$ in $\mathbb{R}^{N+1}_+$ and we write $L^p_m(\mathbb{R}^{N+1}_+)$ for $L^p(\mathbb{R}^{N+1}_+; y^m dxdy)$ and often only L^p_m when $\mathbb{R}^{N+1}_+$ is understood. We omit m in the case of the Lebesgue measure. Similarly $W^{k,p}_m(\mathbb{R}^{N+1}_+) = \{u \in L^p_m(\mathbb{R}^{N+1}_+) : \partial^\alpha u \in L^p_m(\mathbb{R}^{N+1}_+) \quad |\alpha| \leq k\}$. We use $C_c^\infty(\mathbb{R}^{N+1})$ for the space of test functions and $\mathbb{N}_0$ for the natural numbers starting from 0. For $|\theta| \leq \pi$, we denote by Σ_θ the open sector $\{\lambda \in \mathbb{C} : \lambda \neq 0,\ |Arg(\lambda)| < \theta\}$. We use $c^+ = \max\{0, c\}$ and $c^- = -\min\{0, c\}$.

Given $u \in L^1(\mathbb{R}^{N+1})$ we denote by $\hat{u}$ or $\mathcal{F}u$ its Fourier transform defined by $\hat{u}(\xi, \eta) = C_N \int_{\mathbb{R}^N} u(x, y) e^{ix\cdot\xi} e^{iy\eta}\, dxdy$, $C_N = (2\pi)^{-\frac{N+1}{2}}$. With this choice, the inverse Fourier transform is given by $\mathcal{F}^{-1}u(\xi, \eta) = \mathcal{F}u(-\xi, -\eta)$.

2 The Multiplier $N_{\lambda,\sigma}(\xi, s) = \exp\left(-c \int_{\frac{s}{\sigma}}^{s} \frac{r}{\lambda+Q(\xi,r)}\, dr\right)$

Let us write $\zeta = (\xi, s) \in \mathbb{R}^{N+1}$ where $\xi \in \mathbb{R}^N$ and $s \in \mathbb{R}$ and let $Q(\xi, s) = \sum_{i,j=1}^{N} q_{ij}\xi_i\xi_j + s^2$, with Q satisfying (2).

We define for $\lambda \in \Sigma_\pi$, $\sigma \geq 1$ and $(\xi, s) \in \mathbb{R}^{N+1} \setminus \{0\}$

$$M_{\lambda,\sigma}(\xi, s) := \int_{\frac{s}{\sigma}}^{s} \frac{r}{\lambda + Q(\xi, r)}\, dr, \qquad N_{\lambda,\sigma}(\xi, s) = e^{-cM_{\lambda,\sigma}(\xi,s)}. \tag{3}$$

Let us begin by estimating $M_{\lambda,\sigma}$ and its derivatives. $M_{\lambda,\sigma}$ is even in the last variable which we may assume to be positive. In the following elementary lemma we use the branch of the logarithm with argument in $(-\pi,\pi)$.

Lemma 2.1 *Let $0<a<b<\infty$ and $\lambda\in\Sigma_\pi$. Then*

$$\int_a^b \frac{r}{\lambda+Q(\xi,r)}\,dr=\frac{1}{2}\mathrm{Log}\left(\frac{\lambda+Q(\xi,b)}{\lambda+Q(\xi,a)}\right).$$

and for any $\epsilon>0$ there exists a positive constant C_1, C_2 such that for $\lambda\in\Sigma_{\pi-\epsilon}$

$$-C_1\le \mathrm{Re}\int_a^b \frac{r}{\lambda+Q(\xi,r)}\,dr\le \log\left(\frac{b}{a}\right)+C_2$$

Proof The computation of the integral is elementary. Then we use the

$$c_\epsilon\left(|\lambda|+s\right)\le|\lambda+s|\le|\lambda|+s,\qquad \forall\lambda\in\Sigma_{\pi-\epsilon},\quad s\ge 0.$$

to obtain, with $b=ta$, $t\ge 1$ and for suitable $c_1,c_2>0$

$$c_1\le c_1\frac{|\lambda|+|\xi|^2+t^2a^2}{|\lambda|+|\xi|^2+a^2}\le\left|\frac{\lambda+Q(\xi,b)}{\lambda+Q(\xi,a)}\right|\le c_2\frac{|\lambda|+|\xi|^2+t^2a^2}{|\lambda|+|\xi|^2+a^2}\le c_2t^2=c_2\left(\frac{b}{a}\right)^2.$$

The final estimate follows since

$$\mathrm{Re}\,\mathrm{Log}\left(\frac{\lambda+Q(\xi,b)}{\lambda+Q(\xi,a)}\right)=\log\left|\frac{\lambda+Q(\xi,b)}{\lambda+Q(\xi,a)}\right|.$$

□

Corollary 2.1 *Let $\epsilon>0$. Then there exists a positive constant C such that for every $0<a<b$, $\lambda\in\Sigma_{\pi-\epsilon}$ and $\xi\in\mathbb{R}^N$*

$$\left|\exp\left(-c\int_a^b\frac{r}{\lambda+Q(\xi,r)}\,dr\right)\right|\le C\left(\frac{b}{a}\right)^{c^-}.$$

Proof This follows from the previous lemma, by distinguishing between $c\ge 0$ and $c<0$. □

Let us now consider derivatives of $M_{\lambda,\sigma}$.

Lemma 2.2 *Let $0<\varepsilon\le\pi$, $\lambda\in\Sigma_{\pi-\varepsilon}$, $\sigma\ge 1$. Then for any multi-index $|\alpha|\ge 1$ there exists $C>0$ such that*

$$\left|D_\xi^\alpha M_{\lambda,\sigma}(\xi,s)\right|\le C\frac{1}{\left(|\lambda|+|\xi|^2\right)^{\frac{|\alpha|}{2}}},\qquad \forall(\xi,s)\in\mathbb{R}^{N+1}\setminus\{0\}.$$

Proof We differentiate under the integral sign and using Proposition 7.1 we get

$$\left|D_\xi^\alpha M_{\lambda,\sigma}(\xi,s)\right| \le \int_{-\infty}^{\infty} |r| \left|D_\xi^\alpha\left(\frac{1}{\lambda+Q(\xi,r)}\right)\right| dr \le C\int_0^\infty \frac{r}{\left(|\lambda|+|\xi|^2+r^2\right)^{\frac{|\alpha|}{2}+1}}\, dr$$

$$\le C\int_0^\infty \frac{1}{\left(|\lambda|+|\xi|^2+r^2\right)^{\frac{|\alpha|+1}{2}}}\, dr = \frac{C}{\left(|\lambda|+|\xi|^2\right)^{\frac{|\alpha|}{2}}}\int_0^\infty \frac{1}{\left(1+u^2\right)^{\frac{|\alpha|+1}{2}}}\, du.$$

□

Lemma 2.3 *Let* $0<\varepsilon\le\pi$, $\lambda\in\Sigma_{\pi-\varepsilon}$, $\sigma\ge 1$. *For any multi-index* α, *there exists* $C>0$ *such that*

$$\left|D_\xi^\alpha D_s M_{\lambda,\sigma}(\xi,s)\right| \le C\frac{1}{\left(|\lambda|+|\xi|^2\right)^{\frac{|\alpha|}{2}}}\frac{1}{\left(|\lambda|+s^2\right)^{\frac{1}{2}}}, \qquad \forall(\xi,s)\in\mathbb{R}^{N+1}\setminus\{0\}.$$

Proof Since

$$D_s M_{\lambda,\sigma}(\xi,s) = \frac{s}{\lambda+Q(\xi,s)} - \frac{s}{\sigma^2}\frac{1}{\lambda+Q(\xi,\frac{s}{\sigma})}.$$

using Proposition 7.1 we get

$$\left|D_\xi^\alpha D_s M_{\lambda,\sigma}(\xi,s)\right| \le C\left[\frac{|s|}{\left(|\lambda|+|\xi|^2+s^2\right)^{\frac{|\alpha|}{2}+1}} + \frac{|s|}{\sigma^2}\frac{1}{\left(|\lambda|+|\xi|^2+\frac{s^2}{\sigma^2}\right)^{\frac{|\alpha|}{2}+1}}\right]$$

$$\le C\frac{1}{\left(|\lambda|+|\xi|^2\right)^{\frac{|\alpha|}{2}}}\left[\frac{1}{\left(|\lambda|+|\xi|^2+s^2\right)^{\frac{1}{2}}} + \frac{1}{\sigma}\frac{1}{\left(|\lambda|+|\xi|^2+\frac{s^2}{\sigma^2}\right)^{\frac{1}{2}}}\right]$$

$$\le C\frac{1}{\left(|\lambda|+|\xi|^2\right)^{\frac{|\alpha|}{2}}}\frac{1}{\left(|\lambda|+|\xi|^2+s^2\right)^{\frac{1}{2}}}.$$

□

We can finally prove that the multiplier $N_{\lambda,\sigma}(\xi,s)=e^{-cM_{\lambda,\sigma}(\xi,s)}$ defined in (3) satisfies the hypotheses of the Marcinkiewicz multiplier Theorem.

Proposition 2.1 *Let* $0 < \varepsilon \le \pi$, $\lambda \in \Sigma_{\pi-\varepsilon}$, $\sigma \ge 1$. *Then there exist* $C > 0$ *such that*

$$\sup\left\{|\xi^\alpha s^j D_\xi^\alpha D_s^j N_{\lambda,\sigma}(\xi,s)| : (\xi,s) \in \mathbb{R}^{N+1}\setminus\{0\},\ \alpha\in\{0,1\}^N,\ j\in\{0,1\}\right\}$$
$$\le C\sigma^{c^-}.$$

Proof From Corollary 2.1 we get $\left|N_{\lambda,\sigma}(\xi,s)\right| \le C\sigma^{c^-}$, which is the claim for $|\alpha| = j = 0$.

The case $|\alpha|+j > 0$ follows from Lemmas 2.2 and 2.3 since for any multi-index $\beta = (\alpha_1,\dots,\alpha_N, j)$ as in the statement, the derivative $D^\beta_{(\xi,s)}N_{\lambda,\sigma}(\xi,s)$ is a linear combination of terms of the form

$$N_{\lambda,\sigma}\left(D^{\beta^1}_{\xi,s}M_{\lambda,\sigma}\right)^{k_1}\cdots\left(D^{\beta^{N+1}}_{\xi,s}M_{\lambda,\sigma}\right)^{k_{N+1}},\quad |\beta| = |\beta^1|k_1+\cdots+|\beta^{N+1}|k_{N+1}$$

where $k_i \in \{0,1\}$ and $\beta^i = (\beta^i_1,\dots,\beta^i_{N+1}) \in \mathbb{N}^{N+1}$ are multi-indexes such that $\beta^i_j \in \{0,1\}$. □

Proposition 2.1 and Theorem 7.1 show that the operator $T_{N_{\lambda,\sigma}}$ associated to the multiplier $N_{\lambda,\sigma}(\xi,s)$, that is

$$(\mathcal{F}T_{N_{\lambda,\sigma}}f)(\xi,s) = N_{\lambda,\sigma}(\xi,s)\,\hat{f}(\xi,s),\qquad f\in\mathcal{S}\left(\mathbb{R}^{N+1}\right). \tag{4}$$

is bounded in $L^p(\mathbb{R}^{N+1})$ with

$$\|T_{N_{\lambda,\sigma}}\|_p \le C\sigma^{c^-}.$$

Next we state the parabolic version of Proposition 2.1.

Proposition 2.2 *Let us consider for* $\lambda > 0$, $\sigma \ge 1$ *the multiplier*

$$(\xi,s,\tau)\in\mathbb{R}^{N+2}\setminus\{0\}\mapsto N_{\lambda+i\tau,\sigma}(\xi,s).$$

Then there exist $C = C(\epsilon, Q) > 0$ *such that*

$$\sup\left\{|\xi^\alpha s^j\tau^i D_\xi^\alpha D_s^j D_\tau^i N_{\lambda+i\tau,\sigma}(\xi,s)| : (\xi,s,\tau)\in\mathbb{R}^{N+2}\setminus\{0\},\right.$$
$$\left.\alpha\in\mathbb{N}^N,\ |\alpha|\le k,\ \alpha_i, j\in\{0,1\}\right\}\le C\sigma^{c^-}.$$

Proof For $i = 0$ the claim follows from Proposition 2.1. For $i = 1$ we observe that

$$D_\tau M_{\lambda+i\tau,\sigma}(\xi, s) = -i \int_{\frac{s}{\sigma}}^{s} \frac{r}{(\lambda + i\tau + Q(\xi, r))^2}\, dr.$$

Then proceeding as in the proof of Lemma 2.2 one has

$$\left|D_\tau M_{\lambda+i\tau,\sigma}(\xi, s)\right| \le \frac{C}{\lambda + |\tau| + |\xi|^2} \le \frac{C}{|\tau|}.$$

The required claim for $i = 1$ then follows from the last inequality, Proposition 2.1 and equality

$$D_\xi^\alpha D_s^j D_\tau N_{\lambda+i\tau,\sigma}(\xi, s) = -c\left(D_\xi^\alpha D_s^j N_{\lambda+i\tau,\sigma}(\xi, s)\right) D_\tau M_{\lambda+i\tau,\sigma}(\xi, s).$$

□

3 The Elliptic Problem

Let

$$\mathcal{L} = \sum_{i,j=1}^{N} q_{ij} D_{x_i x_j} + D_{yy} + \frac{c}{y} D_y,$$

in the half-space $\mathbb{R}_+^{N+1} = \{(x, y) : x \in \mathbb{R}^N, y > 0\}$. We impose Neumann boundary conditions when $y = 0$ by considering the Sobolev space

$$W_{\mathcal{N}}^{2,p} = \{u \in W^{2,p}(\mathbb{R}_+^{N+1}) : D_y u(x, 0) = 0\}.$$

By Hardy inequality, $\mathcal{L}$ is continuous from $W_{\mathcal{N}}^{2,p}$ to $L^p(\mathbb{R}_+^{N+1})$. In the following we identify the space $W_{\mathcal{N}}^{2,p}$ with the space of even functions with respect to the y variable in $W^{2,p}(\mathbb{R}^{N+1})$ and we consider the set

$$C := \left\{u \in C_c^\infty\left(\mathbb{R}^{N+1}\right),\ u(x, y) = u(x, -y)\right\}. \tag{5}$$

It is important to note that $\lambda u - \mathcal{L}u$ is even in y, for $u \in C$.

We deduce solvability in $W_{\mathcal{N}}^{2,p}$ by the following a-priori estimates.

Proposition 3.1 *Let* $1 < p < \infty$, $\frac{1}{p} < c+1$, $\lambda \in \Sigma_{\pi-\varepsilon}$. *Then there exists a positive constant* C *such that for every* $u \in W^{2,p}_{\mathcal{N}}$

$$|\lambda|\|u\|_p + |\lambda|^{\frac{1}{2}}\|\nabla u\|_p + \|D^2 u\|_p \le C\|\lambda u - \mathcal{L}u\|_p.$$

The constant C *depends on* $\varepsilon, p, N, \nu, M, c$ *and stays bounded if* $0 < a \le c+1-\frac{1}{p} \le b$.

We first prove some preliminary lemmas involving the multiplier $N_{\lambda,\sigma}$ defined in (3).

Lemma 3.1 *Let* $\lambda \in \Sigma_\pi$, $u \in \mathcal{C}$ *and* $f = \lambda u - \mathcal{L}u$. *Then*

$$\hat{u}(\xi,\eta) = \frac{\hat{f}(\xi,\eta)}{\lambda + Q(\xi,\eta)} - \frac{c}{(\lambda + Q(\xi,\eta))}\int_1^{+\infty} \hat{f}(\xi,\eta\sigma)\frac{\eta^2\sigma}{\lambda+Q(\xi,\eta\sigma)}N_{\lambda,\sigma}(\xi,\eta\sigma)\,d\sigma.$$

Proof Since $D_y u(x,0) = 0$ then

$$\frac{D_y u}{y}(x,y) = \int_0^1 D_{yy}u(x,sy)\,ds$$

and the equation $\lambda u - \mathcal{L}u = f$ becomes

$$\lambda u - \sum_{i,j=1}^N q_{ij} D_{x_i x_j} u - D_{yy}u - c\int_0^1 D_{yy}u(x,sy)\,ds = f. \tag{6}$$

Let us compute the Fourier transform of the term

$$v(x,y) = \int_0^1 D_{yy}u(x,sy)\,ds.$$

Since $u(x,y) = u(x,-y)$, it follows that also $\hat{v}$ satisfies $\hat{v}(\xi,\eta) = \hat{v}(\xi,-\eta)$ and we can therefore restrict to $\eta \ge 0$. Then

$$\begin{aligned}
\hat{v}(\xi,\eta) &= C_N \int_{\mathbb{R}^{N+1}} v(x,y)e^{-ix\cdot\xi}e^{-iy\eta}\,dx\,dy \\
&= C_N \int_{\mathbb{R}^N} e^{-ix\cdot\xi}\,dx \int_0^1 ds \int_{\mathbb{R}} D_{yy}u(x,sy)e^{-iy\eta}\,dy \\
&= -C_N \int_{\mathbb{R}^N} e^{-ix\cdot\xi}\,dx \int_0^1 ds \int_{\mathbb{R}} \frac{\eta^2}{s^2}u(x,sy)e^{-iy\eta}\,dy
\end{aligned}$$

$$= -C_N \eta^2 \int_{\mathbb{R}^N} e^{-ix\cdot\xi}\,dx \int_0^1 ds \frac{1}{s^3} \int_{\mathbb{R}} u(x,z) e^{-iz\frac{\eta}{s}}\,dz$$

$$= -\eta^2 \int_0^1 \hat{u}\left(\xi, \frac{\eta}{s}\right)\frac{ds}{s^3} = -\eta^2 \int_1^{+\infty} \hat{u}(\xi, \eta\sigma)\sigma\,d\sigma = -\int_\eta^{+\infty} \hat{u}(\xi, z) z\,dz.$$

It follows that

$$D_\eta \hat{v}(\xi, \eta) = \eta \hat{u}(\xi, \eta). \tag{7}$$

Taking the Fourier transform of (6) we obtain

$$\lambda \hat{u} + Q(\xi, \eta)\hat{u} + c \int_\eta^{+\infty} \hat{u}(\xi, z) z\,dz = \hat{f}. \tag{8}$$

From (7) and (8), we get that $\hat{v}$ satisfies

$$(\lambda + Q(\xi, \eta))\frac{D_\eta \hat{v}}{\eta} - c\hat{v} = \hat{f} \quad \text{that is} \quad D_\eta \hat{v} - c\frac{\eta \hat{v}}{\lambda + Q(\xi, \eta)} = \frac{\eta \hat{f}}{\lambda + Q(\xi, \eta)}.$$

Setting

$$G_\lambda(\xi, \eta) = \int_0^\eta \frac{r}{\lambda + Q(\xi, r)}\,dr$$

we deduce

$$e^{c\,G_\lambda(\xi,\eta)} D_\eta (e^{-c\,G_\lambda(\xi,\eta)}\hat{v}) = \frac{\eta \hat{f}}{\lambda + Q(\xi, \eta)}$$

and then

$$\hat{v}(\xi, \eta) = -\int_\eta^\infty s\frac{\hat{f}(\xi, s)}{\lambda + Q(\xi, s)} \exp\{-c[G_\lambda(\xi, s) - G_\lambda(\xi, \eta)]\}\,ds.$$

From (7)

$$\hat{u} = \frac{\hat{f}}{\lambda + Q(\xi, \eta)} - \frac{c}{\lambda + Q(\xi, \eta)} \int_\eta^\infty s\frac{\hat{f}(\xi, s)}{\lambda + Q(\xi, s)} \exp\{-c[G_\lambda(\xi, s) - G_\lambda(\xi, \eta)]\}\,ds$$

$$:= \frac{\hat{f}}{\lambda + Q(\xi, \eta)} - \frac{c}{\lambda + Q(\xi, \eta)}\hat{w}(\xi, \eta).$$

To end the proof it is sufficient to observe that

$$\hat{w}(\xi,\eta) = \int_\eta^{+\infty} \hat{f}(\xi,s)\frac{s}{\lambda+Q(\xi,s)}e^{-c[G_\lambda(\xi,s)-G_\lambda(\xi,\eta)]}\,ds$$

$$= \int_1^{+\infty} \hat{f}(\xi,\eta\sigma)\frac{\eta^2\sigma}{\lambda+Q(\xi,\eta\sigma)}e^{-c[G_\lambda(\xi,\eta\sigma)-G_\lambda(\xi,\eta)]}\,d\sigma$$

$$= \int_1^{+\infty} \hat{f}(\xi,\eta\sigma)\frac{\eta^2\sigma}{\lambda+Q(\xi,\eta\sigma)}N_{\lambda,\sigma}(\xi,\eta\sigma)\,d\sigma.$$

□

We define the operator $T_{\lambda,\sigma} \in \mathcal{B}\left(L^p(\mathbb{R}^{N+1})\right)$ through the multiplier

$$\frac{s^2}{\lambda+Q(\xi,s)}N_{\lambda,\sigma}\,(\xi,s)\,,$$

i.e.

$$(\mathcal{F}T_{\lambda,\sigma}f)(\xi,s) = \frac{s^2}{\lambda+Q(\xi,s)}N_{\lambda,\sigma}\,(\xi,s)\,\hat{f}(\xi,s). \tag{9}$$

By Propositions 2.1, 7.2 and Theorem 7.1 we have $\|T_{\lambda,\sigma}\|_p \le C\,\sigma^{c^-}$. Note that, recalling (4), if $\mathcal{L}_0 = \text{Tr}\,(D^2Q)$ then $T_{\lambda,\sigma} = D_{yy}\,(\lambda-\mathcal{L}_0)^{-1}\,T_{N_{\lambda,\sigma}}$.

Lemma 3.2 *Let $\lambda \in \Sigma_{\pi-\varepsilon}$, $u \in C$ and $f = \lambda u - \mathcal{L}u$. Then*

$$u = (\lambda-\mathcal{L}_0)^{-1}\,(f - c\,S_\lambda f)\,,$$

where

$$S_\lambda f(x,y) := \int_1^{+\infty} \frac{1}{\sigma^2}(T_{\lambda,\sigma}f)\left(x,\frac{y}{\sigma}\right)\,d\sigma = \frac{1}{y}\int_0^y (T_{\lambda,\frac{y}{s}}f)(x,s)\,ds. \tag{10}$$

Moreover, if $1 < p < \infty$ and $\frac{1}{p} < c+1$, then the operator S_λ defined by (10) is bounded on $L^p\left(\mathbb{R}^{N+1}\right)$ and

$$\|S_\lambda f\|_p \le C_\varepsilon \|f\|_p, \quad \lambda \in \Sigma_{\pi-\varepsilon}.$$

Proof By Lemma 3.1

$$\hat{u}(\xi,\eta) = \frac{1}{\lambda+Q(\xi,\eta)}\left(\hat{f}(\xi,\eta) - c\hat{w}(\xi,\eta)\right).$$

where

$$\hat{w}(\xi,\eta) = \int_1^{+\infty} \hat{f}(\xi,\eta\sigma)\frac{\eta^2\sigma}{\lambda + Q(\xi,\eta\sigma)} N_{\lambda,\sigma}(\xi,\eta\sigma)\, d\sigma.$$

Taking the inverse transform and recalling (9) we get

$$w(x,y) = C_N \int_{\mathbb{R}^{N+1}} e^{i\xi\cdot x} e^{i\eta y}\, d\xi\, d\eta \int_1^{+\infty} \hat{f}(\xi,\eta\sigma)\frac{\eta^2\sigma}{\lambda + Q(\xi,\eta\sigma)} N_{\lambda,\sigma}(\xi,\eta\sigma)\, d\sigma$$

$$= C_N \int_1^{+\infty} \frac{1}{\sigma}\, d\sigma \int_{\mathbb{R}^{N+1}} e^{i\xi\cdot x} e^{i\eta y} \hat{f}(\xi,\eta\sigma)\frac{\eta^2\sigma^2}{\lambda + Q(\xi,\eta\sigma)} N_{\lambda,\sigma}(\xi,\eta\sigma)\, d\xi\, d\eta$$

$$= C_N \int_1^{+\infty} \frac{d\sigma}{\sigma^2} \int_{\mathbb{R}^{N+1}} e^{i\xi\cdot x} e^{i\frac{s}{\sigma}y} \hat{f}(\xi,s)\frac{s^2}{\lambda + Q(\xi,s)} N_{\lambda,\sigma}\,(\xi,s)\; d\xi\; ds$$

$$= \int_1^{+\infty} \frac{d\sigma}{\sigma^2} \int_{\mathbb{R}^{N+1}} (\mathcal{F}T_{\lambda,\sigma} f)(\xi,s) e^{i\xi\cdot x} e^{i\frac{s}{\sigma}y} d\xi\, ds = \int_1^{+\infty} \frac{1}{\sigma^2}(T_{\lambda,\sigma} f)\left(x, \frac{y}{\sigma}\right)\, d\sigma.$$

This proves the first claim. By the discussion preceding the statement of the present lemma, $T_{\lambda,\sigma}$ is bounded in $L^p(\mathbb{R}^{N+1})$ with $\left\|T_{\lambda,\sigma}\right\|_p \le C\sigma^{c^-}$. Therefore, by Minkowski inequality, since $\frac{1}{p} < c+1$, we get

$$\|w\|_p \le \int_1^{+\infty} \frac{d\sigma}{\sigma^2} \left\|(T_{\lambda,\sigma} f)\left(\cdot, \frac{\cdot}{\sigma}\right)\right\|_p = \int_1^{+\infty} \frac{1}{\sigma^2} \|T_{\lambda,\sigma} f\|_p \sigma^{\frac{1}{p}} d\sigma$$

$$\le \|f\|_p \int_1^{+\infty} \sigma^{\frac{1}{p}+c^- -2} d\sigma = C\|f\|_p.$$

□

Proof (Proposition 3.1) Let $u \in C$, $\lambda \in \Sigma_{\pi-\varepsilon}$, set $f = \lambda u - \mathcal{L}u$. Using Lemma 3.2 for the expression of u and the boundedness of S_λ, the claim follows from Propositions 7.1 and 7.2. The gradient estimate follows from classical interpolation inequalities. All the estimates extend to $W^{2,p}_{\mathcal{N}}$ by density. □

Note that the estimate of $y^{-1}D_y u$ is a consequence of Hardy inequality.

In order to prove existence and uniqueness of the equation $\lambda u - \mathcal{L}u = f$ we apply a simple variant of the continuity method.

Remark 3.1 When we apply the operator S_λ of Lemma 3.2 to functions $f \in L^p(\mathbb{R}^{N+1}_+)$ we mean that $S_\lambda f = S_\lambda \tilde{f}$ where $\tilde{f}$ is the even reflection of f around the plane $y = 0$.

Theorem 3.1 *Let $1 < p < \infty$, $\frac{1}{p} < c+1$, $\epsilon > 0$. Then there exists $C > 0$ such that for $\lambda \in \Sigma_{\pi-\varepsilon}$ the operator $\lambda - \mathcal{L} : W^{2,p}_{\mathcal{N}} \to L^p(\mathbb{R}^{N+1}_+)$ is invertible and*

$$(\lambda - \mathcal{L})^{-1} = (\lambda - \mathcal{L}_0)^{-1} (I - c\, S_\lambda).$$

Moreover

$$\|(\lambda - \mathcal{L})^{-1}\|_p \leq \frac{C}{|\lambda|}, \qquad \|D^2(\lambda - \mathcal{L})^{-1}\|_p \leq C.$$

Proof We consider the spaces

$$X = W^{2,p}_{\mathcal{N}}, \qquad Y = L^p(\mathbb{R}^{N+1}_+)$$

and the one parameter family of operators

$$\mathcal{L}_t = (1-t)\mathcal{L}_0 + t\mathcal{L}_1, \qquad t \in [0,1]$$

with $\mathcal{L}_0 = \lambda - \Delta_x - D_{yy}$, $\mathcal{L}_1 = \lambda - \mathcal{L}$. Clearly $\mathcal{L}_0$ is invertible from X to Y. We Observe that, since $0 \leq t \leq 1$ and $\frac{1}{p} < c+1$, then $\frac{1}{p} < ct+1$. Then by Proposition 3.1 applied to $\mathcal{L}_t$, there exists $c(t) > 0$ such that

$$\|u\|_{W^{2,p}_{\mathcal{N}}} \leq c(t)\|(\lambda - \mathcal{L}_t)u\|_p.$$

By Lemma 8.1 the operator $\mathcal{L}_1 = \lambda - \mathcal{L}$ is invertible and the estimates in the statement follow from Proposition 3.1.

Finally, the identity $(\lambda - \mathcal{L})^{-1} = (\lambda - \mathcal{L}_0)^{-1} (I - c\, S_\lambda)$ follows by density from Lemma 3.2, once the invertibility of $\lambda - \mathcal{L}$ has been proved. □

Corollary 3.1 *Assume that $1 < p, q < \infty$, $\frac{1}{p}, \frac{1}{q} < c+1$ and let $f \in L^p(\mathbb{R}^{N+1}_+) \cap L^q(\mathbb{R}^{N+1}_+)$. If $u \in W^{2,p}_{\mathcal{N}}$, $v \in W^{2,q}_{\mathcal{N}}$ satisfy $\lambda u - \mathcal{L}u = f$, $\lambda v - \mathcal{L}v = f$, $\lambda \in \Sigma_\pi$, then $u = v$.*

Proof The assertion is certainly true if $f = \lambda w - \mathcal{L}w$, with $w \in C$, since in this case both u, v coincide with w. Since C is dense in $W^{2,p}_{\mathcal{N}} \cap W^{2,q}_{\mathcal{N}}$, then $(\lambda - \mathcal{L})(C)$ is dense in $L^p(\mathbb{R}^{N+1}_+) \cap L^q(\mathbb{R}^{N+1}_+)$ (both with respect to the norm which is the sum of the corresponding norms) and the thesis follows by density. □

Let us rephrase the above theorem in the language of semigroup theory.

Corollary 3.2 *If $1 < p < \infty$, $\frac{1}{p} < c+1$, the operator $\mathcal{L}$, endowed with domain $W^{2,p}_{\mathcal{N}}$, generates a bounded analytic semigroup of angle $\pi/2$ in $L^p(\mathbb{R}^{N+1}_+)$.*

3.1 Weighted Spaces

Let $m \in \mathbb{R}$ such that $0 < \frac{m+1}{p} < 1$. We generalize the results above to the spaces $L^p_m := L^p\left(\mathbb{R}^{N+1}_+, y^m dxdy\right)$. We define for $1 < p < \infty$ the weighted Sobolev space

$$W^{2,p}_m = \left\{u \in W^{2,p}_{loc}(\mathbb{R}^{N+1}_+) : u,\ D^2u, \nabla u \in L^p_m\right\},$$

$$W^{2,p}_{\mathcal{N},m} = \{u \in W^{2,p}_m : D_y u(x,0) = 0\},$$

where ∇u, D^2u denote the gradient and the hessian matrix with respect to all the variables. Also in this case the set C defined in (5) is dense in $W^{2,p}_{\mathcal{N},m}$. For a proof of this result, as well as for a detailed description of these spaces, we refer the reader to [7].

If $0 < \frac{m+1}{p} < 1$, the weight $w(y) = y^m$ belongs to the Muckenhoupt class $w \in A_p(\mathbb{R}^{N+1}_+)$. The same methods above yield the following result.

Proposition 3.2 *Let $1 < p < \infty$, $m \in \mathbb{R}$ satisfying $0 < \frac{m+1}{p} < c+1$ and $0 < \frac{m+1}{p} < 1$. Let $\epsilon > 0$. Then there exists a positive constant C such that for every $u \in W^{2,p}_{\mathcal{N},m}$ and $\lambda \in \Sigma_{\pi-\varepsilon}$*

$$|\lambda|\|u\|_{L^p_m} + |\lambda|^{\frac{1}{2}}\|\nabla u\|_{L^p_m} + \|D^2u\|_{L^p_m} \le C\|\lambda u - \mathcal{L}u\|_{L^p_m}.$$

Moreover the operator $\lambda - \mathcal{L} : W^{2,p}_{\mathcal{N},m} \to L^p_m$ is invertible and

$$\|(\lambda - \mathcal{L})^{-1}\|_{L^p_m} \le \frac{C}{|\lambda|}, \qquad \|D^2(\lambda - \mathcal{L})^{-1}\|_{L^p_m} \le C.$$

Proof The proof follows as in the case $m = 0$ and we point out only the (minor) changes. Let $u \in C$; then from Lemma 3.2 and writing $\mathcal{L}_0 = \mathrm{Tr}\left(QD^2u\right)$ we have

$$u = (\lambda - \mathcal{L}_0)^{-1}(f - c\,S_\lambda f),$$

$$S_\lambda f(x,y) = \int_1^{+\infty} \frac{1}{\sigma^2}(T_{\lambda,\sigma} f)\left(x, \frac{y}{\sigma}\right) d\sigma.$$

By Propositions 2.1, 7.2 and Theorem 7.1, if $0 < \frac{m+1}{p} < 1$ then $T_{\lambda,\sigma}$ is bounded in L^p_m with $\|T_{\lambda,\sigma}\|_{L^p_m} \le C\sigma^{c^-}$. Therefore, by Minkowski inequality, since $\frac{m+1}{p} < c+1$, we get

$$\|S_\lambda f\|_{L^p_m} \le \int_1^{+\infty} \frac{d\sigma}{\sigma^2} \left\|(T_{\lambda,\sigma} f)\left(\cdot, \frac{\cdot}{\sigma}\right)\right\|_{L^p_m} = \int_1^{+\infty} \frac{1}{\sigma^2} \|T_{\lambda,\sigma} f\|_{L^p_m} \sigma^{\frac{m+1}{p}} d\sigma$$

$$\le \|f\|_{L^p_m} \int_1^{+\infty} \sigma^{\frac{m+1}{p}+c^- -2} d\sigma = C\|f\|_{L^p_m}$$

that is S_λ is bounded on L^p_m and $\|S_\lambda\|_{L^p_m} \le C$, for all $\lambda \in \Sigma_{\pi-\varepsilon}$. Propositions 7.1 and 7.2 and Theorem 7.1 imply that also $D^2(\lambda - \mathcal{L}_0)^{-1}$, $(\lambda - \mathcal{L}_0)^{-1}$ are bounded on L^p_m. The remaining part of the proof follows identically as in Proposition 3.1 and Theorem 3.1. □

4 Elliptic Regularity

Let us show that the usual elliptic regularity results hold for the singular operator $\mathcal{L}$. We always assume that $1 < p, q < \infty$.

The following lemma allows to prove regularity up to the boundary and will be also used to prove the maximum principle.

Lemma 4.1 *Let $c+1 > 0$ and $\lambda \in \Sigma_\pi$. If $f \in C_c^\infty(\mathbb{R}^{N+1})$ then $S_\lambda f \in C^\infty(\mathbb{R}^{N+1})$ and tends to 0 at infinity with all its derivatives.*

Proof Let us set $w := S_\lambda f$. Then for any multi-index $\alpha \in \mathbb{N}_0^N$ and any $k \in \mathbb{N}_0$ one has, as in Lemma 3.2,

$$\xi^\alpha \eta^k \hat{w}(\xi,\eta) = \int_1^{+\infty} \xi^\alpha \eta^k \hat{f}(\xi,\eta\sigma) \frac{\eta^2\sigma}{\lambda + Q(\xi,\eta\sigma)} N_{\lambda,\sigma}(\xi,\eta\sigma)\, d\sigma$$

$$= \int_1^{+\infty} \sigma^{-k-1} \xi^\alpha (\eta\sigma)^k \hat{f}(\xi,\eta\sigma) \frac{\eta^2\sigma^2}{\lambda + Q(\xi,\eta\sigma)} N_{\lambda,\sigma}(\xi,\eta\sigma)\, d\sigma$$

$$= i^{-|\alpha|-k} \int_1^{+\infty} \sigma^{-k-1} \hat{g}(\xi,\eta\sigma) \frac{\eta^2\sigma^2}{\lambda + Q(\xi,\eta\sigma)} N_{\lambda,\sigma}(\xi,\eta\sigma)\, d\sigma$$

where $g := D_x^\alpha D_y^k f$. Then using Corollary 2.1 we have for some positive constant C

$$|\xi^\alpha||\eta^k||\hat{w}(\xi,\eta)| \le C \int_1^{+\infty} \sigma^{-k-1} |\hat{g}(\xi,\eta\sigma)| \sigma^{c^-}\, d\sigma.$$

Therefore, by Minkowski inequality, since $c+1>0$, we get

$$\|\xi^\alpha \eta^k \hat{w}(\xi,\eta)\|_1 \le C\int_1^{+\infty} \sigma^{-k-1+c^-}\left\|\hat{g}(\cdot,\sigma\cdot)\right\|_1 d\sigma = C\|\hat{g}\|_1 \int_1^{+\infty} \sigma^{-k-2+c^-}d\sigma$$

$$= \frac{C}{k+1-c^-}\|\hat{g}\|_1$$

which by the arbitrariness of α, k concludes the proof. □

The global regularity is formulated in the following proposition where, as usual, we identify functions u defined on $\mathbb{R}^{N+1}_+$ satisfying $D_y u(x,0)=0$, with functions defined on $\mathbb{R}^{N+1}$ and even with respect to y.

Proposition 4.1 *If $\frac{1}{p}<c+1$, $\lambda\in\Sigma_{\pi-\varepsilon}$, $f\in C_c^\infty(\mathbb{R}^{N+1}_+)$ with $f(x,-y)=f(x,y)$, then $u=(\lambda-\mathcal{L})^{-1}f\in C^\infty(\mathbb{R}^{N+1})$ and tends to zero at infinity with all derivatives.*

Proof From Theorem 3.1 we have $u=(\lambda-\mathcal{L}_0)^{-1}(f-cS_\lambda f)$ and the thesis follows from the lemma above and the analogous property of $(\lambda-\mathcal{L}_0)^{-1}$. □

Proposition 4.2 *Let $\frac{1}{p},\frac{1}{q}<c+1$ and $u\in W^{2,p}_{\mathcal{N}}$ be such that $u,\mathcal{L}u\in L^q(\mathbb{R}^{N+1}_+)$. Then $u\in W^{2,q}_{\mathcal{N}}$.*

Proof Let $f=u-\mathcal{L}u\in L^p(\mathbb{R}^{N+1}_+)\cap L^q(\mathbb{R}^{N+1}_+)$ and $v\in W^{2,q}_{\mathcal{N}}$ be such that $v-\mathcal{L}v=f$. By Corollary 3.1, $u=v$. □

We prove now a version of the maximum principle in L^p which also gives the positivity of the solutions for positive data, when $\lambda>0$.

Proposition 4.3 *Let $\frac{1}{p}<c+1$, $f\in L^p(\mathbb{R}^{N+1}_+)$, $\lambda>0$ and $u=(\lambda-\mathcal{L})^{-1}f$. Then*

$$(i)\quad \|u\|_\infty \le \frac{1}{\lambda}\|f\|_\infty; \qquad\qquad (ii)\quad \text{if } f\ge 0 \text{ then } u\ge 0.$$

Proof Assume first that $f\in C_c^\infty(\mathbb{R}^{N+1})$, $f(x,-y)=f(x,y)$. Then

$$u=(\lambda-\mathcal{L}_0)^{-1}(f-c\,S_\lambda f)\in C^\infty(\mathbb{R}^{N+1})$$

by Proposition 4.1 and tends to 0 at ∞. Let (x_0,y_0) be a maximum point for $|u|$ and assume, for example, that $u(x_0,y_0)>0$. If $y_0>0$, then $\mathcal{L}u(x_0,y_0)\le 0$.

If $y_0=0$, then the Hessian matrix with respect to the x-variables is negative semidefinite and

$$\lim_{y\to 0}\left(D_{yy}u(x_0,y)+\frac{c}{y}D_y u(x_0,y)\right)=(c+1)D_{yy}u(x_0,0)\le 0,$$

since $c+1>0$. In all cases $\mathcal{L}u(x_0,y_0)\le 0$ and then

$$\lambda\|u\|_\infty=\lambda u(x_0,y_0)\le \lambda u(x_0,y_0)-\mathcal{L}u(x_0,y_0)=f(x_0,y_0)\le \|f\|_\infty.$$

Similarly, u cannot have a positive maximum if $f\le 0$ and then $u\le 0$ whenever $f\le 0$.

When $f\in L^p(\mathbb{R}^{N+1}_+)$, we first extend f to $\mathbb{R}^{N+1}$ by even reflection in y and choose a sequence of even $(f_n)\subset C_c^\infty(\mathbb{R}^{N+1})$ such that $f_n\to f$ in $L^p(\mathbb{R}^{N+1}_+)$ and $\|f_n\|_\infty\le\|f\|_\infty$ in case (i) or $f_n\ge 0$ in case (ii). Then $u_n=(\lambda-\mathcal{L})^{-1}f_n\to u$ in $L^p(\mathbb{R}^{N+1}_+)$ and the proof follows. □

In what follows, we denote by B_R^+ either an half-ball of radius R, centered on the plane $y=0$ or a ball $B_R(z_0)$ centered at $z_0\in\mathbb{R}^{N+1}_+$ such that $B_{2R}(z_0)\subset\mathbb{R}^{N+1}_+$. Accordingly, we use $\|\cdot\|_{p,R}$ for the L^p norm in B_R^+ and similarly for Sobolev norms.

Corollary 4.1 *Let $\frac{1}{p},\frac{1}{q}<c+1$ and $u\in W^{2,p}_{\mathcal{N}}(B_R^+)$ be such that $\mathcal{L}u\in L^q(B_R^+)$. Then $u\in W^{2,q}(B^+_{\frac{R}{2}})$.*

Proof If $B_{2R}\subset\mathbb{R}^{N+1}_+$ this is standard elliptic regularity, since $\mathcal{L}$ is non-degenerate therein. Let us therefore assume that B_R is centered at $(x_0,0)$ and take a smooth cut-off η supported in B_R, which is equal to 1 in $B_{R/2}$ and is even in y. Then $\eta u\in W^{2,p}_{\mathcal{N}}$ and $\mathcal{L}(\eta u)=\eta\mathcal{L}u+2\sum_{i,j}q_{ij}D_{x_i}uD_{x_j}\eta+u\sum_{i,j}q_{ij}D_{x_ix_j}\eta+\frac{c}{y}uD_y\eta$. Note that since η is even, the term $y^{-1}D_y\eta$ is non-singular. If $p\ge N+1$, by Sobolev embedding the right hand side belongs to $L^q(\mathbb{R}^{n+1}_+)$ and the proof follows by Proposition 4.2. If $p<N+1$, then $\mathcal{L}(\eta u)\in L^r(\mathbb{R}^{N+1}_+)$ where r is the minimum between q and p^*, then $\eta u\in W^{2,r}_{\mathcal{N}}$ and the proof follows by iterating the previous argument a finite number of steps (and choosing intermediate radii). □

The following interior estimates are similar to those for uniformly elliptic operators.

Proposition 4.4 *Let $\frac{1}{p}<c+1$. For every $R>0$ there exists $C=C_R>0$ such that for every $u\in W^{2,p}_{\mathcal{N}}(B_R^+)$*

$$\|u\|_{W^{2,p}(B^+_{\frac{R}{2}})}\le C\left(\|\mathcal{L}u\|_{L^p(B_R^+)}+\|u\|_{L^p(B_R^+)}\right).$$

Proof Assume that B_R^+ is an half-ball of radius R centered on the plane $y=0$, otherwise the result follows from interior regularity for uniformly elliptic operators (see [5, Chapter 2, Sect. 4, Lemma 4]). Set $R_n=R\sum_{k=1}^n 2^{-k}$. Then $R_1=R/2$, $R_\infty=R$, $R_{n+1}-R_n=R2^{-(n+1)}$. Denote by B_n the ball of radius R_n and choose cut-off functions $\eta_n(x,y)\in C_c^\infty(\mathbb{R}^{N+1})$ such that $\eta_n(x,y)=\eta_n(x,-y)$, $0\le\eta_n\le$

1, $\eta_n = 1$ in B_n, supp $\eta_n \subset B_{n+1}$, $|\nabla \eta_n| \le \frac{C}{R} 2^n$, $|D^2 \eta_n| \le \frac{C}{R^2} 4^n$ for some constant $C > 0$. Then also $|y^{-1} D_y \eta_n| \le \frac{C}{R^2} 4^n$, since $D_y \eta_n(x, 0) = 0$. We have

$$\mathcal{L}(\eta_n u) = \eta_n \mathcal{L} u - 2 \sum_{i,j=1}^{N} q_{ij} D_{x_i} \eta_n D_{x_j} u - 2 D_y \eta_n D_y u - u \sum_{i,j=1}^{N} q_{ij} D_{x_i x_j} \eta_n - u D_{yy} \eta_n - c \frac{u}{y} D_y \eta_n$$

and $D_y(\eta_n u)(x, 0) = 0$. From Proposition 3.1 and from interpolative inequalities for the gradient, there exists a positive constant C such that for every $\varepsilon > 0$

$$\begin{aligned} \|\eta_n u\|_{2,p} &\le C \left(\|\mathcal{L}(\eta_n u)\|_p + \|\eta_n u\|_p \right) \\ &\le C \left(\|\mathcal{L} u\|_{p,R} + \frac{2^n}{R} \|\nabla u\|_{p,R_{n+1}} + \frac{4^n}{R^2} \|u\|_{p,R_{n+1}} \right) \\ &\le C(R) \left(\|\mathcal{L} u\|_{p,R} + 2^n \|\nabla(\eta_{n+1} u)\|_p + 4^n \|u\|_{p,R} \right) \\ &\le C \left(\|\mathcal{L} u\|_{p,R} + 4^n \|u\|_{p,R} + 2^n \varepsilon \|\eta_{n+1} u\|_{2,p} + \frac{2^n}{\varepsilon} \|u\|_{p,R} \right). \end{aligned}$$

Setting $\xi := C 2^n \varepsilon$, we get

$$\|\eta_n u\|_{2,p} \le C \left(\|\mathcal{L} u\|_{p,R} + \left(\frac{C 4^n}{\xi} + 4^n \right) \|u\|_{p,R} + \frac{\xi}{C} \|\eta_{n+1} u\|_{2,p} \right).$$

It follows that

$$\xi^n \|\eta_n u\|_{2,p} \le \xi^n C \|\mathcal{L} u\|_{p,R} + C_1 4^n \xi^{n-1} \|u\|_{p,R} + \xi^{n+1} \|\eta_{n+1} u\|_{2,p}. \tag{11}$$

By choosing $\varepsilon = \varepsilon_n$ so that $\xi = \frac{1}{8}$ and summing up the inequalities (11),

$$\sum_{n=1}^{\infty} \xi^n \|\eta_n u\|_{2,p} \le C \|\mathcal{L} u\|_{p,R} \sum_{n=1}^{\infty} \xi^n + C_1 \|u\|_{p,R} \sum_{n=1}^{\infty} 4^n \xi^{n-1} + \sum_{n=1}^{\infty} \xi^{n+1} \|\eta_{n+1} u\|_{2,p}$$

(the series converge since $\|\eta_n u\|_{2,p} \le C 4^n \|u\|_{2,p,R}$). Cancelling equal terms on both sides it follows that

$$\|u\|_{2,p,\frac{R}{2}} \le \|\eta_1 u\|_{2,p} \le C(\|\mathcal{L} u\|_{p,R} + \|u\|_{p,R}).$$

□

Corollary 4.2 *If $\frac{1}{p} < c+1$, $p > N+1$ and $u \in W^{2,p}_{\mathcal{N}}(B_R^+)$ for every ball B_R^+ and $u, \mathcal{L}u \in L^\infty(\mathbb{R}^{N+1}_+)$, then $u \in C^1_b(\mathbb{R}^{N+1}_+)$ and*

$$\|\nabla u\|_\infty \le C\|u\|_\infty^{\frac{1}{2}}\|\mathcal{L}u\|_\infty^{\frac{1}{2}}. \tag{12}$$

Proof Applying the above proposition with $R = 1$ and covering $\mathbb{R}^{N+1}_+$ with appropriate balls B_1^+ we obtain, by Sobolev embedding, $\|\nabla u\|_\infty \le C(\|u\|_\infty + \|\mathcal{L}u\|_\infty)$. Then we apply this inequality to $u_\lambda(x) = u(\lambda x)$, $\lambda > 0$, to get $\|\nabla u\|_\infty \le C(\lambda^{-1}\|u\|_\infty + \lambda\|\mathcal{L}u\|_\infty)$ and minimize over $\lambda > 0$. □

The following Liouville-type theorem is an easy consequence.

Proposition 4.5 *If $\frac{1}{p} < c+1$, $u \in W^{2,p}_{\mathcal{N}}(B_R^+)$ for every ball B_R^+ and $u \in L^\infty$ with $\mathcal{L}u = 0$, then u is a constant.*

We finally prove an improved regularity result for solutions associated with the operator $\mathcal{L}$. We define for any $k \ge 2$ the Sobolev space

$$W^{k,p}_{\mathcal{N}} = \{u \in W^{k,p}(\mathbb{R}^{N+1}_+) : \ D_y u(x,0) = 0\}.$$

In Sect. 3 we observed that for $k = 2$, $W^{2,p}_{\mathcal{N}}$ is identified with the space of even functions with respect to the y variable in $W^{2,p}(\mathbb{R}^{N+1})$ and that the set $C := \left\{u \in C_c^\infty\left(\mathbb{R}^{N+1}\right),\ u(x,y) = u(x,-y)\right\}$ is dense in $W^{2,p}_{\mathcal{N}}$. These properties fail if $k > 2$ since any function $u \in W^{k,p}_{\mathcal{N}}$ even with respect to the y variable satisfies the stronger boundary condition $D_y^{2i+1}u(x,0) = 0$, for any $2i+1 \le k$. Nevertheless we have the following result.

Lemma 4.2 *Let $k \ge 2$. The following properties hold.*

(i) For any $\lambda \in \Sigma_\pi$, the operator $\lambda - \Delta : W^{k,p}_{\mathcal{N}} \to W^{k-2,p}\left(\mathbb{R}^{N+1}_+\right)$ is invertible (here Δ is the Laplacian in the x and y variables).
(ii) The set $\mathcal{D} := \left\{u \in C_c^\infty\left(\mathbb{R}^{N+1}_+\right),\ D_y u(x,0) = 0\right\}$ is dense in $W^{k,p}_{\mathcal{N}}$.

Proof Claim (i) is the classic elliptic regularity up to the boundary for the Neumann boundary problem associated with the Laplacian, see [5, Theorem 2, Chapter 9, Section 3]. Let us prove (ii). Let $u \in W^{k,p}_{\mathcal{N}}$ and $f = u - \Delta u$. Let $f_n \in C_c^\infty\left(\mathbb{R}^{N+1}\right)$ such that $f_n \to f$ in $L^p\left(\mathbb{R}^{N+1}_+\right)$. By claim (i) $u_n = (I-\Delta)^{-1}f_n \in W^{k,p}_{\mathcal{N}}$ for any $k \ge 2$ hence, by Sobolev embeddings, $u_n \in C^\infty\left(\overline{\mathbb{R}^{N+1}_+}\right)$; moreover $D_y u_n(x,0) = 0$ and $u_n \to u$ in $W^{k,p}_{\mathcal{N}}$. This proves that u can be approximated with functions in $C^\infty\left(\overline{\mathbb{R}^{N+1}_+}\right) \cap W^{k,p}_{\mathcal{N}}$. A standard cut-off argument at infinity concludes the proof. □

We now turn to higher order regularity. The following result implies that if $u \in W^{2,p}_{\mathcal{N}}$ solves $\lambda u - \mathcal{L}u = f \in W^{k,p}$, then $u \in W^{k+2,p}$.

Proposition 4.6 *Let $1 < p < \infty$, satisfying $0 < \frac{1}{p} < c+1$, $k \in \mathbb{N}_0$. Let $\lambda \in \Sigma_\pi$. Then there exists a positive constant C such that for $u \in W^{k+2,p}_{\mathcal{N}}$*

$$\|u\|_{W^{k+2,p}_{\mathcal{N}}} \le C\|\lambda u - \mathcal{L}u\|_{W^{k,p}(\mathbb{R}^{N+1}_+)}.$$

Moreover the operator $\lambda - \mathcal{L} : W^{k+2,p}_{\mathcal{N}} \to W^{k,p}(\mathbb{R}^{N+1}_+)$ is invertible.

Proof Let us prove the first claim. The case $k = 0$ is Theorem 3.1 and Proposition 3.1. It is then enough to prove the case $k = 1$ since the general case follows by an easy induction argument. By Lemma 4.2(ii), we can restrict to $u \in \mathcal{D}$. By Proposition 3.1 we have $\|u\|_{W^{2,p}_{\mathcal{N}}} \le C\|\lambda u - \mathcal{L}u\|_p$ and therefore we need only to estimate the third order derivatives. For $i = 1, \dots, N$ we have $D_{x_i}\mathcal{L}u = \mathcal{L}D_{x_i}u$. Then by Proposition 3.1 again applied to $D_{x_i}u$ we get

$$\begin{aligned}\|D_{x_i}u\|_{W^{2,p}_{\mathcal{N}}} &\le C\|\lambda D_{x_i}u - \mathcal{L}D_{x_i}u\|_p = C\|D_{x_i}(\lambda u - \mathcal{L}u)\|_p \\ &\le C\|\lambda u - \mathcal{L}u\|_{W^{1,p}(\mathbb{R}^{N+1}_+)}.\end{aligned}$$

It remains to estimates $D_{yyy}u$. With $f = \lambda u - \mathcal{L}u$ we write

$$y^{-c}D_y\left(y^c D_y u\right) = D_{yy}u + \frac{c}{y}D_y u = \lambda u - \sum_{i,j=1}^{N} q_{ij}D_{x_ix_j}u - f := g$$

where by the previous estimates g satisfies $\|D_y g\|_p \le C\|f\|_{W^{1,p}_{\mathcal{N}}}$. Integrating the latter equality we obtain

$$D_y u(x,y) = y^{-c}\int_0^y g(x,s)s^c\,ds = y\int_0^1 g(x,ty)t^c\,dt.$$

Then $D_{yy}u(x,y) = g(x,y) - c\frac{D_y u(x,y)}{y} = g(x,y) - c\int_0^1 g(x,ty)t^c\,dt$ and

$$D_{yyy}u(x,y) = D_y g(x,y) - c\int_0^1 D_y g(x,ty)t^{c+1}\,dt.$$

This implies, using Minkowski inequality for integrals,

$$\begin{aligned}\|D_{yyy}u\|_p &\le \|D_y g\|_p + \int_0^1 \|D_y g(\cdot, t\,\cdot)\|_p t^{c+1}\,dt = \|D_y g\|_p\left(1 + \int_0^1 t^{c+1-\frac{1}{p}}\,dt\right) \\ &\le C\|f\|_{W^{1,p}(\mathbb{R}^{N+1}_+)}.\end{aligned}$$

The invertibility of $\lambda - \mathcal{L} : W^{k+2,p}_{\mathcal{N}} \to W^{k,p}(\mathbb{R}^{N+1}_+)$ follows by applying (i) of Lemma 4.2 and the method of continuity of Lemma 8.1 to the family of operators $\mathcal{L}_t := (1-t)\Delta + t\mathcal{L}$, $0 \le t \le 1$, after noticing that the generation condition for $\mathcal{L}_t$, namely $\frac{1}{p} < ct + 1$, follows from $\frac{1}{p} < c + 1$. □

The following higher order interior estimates follow.

Proposition 4.7 *Let* $\frac{1}{p} < c + 1$. *For every* $R > 0$ *there exists* $C = C_R > 0$ *such that for every* $u \in W^{k+2,p}_{\mathcal{N}}(B^+_R)$

$$\|u\|_{W^{k+2,p}(B^+_{\frac{R}{2}})} \le C\left(\|\mathcal{L}u\|_{W^{k,p}(B^+_R)} + \|u\|_{L^p(B^+_R)}\right).$$

Proof Assume that $k = 1$ and that B^+_R is an half-ball of radius R centered on the plane $y = 0$, otherwise the result follows from higher order interior regularity for uniformly elliptic operators (see [5, Theorem 7, Chapter 1]). Given $R > 0$ and $x_0 \in \mathbb{R}^N$ we choose cut-off functions $\eta(x, y) \in C_c^\infty(\mathbb{R}^{N+1})$ such that $\eta(x, y) = \eta(x, -y)$, $0 \le \eta \le 1$, $\eta = 1$ in $B_{\frac{R}{2}}$, $\operatorname{supp} \eta \subset B_{\frac{3R}{4}}$. We have

$$\mathcal{L}(\eta u) = \eta\mathcal{L}u + 2\sum_{i,j=1}^N q_{ij} D_{x_i}\eta D_{x_j}u + 2D_y\eta D_y u + u\sum_{i,j=1}^N q_{ij} D_{x_i x_j j}\eta$$
$$+ uD_{yy}\eta + c\frac{u}{y}D_y\eta := g$$

with $\frac{D_y\eta}{y} \in C_c^\infty(\mathbb{R}^{N+1})$ and $D_y(\eta u)(x, 0) = 0$. It follows that $g \in W^{1,p}(\mathbb{R}^{N+1}_+)$ and we can apply Proposition 4.6 to ηu to obtain

$$\|u\|_{W^{3,p}(B^+_{\frac{R}{2}})} \le \|\eta u\|_{W^{3,p}_{\mathcal{N}}} \le C\left(\|\mathcal{L}u\|_{W^{1,p}(B^+_R)} + \|u\|_{W^{2,p}(B^+_{\frac{3}{4}R})}\right)$$

and one then applies Proposition 4.4.

The general case follows by iterating this argument. □

5 The Parabolic Problem

For $\lambda > 0$ we consider the problem

$$\begin{cases} \lambda u(t,x,y) + \partial_t u(t,x,y) - \mathcal{L}u(t,x,y) = f(t,x,y), & t \in \mathbb{R}, \quad (x,y) \in \mathbb{R}^{N+1}_+, \\ D_y u(t,x,0) = 0, & t \in \mathbb{R}, \quad x \in \mathbb{R}^N, \end{cases} \tag{13}$$

where $f \in L^p\left(\mathbb{R} \times \mathbb{R}_+^{N+1}\right)$ and $\mathcal{L}$ is the elliptic operator of the previous sections, with Neumann boundary conditions. Accordingly, we define for $1 < p < \infty$

$$W_p^{1,2} = \left\{u \in L^p(\mathbb{R} \times \mathbb{R}_+^{N+1}) : \partial_t u, \nabla u, D^2 u \in L^p(\mathbb{R} \times \mathbb{R}_+^{N+1})\right\},$$

$$W_{p,\mathcal{N}}^{1,2} = \left\{u \in W_p^{1,2} : D_y u(t, x, 0) = 0\right\},$$

where ∇u, $D^2 u$ denote the gradient and the Hessian matrix with respect to all the space variables.

The same methods used in Sect. 3 prove the solvability of (13).

For $\sigma \geq 1$, let $\tilde{N}_{\lambda,\sigma}(\tau, \xi, s) = N_{\lambda+i\tau,\sigma}(\xi, s)$, see (3). We define the operator $\tilde{T}_{\lambda,\sigma} \in \mathcal{B}\left(L^p(\mathbb{R}^{N+2})\right)$ through the multiplier $\frac{s^2}{\lambda+i\tau+Q(\xi,s)}\tilde{N}_{\lambda,\sigma}(\tau, \xi, s)$, that is

$$(\mathcal{F}\tilde{T}_{\lambda,\sigma} f)(\tau, \xi, s) = \frac{s^2}{\lambda + i\tau + Q(\xi, s)}\tilde{N}_{\lambda,\sigma}(\tau, \xi, s)\, \hat{f}(\tau, \xi, s) \tag{14}$$

where this time $\mathcal{F}$ denotes the Fourier transform in $\mathbb{R}^{N+2}$. By Propositions 2.2, (18) and Theorem 7.1, we have $\|\tilde{T}_{\lambda,\sigma}\|_p \leq C\sigma^{c^-}$.

Proposition 5.1 *Let $1 < p < \infty$, $\frac{1}{p} < c + 1$. Then there exists $C > 0$ such that for $\lambda > 0$, $f \in L^p\left(\mathbb{R} \times \mathbb{R}_+^{N+1}\right)$ the equation $\lambda u + \partial_t u - \mathcal{L}u = f$ admits a unique solution u in $W_{p,\mathcal{N}}^{1,2}$. Moreover*

$$\lambda\|u\|_p + \lambda^{\frac{1}{2}}\|\nabla u\|_p + \|D^2 u\|_p + \|\partial_t u\|_p \leq C\|\lambda u + \partial_t u - \mathcal{L}u\|_p.$$

Proof The proof follows similarly as in the elliptic case and we point out only the (minor) changes.

As in Lemma 3.1, taking the Fourier transform of (13) with respect to (t, x, y) with co-variables (τ, ξ, η) we obtain for $u \in C_c^\infty(\mathbb{R}^{N+2})$, even with respect to y

$$\hat{u}(\tau, \xi, \eta) = \frac{\hat{f}(\tau, \xi, \eta)}{\lambda + i\tau + Q(\xi, \eta)} - \frac{c}{\lambda + i\tau + Q(\xi, \eta)} \int_1^{+\infty} \hat{f}(\tau, \xi, \eta\sigma)\frac{\eta^2\sigma}{\lambda + i\tau + Q(\xi, \eta\sigma)}\tilde{N}_{\lambda,\sigma}(\tau, \xi, \eta\sigma)\, d\sigma.$$

Then, as in Lemma 3.2, we write

$$u = (\lambda + \partial_t - \mathcal{L}_0)^{-1}\left(f - c\,\tilde{S}_\lambda f\right),$$

where $\mathcal{L}_0 = \mathrm{Tr}\,\left(QD^2u\right)$ and $\tilde{S}_\lambda$ is defined by

$$\tilde{S}_\lambda f(t,x,y) = \int_1^{+\infty} \frac{1}{\sigma^2}(\tilde{T}_{\lambda,\sigma} f)\left(t,x,\frac{y}{\sigma}\right)\, d\sigma.$$

If $\frac{1}{p} < c+1$, then the operator $\tilde{S}_\lambda$ is bounded on $L^p\left(\mathbb{R}^{N+2}\right)$ and

$$\|\tilde{S}_\lambda f\|_p \le C\|f\|_p, \quad \lambda > 0.$$

The estimates in the statement follow as in Proposition 3.1 from the boundedness of the multipliers M_3^λ, M_4^λ, M_5^λ stated in (18). The invertibility of $\lambda+\partial_t-\mathcal{L} : W^{1,2}_{p,\mathcal{N}} \to L^p(\mathbb{R}\times\mathbb{R}_+^{N+1})$ is proved through the continuity method, as in Theorem 3.1. □

6 Comments on the Associated Semigroup

Assume that $1 < p < \infty$, $\frac{1}{p} < c+1$ and let $(T_p(z))_{z\in\Sigma_{\pi/2}}$ be the semigroup generated by $(\mathcal{L}, W^{2,p}_{\mathcal{N}})$ in $L^p(\mathbb{R}_+^{N+1})$, given by Corollary 3.2. We observe that if $p < q < \infty$, then $T_p(z)f = T_q(z)f$ if $f \in L^p(\mathbb{R}_+^{N+1}) \cap L^q(\mathbb{R}_+^{N+1})$, by Corollary 3.1 and that $T_p(t)f \ge 0$ when $f \ge 0$ and $\|T_p(t)f\|_\infty \le \|f\|_\infty$, for $t \ge 0$, by Proposition 4.3.

The generation condition $\frac{1}{p} < c+1$ cannot be improved since it is precise for the $1d$ Bessel operator $D_{yy} + \frac{c}{y}D_y$, see [8, Theorem 4.2], [9, Proposition 3.2], [10, Theorem 5.1].

The situation in the weighted spaces L^p_m, treated briefly in Section 3.1, is however not completely satisfactory. The precise generation condition is $0 < \frac{m+1}{p} < c+1$, see [9], while we obtain $0 < \frac{m+1}{p} < c+1$ and $0 < \frac{m+1}{p} < 1$ in Proposition 3.2. The extra condition we need here ensures that the weight y^m belongs to $\mathcal{A}_p(\mathbb{R}_+^{N+1})$, which is essential to apply Theorem 7.1. In the simpler situation $\mathcal{L} = \Delta_x + D_{yy} + \frac{c}{y}D_y$, the operators Δ_x and $B_y = D_{yy} + \frac{c}{y}D_y$ commute and one can obtain a semigroup by composing the semigroups generated by the Laplacian and the Bessel operator. The closedness of $\mathcal{L}$ on the intersection of the domains of Δ_x and B_y is however much harder and our proof in [9] relies on tools from operator valued harmonic analysis.

Another remark concerns the contractivity of the semigroup as well as L^p-L^q estimates. From [10] (see Proposition 9.3 for the $1d$ case and proceed as in Lemma 9.1) it follows that $T_p(t)_{t\ge 0}$ is never contractive with respect to the Lebesgue measure, if $p < \infty$. On the other hand it is contractive in the sup-norm, by the maximum principle.

By the domain characterization and the semigroup law, it follows that $T_p(t)$, $t > 0$, maps L^p into L^∞, hence in L^q for every $q > p$. This follows because $T_p(t)$

is analytic and hence maps L^p into $W^{2,p} \subset L^q$ for certain $q > p$, by Sobolev or Morrey embedding. Starting from L^q and using the consistency, after a finite number k of steps, depending only on p, $T_p(kt)f \in L^\infty$, which is the claim since t is arbitrary. This yields the existence of a kernel for the semigroup, which is however not bounded in general, since it does not map L^1 to L^∞, see [10, Theorem 2.6] for the explicit expression of the heat kernel in $1d$.

Next note that, since $I_{s^{-1}}\mathcal{L}I_s = s^2\mathcal{L}$, where $I_s f(x, y) = f(sx, sy)$ for $s > 0$, then $I_{s^{-1}}T(t)I_s = T_p(s^2t)$ and $T(t) = I_sT(1)I_{s^{-1}}$ with $s = t^{-1/2}$ and then $\|T(t)\|_{p,q} = \|T(1)\|_{p,q}t^{-\frac{N}{2}\left(\frac{1}{p}-\frac{1}{q}\right)}$.

As a final remark, we observe that the same proof of [9, Proposition 8.15] shows that the spectrum of $(\mathcal{L}, W^{2,p}_{\mathcal{N}})$ is the half-line $]-\infty, 0]$.

7 Appendix 1: A Review of Some Classic Multipliers

We state a version of the classical Marcinkiewicz multiplier theorem in $L^p\left(\mathbb{R}^k\right)$, see e.g. [11, Theorem 6', Section IV] for the Lebesgue measure and [6, Theorem 3] for the weighted version.

Theorem 7.1 *Let $1 < p < \infty$, $M \in C^k(\mathbb{R}^k \setminus \{0\})$ be such that*

$$\|M\|_{\mathcal{M}} := \sup\left\{|\zeta^\alpha D^\alpha_\zeta M(\zeta)| : \zeta \in \mathbb{R}^k \setminus \{0\},\ |\alpha| \le k,\ \alpha_i \in \{0, 1\}\right\} < +\infty.$$

Then for any $w \in A_p\left(\mathbb{R}^k\right)$, the operator $T_M = \mathcal{F}^{-1}M\mathcal{F}$ is bounded in $L^p(\mathbb{R}^k, w)$ and $\|T_M\|_{L^p(\mathbb{R}^k,w)} \le C\|M\|_{\mathcal{M}}$ with $C = C(p, k, w) > 0$.

The following lemma allows to verify a stronger condition for multipliers depending on quadratic forms.

Lemma 7.1 *Let $1 < p < \infty$, $m \in C^k(\mathbb{R}_+)$ be such that*

$$\sup\left\{s^i|m^{(i)}(s)| : s \in \mathbb{R}_+,\ i \le k\right\} < +\infty.$$

Let $A = \left(a_{ij}\right)_{i,j=1,\dots,k} \in \mathbb{R}^{k,k}$ be a positive definite matrix, $A(\zeta) := (A\zeta, \zeta)$ the associated quadratic form and $M(\zeta) = m\big(A(\zeta)\big)$. Then $M \in C^k(\mathbb{R}^k \setminus \{0\})$ and

$$\sup\left\{|\zeta|^{|\alpha|}|D^\alpha_\zeta M(\zeta)| : \zeta \in \mathbb{R}^N \setminus \{0\},\ |\alpha| \le k\right\}$$
$$\le C \sup\left\{s^i|m^{(i)}(s)| : s \in \mathbb{R}_+,\ i \le k\right\}.$$

Proof Let us first observe that for any multi-index α with $0 < |\alpha| \le k$ one has

$$D_\zeta^\alpha M(\zeta) = \sum_{\substack{i=0,\dots,|\alpha| \\ 2i-|\alpha|\ge 0}} h_{i,\alpha}(\zeta) m^{(i)}(A(\zeta)) \tag{15}$$

where $h_{i,\alpha} \in C^\infty(\mathbb{R}^k \setminus \{0\})$ is a homogeneous polynomials of degree $2i - |\alpha| \ge 0$. Obviously (15) is valid for $|\alpha| = 1$ since $\nabla M(\zeta) = 2m'(A(\zeta))\, A \cdot \zeta$ and follows by induction, since the derivatives of $h_{i,\alpha}$ are homogeneous polynomials of degree $2i - |\alpha| - 1$. Furthermore, from the homogeneity, there are constants $c_{i,\alpha} > 0$ such that

$$|h_{i,\alpha}(\zeta)| \le c_{i,\alpha} |\zeta|^{2i-|\alpha|}.$$

This implies

$$|D_\zeta^\alpha M(\zeta)| \le C(k, A, \alpha) \sum_{\substack{i=0,\dots,|\alpha| \\ 2i-|\alpha|\ge 0}} |\zeta|^{2i-|\alpha|} |m^{(i)}(A(\zeta))|. \tag{16}$$

Let $\nu > 0$ such that $A(\zeta) \ge \nu |\zeta|^2$. Then the previous relations imply

$$|\zeta|^{|\alpha|} |D_\zeta^\alpha M(\zeta)| \le C \sum_{i=0}^{|\alpha|} |\zeta|^{2i} |m^{(i)}(A(\zeta))| \le C \sum_{i=0}^{|\alpha|} |A(\zeta)|^i |m^{(i)}(A(\zeta))|.$$

□

We consider for $\lambda \in \Sigma_\pi$ the following multipliers

$$M_1^\lambda(\zeta) = \frac{1}{\lambda + A(\zeta)}, \qquad M_2^\lambda(\zeta) = \frac{\zeta_i \zeta_j}{\lambda + A(\zeta)}, \quad i, j = 1, \dots, k.$$

If $\mathcal{A} = \mathrm{tr}\left(AD^2\right)$ is the second order elliptic differential operator associated with A, then M_1^λ, M_2^λ, are the Fourier multipliers of respectively $(\lambda - \mathcal{A})^{-1}$ and $D_{x_i x_j}(\lambda - \mathcal{A})^{-1}$. By classical results M_1^λ, M_2^λ satisfy the hypothesis of Theorem 7.1. For the sake of completeness we give below a short proof of this result. We remark preliminarily that for any $0 < \varepsilon \le \pi$ there is $C_\epsilon > 0$ such that

$$C_\epsilon (|\lambda| + s) \le |\lambda + s| \le |\lambda| + s, \qquad \forall \lambda \in \Sigma_{\pi-\epsilon}, \quad s \ge 0. \tag{17}$$

Proposition 7.1 *Let $0 < \varepsilon \le \pi$. For any multi-index $\alpha \in \mathbb{N}^k$ there exists $C = C(\epsilon, A, |\alpha|) > 0$ such that*

$$\left| D_\zeta^\alpha \left(\frac{1}{\lambda + A(\zeta)} \right) \right| \le C \frac{1}{(|\lambda| + |\zeta|^2)^{\frac{|\alpha|}{2}+1}}, \qquad \forall \zeta \in \mathbb{R}^k \setminus \{0\},\ \lambda \in \Sigma_{\pi-\varepsilon}.$$

Proof Setting $m(s) = (\lambda + s)^{-1}$ one has $M_1^\lambda(\zeta) = (\lambda + A(\zeta, \eta))^{-1} = m(A(\zeta))$. Then inequalities (16) and (17) and the positive definiteness of A imply that, for some $C > 0$,

$$|D_\zeta^\alpha M_1^\lambda(\zeta)| \le C \sum_{\substack{i=0,\dots,|\alpha| \\ 2i-|\alpha|\ge 0}} |\zeta|^{2i-|\alpha|} \frac{1}{|\lambda + A(\zeta)|^{i+1}}$$

$$\le C \sum_{\substack{i=0,\dots,|\alpha| \\ 2i-|\alpha|\ge 0}} \left(|\lambda| + |\zeta|^2\right)^{i-\frac{|\alpha|}{2}} \frac{1}{\left(|\lambda| + |\zeta|^2\right)^{i+1}} = C \frac{1}{\left(|\lambda| + |\zeta|^2\right)^{\frac{|\alpha|}{2}+1}}.$$

□

Proposition 7.2 *Let $0 < \varepsilon \le \pi$. For any multi-index $\alpha = (\alpha_1, \dots, \alpha_k) \in \mathbb{N}^k$ such that $\alpha_i \in \{0, 1\}$ there exist $C = C(\varepsilon, A, |\alpha|) > 0$ such that for every $i, j = 1, \dots, k$*

$$\left|D_\zeta^\alpha \left(\frac{\zeta_i \zeta_j}{\lambda + A(\zeta)}\right)\right| \le C \frac{1}{\left(|\lambda| + |\zeta|^2\right)^{\frac{|\alpha|}{2}}}, \qquad \forall \zeta \in \mathbb{R}^k \setminus \{0\},\ \lambda \in \Sigma_{\pi-\varepsilon}.$$

Proof Suppose for example that $i = j$. If $\alpha_i = 0$ then

$$D_\zeta^\alpha \left(\frac{\zeta_i^2}{\lambda + A(\zeta)}\right) = \zeta_i^2 D_\zeta^\alpha \left(\frac{1}{\lambda + A(\zeta)}\right)$$

and Proposition 7.1 yields

$$\left|D_\zeta^\alpha \left(\frac{\zeta_i^2}{\lambda + A(\zeta)}\right)\right| \le C \frac{|\zeta_i|^2}{\left(|\lambda| + |\zeta|^2\right)^{\frac{|\alpha|}{2}+1}} \le C \frac{1}{\left(|\lambda| + |\zeta|^2\right)^{\frac{|\alpha|}{2}}}.$$

which is the claim. If $\alpha_i = 1$, let $\beta = (\beta_1, \dots, \beta_k)$ such that $\beta_r = \alpha_r$ for $r \ne i$ and $\beta_i = 0$; in particular $|\beta| = |\alpha| - 1$. Then we get

$$D_\zeta^\alpha \left(\frac{\zeta_i^2}{\lambda + A(\zeta, \eta)}\right) = \zeta_i^2 D_\zeta^\alpha \left(\frac{1}{\lambda + A(\zeta, \eta)}\right) + 2\zeta_i D_\zeta^\beta \left(\frac{1}{\lambda + A(\zeta, \eta)}\right)$$

and the claim follows similarly as before from Proposition 7.1. The case $i \ne j$ is similar. □

We can also consider the parabolic counterparts of M_1^λ, M_2^λ, namely, for $\lambda > 0$

$$M_3^\lambda(\zeta,\tau) = \frac{1}{\lambda + i\tau + A(\zeta)}, \qquad M_4^\lambda(\zeta,\tau) = \frac{\zeta_i\zeta_j}{\lambda + i\tau + A(\zeta)}, \quad i, j = 1, \dots, k,$$

$$M_5^\lambda(\zeta,\tau) = \frac{i\tau}{\lambda + i\tau + A(\zeta)},$$

which are the multipliers of $(\lambda + \partial_t - \mathcal{A})^{-1}$, $D_{x_ix_j}(\lambda + \partial_t - \mathcal{A})^{-1}$, $\partial_t(\lambda + \partial_t - \mathcal{A})^{-1}$, respectively. As before, M_3^λ, M_4^λ, M_5^λ satisfy the hypothesis of Theorem 7.1 with

$$\|M_3^\lambda\|_{\mathcal{M}} \le \frac{C}{\lambda}, \qquad \|M_4^\lambda\|_{\mathcal{M}} \le C \qquad \|M_5^\lambda\|_{\mathcal{M}} \le C. \tag{18}$$

As an example we give below a short proof for M_5^λ.

Proposition 7.3 *Let $\lambda > 0$. For any multi-index $\alpha = (\alpha_1, \dots, \alpha_k) \in \mathbb{N}^k$ such that $\alpha_i \in \{0, 1\}$ and for any $j \in \{0, 1\}$ there exist $C = C(A, |\alpha|) > 0$ such that*

$$\left| D_\zeta^\alpha D_\tau^j \left(\frac{i\tau}{\lambda + i\tau + A(\zeta)} \right) \right| \le C \frac{1}{\left(\lambda + |\tau| + |\zeta|^2\right)^{\frac{|\alpha|}{2} + j}}, \qquad \forall (\zeta, \tau) \in \mathbb{R}^{k+1} \setminus \{0\}.$$

Proof Using Proposition 7.1 with λ replaced by $\lambda + i\tau$, one has

$$\left| D_\zeta^\alpha \left(\frac{i\tau}{\lambda + i\tau + A(\zeta)} \right) \right| \le C \frac{|\tau|}{\left(\lambda + |\tau| + |\zeta|^2\right)^{\frac{|\alpha|}{2} + 1}} \le C \frac{1}{\left(\lambda + |\tau| + |\zeta|^2\right)^{\frac{|\alpha|}{2}}},$$

which is the claim for $j = 0$. If $j = 1$ then

$$D_\zeta^\alpha D_\tau \left(\frac{i\tau}{\lambda + i\tau + A(\zeta)} \right) = i D_\zeta^\alpha \left(\frac{1}{\lambda + i\tau + A(\zeta)} \right) + \tau D_\zeta^\alpha \left(\frac{1}{(\lambda + i\tau + A(\zeta))^2} \right).$$

The first term in the above equation has been estimated in the above inequality. To treat the second we observe that $(\lambda + i\tau + A(\zeta))^{-2} = m(A(\zeta))$ where $m(s) = (\lambda + i\tau + s)^{-2}$ and, as in Proposition 7.1, from inequality (16) we obtain

$$\left| D_\zeta^\alpha \left(\frac{1}{(\lambda + i\tau + A(\zeta))^2} \right) \right| \le C \frac{1}{\left(\lambda + |\tau| + |\zeta|^2\right)^{\frac{|\alpha|}{2} + 2}}.$$

□

8 Appendix 2: The Method of Continuity

We prove below a slightly stronger version of the well known "continuity method". The novelty consists in the fact that the constants $c(t)$ below are not supposed to be independent of t.

Lemma 8.1 *Let X, Y be two Banach spaces and let* $\{L_t\}_{t\in[0,1]} \subseteq \mathcal{B}(X, Y)$ *a family of bounded operators. Let us assume that*

(i) the map $\mathcal{L}\colon [0,1] \to \mathcal{B}(X,Y), \quad t \mapsto L_t$ *is continuous;*
(ii) for every $t \in [0, 1]$ *there is* $c(t) > 0$ *such that*

$$\|L_t(x)\| \geq c(t)\ \|x\|, \quad \forall x \in X. \tag{19}$$

Then if L_0 *is invertible also* L_1 *is invertible.*

Proof It is enough to prove that the constants $c(t)$ in (19) can be made independent of t. The rest follows from the usual continuity method.

Given $t_0 \in [0, 1]$ we fix $\delta > 0$ such that $\|L_t - L_{t_0}\| < \frac{c(t_0)}{2}$ for $t \in I(t_0) := (t_0 - \delta, t_0 + \delta)$. Then for any $x \in X$, $t \in I(t_0)$ one has

$$\|L_t(x)\| \geq \|L_{t_0}(x)\| - \|(L_t - L_{t_0})(x)\| \geq c(t_0)\|x\| - \frac{c(t_0)}{2}\|x\| = \frac{c(t_0)}{2}\|x\|.$$

By compactness we find $t_1 \ldots t_n$ such that $[0, 1] = \bigcup_{i=1}^{n} I(t_i)$ and

$$\|L_t(x)\| \geq \frac{c(t_i)}{2}\|x\|, \qquad t \in I(t_i).$$

The proof follows by taking $c := \min\left\{\frac{c(t_1)}{2} \ldots \frac{c(t_n)}{2}\right\}$. □

References

1. Dong, H., Phan, T.: On parabolic and elliptic equations with singular or degenerate coefficients. Online preprint: https://doi.org/10.48550/arXiv.2007.04385 (2020)
2. Dong, H., Phan, T.: Weighted mixed-norm L_p-estimates for elliptic and parabolic equations in non-divergence form with singular coefficients. Rev. Mat. Iber. **37**, 04 (2020)
3. Dong, H., Phan, T.: Parabolic and elliptic equations with singular or degenerate coefficients: The Dirichlet problem. Trans. Am. Math. Soc. **374**, 09 (2021)
4. Dong, H., Phan, T.: Weighted mixed-norm L_p estimates for equations in non-divergence form with singular coefficients: the Dirichlet problem. Online preprint: https://doi.org/10.48550/arXiv.2103.08033 (2021)
5. Krylov, N.V.: Lectures on Elliptic and Parabolic Equations in Sobolev Spaces. vol. 96 of Graduate Studies in Mathematics. Amer. Math. Soc. (2008)

6. Kurtz, D.S.: Littlewood-Paley and multiplier theorems on weighted L^p spaces. Trans. Am. Math. Soc. **259**(1), 235–254, (1980)
7. Metafune, G., Negro, L., Spina, C.: Anisotropic sobolev spaces with weights. Tokyo J. Math. **46**(2), (2023)
8. Metafune, G., Negro, L., Spina, C.: Degenerate operators on the half-line. J. Evol. Equations **22**, 60 (2022)
9. Metafune, G., Negro, L., Spina, C.: L^p estimates for the Caffarelli-Silvestre extension operators. J. Differential Equations **316**, 290–345 (2022)
10. Metafune, G., Negro, L., Spina, C.: A unified approach to degenerate problems in the half-space. J. Differential Equations **351**, 63–99 (2023)
11. Stein, E.M.: Singular Integrals and Differentiability Properties of Functions. (PMS-30), vol. 30 of Princeton Mathematical Series. Princeton University Press (1970)

Anisotropic (p, q)-Equations with Convex and Negative Concave Terms

Nikolaos S. Papageorgiou, Dušan D. Repovš, and Calogero Vetro

Abstract We consider a parametric Dirichlet problem driven by the anisotropic (p, q)-Laplacian and with a reaction which exhibits the combined effects of a superlinear (convex) term and of a negative sublinear term. Using variational tools and critical groups we show that for all small values of the parameter, the problem has at least three nontrivial smooth solutions, two of which are of constant sign (positive and negative).

Keywords Variable Lebesgue and Sobolev spaces · Variable *(p,q)*-operator · Regularity theory · Local minimizer · Critical point theory

1 Introduction

Let $\Omega \subseteq \mathbb{R}^N$ be a bounded domain with a C^2-boundary $\partial\Omega$. In this paper we study the following parametric anisotropic Dirichlet problem

$$\begin{cases} -\Delta_{p(z)}u(z) - \Delta_q u(z) = f(z, u(z)) - \lambda|u(z)|^{\tau(z)-2}u(z) & \text{in } \Omega, \\ u\Big|_{\partial\Omega} = 0, \ 1 < \tau(z) < q < p(z) < N \text{ for all } z \in \overline{\Omega}, \ \lambda > 0. \end{cases} \qquad (P_\lambda)$$

N. S. Papageorgiou
Department of Mathematics, National Technical University, Athens, Greece
e-mail: npapg@math.ntua.gr

D. D. Repovš (✉)
Faculty of Education and Faculty of Mathematics and Physics, University of Ljubljana, Institute of Mathematics, Physics and Mechanics, Ljubljana, Slovenia
e-mail: dusan.repovs@guest.arnes.si

C. Vetro
Department of Mathematics and Computer Science, University of Palermo, Palermo, Italy
e-mail: calogero.vetro@unipa.it

A. M. Candela et al. (eds.), *Recent Advances in Mathematical Analysis*,
Trends in Mathematics, https://doi.org/10.1007/978-3-031-20021-2_21

Given $r \in C^{0,1}(\overline{\Omega})$ (= the space of Lipschitz continuous functions on $\overline{\Omega}$) with $1 < r_- = \min_{\overline{\Omega}} r$, by $\Delta_{r(z)}$ we denote the anisotropic r-Laplacian defined by

$$\Delta_{r(z)}u = \operatorname{div}\,(|\nabla u|^{r(z)-2}\nabla u) \text{ for all } u \in W_0^{1,r(z)}(\Omega)(\text{see Sect. 2}).$$

If $r(\cdot)$ is constant, then we have the standard r-Laplacian denoted by Δ_r. In problem (P_λ) above, we have the sum of two such operators, one with variable exponent and the other with constant exponent. In the reaction (the right hand side of (P_λ)), we have the combined effects of two distinct nonlinearities.

One is the Carathéodory function $f(z, x)$ (that is, for all $x \in \mathbb{R}$, $z \to f(z, x)$ is measurable and for a.a. $z \in \Omega$, $x \to f(z, x)$ is continuous). We assume that $f(z, \cdot)$ is $(p_+ - 1)$-superlinear ($p_+ = \max_{\overline{\Omega}} p$) but it needs not satisfy the (common in such cases) Ambrosetti-Rabinowitz condition, see also Papageorgiou-Rădulescu-Repovš [19] (Robin problem). This term represents a "convex" contribution to the reaction.

The other nonlinearity is the parametric function $x \to -\lambda|x|^{\tau(z)-2}x$ with $\tau \in C(\overline{\Omega})$ such that $1 < \tau(z) < q$ for all $z \in \overline{\Omega}$. Therefore this term is $(q - 1)$-sublinear ("concave" term). Thus the reaction of (P_λ) corresponds to a "concave-convex" problem, but with an essential difference. The concave (sublinear) term enters in the equation with a negative sign and this changes the geometry of the problem.

In the past, problems with a negative concave term were studied by Perera [25], de Paiva-Massa [3], Papageorgiou-Rădulescu-Repovš [15] (Robin problems) for semilinear equations driven by the Laplacian, and by Papageorgiou-Winkert [12] for resonant $(p, 2)$-equations. All the aforementioned works deal with isotropic equations and the perturbation $f(z, \cdot)$ is $(p - 1)$-linear.

Using variational tools from the critical point theory and critical groups (see Sect. 2), we show that for all sufficiently small $\lambda > 0$, problem (P_λ) has at least three nontrivial smooth solutions. Two of these solutions have constant sign (one is positive and the other negative). It is an interesting open question, whether this multiplicity theorem still holds when the exponent q is also variable and whether we can show that the third solution is nodal (sign-changing).

For the hypotheses H_0 and H_1 involved in our theorem, we refer to Sect. 2. Also $C_+ = \{u \in C_0^1(\overline{\Omega}) : u(z) \geq 0 \text{ for all } z \in \overline{\Omega}\}$.

Theorem 1.1 *If hypotheses H_0 and H_1 hold, then for all sufficiently small $\lambda > 0$, problem (P_λ) has at least three nontrivial solutions $u_0 \in C_+ \setminus \{0\}$, $v_0 \in (-C_+) \setminus \{0\}$, and $y_0 \in C_0^1(\overline{\Omega}) \setminus \{0\}$.*

To have a more complete picture of the relevant literature, we mention that the standard isotropic concave-convex problems (the concave term having a positive sign), were first considered by Ambrosetti-Brezis-Cerami [1] for semilinear equations driven by Dirichlet Laplacian. Their work was extended to nonlinear equations driven by the p-Laplacian by Garcia Azorero-Peral Alonso-Manfredi [8]. Since then appeared several works with further generalizations. Just to quote a few we mention the works of Gasiński-Papageorgiou [10, 11], Papageorgiou-Repovš-Vetro [20, 23], Papageorgiou-Vetro-Vetro [21, 24], Papageorgiou-Winkert [13], and the recent papers of Papageorgiou-Qin-Rădulescu [17] and Papageorgiou-Rădulescu-Repovš [18] on anisotropic equations. In all these works the concave term enters in the equation with a positive sign and this permits the use of the strong maximum principle which provides more structural information concerning the solution. This extra information allows us to use the result relating Sobolev and Hölder minimizers. In the present setting this is no longer possible and the geometry changes requiring a new approach.

2 Preliminaries

The analysis of problem (P_λ) uses variable Lebesgue and Sobolev spaces. A detailed presentation of these spaces can be found in the books of Cruz Uribe-Fiorenza [2] and of Diening-Hajulehto-Hästö-Růžička [4].

Let $E_1 = \{r \in C(\overline{\Omega}) \,:\, 1 < r_- = \min_{\overline{\Omega}} r\}$. In general, for any $r \in E_1$, we set

$$r_- = \min_{\overline{\Omega}} r \text{ and } r_+ = \max_{\overline{\Omega}} r.$$

Also let $M(\Omega) = \{u \,:\, \Omega \to \mathbb{R} \text{ measurable}\}$. We identify two such functions which differ only on a Lebesgue null set. Given $r \in E_1$, we define the variable Lebesgue space $L^{r(z)}(\Omega)$ by

$$L^{r(z)}(\Omega) = \left\{u \in M(\Omega) : \int_\Omega |u(z)|^{r(z)} dz < +\infty\right\}.$$

This space is equipped with the so-called "Luxemburg norm", defined by

$$\|u\|_{r(z)} = \inf\left\{\lambda > 0 : \int_\Omega \left(\frac{|u(z)|}{\lambda}\right)^{r(z)} dz \leq 1\right\}.$$

The space $L^{r(z)}(\Omega)$ endowed with this norm becomes a Banach space which is separable and uniformly convex (hence reflexive) (see [4], p. 67). For $r \in E_1$ by $r'(\cdot)$ we denote the variable conjugate exponent to $r(\cdot)$, that is, $\frac{1}{r(z)} + \frac{1}{r'(z)} = 1$ for all $z \in \overline{\Omega}$. Evidently, $r' \in E_1$ and

$$(L^{r(z)}(\Omega))^* = L^{r'(z)}(\Omega).$$

Moreover, we have a Hölder-type inequality, namely

$$\int_\Omega |u(z)v(z)|dz \le \left[\frac{1}{r_-} + \frac{1}{r'_-}\right] \|u\|_{r(z)}\|v\|_{r'(z)}$$

for all $u \in L^{r(z)}(\Omega)$, $v \in L^{r'(z)}(\Omega)$ (see [2], p. 27). In addition, if $r, \widehat{r} \in E_1$ and $r(z) \le \widehat{r}(z)$ for all $z \in \overline{\Omega}$, then $L^{\widehat{r}(z)}(\Omega) \hookrightarrow L^{r(z)}(\Omega)$ continuously (see [2], pp. 37–38).

Using the variable Lebesgue spaces, we can define the corresponding variable Sobolev spaces. Taken $r \in E_1$, then

$$W^{1,r(z)}(\Omega) = \left\{u \in L^{r(z)}(\Omega) : |\nabla u| \in L^{r(z)}(\Omega)\right\},$$

where ∇u denotes the weak gradient of u. This space is equipped with the norm

$$\|u\|_{1,r(z)} = \|u\|_{r(z)} + \|\nabla u\|_{r(z)} \text{ for all } u \in W^{1,r(z)}(\Omega),$$

with $\|\nabla u\|_{r(z)} = \||\nabla u|\|_{r(z)}$. If $r \in E_1 \cap C^{0,1}(\overline{\Omega})$, then we define also

$$W_0^{1,r(z)}(\Omega) = \overline{C_c^\infty(\Omega)}^{\|\cdot\|_{1,r(z)}}.$$

Both $W^{1,r(z)}(\Omega)$ and $W_0^{1,r(z)}(\Omega)$ are Banach spaces which are separable and uniformly convex (thus reflexive) (see [4], p. 245). The critical Sobolev exponent $r^*(\cdot)$ is defined by

$$r^*(z) = \begin{cases} \dfrac{Nr(z)}{N - r(z)} & \text{if } r(z) < N, \\ +\infty & \text{if } N \le r(z). \end{cases}$$

For $r, p \in C(\overline{\Omega})$ with $1 < r_-, p_+ < N$ and $1 \le p(z) \le r^*(z)$ for all $z \in \overline{\Omega}$ (resp. $1 \le p(z) < r^*(z)$ for all $z \in \overline{\Omega}$), then we have

$$W^{1,r(z)}(\Omega) \hookrightarrow L^{p(z)}(\Omega) \text{ continuously}$$

$$(\text{resp. } W^{1,r(z)}(\Omega) \hookrightarrow L^{p(z)}(\Omega) \text{ compactly}),$$

(see [4], p. 259). The same embeddings are also valid for $W_0^{1,r(z)}(\Omega)$. We mention that on $W_0^{1,r(z)}(\Omega)$ ($r \in C^{0,1}(\overline{\Omega})$), the Poincaré inequality holds. Recall that the Poincaré inequality says that there exists $c = c(\Omega) > 0$ such that $\|u\|_{r(z)} \le c\|\nabla u\|_{r(z)}$ for all $u \in W_0^{1,r(z)}(\Omega)$ (see [4], p. 249). So, on $W_0^{1,r(z)}(\Omega)$ we can use the following norm

$$\|u\| = \|\nabla u\|_{r(z)} \text{ for all } u \in W_0^{1,r(z)}(\Omega).$$

In what follows, we shall denote by $\rho_r(\cdot)$ the modular function

$$\rho_r(u) = \int_\Omega |u(z)|^{r(z)} dz \text{ for all } u \in L^{r(z)}(\Omega).$$

If $u \in W^{1,r(z)}(\Omega)$ or $u \in W_0^{1,r(z)}(\Omega)$, then $\rho_r(\nabla u) = \rho_r(|\nabla u|)$. The norm $\|\cdot\|_{r(z)}$ and the modular function $\rho_r(\cdot)$ are closely related (see [6], Proposition 2.1).

Proposition 2.1 *If $r \in E_1$ and $u \in L^{r(z)}(\Omega) \setminus \{0\}$, then the following statements hold:*

(a) $\|u\|_{r(z)} = \theta \Leftrightarrow \rho_r\left(\frac{u}{\theta}\right) = 1$ *for all* $\theta > 0$*;*
(b) $\|u\|_{r(z)} < 1$ *(resp.* $= 1, > 1) \Leftrightarrow \rho_r(u) < 1$ *(resp.* $= 1, > 1)$*;*
(c) $\|u\|_{r(z)} < 1 \Rightarrow \|u\|_{r(z)}^{r_+} \leq \rho_r(u) \leq \|u\|_{r(z)}^{r_-}$*;*
(d) $\|u\|_{r(z)} > 1 \Rightarrow \|u\|_{r(z)}^{r_-} \leq \rho_r(u) \leq \|u\|_{r(z)}^{r_+}$*;*
(e) $\|u\|_{r(z)} \to 0$ *(resp.* $\|u\|_{r(z)} \to +\infty) \Leftrightarrow \rho_r(u) \to 0$ *(resp.* $\rho_r(u) \to +\infty)$*.*

We know that for $r \in E_1 \cap C^{0,1}(\overline{\Omega})$, we have

$$W_0^{1,r(z)}(\Omega)^* = W^{-1,r'(z)}(\Omega) \quad \text{(see [4], pp. 378–379).}$$

Consider the operator $A_{r(z)} : W_0^{1,r(z)}(\Omega) \to W^{-1,r'(z)}(\Omega)$ defined by

$$\langle A_{r(z)}(u), h\rangle = \int_\Omega |\nabla u(z)|^{r(z)-2}(\nabla u, \nabla h)_{\mathbb{R}^N} dz \text{ for all } u, h \in W_0^{1,r(z)}(\Omega), \qquad (1)$$

where $(\cdot,\cdot)_{\mathbb{R}^N}$ is the inner product in $\mathbb{R}^N$. This operator has the following properties (see [7], Proposition 2.9).

Proposition 2.2 *If $r \in E_1 \cap C^{0,1}(\overline{\Omega})$, then the operator $A_{r(z)}(\cdot)$ is bounded (that is, it maps bounded sets to bounded sets), continuous, strictly monotone (thus also maximal monotone) and of type $(S)_+$ (that is, $u_n \xrightarrow{w} u$ in $W_0^{1,r(z)}(\Omega)$ and $\limsup_{n\to\infty}\langle A_{r(z)}(u_n), u_n - u\rangle \leq 0$ imply that $u_n \to u$ in $W_0^{1,r(z)}(\Omega)$).*

Let X be a Banach space, $\varphi \in C^1(X,\mathbb{R})$ and $c \in \mathbb{R}$. We set

$$K_\varphi = \{u \in X : \varphi'(u) = 0\} \quad \text{(the critical set of } \varphi\text{)},$$
$$\varphi^c = \{u \in X : \varphi(u) \leq c\}.$$

Let (Y_1, Y_2) be a topological pair such that $Y_2 \subseteq Y_1 \subseteq X$ and $k \in \mathbb{N}_0$. By $H_k(Y_1, Y_2)$ we denote the kth-relative singular homology group with integer coefficients. If $u \in K_\varphi$ is isolated and $c = \varphi(u)$, then the critical groups of φ at u are defined by

$$C_k(\varphi, u) = H_k(\varphi^c \cap U, \varphi^c \cap U \setminus \{u\}) \text{ for all } k \in \mathbb{N}_0,$$

with U a neighborhood of u such that $K_\varphi \cap \varphi^c \cap U = \{u\}$ (see [16], Chapter 6). The excision property of singular homology implies that the above definition is independent of the choice of the isolating neighborhood U. For details we refer to Papageorgiou-Rădulescu-Repovš [16], Chapter 6, where the reader can find explicit computations of the critical groups for various kinds of critical points.

3 Conditions and Hypotheses

Definition 3.1 We say that $\varphi \in C^1(X, \mathbb{R})$ satisfies the *C-condition*, if it has the following property: Every sequence $\{u_n\}_{n\in\mathbb{N}} \subseteq X$ such that

- $\{\varphi(u_n)\}_{n\in\mathbb{N}} \subseteq \mathbb{R}$ is bounded; and
- $(1 + \|u_n\|_X)\varphi'(u_n) \to 0$ in X^* as $n \to \infty$ (X^* denotes the dual of X),

admits a strongly convergent subsequence (see [16], p. 366).

Our hypotheses on the data of problem (P_λ) will be the following:

H_0: $p \in C^{0,1}(\overline{\Omega})$, $\tau \in C(\overline{\Omega})$ and $1 < \tau(z) < q < p(z) < N$ for all $z \in \overline{\Omega}$.
H_1: $f : \Omega \times \mathbb{R} \to \mathbb{R}$ is a Carathéodory function such that $f(z, 0) = 0$ for a.a. $z \in \Omega$ and

(i) $|f(z, x)| \le a(z)[1 + |x|^{r(z)-1}]$ for a.a. $z \in \Omega$, all $x \in \mathbb{R}$, with $a \in L^\infty(\Omega)$, $r \in C(\overline{\Omega})$ with $p(z) < r(z) < p_-^*$ for all $z \in \overline{\Omega}$;

(ii) if $F(z, x) = \int_0^x f(z, s)ds$, then $\lim\limits_{x\to\pm\infty} \frac{F(z,x)}{|x|^{p_+}} = +\infty$ uniformly for a.a. $z \in \Omega$;

(iii) there exist $\mu \in C(\overline{\Omega})$ with $\mu(z) \in \left((r_+ - p_-)\frac{N}{p_-}, p_-^*\right)$ for all $z \in \overline{\Omega}$, $\tau_+ < \mu_-$ and a constant $\beta_0 > 0$ such that

$$\beta_0 \le \liminf_{x\to\pm\infty} \frac{f(z, x)x - p_+ F(z, x)}{|x|^{\mu(z)}} \text{ uniformly for a.a. } z \in \Omega;$$

(iv) there exist $\eta \in L^\infty(\Omega)$ and $\widehat{\eta} > 0$ such that

$$\widehat{\lambda}_1(q) \le \eta(z) \text{ for a.a. } z \in \Omega,\ \eta \not\equiv \widehat{\lambda}_1(q),$$

$$\eta(z) \le \liminf_{x\to 0} \frac{qF(z, x)}{|x|^q} \le \limsup_{x\to 0} \frac{qF(z, x)}{|x|^q} \le \widehat{\eta} \text{ uniformly for a.a. } z \in \Omega,$$

(by $\widehat{\lambda}_1(q)$ we denote the principal eigenvalue of $(-\Delta_q, W_0^{1,q}(\Omega))$; we know $\widehat{\lambda}_1(q) > 0$, see [9], p. 741).

Remark 3.1 Hypotheses H_1 (ii), (iii) imply that for a.a. $z \in \Omega$, $f(z, \cdot)$ is $(p_+ - 1)$-superlinear. We do not employ the AR-condition and this way we incorporate in our framework superlinear nonlinearities with "slower" growth as $x \to \pm\infty$. The following function satisfies hypothesis H_1 but it fails to satisfy the AR-condition:

$$f(z, x) = \begin{cases} \eta[|x|^{q-2}x - |x|^{\theta(z)-2}x] & \text{if } |x| \leq 1, \\ |x|^{p_+-2}x \ln |x| & \text{if } 1 < |x|, \end{cases}$$

with $\theta \in C(\overline{\Omega})$ and $q < \theta(z)$ for all $z \in \overline{\Omega}$.

For $\lambda > 0$, let $\varphi_\lambda : W_0^{1,p(z)}(\Omega) \to \mathbb{R}$ be the energy functional for problem (P_λ) defined by

$$\varphi_\lambda(u) = \int_\Omega \frac{1}{p(z)} |\nabla u(z)|^{p(z)} dz + \frac{1}{q} \|\nabla u\|_q^q + \int_\Omega \frac{\lambda}{\tau(z)} |u(z)|^{\tau(z)} dz - \int_\Omega F(z, u) dz$$

for all $u \in W_0^{1,p(z)}(\Omega)$. Evidently, $\varphi_\lambda \in C^1(W_0^{1,p(z)}(\Omega))$.

We also introduce the positive and negative truncations of $\varphi_\lambda(\cdot)$, namely the C^1-functionals $\varphi_\lambda^\pm : W_0^{1,p(z)}(\Omega) \to \mathbb{R}$ defined by

$$\varphi_\lambda^\pm(u) = \int_\Omega \frac{1}{p(z)} |\nabla u(z)|^{p(z)} dz + \frac{1}{q} \|\nabla u\|_q^q + \int_\Omega \frac{\lambda}{\tau(z)} (u^\pm(z))^{\tau(z)} dz - \int_\Omega F(z, \pm u^\pm) dz$$

for all $u \in W_0^{1,p(z)}(\Omega)$. Recall $u^+ = \max\{u, 0\}$, $u^- = \max\{-u, 0\}$.

We can show that the functionals $\varphi_\lambda^\pm(\cdot)$ and $\varphi_\lambda(\cdot)$ satisfy the C-condition.

Proposition 3.1 *If hypotheses H_0 and H_1 hold and $\lambda > 0$, then the functionals $\varphi_\lambda^\pm(\cdot)$ and $\varphi_\lambda(\cdot)$ satisfy the C-condition.*

Proof We shall present the proof for the functional $\varphi_\lambda^+(\cdot)$, the proofs for $\varphi_\lambda^-(\cdot)$ and $\varphi_\lambda(\cdot)$ are similar. So, consider a sequence $\{u_n\}_{n\in\mathbb{N}} \subseteq W_0^{1,p(z)}(\Omega)$ such that

$$|\varphi_\lambda^+(u_n)| \leq c_1 \text{ for some } c_1 > 0 \text{ and all } n \in \mathbb{N}, \tag{2}$$

$$(1 + \|u_n\|)(\varphi_\lambda^+)'(u_n) \to 0 \text{ in } W^{-1,p'(z)}(\Omega) \text{ as } n \to \infty. \tag{3}$$

Referring to (1), by (3) we have

$$\left| \langle A_{p(z)}(u_n), h\rangle + \langle A_q(u_n), h\rangle + \int_\Omega \lambda (u_n^+)^{\tau(z)-1} h dz - \int_\Omega f(z, u_n^+) h dz \right| \leq \frac{\varepsilon_n \|h\|}{1 + \|u_n\|} \tag{4}$$

for all $h \in W_0^{1,p(z)}(\Omega)$, with $\varepsilon_n \to 0^+$.

In (4) we choose $h = -u_n^- \in W_0^{1,p(z)}(\Omega)$ and obtain

$$\begin{aligned}
&\rho_p(\nabla u_n^-) \le \varepsilon_n \text{ for all } n \in \mathbb{N}, \\
\Rightarrow\quad & u_n^- \to 0 \text{ in } W_0^{1,p(z)}(\Omega) \text{ as } n \to \infty \text{ (see Proposition 2.1).}
\end{aligned} \tag{5}$$

From (2) and (5) we have

$$\rho_p(\nabla u_n^+) + \frac{p_+}{q}\|\nabla u_n^+\|_q^q + \int_\Omega \frac{\lambda p_+}{\tau(z)}(u_n^+)^{\tau(z)}dz - \int_\Omega p_+ F(z, u_n^+)dz \le c_2 \tag{6}$$

for some $c_2 > 0$ and all $n \in \mathbb{N}$.

Also, if in (4) we use the test function $h = u_n^+ \in W_0^{1,p(z)}(\Omega)$, we obtain

$$-\rho_p(\nabla u_n^+) - \|\nabla u_n^+\|_q^q - \int_\Omega \lambda(u_n^+)^{\tau(z)}dz + \int_\Omega f(z, u_n^+)u_n^+ dz \le \varepsilon_n \tag{7}$$

for all $n \in \mathbb{N}$. We add (6) and (7) and obtain

$$\int_\Omega [f(z, u_n^+)u_n^+ - p_+ F(z, u_n^+)]dz \le c_3 \text{ for some } c_3 > 0 \text{ and all } n \in \mathbb{N}.$$

From hypothesis $H_1\,(iii)$ we see that we can always assume that $\mu_- < r_-$. Hypotheses $H_1\,(i), (iii)$ imply that there exist $\widehat{\beta}_0 \in (0, \beta_0)$ and $c_4 > 0$ such that

$$\widehat{\beta}_0|x|^{\mu_-} - c_4 \le f(z, x)x - p_+ F(z, x) \text{ for a.a. } z \in \Omega \text{ and all } x \in \mathbb{R}. \tag{8}$$

We use (8) in (7) and obtain

$$\begin{aligned}
&\|u_n^+\|_{\mu_-}^{\mu_-} \le c_5 \text{ for some } c_5 > 0 \text{ and all } n \in \mathbb{N}, \\
\Rightarrow\quad & \{u_n^+\}_{n\in\mathbb{N}} \subseteq L^{\mu_-}(\Omega) \text{ is bounded.}
\end{aligned} \tag{9}$$

Recall that $\mu_- < r_- \le r_+ < p_-^*$. So, we can find $t \in (0, 1)$ such that

$$\frac{1}{r_+} = \frac{1-t}{\mu_-} + \frac{t}{p_-^*}. \tag{10}$$

Using the interpolation inequality (see Papageorgiou-Winkert [14], p. 116), we have

$$\begin{aligned}
&\|u_n^+\|_{r_+} \le \|u_n\|_{\mu_-}^{1-t}\|u_n\|_{p_-^*}^t \text{ for all } n \in \mathbb{N}, \\
\Rightarrow\quad & \|u_n^+\|_{r_+}^{r_+} \le c_6\|u_n^+\|^{tr_+} \text{ for some } c_6 > 0,\ \text{all } n \in \mathbb{N} \text{ (see (9)).}
\end{aligned} \tag{11}$$

Also, from (4) with $h = u_n^+ \in W_0^{1,p(z)}(\Omega)$, we have

$$\rho_p(\nabla u_n^+) \le c_7 + \int_\Omega f(z, u_n^+) u_n^+ dz \text{ for some } c_7 > 0 \text{ and all } n \in \mathbb{N}.$$

Without loss of generality, we may assume that $\|u_n^+\| \ge 1$. Using hypothesis $H_1(i)$ and Proposition 2.1, we have

$$\begin{aligned} \|u_n^+\|^{p_-} &\le c_8[1 + \|u_n^+\|_{r_+}^{r_+}] \text{ for some } c_8 > 0, \\ &\le c_9[1 + \|u_n^+\|^{tr_+}] \text{ for some } c_9 > 0 \text{ and all } n \in \mathbb{N} \text{ (see (11))}. \end{aligned} \tag{12}$$

From (10) we have

$$\begin{aligned} & tr_+ = \frac{p_-^*(r_+ - \mu_-)}{p_-^* - \mu_-} < p_- \text{ (see hypothesis } H_1\,(iii)), \\ \Rightarrow\quad & \{u_n^+\}_{n \in \mathbb{N}} \subseteq W_0^{1,p(z)}(\Omega) \text{ is bounded (see (12))}, \\ \Rightarrow\quad & \{u_n\}_{n \in \mathbb{N}} \subseteq W_0^{1,p(z)}(\Omega) \text{ is bounded (see (5))}. \end{aligned}$$

So, we may assume that

$$u_n \xrightarrow{w} u \text{ in } W_0^{1,p(z)}(\Omega) \text{ and } u_n \to u \text{ in } L^{r(z)}(\Omega). \tag{13}$$

In (4) we choose $h = u_n - u \in W_0^{1,p(z)}(\Omega)$, pass to the limit as $n \to \infty$, and use (13). We obtain

$$\begin{aligned} & \lim_{n\to\infty} \left[\langle A_{p(z)}(u_n), u_n - u\rangle + \langle A_q(u_n), u_n - u\rangle\right] = 0, \\ \Rightarrow\quad & \limsup_{n\to\infty} \left[\langle A_{p(z)}(u_n), u_n - u\rangle + \langle A_q(u), u_n - u\rangle\right] \le 0, \\ & \qquad\qquad\qquad \text{(since } A_q(\cdot) \text{ is monotone)}, \\ \Rightarrow\quad & \limsup_{n\to\infty} \langle A_{p(z)}(u_n), u_n - u\rangle \le 0 \quad \text{(see (13))}, \\ \Rightarrow\quad & u_n \to u \text{ in } W_0^{1,p(z)}(\Omega) \text{ (see Proposition 2.2)}. \end{aligned}$$

This proves that the functional $\varphi_\lambda^+(\cdot)$ satisfies the C-condition. In a similar fashion we show that $\varphi_\lambda^-(\cdot)$ and $\varphi_\lambda(\cdot)$ also satisfy the C-condition. □

4 Auxiliary Propositions

We shall prove two propositions needed for the proof of the main result.

Proposition 4.1 *If hypotheses H_0 and H_1 hold and $\lambda > 0$, then there exist $\rho_0, c_0 > 0$ such that $\varphi_\lambda^\pm(u) \geq c_0 > 0$ for all $u \in W_0^{1,p(z)}(\Omega)$, $\|u\| = \rho_0$.*

Proof On account of hypothesis H_1 (iv), we have

$$\lim_{x \to 0^+} \frac{F(z,x)}{x^{\tau(z)}} = \lim_{x \to 0^+} \left[\frac{F(z,x)}{x^q} x^{q-\tau(z)} \right] = 0 \text{ (recall that } \tau_+ < q). \tag{14}$$

Then (14) and hypothesis H_1 (i) imply that given $\varepsilon > 0$, we can find $c_{10} = c_{10}(\varepsilon) > 0$ such that

$$F_+(z,x) \leq \frac{\varepsilon}{\tau_+}|x|^{\tau(z)} + c_{10}|x|^{r_-} \text{ for a.a. } z \in \Omega \text{ and all } x \in \mathbb{R}.$$

For $u \in W_0^{1,p(z)}(\Omega)$ with $\|u\| \leq 1$, we have

$$\varphi_\lambda^+(u) \geq \frac{1}{p_+}\rho_p(\nabla u) + \frac{1}{\tau_+}[\lambda - \varepsilon]\rho_\tau(u) - c_{11}\|u\|^{r_-} \text{ for some } c_{11} > 0$$

(since $\|u\|_{\tau(z)} \leq 1$). Choosing $\varepsilon \in (0, \lambda)$ and recalling that $\|u\| \leq 1$, we have

$$\varphi_\lambda^+(u) \geq \frac{1}{p_+}\|u\|^{p_+} - c_{11}\|u\|^{r_-} \text{ (see Proposition 2.1).}$$

Recall that $p_+ < r_-$. So, by choosing $\rho_0 \in (0, 1)$ sufficiently small, we obtain

$$\varphi_\lambda^+(u) \geq c_0 > 0 \text{ for all } u \in W_0^{1,p(z)}(\Omega), \|u\| = \rho_0.$$

Similarly for $\varphi_\lambda^-(\cdot)$. □

Recall that $\widehat{\lambda}_1(q) > 0$ is the principal eigenvalue of $(-\Delta_q, W_0^{1,q}(\Omega))$. Also, by $\widehat{u}_1 = \widehat{u}_1(q)$ we denote the corresponding positive L^q-normalized (that is, $\|\widehat{u}_1\|_q = 1$) eigenfunction. We know that $\widehat{u}_1 \in C_0^1(\overline{\Omega})$ and $\widehat{u}_1(z) > 0$ for all $z \in \Omega$ (see [9], Theorem 6.2.9, p. 739).

Proposition 4.2 *If hypotheses H_0 and H_1 hold, then there exist $\lambda^* > 0$ and $t_\pm > 0$ such that $\varphi_\lambda^\pm(\pm t_\pm \widehat{u}_1) < 0$ for all $\lambda \in (0, \lambda^*)$.*

Proof On account of hypotheses H_1 (i), (iv), given $\varepsilon > 0$, we can find $c_{12} = c_{12}(\varepsilon) > 0$ such that

$$F_+(z,x) \geq \frac{1}{q}[\eta(z) - \varepsilon]|x|^q - c_{12}|x|^{r_-} \text{ for a.a. } z \in \Omega \text{ and all } x \geq 0.$$

Then for $t \in (0, 1]$ we have

$$\varphi_\lambda^+(t\widehat{u}_1) \leq \frac{t^{p_-}}{p_-}\rho_p(\nabla\widehat{u}_1) + \frac{t^q}{q}\left[\int_\Omega (\widehat{\lambda}_1(q) - \eta(z))\widehat{u}_1^q dz + \varepsilon\right] + \frac{\lambda t^{\tau_-}}{\tau_-}\rho_\tau(\widehat{u}_1) + c_{12}t^{r_-}\|\widehat{u}_1\|_{r_-}^{r_-}.$$

As we have mentioned earlier, $\widehat{u}_1(z) > 0$ for all $z \in \Omega$. This fact, combined with hypothesis H_1 (iv), implies that

$$\widehat{\mu} = \int_\Omega (\eta(z) - \widehat{\lambda}_1(q))\widehat{u}_1^q dz > 0.$$

So, choosing $\varepsilon \in (0, \widehat{\mu})$, we obtain

$$\begin{aligned}\varphi_\lambda(t\widehat{u}_1) &\leq c_{13}[t^{p_-} + \lambda t^{\tau_-}] - c_{14}t^q \text{ for some } c_{13}, c_{14} > 0\\ &= [c_{13}(t^{p_- - q} + \lambda t^{\tau_- - q}) - c_{14}]t^q.\end{aligned} \tag{15}$$

Consider the function

$$\xi_\lambda(t) = t^{p_- - q} + \lambda t^{\tau_- - q} \text{ for } t > 0.$$

Since $\tau_- < q < p_-$, we see that

$$\xi_\lambda(t) \to +\infty \text{ as } t \to 0^+ \text{ and as } t \to +\infty.$$

Therefore there exists $t_+ > 0$ such that

$$\begin{aligned}&\xi_\lambda(t_+) = \inf\{\xi_\lambda(t) : t > 0\},\\ \Rightarrow\ &\xi_\lambda'(t_+) = 0,\\ \Rightarrow\ &(p_- q)t_+^{p_- - \tau_-} = \lambda(q - \tau_-),\\ \Rightarrow\ &t_+ = \left[\frac{\lambda(q - \tau_-)}{p_- - q}\right]^{\frac{1}{p_- - \tau_-}}.\end{aligned} \tag{16}$$

Using (16), we see that

$$\xi_\lambda(t_+) \to 0^+ \text{ as } \lambda \to 0^+.$$

Hence we can find $\lambda_1^* > 0$ such that

$$\xi_\lambda(t_+) < \frac{c_{14}}{c_{13}} \text{ for all } \lambda \in (0, \lambda_1^*),$$
$$\Rightarrow \quad \varphi_\lambda^+(t_+\widehat{u}_1) < 0 \text{ for all } \lambda \in (0, \lambda_1^*) \text{ (see (15))}.$$

Similarly working with $\varphi_\lambda^-(\cdot)$, we produce $\lambda_2^* > 0$ and $t_- > 0$ such that

$$\varphi_\lambda^-(-t_-\widehat{u}_1) < 0 \text{ for all } \lambda \in (0, \lambda_2^*).$$

Finally let $\lambda^* = \min\{\lambda_1^*, \lambda_2^*\}$. □

Remark 4.1 We can always choose $\lambda^* > 0$ small so that

$$t_\pm = t_\pm(\lambda) \in (0, \rho_0) \text{ for all } \lambda \in (0, \lambda^*) (\rho_0 > 0 \text{ is as in Proposition 4.1}). \quad (17)$$

5 Proof of Main Theorem

We shall break down the proof of Theorem 1.1 into two steps (5.1 and 5.2).

5.1 Existence of Two Solutions

First, we shall produce two nontrivial constant sign solutions. In what follows, we shall denote $C_+ = \{u \in C_0^1(\overline{\Omega}) : 0 \le u(z) \text{ for all } z \in \overline{\Omega}\}$.

Proposition 5.1 *If hypotheses H_0 and H_1 hold and $\lambda \in (0, \lambda^*)$, then problem (P_λ) has at least two constant sign solutions $u_0 \in C_+ \setminus \{0\}$, $v_0 \in (-C_+) \setminus \{0\}$ and both are local minimizers of the energy functional $\varphi_\lambda(\cdot)$.*

Proof We introduce the closed ball

$$\overline{B}_{\rho_0} = \{u \in W_0^{1,p(z)}(\Omega) : \|u\| \le \rho_0\}$$

with $\rho_0 > 0$ as in Proposition 4.1 and consider the minimization problem

$$\inf\{\varphi_\lambda^+(u) : u \in \overline{B}_{\rho_0}\} = m_\lambda^+. \quad (18)$$

The anisotropic Sobolev embedding theorem (see Sect. 2), implies that $\varphi_\lambda^+(\cdot)$ is sequentially weakly lower semicontinuous. Also the reflexivity of $W_0^{1,p(z)}(\Omega)$ and the Eberlein-Smulian theorem (see [14], p. 221) imply that $\overline{B}_{\rho_0}$ is sequentially weakly compact. So, by the Weierstrass-Tonelli theorem (see [14], p. 78), we can find $u_0 \in \overline{B}_{\rho_0}$ such that

$$\begin{aligned} &\varphi_\lambda^+(u_0) = m_\lambda^+ \leq \varphi_\lambda^+(t_+\widehat{u}_1) < 0 = \varphi_\lambda^+(0) \\ &\text{(see (17), (18) and Proposition 4.2),} \\ \Rightarrow\ & u_0 \neq 0. \end{aligned} \tag{19}$$

From (19) and Proposition 4.1, we have

$$0 < \|u_0\| < \rho_0.$$

Hence we have

$$\begin{aligned} &(\varphi_\lambda^+)'(u_0) = 0, \\ \Rightarrow\ & \langle A_{p(z)}(u_0), h\rangle + \langle A_q(u_0), h\rangle = \int_\Omega f(z, u_0^+)hdz - \lambda \int_\Omega (u_0^+)^{\tau(z)-1}hdz \end{aligned} \tag{20}$$

for all $h \in W_0^{1,p(z)}(\Omega)$. In (20) we choose $h = -u_0^- \in W_0^{1,p(z)}(\Omega)$ and obtain

$$\begin{aligned} &\rho_p(\nabla u_0^-) + \|\nabla u_0^-\|_q^q = 0, \\ \Rightarrow\ & u_0 \geq 0,\ u_0 \neq 0. \end{aligned}$$

By Papageorgiou-Rădulescu-Zhang [22, Proposition A.1], we know that $u_0 \in L^\infty(\Omega)$. Then the anisotropic regularity theory (see Fan [5, Theorem 1.3] and Tan-Fang [26, Corollary 3.1]) implies $u_0 \in C_+ \setminus \{0\}$. So, we have produced a positive smooth solution of (P_λ) for $\lambda \in (0, \lambda^*)$. Similarly working with functional $\varphi_\lambda^-(\cdot)$, we produce a negative solution v_0 of (P_λ) ($\lambda \in (0, \lambda^*)$) such that

$$v_0 \in (-C_+) \setminus \{0\}.$$

Finally, we show that u_0 and v_0 are both local minimizers of the energy functional $\varphi_\lambda(\cdot)$. We shall present the proof for u_0, the proof for v_0 is similar. From the first part of the proof, we know that u_0 is a local $C_0^1(\overline{\Omega})$-minimizer of $\varphi_\lambda^+(\cdot)$. So, we can find $\rho_1 > 0$ such that

$$\varphi_\lambda^+(u_0) \leq \varphi_\lambda^+(u) \text{ for all } u \in \overline{B}_{\rho_1}^{C_0^1}(u_0) = \{u \in C_0^1(\overline{\Omega}) : \|u - u_0\|_{C_0^1(\overline{\Omega})} \leq \rho_1\}. \tag{21}$$

For $u \in \overline{B}_{\rho_1}^{C_0^1}(u_0)$ we have

$$\begin{aligned}
&\varphi_\lambda(u) - \varphi_\lambda(u_0) \\
&= \varphi_\lambda(u) - \varphi_\lambda^+(u_0) \text{ (since } \varphi_\lambda|_{C_+} = \varphi_\lambda^+|_{C_+}) \\
&\geq \varphi_\lambda(u) - \varphi_\lambda^+(u) \text{ (see (21))} \\
&\geq \frac{\lambda}{\tau_+} \int_\Omega [|u|^{\tau(z)} - (u^+)^{\tau(z)}]dz - \int_\Omega [F(z,u) - F(z,u^+)]dz \\
&= \frac{\lambda}{\tau_+} \rho_\tau(u^-) - \int_\Omega F(z,-u^-)dz.
\end{aligned} \tag{22}$$

On account of hypotheses H_1 (i), (iv) we can find $c_{15} > 0$ such that

$$F(z,x) \leq c_{15}[|x|^q + |x|^{r_+}] \text{ for a.a. } z \in \Omega \text{ and all } x \in \mathbb{R}. \tag{23}$$

Using (23) in (22), we obtain

$$\begin{aligned}
&\varphi_\lambda(u) - \varphi_\lambda(u_0) \\
&\geq \frac{\lambda}{\tau_+} \rho_\tau(u^-) - c_{15} \int_\Omega [(u^-)^q + (u^-)^{r_+}]dz \\
&\geq \frac{\lambda}{\tau_+} \rho_\tau(u^-) - c_{15} \int_\Omega [\|u^-\|_\infty^{q-\tau(z)} + \|u^-\|_\infty^{r_+-\tau(z)}](u^-)^{\tau(z)}dz.
\end{aligned} \tag{24}$$

Recall that $u_0 \in C_+ \setminus \{0\}$ and $u \in \overline{B}_{\rho_1}^{C_0^1}(u_0)$. So, by choosing $\rho_1 > 0$ even smaller if necessary, we can have that $\|u^-\|_\infty \leq 1$. Hence

$$\|u^-\|_\infty^{q-\tau(z)} \leq \|u^-\|_\infty^{q-\tau_+}, \ \|u^-\|_\infty^{r_+-\tau(z)} \leq \|u^-\|_\infty^{r_+-\tau_+}. \tag{25}$$

We return to (24) and use (25). We obtain

$$\varphi_\lambda(u) - \varphi_\lambda(u_0) \geq \left[\frac{\lambda}{\tau_+} - c_{15}(\|u^-\|_\infty^{q-\tau_+} + \|u^-\|_\infty^{r_+-\tau_+})\right] \rho_\tau(u^-).$$

Note that $\|u^-\|_\infty \to 0^+$ as $\rho_1 \to 0^+$. Therefore we can choose $\rho_1 > 0$ so small that

$$\varphi_\lambda(u) \geq \varphi_\lambda(u_0) \ \text{ for all } u \in \overline{B}_{\rho_1}^{C_0^1}(u_0).$$

This means that u_0 is a local $C_0^1(\overline{\Omega})$-minimizer of $\varphi_\lambda(\cdot)$. Then Proposition A.3 of Papageorgiou-Rădulescu-Zhang [22], implies that u_0 is a local $W_0^{1,p(z)}(\Omega)$-

minimizer of $\varphi_\lambda(\cdot)$. Similarly we show that $v_0 \in (-C_+) \setminus \{0\}$ is a local minimizer of the energy functional $\varphi_\lambda(\cdot)$. □

Proposition 5.2 *If hypotheses H_0 and H_1 hold and $\lambda > 0$, then $u = 0$ is a local minimizer of the energy functional $\varphi_\lambda(\cdot)$.*

Proof Let $u \in C_0^1(\overline{\Omega})$ with $\|u\|_{C_0^1(\overline{\Omega})} \leq 1$. We have

$$\begin{aligned}\varphi_\lambda(u) - \varphi_\lambda(0) &= \varphi_\lambda(u)\\ &\geq \frac{\lambda}{\tau_+}\rho_\tau(u) - \int_\Omega F(z, u)dz\\ &\geq \left[\frac{\lambda}{\tau_+} - c_{15}(\|u\|_\infty^{q-\tau_+} + \|u\|_\infty^{r_+-\tau_+})\right]\rho_\tau(u) \text{ (see (23)).}\end{aligned}$$

Choosing $\rho > 0$ small enough, we see that

$$\begin{aligned}&\varphi_\lambda(u) \geq 0 = \varphi_\lambda(0) \text{ for all } u \in \overline{B}_\rho^{C_0^1}(0),\\ \Rightarrow\quad & u = 0 \text{ is a local } C_0^1(\overline{\Omega})\text{-minimizer of } \varphi_\lambda(\cdot),\\ \Rightarrow\quad & u = 0 \text{ is a local } W_0^{1,p(z)}(\Omega) - \text{minimizer of } \varphi_\lambda(\cdot)\text{(see [22]).}\end{aligned}$$

□

5.2 *Existence of Third Solution*

Now we are ready to produce the third nontrivial solution for problem (P_λ), $\lambda \in (0, \lambda^*)$.

Proposition 5.3 *If hypotheses H_0 and H_1 hold and $\lambda \in (0, \lambda^*)$, then problem (P_λ) has the third solution $y_0 \in C_0^1(\overline{\Omega})$ and $y_0 \notin \{0, u_0, v_0\}$.*

Proof From the anisotropic regularity theory (see [5], [26]), we have that $K_{\varphi_\lambda} \subseteq C_0^1(\overline{\Omega})$. Since the critical points of $\varphi_\lambda(\cdot)$ are the weak solutions of (P_λ), we may assume that K_{φ_λ} is finite or otherwise we would already have an infinity of nontrivial smooth solutions for (P_λ) and so we would be done. Then Proposition 5.2 and [16, Theorem 5.7.6, p. 449], imply that we can find $\widehat{\rho} > 0$ such that

$$\varphi_\lambda(0) = 0 < \inf\{\varphi_\lambda(u) : \|u\| = \widehat{\rho}\} = \widehat{m}_\lambda. \tag{26}$$

Also, if $u \in C_+$ with $u(z) > 0$ for all $z \in \Omega$, then on account of hypothesis H_1 (ii), we have

$$\varphi_\lambda(tu) \to -\infty \text{ as } t \to +\infty. \tag{27}$$

Then (26), (27) and Proposition 3.1, permit the use of the Mountain Pass Theorem (see [16], p. 401). So, we can find $y_0 \in W_0^{1,p(z)}(\Omega)$ such that

$$\begin{aligned} & y_0 \in K_{\varphi_\lambda},\ \varphi_\lambda(0) = 0 < \widehat{m}_\lambda \leq \varphi_\lambda(y_0), \\ \Rightarrow\quad & y_0 \neq 0. \end{aligned}$$

Moreover, [16, Corollary 6.6.9, p. 533] implies that

$$C_1(\varphi_\lambda, y_0) \neq 0. \tag{28}$$

On the other hand from Proposition 5.1, we infer that

$$C_k(\varphi_\lambda, u_0) = C_k(\varphi_\lambda, v_0) = \delta_{k,0}\mathbb{Z} \text{ for all } k \in \mathbb{N}_0. \tag{29}$$

Comparing (28) and (29), we conclude that

$$y_0 \neq u_0,\ y_0 \neq v_0.$$

The anisotropic regularity theory implies that $y_0 \in C_0^1(\overline{\Omega})$.

□

This also completes the proof of Theorem 1.1. □

Acknowledgments The authors thank the referee for his/her remarks. Repovš was supported by the Slovenian Research Agency grants P1-0292, J1-4031, J1-4001, N1-0278 and N1-0114.

References

1. Ambrosetti, A., Brezis, H., Cerami, G.: Combined effects of concave and convex nonlinearities in some elliptic problems. J. Funct. Anal. **122**, 519–543 (1994)
2. Cruz-Uribe, D.V., Fiorenza, A.: Variable Lebesgue Spaces: Foundations and Harmonic Analysis. Springer, Birkhäuser, Basel (2013)
3. de Paiva, F.O., Massa, E.: Multiple solutions for some elliptic equations with a nonlinearity concave at the origin. Nonlinear Anal. **66**, 2940–2946 (2007)
4. Diening, L., Harjulehto, P., Hästö, P., Růžička, M.: Lebesgue and Sobolev Spaces with Variable Exponents. Lecture Notes in Math., vol. 2017. Springer-Verlag, Heidelberg (2011)
5. Fan, X.: Global $C^{1,\alpha}$-regularity for variable exponent elliptic equations in divergence form. J. Differential Equations **235**, 397–417 (2007)
6. Fan, X., Deng, S.-G.: Multiplicity of positive solutions for a class of inhomogeneous Neumann problems involving the $p(x)$-Laplacian. Nonlinear Differ. Equ. Appl. **16**, 255–271 (2009)
7. Fan, X., Zhao, Y.: Nodal solutions of $p(x)$-Laplacian equations. Nonlinear Anal. **67**, 2859–2868 (2007)
8. García Azorero, J.P., Peral Alonso, I., Manfredi, J.J.: Sobolev versus Hölder local minimizers and global multiplicity for some quasilinear elliptic equations. Commun. Contemp. Math. **2**, 385–404 (2000)

9. Gasiński, L., Papageorgiou, N.S.: Nonlinear Analysis. Ser. Math. Anal. Appl., vol. 9. Chapman and Hall/CRC Press, Boca Raton (2006)
10. Gasiński, L., Papageorgiou, N.S.: Positive solutions for the Robin p-Laplacian problem with competing nonlinearities. Adv. Calc. Var. **12**, 31–56 (2019)
11. Gasiński, L., Papageorgiou, N.S.: Multiple solutions for $(p, 2)$-equations with resonance and concave terms. Results Math. **74**:79, pp. 34 (2019)
12. Papageorgiou, N.S., Winkert, P.: Resonant $(p, 2)$-equations with concave terms. Appl. Anal. **94**, 341–359 (2015)
13. Papageorgiou, N.S., Winkert, P.: Positive solutions for nonlinear nonhomogeneous Dirichlet problems with concave-convex nonlinearities. Positivity **20**, 945–979 (2016)
14. Papageorgiou, N.S., Winkert, P.: Applied Nonlinear Functional Analysis. W. de Gruyter, Berlin (2018)
15. Papageorgiou, N.S., Rădulescu, V.D., Repovš, D.D.: Asymmetric Robin problems with indefinite potential and concave terms. Adv. Nonlin. Stud. **19**, 69–87 (2019)
16. Papageorgiou, N.S., Rădulescu, V.D., Repovš, D.D.: Nonlinear Analysis - Theory and Methods. Springer Monographs in Mathematics. Springer, Cham (2019)
17. Papageorgiou, N.S., Qin, D., Rădulescu, V.D.: Anisotropic double-phase problems with indefinite potential: multiplicity of solutions. Anal. Math. Phys. **10**:63, pp. 37 (2020)
18. Papageorgiou, N.S., Rădulescu, V.D., Repovš, D.D.: Anisotropic equations with indefinite potential and competing nonlinearities. Nonlinear Anal. **201**:111861, pp. 24 (2020)
19. Papageorgiou, N.S., Rădulescu, V.D., Repovš, D.D.: Superlinear perturbations of the eigenvalue problem for the Robin Laplacian plus an indefinite and unbounded potential. Results Math. **75**:116, pp. 22 (2020)
20. Papageorgiou, N.S., Repovš, D.D., Vetro, C.: Nonlinear nonhomogeneous Robin problems with almost critical and partially concave reaction. J. Geom. Anal. **30**, 1774–1803 (2020)
21. Papageorgiou, N.S., Vetro, C., Vetro, F.: Multiple solutions with sign information for a $(p, 2)$-equation with combined nonlinearities. Nonlinear Anal. **192**:111716, pp. 25 (2020)
22. Papageorgiou, N.S., Rădulescu, V.D., Zhang, Y.: Anisotropic singular double phase Dirichlet problem. Discrete Contin. Dyn. Syst. Ser. S **14**, 4465–4502 (2021)
23. Papageorgiou, N.S., Repovš, D.D., Vetro, C.: Constant sign and nodal solutions for parametric anisotropic $(p, 2)$-equations. Appl. Anal. (2021). https://doi.org/10.1080/00036811.2021.1971199
24. Papageorgiou, N.S., Vetro, C., Vetro, F.: Multiple solutions for parametric double phase Dirichlet problems. Commun. Contemp. Math. **23**:2050006, pp. 18 (2021)
25. Perera, K.: Multiplicity results for some elliptic problems with concave nonlinearities. J. Differential Equations **140**, 133–141 (1997)
26. Tan, Z., Fang, F.: Orlicz-Sobolev versus Hölder local minimizer and multiplicity results for quasilinear elliptic equations. J. Math. Anal. Appl. **402**, 348–370 (2013)

Implicit Coupled k-Generalized ψ-Hilfer Fractional Differential Systems with Terminal Conditions in Banach Spaces

Abdelkrim Salim, Mouffak Benchohra, and Jamal Eddine Lazreg

Abstract This chapter deals with some existence results for a class of coupled systems for implicit nonlinear k-generalized ψ-Hilfer fractional differential equations with terminal conditions. The tools employed for this study are the fixed point theorem of Mönch combined with the technique of measure of noncompactness. Furthermore, an example is provided to illustrate of our results.

Keywords ψ-Hilfer fractional derivative · k-Generalized ψ-Hilfer fractional derivative · Banach spaces · Coupled systems · Existence · Measure of noncompactness

1 Introduction

In recent years, fractional calculus has proven to be a highly essential approach for dealing with the complexity structures observed in a range of diverse areas. Its theory and application are extensive, and it is concerned with the expansion of integer order differentiation and integration of a function to non-integer order. The reader is directed to the publications [1–3, 10, 12, 20, 21, 23, 24], for more details.

Several papers and books have lately been published in which the authors addressed the existence, stability, and uniqueness of solutions for diverse systems with fractional differential equations and inclusions using various fractional derivatives and some types of conditions. One may see the papers [7, 8, 14, 16, 27, 33], and the references therein.

In [22, 25, 26], we recently presented the definition of the k-generalized ψ-Hilfer fractional derivative. It is regarded to be an expansion of previous fractional operators, such as the ψ-Hilfer fractional derivative introduced by Sousa et al.

A. Salim
Hassiba Benbouali University, Chlef, Algeria

M. Benchohra (✉) · J. E. Lazreg
Djillali Liabes University, Sidi Bel-Abbes, Algeria

A. M. Candela et al. (eds.), *Recent Advances in Mathematical Analysis*, Trends in Mathematics, https://doi.org/10.1007/978-3-031-20021-2_22

in [30]. Form more studies by the previous authors, we refer the readers to the papers [28, 29, 31, 32] and the references therein. We also outlined a number of its properties considering the special functions presented by Diaz et al. in [13], as well as the results and generalizations for various fractional operators based on these characteristics found in [11, 18].

In [4], Abbas et al. considered the coupled system with the Hilfer-Hadamard fractional derivative and initial integral conditions that follows:

$$\begin{cases} \left({}^H D_1^{\alpha_1,\beta_1} u\right)(t) = g_1(t, u(t), v(t)), \\ \left({}^H D_1^{\alpha_2,\beta_2} v\right)(t) = g_2(t, u(t), v(t)), \end{cases} ; t \in [1, T],$$

with the following initial conditions

$$\begin{cases} \left({}^H I_1^{1-\gamma_1} u\right)(1) = \psi_1, \\ \left({}^H I_1^{1-\gamma_2} v\right)(1) = \psi_2, \end{cases}$$

where $T > 1, \alpha_i \in (0, 1), \beta_i \in [0, 1], \gamma_i = \alpha_i + \beta_i - \alpha_i\beta_i, \psi_i \in E, g_i : [1, T] \times E \times E \to E; i = 1, 2$ are given functions, ${}^H I_1^{1-\gamma_i}$ is the left-sided mixed Hadamard integral of order $1 - \gamma_i$, and ${}^H D_1^{\alpha_i,\beta_i}$ is the Hilfer-Hadamard fractional derivative of order α_i and type $\beta_i; i = 1, 2$. They employed the technique that relies on the concept of the measure of noncompactness and the fixed point theory.

In [6], Ahmed et al. considered the following switched coupled implicit ψ-Hilfer fractional differential system:

$$\begin{cases} {}_H D_{a^+}^{p,q;\psi} u(t) = f\left(t, u(t), {}_H D_{a^+}^{p,q;\psi} v(t)\right), & t \in \mathcal{J} = (a, b], \\ {}_H D_{a^+}^{p,q;\psi} v(t) = g\left(t, {}_H D_{a^+}^{p,q;\psi} u(t), v(t)\right), & \gamma = p + q - pq, \\ I_{a^+}^{1-\gamma;\psi} u(t)\Big|_{t=a} = u_a, \ I_{a^+}^{1-\gamma;\psi} v(t)\Big|_{t=a} = v_a, & u_a, v_a \in \mathbb{R}, \end{cases}$$

where ${}_H D_{a^+}^{p,q;\psi}$ represent the ψ-Hilfer fractional derivative of order p and type q with $p \in (0, 1), q \in (0, 1]$. $I_{a^+}^{1-\gamma;\psi}$ denote the ψ-Riemann-Liouville fractional integral of order $1 - \gamma$. Moreover $f, g : \mathcal{J} \times \mathcal{X} \times \mathcal{X} \to \mathcal{X}$ are continuous and nonlinear functions on a Banach space $\mathcal{X}$. The linear function $\psi : \mathcal{J} \to \mathbb{R}$ satisfies $\psi'(t) \neq 0, \ t \in \mathcal{J}$. The tools employed for this study are the Banach contraction principle and Schauder's fixed point theorem.

Using the fixed point techniques of Banach and Krasnoselskii, in [5], Abdo et al. proved some existence, uniqueness, and Ulam–Hyers stability results of the following coupled system for generalized Hilfer fractional derivative:

$$\begin{cases} D_{a^+}^{\theta_1,\eta_1;\psi} y(t) = f_1(t, x(t)), a < t \leq T, a > 0, \\ D_{a^+}^{\theta_2,\eta_2;\psi} x(t) = f_2(t, y(t)), a < t \leq T, a > 0, \\ y(T) = w_1 \in \mathbb{R}, \\ x(T) = w_2 \in \mathbb{R}, \end{cases}$$

where $0 < \theta_i < 1, 0 \leq \eta_i \leq 1, D_{a^+}^{\theta_i,\eta_i\psi} (i = 1, 2)$ is the ψ-Hilfer fractional derivative of order θ_i and type η_i with respect to ψ and $f : (a, T] \times \mathbb{R} \to \mathbb{R}$ is a given function. The tools employed for their demonstrations are the fixed point techniques of Banach and Krasnoselskii.

Inspired by the abovementioned papers and with the goal of extending previous outcomes in mind, in this work, we investigate the existence results for a class of coupled systems of nonlinear implicit k-generalized ψ-Hilfer type fractional differential equation and terminal conditions as follows:

$$\begin{cases} \left({}_k^H\mathcal{D}_{\theta_1+}^{\lambda_1,r_1;\psi} x\right)(t) = f_1\left(t, x(t), y(t), \left({}_k^H\mathcal{D}_{\theta_1+}^{\lambda_1,r_1;\psi} x\right)(t), \left({}_k^H\mathcal{D}_{\theta_1+}^{\lambda_2,r_2;\psi} y\right)(t)\right), \\ \left({}_k^H\mathcal{D}_{\theta_1+}^{\lambda_2,r_2;\psi} y\right)(t) = f_2\left(t, x(t), y(t), \left({}_k^H\mathcal{D}_{\theta_1+}^{\lambda_1,r_1;\psi} x\right)(t), \left({}_k^H\mathcal{D}_{\theta_1+}^{\lambda_2,r_2;\psi} y\right)(t)\right), \end{cases} \tag{1}$$

where $t \in (\theta_1, \theta_2]$, with the terminal conditions

$$\begin{cases} x(\theta_2) = \alpha, \\ y(\theta_2) = \beta, \end{cases} \tag{2}$$

where for $i = 1, 2$, ${}_k^H\mathcal{D}_{\theta_1+}^{\lambda_i,r_i;\psi}$ and $\mathcal{J}_{\theta_1+}^{k(1-\eta_i),k;\psi}$ are, respectively, the k-generalized ψ-Hilfer fractional derivative of order $\lambda_i \in (0, k)$ and type $r_i \in [0, 1]$, and k-generalized ψ-fractional integral of order $k(1 - \eta_i)$, where $\eta_i = \frac{1}{k}(r_i(k - \lambda_i) + \lambda_i)$, $k > 0, \alpha, \beta \in E$, and $f_i : [\theta_1, \theta_2] \times E^4 \longrightarrow E$ are given functions, where $(E, \|\cdot\|)$ is a Banach space.

The chapter is arranged as follows: The second section introduces various notations, preceding facts, and auxiliary results. Section 3 presents our major results for problem (1) and (2) that are founded on the fixed point theorem Mönch combined with the technique of measure of noncompactness. In the last part, we present some illustrations to demonstrate the practicability of our results.

2 Preliminaries

We will begin by introducing the weighted spaces, notations, definitions, and fundamental notions that will be employed in this study. More details on the used norms and definitions, one can see [15, 22, 26].

Let $0 < \theta_1 < \theta_2 < \infty$, $J = [\theta_1, \theta_2]$, and let $\lambda \in (0, k)$, $r \in [0, 1]$, $k > 0$ and $\eta = \frac{1}{k}(r(k - \lambda) + \lambda)$.

By $C(J, E)$ we denote the Banach space of all continuous functions from J into E with the norm

$$\|x\|_\infty = \sup\{\|x(t)\| : t \in J\}.$$

Let $AC^n(J, E)$ and $C^n(J, E)$ be the spaces of n-times absolutely continuous and n-times continuously differentiable functions on J, respectively.

Consider the weighted Banach space

$$C_{\eta;\psi}(J) = \left\{x : (\theta_1, \theta_2] \to E : t \to \Phi_\eta^\psi(t, \theta_1)x(t) \in C(J, E)\right\},$$

where $\Phi_\eta^\psi(t, \theta_1) = (\psi(t) - \psi(\theta_1))^{1-\eta}$, with the norm

$$\|x\|_{C_{\eta;\psi}} = \sup_{t\in J} \left\|\Phi_\eta^\psi(t, \theta_1)x(t)\right\|,$$

and

$$C^n_{\eta;\psi}(J) = \left\{x \in C^{n-1}(J) : x^{(n)} \in C_{\eta;\psi}(J)\right\}, n \in \mathbb{N},$$
$$C^0_{\eta;\psi}(J) = C_{\eta;\psi}(J),$$

with the norm

$$\|x\|_{C^n_{\eta;\psi}} = \sum_{i=0}^{n-1} \|x^{(i)}\|_\infty + \|x^{(n)}\|_{C_{\eta;\psi}}.$$

Now, let us consider the Banach space

$$\mathcal{F}_{\eta_1,\eta_2} := C_{\eta_1;\psi}(J) \times C_{\eta_2;\psi}(J),$$

with the norm

$$\|(x, y)\|_{\mathcal{F}_{\eta_1,\eta_2}} = \max\left\{\|x\|_{C_{\eta_1;\psi}}, \|y\|_{C_{\eta_2;\psi}}\right\},$$

where $0 < \eta_1, \eta_2 \leq 1$.

By $L^1(J)$, we denote the space of Bochner–integrable functions $f : J \longrightarrow E$ with the norm

$$\|f\|_1 = \int_{\theta_1}^{\theta_2} \|f(t)\|dt.$$

Definition 2.1 ([13]) The k-gamma function is defined by

$$\Gamma_k(\theta) = \int_0^{\infty} t^{\theta-1}e^{-\frac{t^k}{k}}dt, \theta > 0.$$

When $k \to 1$ then $\Gamma(\theta) = \Gamma_k(\theta)$, and some other useful relations are $\Gamma_k(\theta) = k^{\frac{\theta}{k}-1}\Gamma\left(\frac{\theta}{k}\right)$, $\Gamma_k(\theta + k) = \theta\Gamma_k(\theta)$ and $\Gamma_k(k) = 1$. Moreover, the k-beta function is given as

$$B_k(\theta, \gamma) = \frac{1}{k}\int_0^1 t^{\frac{\theta}{k}-1}(1-t)^{\frac{\gamma}{k}-1}dt,$$

so that $B_k(\theta, \gamma) = \frac{1}{k}B\left(\frac{\theta}{k}, \frac{\gamma}{k}\right)$ and $B_k(\theta, \gamma) = \frac{\Gamma_k(\theta)\Gamma_k(\gamma)}{\Gamma_k(\theta+\gamma)}$.

Now, we give the definition to the integral fractional operator used throughout this paper and some of its properties.

Definition 2.2 (k-Generalized ψ-Fractional Integral [19]) Let $\psi(t) > 0$ be an increasing function on $(\theta_1, \theta_2]$ and $\psi'(t) > 0$ be continuous on (θ_1, θ_2), and $\lambda > 0$. The generalized k-fractional integral operators of a function $\varkappa$ of order λ is defined by:

$$\mathcal{J}_{\theta_1+}^{\lambda,k;\psi}\varkappa(t) = \int_{\theta_1}^{t} \bar{\Phi}_{\lambda}^{k,\psi}(t,s)\psi'(s)\varkappa(s)ds,$$

with $k > 0$ and $\bar{\Phi}_{\lambda}^{k,\psi}(t,s) = \dfrac{(\psi(t)-\psi(s))^{\frac{\lambda}{k}-1}}{k\Gamma_k(\lambda)}$.

Theorem 2.1 ([22, 25]) *Let $\varkappa \in L^1(J)$ and take $\lambda > 0$ and $k > 0$. Then $\mathcal{J}_{\theta_1+}^{\lambda,k;\psi}\varkappa \in C([\theta_1, \theta_2], \mathbb{R})$.*

Lemma 2.1 ([22, 25]) *Let $\lambda > 0$, $\varrho > 0$ and $k > 0$. Then, the semigroup properties that follow are met:*

$$\mathcal{J}_{\theta_1+}^{\lambda,k;\psi}\mathcal{J}_{\theta_1+}^{\varrho,k;\psi}\varkappa(t) = \mathcal{J}_{\theta_1+}^{\lambda+\varrho,k;\psi}\varkappa(t) = \mathcal{J}_{\theta_1+}^{\varrho,k;\psi}\mathcal{J}_{\theta_1+}^{\lambda,k;\psi}\varkappa(t).$$

Lemma 2.2 ([22, 25]) *Let $\lambda, \varrho > 0$ and $k > 0$. Then, we obtain*

$$\mathcal{J}_{\theta_1+}^{\lambda,k;\psi}\bar{\Phi}_{\varrho}^{k,\psi}(t,\theta_1) = \bar{\Phi}_{\lambda+\varrho}^{k,\psi}(t,\theta_1).$$

Theorem 2.2 ([22, 25]) *Let* $0 < \theta_1 < \theta_2 < \infty, \lambda, \varrho > 0, 0 < \eta = \frac{1}{k}(\varrho(k-\lambda)+\lambda) \leq 1$, $k > 0$ *and* $x \in C_{\eta;\psi}(J)$. *If* $\frac{\lambda}{k} > 1-\eta$, *then*

$$\left(\mathcal{J}_{\theta_1+}^{\lambda,k;\psi} x\right)(\theta_1) = \lim_{t\to\theta_1{}^+} \left(\mathcal{J}_{\theta_1+}^{\lambda,k;\psi} x\right)(t) = 0.$$

Definition 2.3 (k-Generalized ψ-Hilfer Derivative [22, 25]) Let $m-1 < \frac{\lambda}{k} \leq m$ with $m \in \mathbb{N}$, $-\infty \leq \theta_1 < \theta_2 \leq \infty$ and $\varkappa, \psi \in C^m([\theta_1, \theta_2], \mathbb{R})$ two functions where ψ is increasing and $\psi'(t) \neq 0$, for all $t \in J$. The k-generalized ψ-Hilfer fractional derivatives ${}_k^H\mathcal{D}_{\theta_1+}^{\lambda,\varrho;\psi}(\cdot)$ of a function $\varkappa$ of order λ and type $0 \leq \varrho \leq 1$, with $k > 0$ is given by:

$$\begin{aligned} {}_k^H\mathcal{D}_{\theta_1+}^{\lambda,\varrho;\psi} \varkappa(t) &= \left(\mathcal{J}_{\theta_1+}^{\varrho(km-\lambda),k;\psi} \left(\frac{1}{\psi'(t)}\frac{d}{dt}\right)^m \left(k^m \mathcal{J}_{\theta_1+}^{(1-\varrho)(km-\lambda),k;\psi} \varkappa\right)\right)(t) \\ &= \left(\mathcal{J}_{\theta_1+}^{\varrho(km-\lambda),k;\psi} \delta_\psi^m \left(k^m \mathcal{J}_{\theta_1+}^{(1-\varrho)(km-\lambda),k;\psi} \varkappa\right)\right)(t), \end{aligned}$$

where $\delta_\psi^m = \left(\frac{1}{\psi'(t)}\frac{d}{dt}\right)^m$.

Lemma 2.3 ([22, 25]) *Let* $t > \theta_1$, $0 < \frac{\lambda}{k} < 1, 0 \leq \varrho \leq 1, k > 0$. *Then for* $0 < \eta < 1$; $\eta = \frac{1}{k}(\varrho(k-\lambda)+\lambda)$, *we have*

$$\left[{}_k^H\mathcal{D}_{\theta_1+}^{\lambda,\varrho;\psi} \left(\Phi_\eta^\psi(s,\theta_1)\right)^{-1}\right](t) = 0.$$

Theorem 2.3 ([22, 25]) *If* $\varkappa \in C_{\eta;\psi}^m[\theta_1, \theta_2]$, $m-1 < \frac{\lambda}{k} < m$, $0 \leq \varrho \leq 1$, *where* $m \in \mathbb{N}$ *and* $k > 0$, *then*

$$\begin{aligned} &\left(\mathcal{J}_{\theta_1+}^{\lambda,k;\psi}\, {}_k^H\mathcal{D}_{\theta_1+}^{\lambda,\varrho;\psi} \varkappa\right)(t) \\ &= \varkappa(t) - \sum_{i=1}^m \frac{(\psi(t)-\psi(\theta_1))^{\eta-i}}{k^{i-m}\Gamma_k(k(\eta-i+1))} \left\{\delta_\psi^{m-i} \left(\mathcal{J}_{\theta_1+}^{k(m-\eta),k;\psi} \varkappa(\theta_1)\right)\right\}, \end{aligned}$$

where

$$\eta = \frac{1}{k}(\varrho(km-\lambda)+\lambda).$$

If $m = 1$, *we have*

$$\left(\mathcal{J}_{\theta_1+}^{\lambda,k;\psi}\, {}_k^H\mathcal{D}_{\theta_1+}^{\lambda,\varrho;\psi}\varkappa\right)(t) = \varkappa(t) - \frac{(\psi(t)-\psi(\theta_1))^{\eta-1}}{\Gamma_k(\varrho(k-\lambda)+\lambda)}\mathcal{J}_{\theta_1+}^{(1-\varrho)(k-\lambda),k;\psi}\varkappa(\theta_1).$$

Lemma 2.4 ([22, 25]) *Let* $\lambda > 0, 0 \le \varrho \le 1$, *and* $x \in C^1_{\eta;\psi}(J)$, *where* $k > 0$. *Then for* $t \in (\theta_1, \theta_2]$, *we have*

$$\left({}_k^H\mathcal{D}_{\theta_1+}^{\lambda,\varrho;\psi}\, \mathcal{J}_{\theta_1+}^{\lambda,k;\psi}x\right)(t) = x(t).$$

Definition 2.4 ([9]) let X be a Banach space and let Ω_X be the family of bounded subsets of X. The Kuratowski measure of noncompactness is the map $\mu : \Omega_X \longrightarrow [0,\infty)$ defined by

$$\mu(M) = inf\left\{\epsilon > 0 : M \subset \bigcup_{j=1}^{m} M_j, diam(M_j) \le \epsilon\right\},$$

where $M \in \Omega_X$. The map μ satisfies the following properties:

- $\mu(M) = 0 \Leftrightarrow \overline{M}$ is compact (M is relatively compact).
- $\mu(M) = \mu(\overline{M})$.
- $M_1 \subset M_2 \Rightarrow \mu(M_1) \le \mu(M_2)$.
- $\mu(M_1 + M_2) \le \mu(B_1) + \mu(B_2)$.
- $\mu(cM) = |c|\mu(M), c \in \mathbb{R}$.
- $\mu(convM) = \mu(M)$.

Theorem 2.4 (Mönch's Fixed Point Theorem [17]) *Let D be closed, bounded and convex subset of a Banach space X and let T be a continuous mapping of D into itself. If the implication*

$$V = \overline{conv}T(V), or\, V = T(V) \cup \{0\} \Rightarrow \mu(V) = 0, \tag{3}$$

holds for every subset V of D, then T has a fixed point.

3 Existence of Solutions

Firstly, we provide the following theorem in order to convert our system (1) and (2) into a coupled system of fractional integral equations.

Theorem 3.1 *Let* $\eta = \dfrac{\varrho(k-\lambda)+\lambda}{k}$*, where* $k > 0$*,* $0 < \lambda < k$*,* $0 \le \varrho \le 1$*, and let* $\varphi(\cdot) \in C(J, E)$*. The function* x *satisfies the terminal value problem:*

$$\left({}_{k}^{H}\mathcal{D}_{\theta_1+}^{\lambda,\varrho;\psi} x\right)(t) = \varphi(t), \quad t \in (\theta_1, \theta_2], \tag{4}$$

$$x(\theta_2) = \alpha, \tag{5}$$

if and only if it verifies the following integral equation:

$$x(t) = \frac{\Phi_{\eta}^{\psi}(\theta_2,\theta_1)\left(\alpha - \left(\mathcal{J}_{\theta_1+}^{\lambda,k;\psi}\varphi\right)(\theta_2)\right)}{\Phi_{\eta}^{\psi}(t,\theta_1)} + \left(\mathcal{J}_{\theta_1+}^{\lambda,k;\psi}\varphi\right)(t), \quad t \in (\theta_1, \theta_2], \tag{6}$$

where $\alpha \in E$.

Proof Assume that x satisfies Eqs. (4) and (5). We apply $\mathcal{J}_{\theta_1+}^{\lambda,k;\psi}(\cdot)$ on both sides of (4) to obtain

$$\left(\mathcal{J}_{\theta_1+}^{\lambda,k;\psi}\,{}_{k}^{H}\mathcal{D}_{\theta_1+}^{\lambda,\varrho;\psi} x\right)(t) = \left(\mathcal{J}_{\theta_1+}^{\lambda,k;\psi}\varphi\right)(t),$$

and using Theorem 2.3, we get

$$x(t) = \frac{\mathcal{J}_{\theta_1+}^{k(1-\eta),k;\psi} x(\theta_1)}{\Phi_{\eta}^{\psi}(t,\theta_1)\Gamma_k(k\eta)} + \left(\mathcal{J}_{\theta_1+}^{\lambda,k;\psi}\varphi\right)(t). \tag{7}$$

Using Eq. (5), we obtain

$$x(t) = \frac{\Phi_{\eta}^{\psi}(\theta_2,\theta_1)\left(\alpha - \left(\mathcal{J}_{\theta_1+}^{\lambda,k;\psi}\varphi\right)(\theta_2)\right)}{\Phi_{\eta}^{\psi}(t,\theta_1)} + \left(\mathcal{J}_{\theta_1+}^{\lambda,k;\psi}\varphi\right)(t),$$

with $t \in (\theta_1, \theta_2]$, that is x verifies (6).

For the converse, let us now prove that if x satisfies Eq. (6), then it verifies (4) and (5). We apply ${}_{k}^{H}\mathcal{D}_{\theta_1+}^{\lambda,\varrho;\psi}(\cdot)$ on (6) to get

$$\begin{aligned}\left({}_{k}^{H}\mathcal{D}_{\theta_1+}^{\lambda,\varrho;\psi} x\right)(t) = {}_{k}^{H}\mathcal{D}_{\theta_1+}^{\lambda,\varrho;\psi}&\left(\frac{\Phi_{\eta}^{\psi}(\theta_2,\theta_1)\left(\alpha - \left(\mathcal{J}_{\theta_1+}^{\lambda,k;\psi}\varphi\right)(\theta_2)\right)}{\Phi_{\eta}^{\psi}(t,\theta_1)}\right)\\ &+ \left({}_{k}^{H}\mathcal{D}_{\theta_1+}^{\lambda,\varrho;\psi}\mathcal{J}_{\theta_1+}^{\lambda,k;\psi}\varphi\right)(t).\end{aligned}$$

By Lemmas 2.3 and 2.4, we get (4). We can also set $t = \theta_2$ to easily obtain the condition (5). This completes the proof. □

As a consequence of Theorem 3.1, we have the following result:

Lemma 3.1 *Let* $i = 1, 2$, $\eta_i = \dfrac{r_i(k-\lambda_i)+\lambda_i}{k}$ *where* $0 < \lambda_i < k$ *and* $0 \le r_i \le 1$, *and let* $f_i : J \times E^4 \to E$ *be continuous functions. Then* $(x, y) \in \mathcal{F}_{\eta_1,\eta_2}$ *satisfies the coupled system (1) and (2) if and only if* (x, y) *is the fixed point of the operator* $\mathcal{H} : \mathcal{F}_{\eta_1,\eta_2} \to \mathcal{F}_{\eta_1,\eta_2}$ *defined by:*

$$\mathcal{H}(x, y)(t) = (\mathcal{H}_1(x, y)(t), \mathcal{H}_2(x, y)(t)), \quad t \in (\theta_1, \theta_2], \tag{8}$$

where $\mathcal{H}_1$ *and* $\mathcal{H}_2$ *are the operators defined for* $t \in (\theta_1, \theta_2]$, *as follows:*

$$\mathcal{H}_1(x, y)(t) = \frac{\Phi^{\psi}_{\eta_1}(\theta_2, \theta_1)\left(\alpha - \left(\mathcal{J}^{\lambda_1,k;\psi}_{\theta_1+}\varphi_1(s)\right)(\theta_2)\right)}{\Phi^{\psi}_{\eta_1}(t, \theta_1)} + \left(\mathcal{J}^{\lambda_1,k;\psi}_{\theta_1+}\varphi_1(s)\right)(t), \tag{9}$$

and

$$\mathcal{H}_2(x, y)(t) = \frac{\Phi^{\psi}_{\eta_2}(\theta_2, \theta_1)\left(\beta - \left(\mathcal{J}^{\lambda_2,k;\psi}_{\theta_1+}\varphi_2(s)\right)(\theta_2)\right)}{\Phi^{\psi}_{\eta_2}(t, \theta_1)} + \left(\mathcal{J}^{\lambda_2,k;\psi}_{\theta_1+}\varphi_2(s)\right)(t), \tag{10}$$

where for $i = 1, 2$, $\varphi_i \in C(J, E)$ satisfy the following system of functional equations:

$$\begin{cases} \varphi_1(t) = f_1(t, x(t), y(t), \varphi_1(t), \varphi_2(t)), \\ \varphi_2(t) = f_2(t, x(t), y(t), \varphi_1(t), \varphi_2(t)). \end{cases}$$

We may employ Theorem 2.1 to easily demonstrate that for $(x, y) \in \mathcal{F}_{\eta_1,\eta_2}$, we have $\mathcal{H}(x, y) \in \mathcal{F}_{\eta_1,\eta_2}$, where $\mathcal{H}$ is the operator defined in (8).

The Hypotheses

(Cd_1) The functions $f_i : J \times E^4 \to E; i = 1, 2$, are continuous.

(Cd_2) There exist constants $\zeta_i, \varpi_i, \bar{\zeta}_i, \bar{\varpi}_i > 0$ such that $0 < \bar{\zeta}_1 < 1, 0 < \bar{\varpi}_2 < 1$ and

$$\begin{aligned}&\|f_1(t, x_1, y_1, w_1, z_1) - f_1(t, x_2, y_2, w_2, z_2)\| \\ &\le \zeta_1 \Phi^{\psi}_{\eta_1}(t, \theta_1)\|x_1 - x_2\| + \varpi_1 \Phi^{\psi}_{\eta_2}(t, \theta_1)\|y_1 - y_2\| + \bar{\zeta}_1\|w_1 - w_2\| + \bar{\varpi}_1\|z_1 - z_2\|\end{aligned}$$

and

$$\|f_2(t,x_1,y_1,w_1,z_1)-f_2(t,x_2,y_2,w_2,z_2)\|$$
$$\leq \zeta_2\Phi^{\psi}_{\eta_1}(t,\theta_1)\|x_1-x_2\|+\varpi_2\Phi^{\psi}_{\eta_2}(t,\theta_1)\|y_1-y_2\|+\bar{\zeta}_2\|w_1-w_2\|+\bar{\varpi}_2\|z_1-z_2\|$$

for any $x_i, y_i, w_i, z_i \in \mathbb{R}$ and $t \in (\theta_1, \theta_2]$, where $i = 1, 2$.

(Cd_3) For each bounded sets Ω and for each $t \in J$, the following inequalities hold:

$$\mu\left(f_1\left(t,\Omega,\Omega,{}^{H}_{k}\mathcal{D}^{\lambda_1,r_1;\psi}_{\theta_1+}\Omega,{}^{H}_{k}\mathcal{D}^{\lambda_2,r_2;\psi}_{\theta_1+}\Omega\right)\right)\leq \zeta_3\Phi^{\psi}_{\eta_1}(t,\theta_1)\mu(\Omega)$$

and

$$\mu\left(f_2\left(t,\Omega,\Omega,{}^{H}_{k}\mathcal{D}^{\lambda_1,r_1;\psi}_{\theta_1+}\Omega,{}^{H}_{k}\mathcal{D}^{\lambda_2,r_2;\psi}_{\theta_1+}\Omega\right)\right)\leq \zeta_4\Phi^{\psi}_{\eta_2}(t,\theta_1)\mu(\Omega),$$

where

$${}^{H}_{k}\mathcal{D}^{\lambda_i,r_i;\psi}_{\theta_1+}\Omega=\left\{{}^{H}_{k}\mathcal{D}^{\lambda_i,r_i;\psi}_{\theta_1+}w : w\in\Omega\right\};\, i=1,2.$$

We can now declare and demonstrate our existence result for problem (1) and (2). The argument is based on Mönch's fixed point Theorem [17].

Theorem 3.2 *Suppose that the hypotheses* (Cd_1)-(Cd_3) *hold. If*

$$\ell=\max\{\ell_1,\ell_2,\ell_3\}<1, \tag{11}$$

where

$$\ell_1=\frac{\zeta_3\left(\psi(\theta_2)-\psi(\theta_1)\right)^{1-\eta_1+\frac{\lambda_1}{k}}}{\Gamma_k(\lambda_1+k)},$$

$$\ell_2=\frac{\zeta_4\left(\psi(\theta_2)-\psi(\theta_1)\right)^{1-\eta_2+\frac{\lambda_2}{k}}}{\Gamma_k(\lambda_2+k)},$$

and

$$\ell_3=\frac{\bar{\zeta}_2\bar{\varpi}_1}{(1-\bar{\varpi}_2)(1-\bar{\zeta}_1)}.$$

Then problem (1) and (2) has at least one solution in $\mathcal{F}_{\eta_1,\eta_2}$.

Proof Now, we will employ Mönch's fixed point theorem to demonstrate that the operator $\mathcal{H}$ defined in (8) has a fixed point. The proof will be given in several steps.

Step 1 We show that the operator $\mathcal{H}$ defined in (8), transforms the ball $B_\delta = B(0,\delta) = \left\{(x,y)\in\mathcal{F}_{\eta_1,\eta_2} : \|(x,y)\|_{\mathcal{F}_{\eta_1,\eta_2}} \le \delta\right\}$ into itself. Let δ a positive constant such that

$$\delta \ge \frac{\delta_3}{1-\delta_1-\delta_2},$$

where

$$\delta_1 = \max\{\mathcal{L}_1, \mathcal{L}_2\},\ \delta_2 = \max\{\mathcal{L}_3, \mathcal{L}_4\} \text{ and } \delta_3 = \mathcal{L}_5 + \mathcal{L}_6,$$

such that

$$\mathcal{L}_1 = \left[\frac{\zeta_2\bar{\varpi}_1 + \zeta_1(1-\bar{\varpi}_2)}{(1-\bar{\varpi}_2)(1-\bar{\zeta}_1)-\bar{\zeta}_2\bar{\varpi}_1}\right]\frac{2(\psi(\theta_2)-\psi(\theta_1))^{1-\eta_1+\frac{\lambda_1}{k}}}{\Gamma_k(\lambda_1+k)},$$

$$\mathcal{L}_2 = \left[\frac{\zeta_2(1-\bar{\zeta}_1) + \zeta_1\bar{\zeta}_2}{(1-\bar{\zeta}_1)(1-\bar{\varpi}_2)-\bar{\varpi}_1\bar{\zeta}_2}\right]\frac{2(\psi(\theta_2)-\psi(\theta_1))^{1-\eta_2+\frac{\lambda_2}{k}}}{\Gamma_k(\lambda_2+k)},$$

$$\mathcal{L}_3 = \left[\frac{\varpi_2\bar{\varpi}_1 + \varpi_1(1-\bar{\varpi}_2)}{(1-\bar{\varpi}_2)(1-\bar{\zeta}_1)-\bar{\zeta}_2\bar{\varpi}_1}\right]\frac{2(\psi(\theta_2)-\psi(\theta_1))^{1-\eta_1+\frac{\lambda_1}{k}}}{\Gamma_k(\lambda_1+k)},$$

$$\mathcal{L}_4 = \left[\frac{\varpi_2(1-\bar{\zeta}_1) + \varpi_1\bar{\zeta}_2}{(1-\bar{\zeta}_1)(1-\bar{\varpi}_2)-\bar{\varpi}_1\bar{\zeta}_2}\right]\frac{2(\psi(\theta_2)-\psi(\theta_1))^{1-\eta_2+\frac{\lambda_2}{k}}}{\Gamma_k(\lambda_2+k)},$$

$$\mathcal{L}_5 = \left[\frac{f^*\bar{\varpi}_1 + f^*(1-\bar{\varpi}_2)}{(1-\bar{\varpi}_2)(1-\bar{\zeta}_1)-\bar{\zeta}_2\bar{\varpi}_1}\right]\frac{2(\psi(\theta_2)-\psi(\theta_1))^{1-\eta_1+\frac{\lambda_1}{k}}}{\Gamma_k(\lambda_1+k)} + \Phi^{\psi}_{\eta_1}(\theta_2,\theta_1)\|\alpha\|,$$

$$\mathcal{L}_6 = \left[\frac{f^*(1-\bar{\zeta}_1) + f^*\bar{\zeta}_2}{(1-\bar{\zeta}_1)(1-\bar{\varpi}_2)-\bar{\varpi}_1\bar{\zeta}_2}\right]\frac{2(\psi(\theta_2)-\psi(\theta_1))^{1-\eta_2+\frac{\lambda_2}{k}}}{\Gamma_k(\lambda_2+k)} + \Phi^{\psi}_{\eta_2}(\theta_2,\theta_1)\|\beta\|,$$

and

$$f^* = \sup_{t\in J}\|f(t,0,0,0,0)\|.$$

For each $t\in(\theta_1,\theta_2]$, (9) and (10) imply that we have

$$\|\mathcal{H}_1(x,y)(t)\|$$
$$\le \frac{\Phi^{\psi}_{\eta_1}(\theta_2,\theta_1)\left(\|\alpha\| + \left(\mathcal{J}^{\lambda_1,k;\psi}_{\theta_1+}\|\varphi_1(s)\|\right)(\theta_2)\right)}{\Phi^{\psi}_{\eta_1}(t,\theta_1)} + \left(\mathcal{J}^{\lambda_1,k;\psi}_{\theta_1+}\|\varphi_1(s)\|\right)(t), \qquad (12)$$

and

$$\|\mathcal{H}_2(x,y)(t)\|$$
$$\leq \frac{\Phi^{\psi}_{\eta_2}(\theta_2,\theta_1)\left(\|\beta\|+\left(\mathcal{J}^{\lambda_2,k;\psi}_{\theta_1+}\|\varphi_2(s)\|\right)(\theta_2)\right)}{\Phi^{\psi}_{\eta_2}(t,\theta_1)}+\left(\mathcal{J}^{\lambda_2,k;\psi}_{\theta_1+}\|\varphi_2(s)\|\right)(t). \tag{13}$$

By the hypothesis (Cd_2), for $t\in(\theta_1,\theta_2]$, we have

$$\begin{aligned}\|\varphi_1(t)\| &= \|f_1(t,x(t),y(t),\varphi_1(t),\varphi_2(t))-f(t,0,0,0,0)\|+\|f(t,0,0,0,0)\|\\ &\leq \zeta_1\Phi^{\psi}_{\eta_1}(t,\theta_1)\|x(t)\|+\varpi_1\Phi^{\psi}_{\eta_2}(t,\theta_1)\|y(t)\|+\bar{\zeta}_1\|\varphi_1(t)\|+\bar{\varpi}_1\|\varphi_2(t)\|+f^*,\end{aligned}$$

which implies that

$$\|\varphi_1(t)\|\leq\frac{\zeta_1}{1-\bar{\zeta}_1}\|x\|_{C_{\eta_1;\psi}}+\frac{\varpi_1}{1-\bar{\zeta}_1}\|y\|_{C_{\eta_2;\psi}}+\frac{\bar{\varpi}_1}{1-\bar{\zeta}_1}\|\varphi_2(t)\|+\frac{f^*}{1-\bar{\zeta}_1}.$$

Similarly, one can find that

$$\|\varphi_2(t)\|\leq\frac{\zeta_2}{1-\bar{\varpi}_2}\|x\|_{C_{\eta_1;\psi}}+\frac{\varpi_2}{1-\bar{\varpi}_2}\|y\|_{C_{\eta_2;\psi}}+\frac{\bar{\zeta}_2}{1-\bar{\varpi}_2}\|\varphi_1(t)\|+\frac{f^*}{1-\bar{\varpi}_2}.$$

Therefore

$$\begin{aligned}\|\varphi_1(t)\|\leq{}&\frac{\zeta_1}{1-\bar{\zeta}_1}\|x\|_{C_{\eta_1;\psi}}+\frac{\varpi_1}{1-\bar{\zeta}_1}\|y\|_{C_{\eta_2;\psi}}+\frac{f^*}{1-\bar{\zeta}_1}\\ &+\frac{\zeta_2\bar{\varpi}_1}{(1-\bar{\varpi}_2)(1-\bar{\zeta}_1)}\|x\|_{C_{\eta_1;\psi}}\frac{\varpi_2\bar{\varpi}_1}{(1-\bar{\varpi}_2)(1-\bar{\zeta}_1)}\|y\|_{C_{\eta_2;\psi}}\\ &+\frac{\bar{\zeta}_2\bar{\varpi}_1}{(1-\bar{\varpi}_2)(1-\bar{\zeta}_1)}\|\varphi_1(t)\|+\frac{f^*\bar{\varpi}_1}{(1-\bar{\varpi}_2)(1-\bar{\zeta}_1)},\end{aligned}$$

then

$$\begin{aligned}\|\varphi_1(t)\|\leq{}&\frac{\zeta_2\bar{\varpi}_1+\zeta_1(1-\bar{\varpi}_2)}{(1-\bar{\varpi}_2)(1-\bar{\zeta}_1)-\bar{\zeta}_2\bar{\varpi}_1}\|x\|_{C_{\eta_1;\psi}}+\frac{\varpi_2\bar{\varpi}_1+\varpi_1(1-\bar{\varpi}_2)}{(1-\bar{\varpi}_2)(1-\bar{\zeta}_1)-\bar{\zeta}_2\bar{\varpi}_1}\|y\|_{C_{\eta_2;\psi}}\\ &+\frac{f^*\bar{\varpi}_1+f^*(1-\bar{\varpi}_2)}{(1-\bar{\varpi}_2)(1-\bar{\zeta}_1)-\bar{\zeta}_2\bar{\varpi}_1}.\end{aligned}$$

By following the same approach, we can also obtain the following:

$$\|\varphi_2(t)\| \leq \frac{\zeta_2(1-\bar{\zeta}_1)+\zeta_1\bar{\zeta}_2}{(1-\bar{\zeta}_1)(1-\bar{\varpi}_2)-\bar{\varpi}_1\bar{\zeta}_2}\|x\|_{C_{\eta_1;\psi}} + \frac{\varpi_2(1-\bar{\zeta}_1)+\varpi_1\bar{\zeta}_2}{(1-\bar{\zeta}_1)(1-\bar{\varpi}_2)-\bar{\varpi}_1\bar{\zeta}_2}\|y\|_{C_{\eta_2;\psi}}$$
$$+ \frac{f^*(1-\bar{\zeta}_1)+f^*\bar{\zeta}_2}{(1-\bar{\zeta}_1)(1-\bar{\varpi}_2)-\bar{\varpi}_1\bar{\zeta}_2}.$$

Thus, By the hypothesis (Cd_2), for $t \in (\theta_1, \theta_2]$, we have

$$\|\Phi_{\eta_1}^{\psi}(t,\theta_1)\mathcal{H}_1(x,y)(t)\| \leq \Phi_{\eta_1}^{\psi}(\theta_2,\theta_1)\left(\|\alpha\| + \left(\mathcal{J}_{\theta_1+}^{\lambda_1,k;\psi}\|\varphi_1(s)\|\right)(\theta_2)\right)$$
$$+ \Phi_{\eta_1}^{\psi}(t,\theta_1)\left(\mathcal{J}_{\theta_1+}^{\lambda_1,k;\psi}\|\varphi_1(s)\|\right)(t),$$

and

$$\|\Phi_{\eta_2}^{\psi}(t,\theta_1)\mathcal{H}_2(x,y)(t)\| \leq \Phi_{\eta_2}^{\psi}(\theta_2,\theta_1)\left(\|\beta\| + \left(\mathcal{J}_{\theta_1+}^{\lambda_2,k;\psi}\|\varphi_2(s)\|\right)(\theta_2)\right)$$
$$+ \Phi_{\eta_2}^{\psi}(t,\theta_1)\left(\mathcal{J}_{\theta_1+}^{\lambda_2,k;\psi}\|\varphi_2(s)\|\right)(t),$$

which implies

$$\|\Phi_{\eta_1}^{\psi}(t,\theta_1)\mathcal{H}_1(x,y)(t)\|$$
$$\leq \left[\frac{\zeta_2\bar{\varpi}_1+\zeta_1(1-\bar{\varpi}_2)}{(1-\bar{\varpi}_2)(1-\bar{\zeta}_1)-\bar{\zeta}_2\bar{\varpi}_1}\|x\|_{C_{\eta_1;\psi}} + \frac{\varpi_2\bar{\varpi}_1+\varpi_1(1-\bar{\varpi}_2)}{(1-\bar{\varpi}_2)(1-\bar{\zeta}_1)-\bar{\zeta}_2\bar{\varpi}_1}\|y\|_{C_{\eta_2;\psi}}\right.$$
$$\left.+ \frac{f^*\bar{\varpi}_1+f^*(1-\bar{\varpi}_2)}{(1-\bar{\varpi}_2)(1-\bar{\zeta}_1)-\bar{\zeta}_2\bar{\varpi}_1}\right] \times \left[\Phi_{\eta_1}^{\psi}(\theta_2,\theta_1)\left(\mathcal{J}_{\theta_1+}^{\lambda_1,k;\psi}(1)\right)(\theta_2)\right.$$
$$\left.+ \Phi_{\eta_1}^{\psi}(t,\theta_1)\left(\mathcal{J}_{\theta_1+}^{\lambda_1,k;\psi}(1)\right)(t)\right] + \Phi_{\eta_1}^{\psi}(\theta_2,\theta_1)\|\alpha\|,$$

and

$$\|\Phi_{\eta_2}^{\psi}(t,\theta_1)\mathcal{H}_2(x,y)(t)\|$$
$$\leq \left[\frac{\zeta_2(1-\bar{\zeta}_1)+\zeta_1\bar{\zeta}_2}{(1-\bar{\zeta}_1)(1-\bar{\varpi}_2)-\bar{\varpi}_1\bar{\zeta}_2}\|x\|_{C_{\eta_1;\psi}} + \frac{\varpi_2(1-\bar{\zeta}_1)+\varpi_1\bar{\zeta}_2}{(1-\bar{\zeta}_1)(1-\bar{\varpi}_2)-\bar{\varpi}_1\bar{\zeta}_2}\|y\|_{C_{\eta_2;\psi}}\right.$$
$$\left.+ \frac{f^*(1-\bar{\zeta}_1)+f^*\bar{\zeta}_2}{(1-\bar{\zeta}_1)(1-\bar{\varpi}_2)-\bar{\varpi}_1\bar{\zeta}_2}\right] \times \left[\Phi_{\eta_2}^{\psi}(\theta_2,\theta_1)\left(\mathcal{J}_{\theta_1+}^{\lambda_2,k;\psi}(1)\right)(\theta_2)\right.$$
$$\left.+ \Phi_{\eta_2}^{\psi}(t,\theta_1)\left(\mathcal{J}_{\theta_1+}^{\lambda_2,k;\psi}(1)\right)(t)\right] + \Phi_{\eta_2}^{\psi}(\theta_2,\theta_1)\|\beta\|.$$

Lemma 2.2 implies

$$
\begin{aligned}
&\|\Phi_{\eta_1}^{\psi}(t,\theta_1)\mathcal{H}_1(x,y)(t)\| \\
&\leq \left[\frac{\zeta_2\bar{\varpi}_1+\zeta_1(1-\bar{\varpi}_2)}{(1-\bar{\varpi}_2)(1-\bar{\zeta}_1)-\bar{\zeta}_2\bar{\varpi}_1}\|x\|_{C_{\eta_1;\psi}}+\frac{\varpi_2\bar{\varpi}_1+\varpi_1(1-\bar{\varpi}_2)}{(1-\bar{\varpi}_2)(1-\bar{\zeta}_1)-\bar{\zeta}_2\bar{\varpi}_1}\|y\|_{C_{\eta_2;\psi}}\right. \\
&\quad\left.+\frac{f^*\bar{\varpi}_1+f^*(1-\bar{\varpi}_2)}{(1-\bar{\varpi}_2)(1-\bar{\zeta}_1)-\bar{\zeta}_2\bar{\varpi}_1}\right]\frac{2(\psi(\theta_2)-\psi(\theta_1))^{1-\eta_1+\frac{\lambda_1}{k}}}{\Gamma_k(\lambda_1+k)}+\Phi_{\eta_1}^{\psi}(\theta_2,\theta_1)\|\alpha\|,
\end{aligned}
$$

and

$$
\begin{aligned}
&\|\Phi_{\eta_2}^{\psi}(t,\theta_1)\mathcal{H}_2(x,y)(t)\| \\
&\leq \left[\frac{\zeta_2(1-\bar{\zeta}_1)+\zeta_1\bar{\zeta}_2}{(1-\bar{\zeta}_1)(1-\bar{\varpi}_2)-\bar{\varpi}_1\bar{\zeta}_2}\|x\|_{C_{\eta_1;\psi}}+\frac{\varpi_2(1-\bar{\zeta}_1)+\varpi_1\bar{\zeta}_2}{(1-\bar{\zeta}_1)(1-\bar{\varpi}_2)-\bar{\varpi}_1\bar{\zeta}_2}\|y\|_{C_{\eta_2;\psi}}\right. \\
&\quad\left.+\frac{f^*(1-\bar{\zeta}_1)+f^*\bar{\zeta}_2}{(1-\bar{\zeta}_1)(1-\bar{\varpi}_2)-\bar{\varpi}_1\bar{\zeta}_2}\right]\frac{2(\psi(\theta_2)-\psi(\theta_1))^{1-\eta_2+\frac{\lambda_2}{k}}}{\Gamma_k(\lambda_2+k)}+\Phi_{\eta_2}^{\psi}(\theta_2,\theta_1)\|\beta\|.
\end{aligned}
$$

Thus

$$
\begin{aligned}
&\|\mathcal{H}_1(x,y)(t)\|_{C_{\eta_1;\psi}} \\
&\leq \delta_1\|x\|_{C_{\eta_1;\psi}}+\delta_2\|y\|_{C_{\eta_2;\psi}} \\
&\quad+\left[\frac{f^*\bar{\varpi}_1+f^*(1-\bar{\varpi}_2)}{(1-\bar{\varpi}_2)(1-\bar{\zeta}_1)-\bar{\zeta}_2\bar{\varpi}_1}\right]\frac{2(\psi(\theta_2)-\psi(\theta_1))^{1-\eta_1+\frac{\lambda_1}{k}}}{\Gamma_k(\lambda_1+k)}+\Phi_{\eta_1}^{\psi}(\theta_2,\theta_1)\|\alpha\|,
\end{aligned}
$$

and

$$
\begin{aligned}
&\|\mathcal{H}_2(x,y)(t)\|_{C_{\eta_2;\psi}} \\
&\leq \delta_1\|x\|_{C_{\eta_1;\psi}}+\delta_2\|y\|_{C_{\eta_2;\psi}} \\
&\quad+\left[\frac{f^*(1-\bar{\zeta}_1)+f^*\bar{\zeta}_2}{(1-\bar{\zeta}_1)(1-\bar{\varpi}_2)-\bar{\varpi}_1\bar{\zeta}_2}\right]\frac{2(\psi(\theta_2)-\psi(\theta_1))^{1-\eta_2+\frac{\lambda_2}{k}}}{\Gamma_k(\lambda_2+k)}+\Phi_{\eta_2}^{\psi}(\theta_2,\theta_1)\|\beta\|.
\end{aligned}
$$

Thus, for each $t\in(\theta_1,\theta_2]$ we get

$$
\|\mathcal{H}(x,y)\|_{\mathcal{F}_{\eta_1,\eta_2}}\leq(\delta_1+\delta_2)\delta+\delta_3\leq\delta.
$$

Step 2 The operator $\mathcal{H} : B_\delta \to B_\delta$ is continuous.

Let $\{(x_n, y_n)\}$ be a sequence where $(x_n, y_n) \longrightarrow (x, y)$ in $\mathcal{F}_{\eta_1,\eta_2}$. For each $t \in (\theta_1, \theta_2]$, we have

$$\begin{aligned}
&\|\mathcal{H}_1(x_n, y_n)(t) - \mathcal{H}_1(x, y)(t)\| \\
&\leq \frac{\Phi^{\psi}_{\eta_1}(\theta_2, \theta_1)\left(\mathcal{J}^{\lambda_1,k;\psi}_{\theta_1+}\|\varphi_{1,n}(s) - \varphi_1(s)\|\right)(\theta_2)}{\Phi^{\psi}_{\eta_1}(t, \theta_1)} + \left(\mathcal{J}^{\lambda_1,k;\psi}_{\theta_1+}\|\varphi_{1,n}(s) - \varphi_1(s)\|\right)(t),
\end{aligned}$$

and

$$\begin{aligned}
&\|\mathcal{H}_2(x_n, y_n)(t) - \mathcal{H}_2(x, y)(t)\| \\
&\leq \frac{\Phi^{\psi}_{\eta_2}(\theta_2, \theta_1)\left(\mathcal{J}^{\lambda_2,k;\psi}_{\theta_1+}\|\varphi_{2,n}(s) - \varphi_2(s)\|\right)(\theta_2)}{\Phi^{\psi}_{\eta_2}(t, \theta_1)} + \left(\mathcal{J}^{\lambda_2,k;\psi}_{\theta_1+}\|\varphi_{2,n}(s) - \varphi_2(s)\|\right)(t),
\end{aligned}$$

where for $i = 1, 2$, φ_i and $\varphi_{i,n}$ satisfy the following systems of functional equations:

$$\begin{cases}
\varphi_1(t) = f_1\left(t, x(t), y(t), \varphi_1(t), \varphi_2(t)\right), \\
\varphi_2(t) = f_2\left(t, x(t), y(t), \varphi_1(t), \varphi_2(t)\right),
\end{cases}$$

and

$$\begin{cases}
\varphi_{1,n}(t) = f_1\left(t, x_n(t), y_n(t), \varphi_{1,n}(t), \varphi_{2,n}(t)\right), \\
\varphi_{2,n}(t) = f_2\left(t, x_n(t), y_n(t), \varphi_{1,n}(t), \varphi_{2,n}(t)\right).
\end{cases}$$

Since $(x_n, y_n) \to (x, y)$, then

$$\varphi_{1,n}(t) \to \varphi_1(t)$$

and

$$\varphi_{2,n}(t) \to \varphi_2(t),$$

as $n \to \infty$ for each $t \in (\theta_1, \theta_2]$, and since f_1 and f_2 are continuous, then we have

$$\|\mathcal{H}_1(x_n, y_n) - \mathcal{H}_1(x, y)\|_{C_{\eta_1;\psi}} \to 0 \text{ and } \|\mathcal{H}_2(x_n, y_n) - \mathcal{H}_2(x, y)\|_{C_{\eta_2;\psi}} \to 0 \text{ as } n \to \infty.$$

Thus, for each $t \in (\theta_1, \theta_2]$, we get

$$\|\mathcal{H}(x_n, y_n) - \mathcal{H}(x, y)\|_{\mathcal{F}_{\eta_1,\eta_2}} \to 0 \text{ as } n \to \infty.$$

Consequently, $\mathcal{H}$ is continuous.

Step 3 $\mathcal{H}(B_\delta)$ is bounded and equicontinuous.

Since $\mathcal{H}(B_\delta) \subset B_\delta$ and B_δ is bounded, then $\mathcal{H}(B_\delta)$ is bounded.

Let $\tau_1, \tau_2 \in (\theta_1, \theta_2]$, $\tau_1 < \tau_2$ and let $(x, y) \in B_\delta$. Thus

$$\begin{aligned}
&\left\|\Phi^{\psi}_{\eta_1}(\tau_1,\theta_1)\mathcal{H}_1(x,y)(\tau_1) - \Phi^{\psi}_{\eta_1}(\tau_2,\theta_1)\mathcal{H}_1(x,y)(\tau_2)\right\| \\
&\leq \left\|\Phi^{\psi}_{\eta_1}(\tau_1,\theta_1)\left(\mathcal{J}^{\lambda_1,k;\psi}_{\theta_1^+}\|\varphi_1(s)\|\right)(\tau_1) - \Phi^{\psi}_{\eta_1}(\tau_2,\theta_1)\left(\mathcal{J}^{\lambda_1,k;\psi}_{\theta_1^+}\|\varphi_1(s)\|\right)(\tau_2)\right\| \\
&\leq \int_{\theta_1}^{\tau_1}\left\|\Phi^{\psi}_{\eta_1}(\tau_1,\theta_1)\bar{\Phi}^{k,\psi}_{\lambda_1}(\tau_1,s) - \Phi^{\psi}_{\eta_1}(\tau_2,\theta_1)\bar{\Phi}^{k,\psi}_{\lambda_1}(\tau_2,s)\right\|\|\psi'(s)\varphi_1(s)\|ds \\
&\quad + \left\|\Phi^{\psi}_{\eta_1}(\tau_2,\theta_1)\left(\mathcal{J}^{\lambda_1,k;\psi}_{\tau_1^+}\|\varphi_1(s)\|\right)(\tau_2)\right\|,
\end{aligned}$$

and

$$\begin{aligned}
&\left\|\Phi^{\psi}_{\eta_2}(\tau_1,\theta_1)\mathcal{H}_2(x,y)(\tau_1) - \Phi^{\psi}_{\eta_2}(\tau_2,\theta_1)\mathcal{H}_2(x,y)(\tau_2)\right\| \\
&\leq \left\|\Phi^{\psi}_{\eta_2}(\tau_1,\theta_1)\left(\mathcal{J}^{\lambda_2,k;\psi}_{\theta_1^+}\|\varphi_2(s)\|\right)(\tau_1) - \Phi^{\psi}_{\eta_2}(\tau_2,\theta_1)\left(\mathcal{J}^{\lambda_2,k;\psi}_{\theta_1^+}\|\varphi_2(s)\|\right)(\tau_2)\right\| \\
&\leq \int_{\theta_1}^{\tau_1}\left\|\Phi^{\psi}_{\eta_2}(\tau_1,\theta_1)\bar{\Phi}^{k,\psi}_{\lambda_2}(\tau_1,s) - \Phi^{\psi}_{\eta_2}(\tau_2,\theta_1)\bar{\Phi}^{k,\psi}_{\lambda_2}(\tau_2,s)\right\|\|\psi'(s)\varphi_2(s)\|ds \\
&\quad + \left\|\Phi^{\psi}_{\eta_2}(\tau_2,\theta_1)\left(\mathcal{J}^{\lambda_2,k;\psi}_{\tau_1^+}\|\varphi_2(s)\|\right)(\tau_2)\right\|.
\end{aligned}$$

Lemma 2.2 implies

$$\begin{aligned}
&\left\|\Phi^{\psi}_{\eta_1}(\tau_1,\theta_1)\mathcal{H}_1(x,y)(\tau_1) - \Phi^{\psi}_{\eta_1}(\tau_2,\theta_1)\mathcal{H}_1(x,y)(\tau_2)\right\| \\
&\leq \delta\int_{\theta_1}^{\tau_1}\left\|\Phi^{\psi}_{\eta_1}(\tau_1,\theta_1)\bar{\Phi}^{k,\psi}_{\lambda_1}(\tau_1,s) - \Phi^{\psi}_{\eta_1}(\tau_2,\theta_1)\bar{\Phi}^{k,\psi}_{\lambda_1}(\tau_2,s)\right\|\|\psi'(s)\|ds \\
&\quad + \frac{\delta\Phi^{\psi}_{\eta_1}(\tau_2,\theta_1)\left(\psi(\tau_2)-\psi(\tau_1)\right)^{\frac{\lambda_1}{k}}}{\Gamma_k(\lambda_1+k)},
\end{aligned}$$

and

$$\begin{aligned}
&\left\|\Phi^{\psi}_{\eta_2}(\tau_1,\theta_1)\mathcal{H}_2(x,y)(\tau_1) - \Phi^{\psi}_{\eta_2}(\tau_2,\theta_1)\mathcal{H}_2(x,y)(\tau_2)\right\| \\
&\leq \delta\int_{\theta_1}^{\tau_1}\left\|\Phi^{\psi}_{\eta_2}(\tau_1,\theta_1)\bar{\Phi}^{k,\psi}_{\lambda_2}(\tau_1,s) - \Phi^{\psi}_{\eta_2}(\tau_2,\theta_1)\bar{\Phi}^{k,\psi}_{\lambda_2}(\tau_2,s)\right\|\|\psi'(s)\|ds \\
&\quad + \frac{\delta\Phi^{\psi}_{\eta_2}(\tau_2,\theta_1)\left(\psi(\tau_2)-\psi(\tau_1)\right)^{\frac{\lambda_2}{k}}}{\Gamma_k(\lambda_2+k)}.
\end{aligned}$$

As $\tau_1 \to \tau_2$, the right-hand side of the above inequalities tends to zero. This exhibits the equicontinuity of $\mathcal{H}(B_\delta)$.

Step 4: The implication (3) of Theorem 2.4 holds.

Now let $D = D_1 \cap D_2$ be an equicontinuous subset of B_δ such that $D_i \subset \overline{\mathcal{T}_i(D_i)} \cup \{0\}$; $i = 1, 2$, therefore the functions $t \longrightarrow d_i(t) = \mu(D_i(t))$ are continuous on $(\theta_1, \theta_2]$. By the hypotheses (Cd_3) and the properties of the measure μ, for each $t \in (\theta_1, \theta_2]$, we have

$$\begin{aligned}
\Phi^{\psi}_{\eta_1}(t, \theta_1) d_1(t) &\le \mu\left(\Phi^{\psi}_{\eta_1}(t, \theta_1)(\mathcal{H}_1 D_1)(t) \cup \{0\}\right) \\
&\le \mu\left(\Phi^{\psi}_{\eta_1}(t, \theta_1)(\mathcal{H}_1 D_1)(t)\right) \\
&\le \Phi^{\psi}_{\eta_1}(t, \theta_1)\left(\mathcal{J}^{\lambda_1, k; \psi}_{\theta_1 +} \zeta_3 \Phi^{\psi}_{\eta_1}(t, \theta_1)\mu(D_1(s))\right)(t) \\
&\le \zeta_3 \|d_1\|_{C_{\eta_1;\psi}} \left[\frac{(\psi(\theta_2) - \psi(\theta_1))^{1-\eta_1+\frac{\lambda_1}{k}}}{\Gamma_k(\lambda_1 + k)}\right],
\end{aligned}$$

which implies that

$$\|d_1\|_{C_{\eta_1;\psi}} \le \ell \|d_1\|_{C_{\eta_1;\psi}}.$$

From (11), we get $\|d_1\|_{C_{\eta_1;\psi}} = 0$, that is $d_1(t) = \mu(D_1(t)) = 0$, for each $t \in J$. Similarly, we have

$$\begin{aligned}
\|d_2\|_{C_{\eta_2;\psi}} &\le \frac{\zeta_4 (\psi(\theta_2) - \psi(\theta_1))^{1-\eta_2+\frac{\lambda_2}{k}}}{\Gamma_k(\lambda_2 + k)} \|d_2\|_{C_{\eta_2;\psi}} \\
&\le \ell \|d_2\|_{C_{\eta_2;\psi}},
\end{aligned}$$

that is $d_1(t) = \mu(D_1(t)) = 0$. Thus, $\mu(D(t)) \le \mu(D_1(t)) = 0$ and $\mu(D(t)) \le \mu(D_2(t)) = 0$, which means that $D(t)$ is relatively compact in $E \times E$. In view of the Ascoli-Arzela Theorem, D is relatively compact in B_δ. Applying now Theorem 2.4, we conclude that $\mathcal{H}$ has a fixed point, which is a solution to the system (1) and (2). □

4 An Example

In this section, we illustrate our results; this example may be thought of as a specific case of our problem (1) and (2). Let

$$E = c_0 = \{v = (v_1, v_2, \ldots, v_n, \ldots), v_n \to v \text{ as } n \to \infty\}$$

be the Banach space of real sequences converging to zero, with the norm

$$\|v\| = \sup_{n\geq 1} |v_n|.$$

Suppose that $J = [1, 2]$, $\eta_i = \frac{1}{k}(r_i(k - \lambda_i) + \lambda_i)$; $i = 1, 2$.

Taking $r_1 \to 1$, $r_2 \to 0$, $\lambda_1 = \lambda_2 = \frac{1}{2}$, $k = 1$, $\psi(t) = t$, $\alpha = \beta = (0, 0, \ldots, 0, \ldots)$, $\eta_1 = 1$ and $\eta_2 = \frac{1}{2}$, we obtain a particular case of problem (1) and (2). It is a coupled system of two problems: The first one is a terminal value problem with Caputo fractional derivative, the second is a terminal value problem with Riemann-Liouville fractional derivative. The coupled system is given for $t \in (1, e]$ by:

$$\begin{cases} \left({}_1^H\mathcal{D}_{1^+}^{\frac{1}{2},1;\psi} x\right)(t) = \left({}^C\mathbb{D}_{1^+}^{\frac{1}{2}} x\right)(t) = f_1\left(t, x(t), y(t), \left({}^C\mathbb{D}_{1^+}^{\frac{1}{2}} x\right)(t), \left({}^{RL}\mathbb{D}_{1^+}^{\frac{1}{2}} y\right)(t)\right), \\ \left({}_1^H\mathcal{D}_{1^+}^{\frac{1}{2},0;\psi} y\right)(t) = \left({}^{RL}\mathbb{D}_{1^+}^{\frac{1}{2}} y\right)(t) = f_2\left(t, x(t), y(t), \left({}^C\mathbb{D}_{1^+}^{\frac{1}{2}} x\right)(t), \left({}^{RL}\mathbb{D}_{1^+}^{\frac{1}{2}} y\right)(t)\right), \end{cases} \tag{14}$$

$$\begin{cases} x(2) = (0, 0, \ldots, 0, \ldots), \\ y(2) = (0, 0, \ldots, 0, \ldots), \end{cases} \tag{15}$$

where

$$x = (x_1, x_2, \ldots, x_n, \ldots), \ \ y = (y_1, y_2, \ldots, y_n, \ldots),$$

$$f_1 = (f_{1,1}, f_{1,2}, \ldots, f_{1,n}, \ldots), \ \ f_2 = (f_{2,1}, f_{2,2}, \ldots, f_{2,n}, \ldots),$$

$${}^C\mathbb{D}_{1^+}^{\frac{1}{2}} x = \left({}^C\mathbb{D}_{1^+}^{\frac{1}{2}} x_1, {}^C\mathbb{D}_{1^+}^{\frac{1}{2}} x_2, \ldots, {}^C\mathbb{D}_{1^+}^{\frac{1}{2}} x_n, \ldots\right),$$

$${}^{RL}\mathbb{D}_{1^+}^{\frac{1}{2}} y = \left({}^{RL}\mathbb{D}_{1^+}^{\frac{1}{2}} y_1, {}^{RL}\mathbb{D}_{1^+}^{\frac{1}{2}} y_2, \ldots, {}^{RL}\mathbb{D}_{1^+}^{\frac{1}{2}} y_n, \ldots\right);$$

$$f_{1,n}\left(t, x_n(t), y_n(t), \left({}^C\mathbb{D}_{1^+}^{\frac{1}{2}} x_n\right)(t), \left({}^{RL}\mathbb{D}_{1^+}^{\frac{1}{2}} y_n\right)(t)\right)$$
$$= \frac{\sin(t)\left(1 + |x_n(t)| + \sqrt{t-1}|y_n(t)|\right)}{283e^{t+3}\left(1 + \|x(t)\| + \|y(t)\| + \left\|\left({}^C\mathbb{D}_{1^+}^{\frac{1}{2}} x\right)(t)\right\| + \left\|\left({}^{RL}\mathbb{D}_{1^+}^{\frac{1}{2}} y\right)(t)\right\|\right)}, \ t \in (1, 2],$$

and

$$f_{2,n}\left(t, x_n(t), y_n(t), \left({}^{C}\mathbb{D}^{\frac{1}{2}}_{1^+}x_n\right)(t), \left({}^{RL}\mathbb{D}^{\frac{1}{2}}_{1^+}y_n\right)(t)\right)$$
$$= \frac{\cos(t)\sqrt{t-1}\,(1+|x_n(t)|+|y_n(t)|)}{83e^{t+1}\left(1+\|x(t)\|+\|y(t)\|+\left\|\left({}^{C}\mathbb{D}^{\frac{1}{2}}_{1^+}x\right)(t)\right\|+\left\|\left({}^{RL}\mathbb{D}^{\frac{1}{2}}_{1^+}y\right)(t)\right\|\right)}, \quad t \in (1,2].$$

We have

$$C_{\eta_1,k;\psi}(J) = C_{1,1;\psi}(J) = C(J,\mathbb{R}),$$

and

$$C_{\eta_2,k;\psi}(J) = C_{\frac{1}{2},1;\psi}(J) = \left\{x : (1,2] \to \mathbb{R} : x\sqrt{t-1} \in C(J,\mathbb{R})\right\}.$$

Since it is clear to see that the functions f_1 and f_2 are continuous, then (Cd_1) is verified.

Further, for each $x_1, y_1, x_2, y_2, w_1, w_2, z_1, z_2 \in E$ and $t \in J$, we have

$$\|f_1(t,x_1,y_1,w_1,z_1) - f_1(t,x_2,y_2,w_2,z_2)\|$$
$$\leq \frac{\sin(t)}{283e^{t+3}}|x_1-x_2| + \frac{\sin(t)\sqrt{t-1}}{283e^{t+3}}|y_1-y_2| + \frac{\sin(t)}{283e^{t+3}}\left(|w_1-w_2|+|z_1-z_2|\right),$$

and

$$|f_2(t,x_1,y_1,w_1,z_1) - f_2(t,x_2,y_2,w_2,z_2)|$$
$$\leq \frac{\cos(t)\sqrt{t-1}}{83e^{t+1}}\left(|x_1-x_2|+|y_1-y_2|+|w_1-w_2|+|z_1-z_2|\right).$$

Thus, the condition (Cd_2) is satisfied with

$$\zeta_1 = \varpi_1 = \bar{\zeta}_1 = \bar{\varpi}_1 = \frac{1}{283e^4},$$

and

$$\zeta_2 = \varpi_2 = \bar{\zeta}_2 = \bar{\varpi}_2 = \frac{1}{83e^2}.$$

The hypothesis (Cd_3) is easily verified by taking

$$\zeta_3 = \frac{1}{283e^4}, \text{ and } \zeta_4 = \frac{1}{83e^2}.$$

Also

$$\ell = \max\left\{\frac{2}{283e^4\sqrt{\pi}}, \frac{2}{83e^2\sqrt{\pi}}, \frac{1}{(283e^4-1)(83e^2-1)}\right\} < 1.$$

As all the conditions of Theorem 3.2 are satisfied, then the problem (14) and (15) has at least one solution in $C_{1;\psi}([1,2]) \times C_{\frac{1}{2};\psi}([1,2])$.

References

1. Abbas, S., Benchohra, M., N'Guérékata, G.M.: Topics in Fractional Differential Equations. Springer-Verlag, New York (2012)
2. Abbas, S., Benchohra, M., N'Guérékata, G. M.: Advanced Fractional Differential and Integral Equations. Nova Science Publishers, New York (2014)
3. Abbas, S., Benchohra, M., Graef, J. R. Henderson, J.: Implicit Differential and Integral Equations: Existence and Stability. Walter de Gruyter, London (2018)
4. Abbas, S., Benchohra, M., Lazreg, J.E., Nieto, J.J.: On a coupled system of Hilfer and Hilfer-Hadamard fractional differential equations in Banach spaces. J. Nonlinear Funct. Anal. **2018**, 12pp (2018)
5. Abdo, M.S., Shah, K., Panchal, S.K., Wahash, H.A.: Existence and Ulam stability results of a coupled system for terminal value problems involving ψ-Hilfer fractional operator. Adv. Differ. Equ. **2020**, 21pp. (2020)
6. Ahmad, M., Zada, A., Wang, X.: Existence, Uniqueness and stability of implicit switched coupled fractional differential equations of ψ-Hilfer type. Int. J. Nonlinear Sci. Numer. Simul. **21**, 327–337 (2020)
7. Ali, A., Shah, K., Khan, R.A.: Existence of solution to a coupled system of hybrid fractional differential equations. Bull. Math. Anal. Appl. **9**, 9–18 (2017)
8. Almalahi, M.A., Bazighifan, O., Panchal, S.K., Askar, S.S., Oros, G.I.: Analytical study of two nonlinear coupled hybrid systems involving generalized Hilfer fractional operators. Fractal Fract. **5**, 22pp. (2021)
9. Banas, J., Goebel, K.: Measures of noncompactness in Banach spaces. Marcel Dekker, New York (1980)
10. Benchohra, M., Henderson, J., Ntouyas, S. K.: Impulsive Differential Equations and Inclusions, vol. 2. Hindawi Publishing Corporation, New York (2006)
11. Chu, Y.M., Awan, M.U., Talib, S., Noor, M.A., Noor, K.I.: Generalizations of Hermite-Hadamard like inequalities involving χ_κ-Hilfer fractional integrals. Adv. Differ. Equ. **2020**, 594 (2020)
12. Derbazi, C., Hammouche, H., Salim, A., Benchohra, M.: Measure of noncompactness and fractional Hybrid differential equations with Hybrid conditions. Differ. Equ. Appl. **14**, 145–161 (2022). https://doi.org/10.7153/dea-2022-14-09
13. Diaz, R., Teruel, C.: q, k-Generalized gamma and beta functions. J. Nonlinear Math. Phys. **12**, 118–134 (2005)
14. Guida, K., Hilal, K., Ibnelazyz, L.: Existence of mild solutions for a class of impulsive Hilfer fractional coupled systems. Adv. Math. Phys. **2020**, 12pp. (2020)
15. Kucche, K.D., Mali, A.D.: On the nonlinear (k, ψ)-Hilfer fractional differential equations. Chaos Solitons Fractals **152** (2021)
16. Lin, L., Liu, Y., Zhao, D.: Study on implicit-type fractional coupled system with integral boundary conditions. Mathematics **9**, 15pp. (2021)
17. Monch, H.: BVP for nonlinear ordinary differential equations of second order in Banach spaces. Nonlinear Anal. **4**, 985–999 (1980)

18. Mubeen, S., Habibullah, G.M.: k-Fractional integrals and application. Int. J. Contemp. Math. Sci. **7**, 89–94 (2012)
19. Rashid, S., Aslam Noor, M., Inayat Noor, K., Chu, Y.M.: Ostrowski type inequalities in the sense of generalized $\mathcal{K}$-fractional integral operator for exponentially convex functions. AIMS Math. **5**, 2629–2645 (2020)
20. Salim, A., Benchohra, M., Lazreg, J. ., Henderson, J.: Nonlinear implicit generalized Hilfer-type fractional differential equations with non-instantaneous impulses in Banach spaces. Adv. Theory Nonlinear Anal. Appl. **4**, 332–348 (2020). https://doi.org/10.31197/atnaa.825294
21. Salim, A., Benchohra, M., Graef, J.R., Lazreg, J.E.: Boundary value problem for fractional generalized Hilfer-type fractional derivative with non-instantaneous impulses. Fractal Fract. **5**, 1–21 (2021). https://doi.org/10.3390/fractalfract5010001
22. Salim, A. Benchohra, M., Lazreg, J. E., Karapinar, E.: On k-generalized ψ-Hilfer impulsive boundary value problem with retarded and advanced arguments. J. Math. Ext. **15**, 39 pp. (2021)
23. Salim, A., Benchohra, M., Lazreg, J.E., N'Guérékata, G.: Boundary value problem for nonlinear implicit generalized Hilfer-type fractional differential equations with impulses. Abstr. Appl. Anal. **2021**, 17pp. (2021). https://doi.org/10.1155/2021/5592010
24. Salim, A., Benchohra, M., Graef, J.R., Lazreg, J.E.: Initial value problem for hybrid ψ-Hilfer fractional implicit differential equations. J. Fixed Point Theory Appl. **24**, 14 pp. (2022). https://doi.org/10.1007/s11784-021-00920-x
25. Salim, A., Benchohra, M., Lazreg, J.E., Henderson, J.: On k-generalized ψ-Hilfer boundary value problems with retardation and anticipation. Adv. Theory Nonlinear Anal. Appl. **6**, 173–190 (2022). https://doi.org/10.31197/atnaa.973992
26. Salim, A., Lazreg, J.E., Ahmad, B., Benchohra, M., Nieto, J.J.: A Study on k-Generalized ψ-Hilfer Derivative Operator. Vietnam J. Math. (2022). https://doi.org/10.1007/s10013-022-00561-8
27. Samina, Shah, K., Khan, R.A.: Stability theory to a coupled system of nonlinear fractional hybrid differential equations. Indian J. Pure Appl. Math. **51**, 669–687 (2020)
28. Sousa, J.V.C., Capelas de Oliveira, E.: A Gronwall inequality and the Cauchy-type problem by means of ψ-Hilfer operator. Differential Equations Appl. **11**, 87–106 (2019)
29. Sousa, J.V.C., Capelas de Oliveira, E.: Fractional order pseudoparabolic partial differential equation: Ulam–Hyers stability. Bull. Braz. Math. Soc. **50**, 481–496 (2019)
30. Sousa, J.V.C., de Oliveira, E.C.: On the ψ-Hilfer fractional derivative. Commun. Nonlinear Sci. Numer. Simul. **60**, 72–91 (2018)
31. Sousa, J.V.C., de Oliveira, E.C.: Leibniz type rule: ψ-Hilfer fractional operator. Commun. Nonlinear Sci. Numer. Simul. **77**, 305–311 (2019)
32. Sousa, J.V.C., Frederico, G.S.F., de Oliveira, E.C.: ψ-Hilfer pseudo-fractional operator: new results about fractional calculus. Comput. Appl. Math. **39**(4), Paper No. 254, 33pp. (2020)
33. Wongcharoen, A., Ntouyas, S.K., Tariboon, J.: On coupled systems for Hilfer fractional differential equations with nonlocal integral boundary conditions. J. Math. **2020**, 12pp. (2020)

MIX
Papier aus verantwortungsvollen Quellen
Paper from responsible sources
FSC® C105338

If you have any concerns about our products,
you can contact us on
ProductSafety@springernature.com

In case Publisher is established outside the EU,
the EU authorized representative is:
Springer Nature Customer Service Center GmbH
Europaplatz 3, 69115 Heidelberg, Germany

Printed by Libri Plureos GmbH
in Hamburg, Germany